THE POWDERY MILDEWS

THE
POWDERY MILDEWS

Edited by

D. M. SPENCER

Glasshouse Crops Research Institute, Littlehampton, Sussex, England

1978

ACADEMIC PRESS

London　　New York　　San Francisco

A Subsidiary of Harcourt Brace Jovanovich, Publishers

ACADEMIC PRESS INC. (LONDON) LTD.
24–28 Oval Road
London, NW1

U.S. Edition published by
ACADEMIC PRESS INC.
111 Fifth Avenue,
New York, New York 10003

Library of Congress Catalog Card Number: 77-92827
ISBN: 0–12–656850–2

Text set in 10/12 pt Monotype Times Roman, printed by letterpress
in Great Britain at The Pitman Press, Bath

Contributors

BAINBRIDGE, A. *Plant Pathology Department Rothamsted Experimental Station Harpenden, Herts, England*

BENT, K. J. *Imperial Chemical Industries Ltd., Jealott's Hill Research Station, Bracknell, Berks, England*

BULIT, J. *INRA Bordeaux Station de Pathologie Végétale 33140 Pont de la Maye, France*

BURCHILL, R. T. *National Vegetable Research Station, Wellesbourne, Warwick, England*

BUSHNELL, W. R. *Cereal Rust Laboratory, U.S.D.A., University of Minnesota, St. Paul, Minnesota 55108, U.S.A.*

BUTT, D. J. *East Malling Research Station, Maidstone, Kent, England*

COLE, J. S. *Tobacco Research Board of Rhodesia, Salisbury, Rhodesia*

CORKE, A. T. K. *Long Ashton Research Station, University of Bristol, England*

DIXON, G. R. *National Institute of Agricultural Botany, Huntingdon Road, Cambridge, England*

DRANDAREVSKI, C. A. *Cela Merck, Biolog. Research 6507 Ingelheim, Germany*

ELLINGBOE, Albert H. *Genetics Program, and Department of Botany and Plant Pathology, Michigan State University, East Lansing, Michigan 48824, U.S.A.*

ESHED, Nava *Faculty of Agriculture, Hebrew University of Jerusalem, Rehovot, Israe*

GAY, J. L. *Department of Botany, Imperial College of Science and Technology, Prince Consort Road, London SW7 2BB, England*

HIURA, U. *Institute for Agricultural and Biological Sciences, Okayama University, Kurashiki, Japan*

JENKYN, J. F. *Plant Pathology Department, Rothamsted Experimental Station, Harpenden, Herts, England*

JORDAN, V. W. L. *Long Ashton Research Station, University of Bristol, England*

LAFON, R. *INRA Bordeaux Station de Pathologie Végétale 33140 Pont de la Maye, France*

McINTOSH, R. A. *Plant Breeding Institute, The University of Sydney, N.S.W., Australia 2006*

ROYLE, D. J. *Department of Hop Research, Wye College, Ashford, Kent, England*

SCHWARZBACH, E. *Technische Universität München, Lehrstuhl für Pflanzenbau und Pflanzenzüchtung 8050 Freising-Weihenstephan, West Germany, FRG*

SEGAL, A. *Faculty of Life Sciences, Tel Aviv University, Tel-Aviv, Israel*

SITTERLY, Wayne R. *Clemson University Truck Experiment Station, Charleston, South Carolina, U.S.A.*

SOBEL, Z. *Faculty of Life Sciences, Tel-Aviv University, Tel-Aviv, Israel*

WAHL, I. *Faculty of Life Sciences, Tel-Aviv University, Tel-Aviv, Israel*

WELTZIEN, H. C. *Institut für Pflanzenkrankheiten Universität, Bonn, Germany*

WHEELER, B. E. J. *Imperial College Field Station, Silwood Park, Sunninghill, Ascot, Berks, England*

WOLFE, M. S. *Plant Breeding Institute, Maris Lane, Trumpington, Cambridge, England*

YARWOOD, C. E. *Department of Plant Pathology, University of California, Berkeley, California, U.S.A.*

v

189243

Preface

The aim of this treatise is to present an integrated discussion of our present knowledge of the group of diseases known as the the powdery mildews. It was somewhat disconcerting to find, therefore, that the author of the opening chapter might have chosen "The White Mildews" as a better title for the book. Be that as it may, the intention has been to bring together in one volume the information we now have on the diseases as they affect a wide variety of plants and, at the same time, to examine the progress which has been made in recent years in fundamental studies of some of the fungi involved and their effects on the host plants, all of which can shed light on the problems of disease control.

Each subject is comprehensively treated so as to be informative and also to be thought-provoking, where opinions diverge, or where there are apparent gaps in our knowledge. Although each has been written to be complete in itself, the chapters were prepared with an understanding of their relationship, one to another.

References to powdery mildews were not infrequent before the turn of this century but I think it is true to say that Salmon's monograph in 1900 gave an impetus to their scientific study which has hardly diminished after more than three quarters of a century. Indeed, Salmon's pronouncements on taxonomy still carry weight, though discussions on this subject are by no means over, as judged by the evidence of several contributions especially in Chapter 1.

Until some twenty years ago it was possible for a well-informed worker to review the then current status of our knowledge of the powdery mildews. In the ensuing years the scientific discoveries made through the individual and combined research efforts of plant pathologists, other biologists, chemists, biochemists, physicists, engineers and others, have proliferated in several directions, our information on this group of diseases. There is now such a vast fund of knowledge that it was immediately obvious, when the book was conceived that only a multi-author approach could provide the desirable comprehensive treatment. In order that the broadest coverage, in depth, should be provided for the various subject areas, the contributions were requested from an international selection of authors.

The first part of the book is concerned with subjects of overall importance in relation to the powdery mildews as a group. Subjects such as taxonomy, distribution, epidemiology, host-parasite relations and methods of control are all applicable to powdery mildews in general though some of them are dealt with here with particular reference to individual diseases. With some of these chapters it might be judged that the powdery mildew fungus has been selected as a convenient test organism for use in investigations which have much deeper significance than their contribution to the subject of this book. The last nine chapters deal with the powdery mildews of individual crops or with those of commodity groups. Some of these chapters, therefore, are concerned with such single diseases as those of tobacco, the hop and the vine, whereas others have to be wide-ranging and deal with the large number of diseases of vegetables, ornamentals or tree crops. Between these extremes other chapters discuss a limited number of diseases of bush and soft fruits, cucurbits or root crops. In these nine chapters topics such as history, taxonomy, economic importance, host-parasite relations and disease resistance are mentioned in the context of the particular crops and can thus be seen to supplement the specialized treatment of these subjects in the earlier chapters. Indeed, the chapter on powdery mildews of cereals, which as a group have received more attention than any other group, serves as a bridge between the two parts of the book.

It is hoped that the book will be useful to students of plant pathology and to all those professional workers with an interest in plant diseases generally and not just the powdery mildews. It will serve as a source of information and, hopefully, will stimulate the generation of new ideas and new approaches to researchers in the field of host-parasite interaction. Since a basic understanding of principles is necessary for the best understanding of plant disease problems, the information presented here should also be useful to research and extension workers interested in better plant disease control.

It might be asserted that some of the more "modern" fields of research in which there has been considerable progress, have not been given extensive treatment here. This omission has resulted from a desire on the one hand to avoid excessive overlap between chapters and on the other from an unavoidable withdrawal from the venture of certain authors. Indeed, some of the present contributions were received so long after the agreed deadline that there was a real danger that they too would have to be omitted. This having been said, there is no doubt that the greatest share of the work in the preparation of this book has been borne by the contributors, some of whom most willingly extended their chapters and thus reduced the deficiency in the treatment of some subjects. It is perhaps worth noting here that where there was a possibility of alternative interpretation of certain phrases in some chapters, the editor has introduced changes, only for the sake of clarity. He has not attempted to standardize the style of writing, nor has he questioned the opinions expressed. It is hoped therefore, that the presentation of all the chapters will be recognizable, unmistakably, as that of the respective authors, to whom all credit is due. I am most appreciative of the industriousness, thoroughness and patience shown by all contributors in performing their tasks. The advice and support of Dr. D. Rudd-Jones, Director of the Glasshouse Crops Research Institute in starting this venture is gratefully acknowledged. Finally I want to thank my wife Eileen for the encouragement and practical help which she has so freely given at all stages of the preparation of this book.

APRIL 1978   D. M. SPENCER,
                                   Glasshouse Crops Research Institute,
                                   Rustington, Littlehampton, Sussex

Contents

3. Epidemiology of Powdery Mildews

D. J. BUTT

4. Significance of Wild Relatives of Small Grains and Other Wild Grasses in Cereal Powdery Mildews

I. WAHL, NAVA ESHED, A. SEGAL and Z. SOBEL

5. Genetic Basis of Formae Speciales in *Erysiphe graminis* DC.

U. HIURA

6. The Recent History of the Evolution of Barley Powdery Mildew in Europe

M. S. WOLFE and E. SCHWARZBACH

7. A Genetic Analysis of Host-Parasite Interactions

ALBERT H. ELLINGBOE

8. Accumulations of Solutes in Relation to the Structure and Function of Haustoria in Powdery Mildews

W. R. BUSHNELL and JOHN GAY

9. Breeding for Resistance to Powdery Mildew in the Temperate Cereals

R. A. McINTOSH

10. Chemical Control of Powdery Mildews

K. J. BENT

11. Biology and Pathology of Cereal Powdery Mildews

J. F. JENKYN and A. BAINBRIDGE

12. Powdery Mildews of Beet Crops

C. A. DRANDAREVSKI

13. Powdery Mildews of Bush and Soft Fruits

A. T. K. CORKE and V. W. L. JORDAN

14. Powdery Mildews of Cucurbits

WAYNE R. SITTERLY

15. Powdery Mildew of the Hop

D. J. ROYLE

16. Powdery Mildews of Ornamentals

B. E. J. WHEELER

17. Powdery Mildew of Tobacco

J. S. COLE

18. Powdery Mildews of Tree Crops

R. T. BURCHILL

19. Powdery Mildews of Vegetable and Allied Crops

G. R. DIXON

20. Powdery Mildew of the Vine

J. BULIT and R. LAFON

Chapter 1

History and Taxonomy of Powdery Mildews

C. E. YARWOOD

Department of Plant Pathology, University of California, Berkeley, USA

I. Introduction

Powdery mildews are one of the most conspicuous and most studied groups of parasitic fungi. Previous treatments of the taxonomy and of other aspects of the group which overlap on the present study include those by De Candolle (1805), Leveille (1851), Jaczewski (1896), Salmon (1900), Neger (1901, 1902, 1923), Pollacci (1911), Jorstad (1925), Sawada (1927), Blumer (1933, 1967), Brundza (1934), Homma (1937), Yarwood (1957, 1973), Golovin (1958), Clare (1964), and Hirata (1966). Blumer's (1967) treatment in German replaces Salmon's (1900) treatment in English as the most authoritative general and taxonomic treatment. The longest list of references (1462) and hosts (7606) yet produced is by Hirata (1966) and this is far from complete. Space does not permit a comparable treatment here. Rather I will briefly review the taxonomy and some other aspects of the group, and present keys to genera based on characters of the perithecial and conidial stages.

II. Definition

Powdery mildews or Erysiphaceae are those fungi with white (colourless or hyaline) hyphae, and colourless one-celled ascospores borne in asci enclosed in black non-ostiolate perithecia (cleistothecia) on the surface of living plants. This definition based on the perfect (sexual) stage has limited use because most collections do not contain perithecia. A better working definition is that powdery mildews are those fungi with white superficial hyphae on the aerial parts of living plants, with large one-celled conidia produced terminally on isolated aerial unbranched conidiophores and with haustoria in the epidermal cells of their hosts. A combination of these definitions would be better than either alone and there are some exceptions. A few species of powdery mildews are only known to produce perithecia; many produce only conidia in most situations but most produce first conidia and later perithecia. A few collections have been found only as mycelium as on *Lycium herbarium* (Leveille, 1851) and on *Aristolochia californica* (Gardner and Yarwood, 1974).

As well as being the name given to the organism, the term powdery mildew is also used to describe the disease caused by a member of the Erysiphaceae. Taxonomically, on the basis of their ascus or perfect stage, powdery mildews comprise the family Erysiphaceae in the order Erysiphales, class Pyrenomycetes, and their closest relatives are the Perisporiaceae or dark mildews. On the basis of their conidial or imperfect stage the powdery mildews are usually classified in the family Moniliaceae, order Moniliales, class Deuteromycetes. As with so many fungi, the conidia and ascospores of Erysiphaceae resemble each other in that they are both large, smooth, one-celled and colourless. It is a common belief that the conidial stages of all powdery mildews are similar to each other (Alexopoulos, 1952; Arnaud, 1921) and that therefore they cannot be classified or identified on the basis of the conidial stage. In fact there are many differences in the conidial stages and these have been used for identification (Sawada, 1927; Clare, 1964; Yarwood, 1973). However powdery mildews are logically usually named for the perfect stage, even though the perfect stage may be rarely found. For example, grape powdery mildew is usually called *Uncinula necator*, the name of the perfect stage, though the perfect stage may be absent in the specimen under observation. This is usually justified in part because there is usually only one species of powdery mildew on one species of host (including grape) and certain morphological types of conidiophores are well correlated with certain morphological types of perithecia.

For many years the perfect and imperfect stages of powdery mildews were considered separate organisms and given separate names. For example, the perithecial stage of the powdery mildew of grape was first found and described in the United States and was named *Erysiphe necator* by Schweinitz in 1834 while the conidial stage was first found in England and was named *Oidium tuckeri* by Berkeley in 1847. The realization that these two species were different stages of the same fungus came gradually and as a result of some controversy (see Large, 1940) and is commonly attributed to the Tulasne brothers in 1861, but the perithecial stage of grape powdery mildew was not found in Europe until 1893 by Couderc. Better evidence that the conidial and perithecial stages are of the same fungus comes from production of the perithecial stage by inoculation with single conidia (Homma, 1933) or paired cultures derived from single conidia (Yarwood, 1935).

Mildew has had many meanings such as Erysiphaceae, Peronosporaceae, Uredinales (Knight, 1818; Murray, 1908), insects (Forsyth, 1824), or the fungus growth on old cloth, leather, paper, or wood (Cooke, 1878). According to Large (1940) it may have originated from the German *mehltau* or *meal deau*, from the Gothic *mili* for honey and *tau* for dew. As growths of mildew were common in low lying damp localities such as near water mills, the word was sometimes written mill-dew. Other spellings are meledeaw, myldewe, myldeawe, mildewe, mieldew and mealydew (Murray, 1908).

Other words for powdery mildew have been mould, fen (Hales, 1727), white mould (Wallroth, 1829), egg mildew (anon., 1848), white mildew (Halsted, 1884), field fungus (D'Angremond, 1924), red or brown mould (Large, 1940), blight (Ogilvie, 1942) and white rust (Du Toit, 1948). A good discussion of the meanings of mildew is given by Cooke (1878).

The first or at least an early use of the word as a disease or pest was in the Bible (King James version, I Kings 8:37. 1004 B.C. ?) as follows: "If there be in the land famine, if there be pestilence, blasting, mildew, locust, or if there be caterpillar . . ." In another independent view, Pliny (see Shepherd, 1939) wrote about pest averting sulphur about 1000 B.C. It is likely that the pests he was referring to were, or at least included, the Erysiphaceae, since these were among the first parasitic fungi recognized by man and since these are the pests most readily controlled by sulphur.

When Hales (1727) referred to mould or fen or mouldy fen of hops, he gave us what is likely the first example of a specific disease caused by what we now recognize as a powdery mildew (*Sphaerotheca humuli*) though he did not use the term mildew. He clearly inferred the infectiousness of the fungus though he did not demonstrate it.

An early record of the use of mildew for a disease we can specifically identify was for peach powdery mildew (*Sphaerotheca pannosa*) by Knight (1818) though Knight also used mildew to apply to what we now regard as wheat rust.

The restriction of mildew to Erysiphaceae seemed to prevail for a short time but was never complete. In the extensive writings of Berkeley (1803–1889), specifically Berkeley (1855), the word mildew usually refers to Erysiphaceae though Berkeley (1848) did at least once use mildew for a member of the Peronosporaceae. With the introduction of *Plasmopara viticola* from America to Europe about 1874, mildew was also applied to this and related fungi, but the Peronosporaceae were increasingly called false or downy mildews. False mildew is also used for diseases caused by *Ramularia* (Cauquil and Sement, 1973). Halsted (1884) used mildew for Peronosporaceae and white mildew for Erysiphaceae. Riley (1886) may have been the first to point out clearly the difference between these two large and so clearly different groups of fungi. He applied "downy mildews" to the Peronosporaceae and "powdery mildews" to the Erysiphaceae and these terms are now in general use.

The application of "powdery" to the mildews caused by Erysiphaceae is generally accepted but slightly misleading. Powdery usually refers to something dry which blows around readily. Powdery mildews do blow around readily and it is common to shake a heavily mildewed branch and produce a conspicuous cloud of spores but "in mass" these spores are distinctly wet from their high water content (Jhooty and McKeen, 1965) and therefore sticky, in contrast to spores of a low water content such as conidia of Penicillium, chlamydospores of *Ustilago* or uredospores of *Uromyces*. I believe white mildews would be a better term than powdery mildews for the Erysiphaceae.

III. Characters of Value in Taxonomy

A. Perithecia

Classification of Erysiphaceae, like that of other perfect fungi, is based primarily on characters of the sexual stage and the perithecia of Erysiphaceae offer many structural features of excellent taxonomic value. These are primarily size (50–350 µm diameter); dorsiventrality; attachment to mycelium; surface of occurrence; transparency; number of layers of wall; size of wall cells; number, size and shape of asci; number, size and shape of ascospores; size, number, type, position, septation and colour of appendages; seasonal development and spore discharge. The use of these characters in separation of genera and species is illustrated and discussed by Salmon (1900), Blumer (1933, 1967), Homma (1937) and many others. Number of asci per perithecium and type of appendages, alone, are universally used in the separation of major genera and these characters are so well defined that the Erysiphaceae are the only family of fungi I know of in which diagnosis to genus is a regular and successful exercise for beginning students of mycology and plant pathology. Types of appendages are illustrated in Figs 1–4.

Perithecia are of limited value in taxonomy of powdery mildews because most collections do not contain them. Only 13% of 515 collections listed by Gardner and Yarwood (1974) contained perithecia. Tropical and subtropical regions contain many powdery mildews but perithecia are uncommon there (Hansford, 1961; Bessey, 1961; Clare, 1964; Blumer, 1967). Liabach (1930) indicates that old leaves, a low state of host nutrition, a dry atmosphere and low temperature favour perithecium formation. Continuous culture in the glasshouse may lead to loss of perithecium formation (Mamluk and Weltzien, 1973). Infections of *Puccinia* and *Coleosporium* may favour perithecium formation (Blumer, 1967). Two sexually compatible strains are necessary for perithecium formation in many cases (Smith, 1970) and this could explain the long interval between the discovery of the conidial and perithecial stages of *U. necator* on grape in Europe (Couderc, 1893) and *Erysiphe polygoni* on beet in USA (Coyier *et al.*, 1975). Clearly certain geographic areas favour perithecium formation. The perithecial stage of *Erysiphe cichoracearum* on potato (Menzies, 1950) of *E. cichoracearum* on cucumbers (Randall and Menzies, 1956) and of *E. polygoni* on beet (Coyier *et al.*, 1975) have apparently been found in the USA only in central Washington State. Perithecia of *E. polygoni* on red clover were first found in the USA in the Pacific Northwest (Mains, 1923). Not only the host species but also its variety may be important, thus perithecia of *E. cichoracearum* were found only on certain varieties of cucumber (Randall and Menzies, 1956). Perithecia of *E. polygoni* occur abundantly on *Trifolium longipes* but rarely on *T. pratense* in California.

Some of the characters of Erysiphaceae which might be, but usually are not widely used in taxonomy, are as follows.

B. Mycelium

It seems that Linnaeus (1753) clearly recognized only the white mycelial stage of the fungi to which he gave the name *Mucor erysiphe* and which are

6 C. E. YARWOOD

now understood as the powdery mildews. This white (hyaline or colourless) parasitic mycelium on the surface of living leaves still distinguishes the Erysiphaceae from other fungi. With a few species such as *Sphaerotheca lanestris* and *Sphaerotheca mors-uvae* the mycelium is dark or darkens with age and may become almost black; this is certainly of taxonomic value.

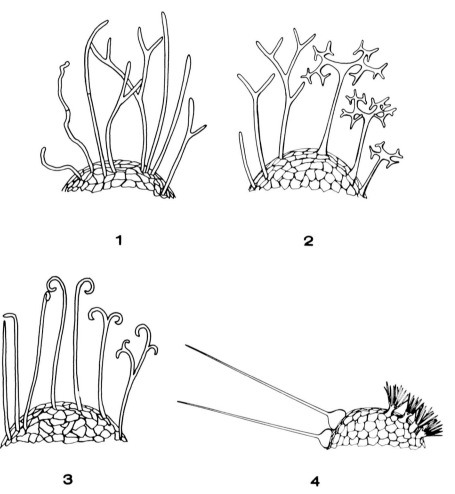

Fig. 1. Hypha-like appendages of *Erysiphe*, *Sphaerotheca*, and *Leveillula*.
Fig. 2. Dichotomously branched appendages of *Microsphaea* and *Podosphaera*.
Fig. 3. Uncinulate appendages of *Uncinula*.
Fig. 4. Aciculate appendages with basal swelling, and brush-like appendages of *Phyllactinia*.

Homma (1937) has used the dark aerial hyphae (special hairs according to her) to characterize the genus *Cystotheca*. However, Berkeley and Curtis (1858) who created the genus *Cystotheca* on what I regard as doubtful grounds, make no mention of special hairs. With *Erysiphe graminis* the mycelium turns buff yellow with age.

Most Erysiphaceae have only external (superficial) mycelium. *Phyllactinia* has mostly external mycelium but sends branches through the stomata to the mesophyll tissue where haustoria are formed. *Leveillula* forms well developed internal mycelium. *Uncinula polychaeta* forms internal mycelium and this could be the logical basis of the genus *Polychaeta* (see Kimbrough and Korf, 1963) but this species also has uncinulate appendages. *S. lanestris* grows mostly on the lower surface of leaves but on some vigorous leaves the mycelium grows through the leaf from the lower to the upper surface on *Quercus agrifolia, Q. borealis, Q. coccinea, Q. douglasii, Q. kelloggii, Q. robur* and *Q. velutina* (Yarwood, unpublished). The mycelium of *E. graminis, E. cichoracearum* and *Sphaerotheca fuliginea* is normally external but has been induced to become internal (endophytic) by removing the epidermis (Salmon, 1906a), by heat treatment (Salmon, 1906a; Jarvis, 1964) and by translocated heat treatment (Jarvis, 1964).

Mycelium is commonly localized and early and sometimes late infections can commonly be seen as discrete colonies, most of which originated from a single spore. Exceptions are overwintering bud infections of *Microsphaera alni* on lilac, *Podosphaera leucotricha* on apple, *S. pannosa* on peach, *U. necator* on grape and several others. When infected buds grow in the spring, all new growth is covered with mycelium and spores produced here are the source of secondary infections. In California at least, bud infections seem to be a more important method of overwintering than are perithecia. Seed infection is reported for *E. polygoni* on pea (Van Hook, 1906) but apparently never adequately confirmed. Powdery mildews are among the relatively few groups of plant pathogens apparently never recorded on roots except for *Erysiphe epigea* on grass roots (Leveille, 1851).

With most Erysiphaceae the mycelium and conidiophores are conspicuous, especially on glabrous leaves (e.g., *Erysiphe trina* on *Quercus agrifolia*) but *E. trina* on pubescent leaves of *Lithocarpus densiflora* is more difficult to recognize. *E. cichoracearum* (conidiophore type) on the pubescent lower surface of *Gnaphalium chilense* is the only powdery mildew which I could not detect and recognize as a powdery mildew with the unaided eye.

C. Conidiophores

Most of the white powdery appearance of Erysiphaceae is due to conidiophores and conidia borne on the mycelium but with some, such as *E. trina* on

Quercus and the pannose mycelium of *S. pannosa* on *Rosa* and *Prunus persica*, there are no or few conidia and the mycelium seems more felty or glistening than powdery. No conidia are known for *Erysiphe aggregata* (Yarwood, 1973), *Podosphaera major* (Blumer, 1967) and *Typhulochaeta* (Clare, 1964). The conidial stage of Erysiphaceae is usually more conspicuous than the perithecial stage and was likely recognized as a fungus and as a pathogen long before the perithecial stage. Yet the conidial stage is rarely used in classification. This is probably because perithecia were recognized soon after taxonomy became fashionable, and because early as well as present schemes of classification usually, by rules of nomenclature, are based on the perfect stage, if present. Also, it is because early investigators as well as some present ones regarded conidial stages of genera and species to be indistinguishable from each other, and because the structure of the conidial stage is not well preserved in dried specimens, while the perithecial stage is.

Conidiophores of Erysiphaceae range in length from about 60 μm (Clare, 1964, gives much lower values) for *M. alni* on the upper surface of oak leaves, to 670 μm for *S. humuli* on the lower surface of strawberry leaves (Yarwood and Gardner, 1970) or even 700 μm for *Leveillula* (Salmon, 1906b). Part of this difference is undoubtedly a character of the fungus and part is due to the physical environment and the host. Chain–forming conidiophores such as *E. cichoracearum* are usually longer than conidiophores where the conidia are borne singly such as *E. polygoni*. The same culture may produce spores singly on one host species and in chains on another (Peries, 1966). With chain forming species the number of cells of the chain may range from three, and thus closely resemble *E. polygoni*, to 12 or more as in *P. leucotricha* (Yarwood, 1937). At high humidity *E. polygoni* type conidiophores may form chains of conidia (Neger, 1901; Salmon, 1900). Conidiophores are commonly longer on the lower leaf surface than on the upper and longer on leaves with hairs than on glabrous leaves (Yarwood and Gardner, 1970). Both chain forming and non-chain forming conidiophores show a diurnal periodicity of development (Childs, 1940; Yarwood, 1936a) but no diurnal periodicity was noted for *E. graminis* (Pady *et al.*, 1969). The presence of wart-like structures on otherwise smooth conidiophores (Crepin, 1922; Golovin, 1958; Kendrick and Carmichael, 1973) is rare and I have not seen them.

The principal types of conidiophores are those with a swollen basal cell *v.* those with unswollen basal cell, those with a twisted basal cell *v.* those not twisted, those with conidia borne singly *v.* those borne in chains, those with fibrosin bodies *v.* those without fibrosin bodies and those with conidia of various shapes. Examples of types of conidiophores and conidia are illustrated in Figs 5–17.

Only two spore stages in the Erysiphaceae—conidial and ascus—are usually now recognized but others have been described. Berkeley (1857) believed *Erysiphe* had five spore stages. A pycnidial fungus described by

Tulasne (1853) has been shown to be a parasite of Erysiphaceae usually called *Cicinnobolus cesatii* or *Ampelomyces quisqualis* (see Rogers, 1959). This parasite tends to suppress powdery mildews in rainy areas and seasons, but has not been found of commercial value for disease control in dry areas where powdery mildews are more severe. Vanha (1903) described and illustrated zoosporangia and zoospores of *E. polygoni* (*Microsphaera betae*) of beet, but these have never been confirmed. He may have confused the explosion of conidia in water with germinating sporangia. "Hunger" conidia were described by Neger (1902). Microconidia were described by Homma (1937) for *Sawadaea tulasni* (*Uncinula aceris*) and by Blumer (1967) for *Uncinula tulasni*. Chlamydospores are described by Foex (1924).

If powdery mildews are identified on the basis of their imperfect stages, they are usually placed in the genus *Oidium* (Hirata, 1966; USDA, 1960) though *Acrosporium* would be more correct (Linder, 1942). This placing of imperfect stages in a single genus and with or without designation of species is based on the incorrect idea that most imperfect stages are indistinguishable from each other (Dash, 1913; Large, 1940; Alexopoulos, 1952). I believe with Foex (1912), Clare (1964) and Sawada (1927) that many powdery mildews can be correctly assigned to genera and species on the basis of their conidial stages but this will require a more careful rendering of structure than is commonly encountered in published illustrations. The common incorrectness of the reported structure of conidiophores is well demonstrated with *U. necator* for which most illustrations are incorrect or misleading (Weltzien and Yarwood, 1963).

While I consider conidial stages of value in the identification of Erysiphaceae, there are serious limitations. The length of conidiophores varies with time of day (Childs, 1940; Yarwood, 1936a), surface and hairiness of leaves (Yarwood and Gardner, 1970), humidity (Neger, 1901; Salmon, 1905), host species (Peries, 1966) and undetermined factors (Yarwood, 1957). I believe that size of perithecia, conidiophores, or conidia should usually not be used even for the separation of species.

D. Conidia

When conidia are used in taxonomy, it is the size of conidia which is most commonly used to characterize species (Allison, 1934; Weltzien, 1963). Size of conidia is fairly constant for a given host in a given environment (Blumer, 1922; Bouwens, 1927) but varies greatly with host nutrition (Neger, 1902), age and vigour of host leaves (Neger, 1902; Fischer, 1957), nutrition of host leaves (Zwirn, 1943), season (Homma, 1937), humidity (Bouwens, 1924; Neger, 1901), host species (Bouwens, 1924; Homma, 1937) and undetermined factors (Yarwood, 1957). I believe size of conidia should usually not be used

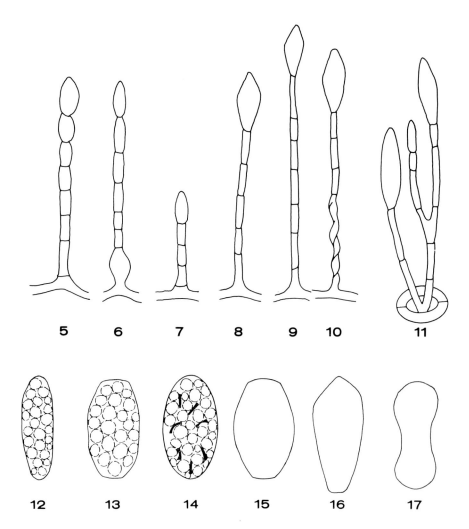

5   6   7   8   9   10   11

12   13   14   15   16   17

in the taxonomy of Erysiphaceae. Rather, shape and the presence of vacuoles, fibrosin bodies and oil drops within the conidia seem to be their best diagnostic characters.

Taxonomy of the conidial stage has proceeded largely independently of the perfect stages. De Candolle (1805), Salmon (1900), Blumer (1933, 1967) and several others either omitted consideration of the conidial stages of Erysiphaceae or placed conidial stages in the genus *Oidium*. The genus *Oidium* was created by Link in 1809 but the original descriptions and illustrations of *Oidium* (lateral spores, etc.) would not include our present concept of the imperfect stage of Erysiphaceae. However, in 1824 Link clearly includes the conidial stage of powdery mildews in the genus *Oidium*, though he also included non-Erysiphaceous fungi. The first genus name applied to the imperfect stage of Erysiphaceae appears to be *Acrosporium* by Nees von Esenbech (1817) and he included an illustration of *Acrosporium monilioides*, which we now recognize as *E. graminis* but he also included some non-Erysiphaceous fungi in this genus. I agree with Ainsworth (1971), Hughes (1958), Kendrick and Carmichael (1973), Linder (1942) and Sumstine (1913) that *Acrosporium* is the correct genus name and *Oidium* is incorrect for imperfect stages of Erysiphaceae, even though *Oidium* is much better established in the literature (Blumer, 1967; Hirata, 1966; USDA, 1960). Though it may be incorrect, the genus name *Oidium* has been applied to some 90 species names of powdery mildews (Hirata, 1966). Should these species be renamed *Acrosporium*? I think not. I recommend that both *Oidium* and *Acrosporium* be avoided when possible and that conidial stages of Erysiphaceae be named for the conidiophore type corresponding to the perfect stage, as done by Yarwood (1937), Clare (1964) and Gardner and Yarwood (1974) though this method is not completely satisfactory. Other generic names applied to

Fig. 5. Conidiophore forming conidia in a chain as with *Erysiphe cichoracearum*, and *Sphaerotheca fuliginea*.
Fig. 6. Conidiophore with bulbous base, as with *Erysiphe graminis*.
Fig. 7. Conidiophore forming conidia singly, as with *Erysiphe polygoni*, *Microsphaera alni*, and *Uncinula necator*.
Figs 8 and 9. Conidiophores of *Phyllactinia* sp.
Fig. 10. Conidiophore of *Phyllactinia subspirales*.
Fig. 11. Conidiophore of *Leveillula taurica*.
Fig. 12. Conidium with vacuoles, characteristic of *Erysiphe graminis*.
Fig. 13. Conidium with vacuoles and flattened sides characteristic of *Microsphaera alni* from *Acer macrophyllum*.
Fig. 14. Conidium with vacuoles and fibrosin bodies, characteristic of *Sphaerotheca fuliginea* from *Cucumis sativus*.
Fig. 15. Barrel-shaped conidia characteristic of *Sphaerotheca lanestris* from *Quercus agrifolia*.
Fig. 16. Pointed conidium characteristic of *Phyllactinia corylea* from *Aesculus californica*.
Fig. 17. Constricted conidium characteristic of *Phyllactinia corylea* from *Quercus agriolia*. (All illustrations modified from Yarwood, 1973.)

conidiophore stages are *Alysidium, Oidiopsis, Oospora, Ovularia, Ovulariopsis, Trichocladia* and *Torula*.

E. Fibrosin bodies

Well developed fibrosin bodies (f.b.) (Zopf, 1887) occur in the conidiophores and conidia of most *Sphaerotheca* and *Podosphaera* and some *Uncinula* spp. but do not occur in *Erysiphe* or *Microsphaera* (Sawada, 1914). They have been emphasized in recent years as a character which distinguishes *S. fuliginea* (with f.b.) from *E. cichoracearum* (without f.b.) which otherwise have similar conidiophores (Ballantyne, 1963; Clare, 1964; Nagy, 1970). I agree with the utility of this character but with Neger (1901) and Blumer (1967) I feel that it may have been overemphasized. The prevalent form of cucurbit powdery mildew in most parts of the world and in California does not have perithecia, but usually has f.b. and has therefore been named *S. fuliginea* (Clare, 1964). Even f.b.-forming cultures do not always have them in every conidium. My culture of cucumber powdery mildew showed 4–8 f.b. in each conidium produced on the upper surface of cotyledons and true leaves in the green-house at about 20°C but on the lower surface of old true leaves, f.b. were sometimes absent. On outdoor squash in Berkeley the percentage of conidia with f.b. has been as low as 20%. The incidence of f.b. on cucumber in the greenhouse was lower at 16°C than at 20, 24, or 28°C. With *S. lanestris* the percentage of conidia with f.b. has been as low as 3% (*Quercus velutina*, 3 May 1971). Most conidia of *S. humuli* from hops in Moscow, Russia, 21 July 1975, did not contain fibrosin bodies.

F. Vacuoles

Vacuoles are the clear, spherical bodies in the protoplasm of conidia and mycelium and are the most conspicuous contents of most powdery mildew conidia. Most conidia contain vacuoles but they vary greatly in size, number and visibility in different species and different environments. In my observations the average number of vacuoles per conidium and their visibility were 3·8 and conspicuous for *S. pannosa* on rose, 9·3 and conspicuous for *S. lanestris* on oak, 15 and moderately conspicuous for *S. fuliginea* on cucumber, and 3·0 and faintly visible for *E. graminis* on barley. On 15 September the average number of vacuoles for *S. lanestris* was 2·5 and many conidia had only one vacuole which almost filled the spore. Sawada (1914) indicated that conidia of *Erysiphe galeopsidis, Podosphaera tridactyla, P. leucotricha, Sawadaea aceris* and *Uncinula clandestina* do not contain vacuoles but I do not agree. I have found no vacuoles in conidia of *Leveillula* on *Mimulus aurantiacus*.

I believe that vacuoles are primarily water and powdery mildews are apparently the only fungi with air borne spores with a high water content (Yarwood, 1950, 1952). Somers and Horsfall (1966) indicate that conidia of powdery mildews have a lower water content than other air borne fungi but I think their data are questionable.

Oil is reported in conidia or ascospores of Erysiphaceae by Tulasne and Tulasne (1861), Yarwood (1950), Kimbrough (1963), McKeen (1970), Yarwood and Gardner (1972) and Hess *et al.* (1974). Conidia of *Uncinula salicis* on *Salix* sp. are almost filled with clear, yellowish globules which are not miscible with water and which I believe to be oil. I have not found oil globules in conidia of *U. necator* or in any other genus of Erysiphaceae. This oil might be a useful taxonomic character but more study is needed.

Particles within vacuoles of conidia of *E. cichoracearum, E. graminis, E. polygoni, Phyllactinia corylea, S. fuliginea, S. humuli, S. lanestris,* and *S. pannosa* showing rapid motion (presumably Brownian movement) are not uncommon though most collections show none when observed in the fresh condition. With conidia of *E. polygoni* from *Trifolium pratense* mounted in water under a coverglass, the percent of spores with moving particles in the vacuoles averaged 4% at about 30 s after mounting, 94% at 100–180 min after mounting and 0% at 24 h after mounting. This development and loss of Brownian motion in vacuoles of Erysiphaceae probably has little taxonomic significance though differences between species in this regard are very great. Of 22 powdery mildews examined as fresh specimens in the USSR Botanic Garden, Moscow, in July 1975, Brownian movement was observed only with *S. humuli* from *Humulus lupulus* and most conidia of this species showed none.

G. Appressoria

Appressoria are the holdfasts which attach the hyphae to the host surface and from which haustoria extend into the lumen of host cells. They are usually terminal on the first germ tube from a conidium but lateral on later vegetative hyphae. They may be large and highly lobed (Kimbrough, 1963) or small and inconspicuous (Salmon, 1900; Clare, 1964). They differ considerably between species but except for Clare (1964) and Zaracovitis (1965) they seem never to have been used as a taxonomic character.

H. Nuclei and chromosomes

Each vegetative cell, conidium and haustorium usually contains one nucleus, with a conspicuous nucleolus. The inner cells of the wall of the perithecium are binucleate (Homma, 1937).

The haploid chromosome numbers of Erysiphaceae, summarized by Clare (1964), are usually 4 for eight species in five genera. However, for *Phyllactinia corylea*, values of 4, 8, and 10 are given by different observers.

I. Discharge of conidia

The mechanism of release of conidia is controversial. Hammarlund (1925) fitted capillary tubes over individual conidiophores and observed spore maturation microscopically. He reported that the rim of attachment of a maturing conidium to the cell below was suddenly broken and that the maturing conidium was forcibly discharged to a distance of several spore lengths by the sudden rounding off of the two adjacent cells by internal pressure. Mendonca (1965) illustrates the torn membrane left after the discharge of conidia. Yarwood (1936a), Childs (1940), and Brodie and Neufeld (1942) concluded that spore liberation was passive though Yarwood (1957) reported sudden displacements of maturing conidia when dish cultures of *E. polygoni* on detached leaves were opened. Leach (1975) found that active discharge of conidia of *S. fuliginea* resulted from the exposure of cultures to infra-red light and decreasing relative humidity. Clerk and Ankara (1969) describe two ways in which the release of conidia of *Phyllactinia corylea* apparently differs from all other powdery mildews. They found that the generative cell was terminal while the generative cell of all other Erysiphaceae is subterminal. They also found that when the conidium is mature the generative cell bends through more than 90° while the conidium is being matured.

J. Germination of conidia

Neger (1902), Clare (1964) and Zaracovitis (1965) have distinguished some Erysiphaceae by the method of germination of the conidia. *E. betae* (*E. polygoni*) formed short germ tubes with lobed appressoria; *S. fuliginea* formed short germ tubes with forked appressoria and *S. macularis* formed long germ tubes without appressoria (Zaracovitis, 1965). *E. polygoni*, *M. alni* and *U. necator* conidia germinate at close to 0% relative humidity (Delp, 1954). Conidia of *S. pannosa* and *Podosphaera leucotricha* require approximately 100% relative humidity for germination on glass (Brodie, 1954). The germ tubes of some species are phototropic (Neger, 1902) though those of *E. polygoni* may be positive or negative (Yarwood, 1957). The host may greatly stimulate the germination of conidia (Jhooty, 1971; Yarwood, 1936a).

    Under most conditions conidia of Erysiphaceae are short-lived (Yarwood *et al.*, 1954). Hammarlund (1925) and Hermansen (1972) found that conidia produced under dry conditions are more infective and/or live longer than conidia produced under humid conditions. Hermansen found that conidia of *E. graminis* produced at 25% relative humidity could live for two years at −20 to −60°C.

For obligate parasites such as Erysiphaceae, the germination of spores is one of the few characters which can be studied apart from the host. With most other characters such as size of spores, type of appressoria, formation of conidia and perithecia, etc. the expression of the fungus could be a result of host influence. This host influence is a major argument for including the name of the host in any discussion of physiology or taxonomy of powdery mildews and for caution in giving different species names to mildews from different hosts.

K. Heterothallism

Heterothallism is reported for *E. cichoracearum* (Yarwood, 1935), *E. graminis* (Hiura, 1962), *E. polygoni*, *Microsphaera penicillata* and *U. necator* (Smith, 1970). *E. polygoni* is homothallic on *Ranunculus* but heterothallic on six other species.

L. Host range

The easiest character of a powdery mildew to recognize is usually the host on which it is found. Host range is not usually used to characterize genera but there are clear host limitations of genera. *Erysiphe*, *Sphaerotheca*, and *Leveillula* generally occur on herbs, while *Microsphaera*, *Podosphaera*, *Phyllactinia*, *Polychaeta* and *Uncinula* usually occur on woody plants. *E. graminis* is found only on grasses and no other genus or species of powdery mildew occurs on grasses. The conidial stages of *E. polygoni* and *M. alni* are usually indistinguishable morphologically but they are usually separable by hosts as indicated above.

Most species of Erysiphaceae are named for a host genus on which they are found though this practice has been logically deplored by Salmon (1900) and others. Of the 90 or more named species of *Oidium* (Hirata, 1966), 84 were named for a host genus. This naming of Erysiphaceae by host genera is defensible in that each race (pathologically distinct culture) is usually restricted to a single host genus or closely related genera and some are restricted to certain clones within a variety (Yarwood, 1936b). Yet each morphologic species of Erysiphaceae usually occurs on several host genera and species. The morphologic species of powdery mildew with the greatest host range is likely *E. graminis* with 609 host species (Hirata, 1966) and *E. polygoni* may be next with 356 host species (Salmon, 1900). The cultural species with the greatest host range may be *S. fuliginea* from cucumber (Stone, 1962; Alcorn, 1969) and *Erysiphe polyphaga* of Hammarlund (1945) may be the same or similar species.

Salmon (1904), Steiner (1908) and Homma (1937) indicate that host range of Erysiphaceae may be modified by host passage but Blumer (1922) and Hammarlund (1925) found otherwise. Stakman and Harrar (1957) conclude that rusts and mildews do not adapt to foreign hosts by host passage. It seems logical that adaptation to hosts, as adaptation to antibiotics (Moyed, 1964), can occur in powdery mildews but if so, the conditions for its occurrence seem as yet unclear.

Usually only one species of Erysiphaceae occurs on one species of host but there are so many exceptions that this generalization might discourage progress and lead to incorrect diagnosis. Only one species is known each for *Allium cepa*, *Beta vulgaris*, *Vitis vinifera*, etc.; two each for *Alnus glutinosa*, *Betula alba* and *Corylus cornuta*; and three each for *Quercus prinus*, *Ribes grossularia* and *Prunus domestica* (Hirata, 1966). *Quercus agrifolia* is the only species on which Yarwood *et al.* (1972) found as many as four species of Erysiphaceae in California but Hirata (1966) reports six species of Erysiphaceae on *Q. serrata* in Japan.

Leaves are the most common site of infection but stems, petals and fruits are commonly invaded. Most Erysiphaceae occur on both surfaces of leaves (amphigynous) but they are usually more abundant on one surface than the other. *E. trina* is almost exclusively epiphyllous while *Phyllactinia* is hypophyllous. Infection may almost completely avoid fruits as with *Sphaerotheca* on most cucurbits or *Podosphaera* on most apples and pears but may preferentially attack the fruits as with *E. aggregata* on female alder catkins or *Podosphaera* on maple keys.

M. Symptoms *v.* Signs

As diseases, powdery mildews are usually diagnosed by the presence, and details of the morphology, of the causal agent (signs); rarely, if ever, by the host response (symptoms). Unlike most diseases, powdery mildews usually cause no symptoms other than a slow decline (or killing) of plant parts and decrease in productivity of their hosts. The only well-defined exception I know is *S. lanestris* on *Quercus*, especially *Quercus agrifolia* (coast live oak) in California. On this host *Sphaerotheca* causes inconspicuous localized infections on the lower surfaces of leaves but also causes conspicuous systemic infections of buds and branches commonly leading to witches brooms. These systemically infected shoots have smaller leaves than normal, are thicker than normal and grow faster than normal but they usually die during the winter. They are similar in general appearance to the witches brooms caused by mycoplasms (Bos, 1970) and it has been suggested that these witches brooms are caused by a virus (Yarwood, 1971) or mycoplasm carried by the *Sphaerotheca*. Hyperplasia of *Malus baccata* due to *Sphaerotheca leucotricha* is indicated by Homma (1937). Girard (1889) indicates that *Erysiphe martii* may cause increase in the size of leaves of *Hypericum* but sterility of the flowers. I have observed slight localized hypertrophies of leaves of *Helianthus*, *Crataegus* and *Humulus* due to mildew infection.

In incompatible (resistant) hosts or tissues, many mildews cause necrosis (hypersensitive reaction) at the point of penetration and this necrosis may extend to non-invaded cells close to the invaded cells. A good example is necrosis of ripening green grape fruits and grape canes caused by *U. necator*.

N. Life history

Powdery mildews may pass a dormant season as perithecia, as mycelium and as chlamydospores. For some there is apparently no dormant season and for some the dormant stage, if any, is unknown. Perithecia are likely agents for the overwintering of *Phyllactinia* on *Aesculus* and *Microsphaera* on *Platanus* and for the oversummering of *E. graminis* on *Hordeum* (Koltin and Kenneth, 1970). Powdery mildews believed to overwinter as mycelium in dormant buds are *Microsphaera* on *Syringa* and *Uncinula* on *Vitis* (Yarwood, 1944) and *Sphaerotheca* on *Rosa* (Price, 1970) even though they form an abundance of perithecia. Chlamydospores are reported to function for the overwintering of mildews on *Quercus* and *Vitis* (Blumer, 1967) but I think this needs confirmation. *E. trina* on *Quercus*, *Microsphaera* on *Euonymus* and other powdery mildews are apparently active throughout the year in California and therefore need no special method of passing the dormant season. The method of overwintering of *Sphaerotheca* on *Cucumis* and *Erysiphe* on *Phaseolus* and several other powdery mildews is obscure. It is possible that they remain active throughout the year in southern areas and are carried north by wind each summer season.

IV. History of Taxonomy

Powdery mildews like many other parasitic fungi were once considered to be excretions of their hosts (Forsyth, 1824; Whetzel, 1918). The first specific naming of a powdery mildew as an organism was probably that of Linnaeus (1753) when he gave the binomial *Mucor erysiphe* to a white fungus on the leaves of *Humulus, Acer, Lamia, Galeopsis* and *Lithospermum*. He did not describe or illustrate either the conidial or perithecial stage. Why he called it *Mucor* or *Erysiphe* is not clear. *Mucor* as a genus was listed by Linnaeus as coordinate with *Agaricus, Boletus, Peziza*, etc. and was probably taken from Micheli (1729) who distinguished *Mucor* from *Puccinia, Lycogala*, etc. Micheli's illustration of *Mucor* on Plate 95 is recognizable as what we now call *Rhizopus. Mucor* was likely a group name for what we would now consider several genera of fungi. *Erysiphe* is usually considered as derived from the Greek *erythros* (red) and its application to powdery mildew may result from the grouping of the powdery mildews with the reddish rust fungi (see *Erysiphe* in Ainsworth, 1971). Tulasne and Tulasne (1861) indicated that mycelium of *E. graminis* may be reddish. No powdery mildew is ever red in my opinion but maturing perithecia are brown, as is easily seen in the maturing perithecia of *E. aggregata* on alder catkins of *Alnus* and Large (1940) writes of red or brown mould in referring to powdery mildew of hops. The name *Mucor* was soon dropped from the names of powdery mildews because it had previously been applied to a clearly distinct fungus but *Erysiphe* was continued as a genus and later as a family name applying to the group.

Presumably following Linnaeus' nomenclature but aware that *Mucor* was invalid, De Candolle (1805, 1815) names 25 new species of *Erysiphe*—more, I believe, than named by any other mycologist before or since—and many if not most of those species names are still in vogue, even though they were named primarily for the host on which they were found.

Leveille (1851) was apparently the first to set up a taxonomic system with a key for the genera of powdery mildews. He divided De Candolle's *Erysiphe* into the genera *Calocladia, Erysiphe, Phyllactinia, Podosphaera, Sphaerotheca* and *Uncinula*. He feared he would be criticized for forming so many new genera but five of these genera are still universally used today and include most species of powdery mildews accepted today.

In 1900 Salmon published a monograph of the Erysiphaceae of the world and his keys to genera and species are probably the most widely used today. Except for changing *Calocladia* to *Microsphaera*, Salmon's key to genera is almost identical to Leveille's. The major criticism I have of Salmon's taxonomy, as that of Leveille before him, is that he practically omitted consideration of the conidial stage. Since Salmon, many new genera of Erysiphaceae have been added, but most have not stood the test of time. Alexopoulos (1952), Bessey (1961) and Clements and Shear (1931) and others, added species with two- and four-celled ascospores but most taxonomists have limited the family to species with one-celled ascospores. Until Salmon's monograph, most studies of powdery mildews were probably taxonomic; since Salmon, most studies have been pathologic, physiologic and genetic. A chronologic review of the major genera of Erysiphaceae by the major contributors to their taxonomy is as follows:

TABLE 1

Major Genera of Erysiphaceae (chronological by authors)

1763	Linnaeus	*Mucor*, 1 species
1796	Persoon	*Sclerotium*, 1 species
1805,		
1815	De Candolle	*Erysiphe*, 25 species mostly named for host genus
1819	Wallroth	*Alphitimorpha*, 25 species
1821	Merat	*Erysiphe*, 12 species
1829	Fries	*Erysiphe*, 16 species
1829	Schlectendal	*Alphitimorpha*, 19 species. First separation into groups based on number of asci and type of appendages.
1851	Leveille	*Calocladia, Erysiphe, Phyllactinia, Podosphaera, Sphaerotheca, Uncinula*, 33 species
1896	Jaczewski	*Apiosporium, Dimerisporium, Erysiphe, Microsphaera, Microthyrium, Phyllactinia, Podosphaera, Sphaerotheca, Uncinula*
1900	Salmon	*Erysiphe, Microsphaera, Phyllactinia, Podosphaera, Sphaerotheca, Uncinula*, 49 species, 11 varieties
1911	Pollacci	*Erysiphe, Microsphaera, Phyllactinia, Podosphaera, Sphaerotheca, Uncinula*, 25 species
1913	Migula	*Erysiphe, Microsphaera, Phyllactinia, Podosphaera, Sphaerotheca, Trichocladia, Uncinula*, 39 species
1915	Ito	*Erysiphe, Microsphaera, Oidiopsis, Phyllactinia, Podosphaera, Sawadaea, Sphaerotheca, Typhulochaeta, Uncinula*
1921	Arnaud	*Erysiphe, Leveillula, Microsphaera, Phyllactinia, Podosphaera, Sphaerotheca, Uncinula*
1937	Homma	*Cystotheca, Erysiphe, Leveillula, Microsphaera, Phyllactinia, Podosphaera, Sawadaea, Sphaerotheca, Typhulochaeta, Uncinula, Uncinulopsis*, 74 species
1958	Golovin	*Arthrocladia, Blumeria, Cystotheca, Erysiphe, Leveillula, Linkomyces, Microsphaera, Phyllactinia, Podosphaera, Sphaerotheca, Trichocladia, Uncinula*
1961	Bessey	*Astomella, Brasilomyces, Chilomyces, Erysiphe, Microsphaera, Leucoconia, Phyllactinia, Podosphaera, Schistodes, Sphaerotheca, Typhulochaeta, Uncinula, Uncinulopsis*
1964	Clare	*Brasilomyces, Cystotheca, Erysiphe, Leveillula, Microsphaera, Phyllactinia Podosphaera, Sawadaea, Sphaerotheca, Typhulochaeta, Uncinula, Uncinulopsis*
1966	Hirata	*Arthrocladia, Brasilomyces, Cystotheca, Erysiphe, Erysiphopsis, Leveillula, Medusosphaera, Microsphaera, Oidium, Oidiopsis, Ovulariopsis, Phyllactinia, Podosphaera, Quierozia, Salmonomyces, Sawadaea, Sphaerotheca, Trichocladia, Typhulochaeta, Uncinula, Uncinulopsis*, 370 species
1967	Blumer	*Brasilomyces, Cystotheca, Erysiphe, Leveillula, Medusosphaera, Microsphaera, Oidium, Phyllactinia, Pleochaeta, Podosphaera, Sphaerotheca, Typhulochaeta, Uncinula*, 140 species

The following table shows all the generic names applied to Erysiphaceae, which I have been able to find, with the names of the authors who are believed to have first used them, together with some justification for inclusion

here. My recommendations for future use are A = accept; R = reject;
D = doubtful.

TABLE 2

Generic Names Applied to Erysiphaceae

Generic name	Author	Justification	Recom-mendation
Acrosporium	(Nees, 1817) = *Oidium*.	Correct name for conidial stages if such is necessary. Also applied to non-Erysiphaceae. See Linder, 1942.	D
Albigo	(Kuntze, 1892)	Contained some present species of *Podosphaera* and *Sphaerotheca*. See Salmon, 1900.	R
Albugo	(Persoon, 1801?)	Used as synonym of *Alphitimorpha* by Wallroth, 1829. Now included in Peronosporales.	R
Alphitimorpha	(Wallroth, 1819)	Genus name for all Erysiphaceae.	R
Alysidium	(Kunze and Schmidt, 1817?)	Given as synonym of *Oidium* by Fries (1829).	R
Anixia	(Fries, 1829)	Included in Karsten's (1885) Erysiphaceae. Now usually in Perisporiaceae.	R
Apiosporium	(Schroeter, 1893?)	Used as genus of Erysiphaceae by Jaczewski (1896).	R
Arthrocladia	(Golovin, 1958)	In part *Microsphaera* and *Calocladia*. Invalid because first used for algae (Duby, 1830). Used as section of genus *Microsphaera* by Blumer, 1967.	R
Arthrocladiella	(Vasil'Kev, 1960?)	Apparently used in place of *Arthrocladia*. See Ainsworth, 1971.	R
Aspergillus	(Micheli, 1729)	Given as synonym of *Oidium* by Fries (1829).	R
Astomella	(Thirumalachar, 1947)	Septate ascospores. Intermediate between Erysiphaceae and Perisporiaceae.	R
Blumeria	(Golovin, 1958)	A genus for present *Erysiphe graminis*. Probably justified, but tentatively rejected in the cause of preservation of established names.	D
Botrytis	(Castagne, 1851)	Listed as synonym of *Sphaerotheca* by Salmon, 1900. Now in Moniliales.	R
Brasilomyces	(Viegas, 1944)	No appendages, 2 asci per perithecium. Similar to *Erysiphe trina*.	R

TABLE 2 (*contd.*)

Generic name	Author	Justification	Recom- mendation
Calocladia	(Leveille, 1851) = *Microsphaera*.	Invalid because first used for algae (Agarth, 1852). Used as section of genus *Erysiphe* by Blumer, 1933.	R
Capnodium	(Montagne, 1849?)	Included in Karsten's (1885) Erysiphaceae. Now in Capnodiales.	R
Chilemyces	(Spegazzini, 1910)	Septate ascospores. Given as genus of Erysiphaceae by Bessey (1961).	R
Crocisporium or *Corcysporium*	(Bonorden, 1861)	Conidial stage of *Erysiphe graminis*? See Salmon, 1905.	R
Cystotheca	(Berkeley and Curtis, 1858)	Single ascus, No or poorly defined appendages, dark hyphae with hairy branches. Not clearly defined as Erysiphaceae. Accepted as valid genus by Blumer (1967).	D
Dematium	(Sprengel, 1806)	Lateral conidia. Given as synonym of *Phyllactinia* by Salmon (1900). Now in Dematiaceae.	D
Dichothrix	(Thiessen, 1912)	Septate ascospores. Given as synonym of Erysiphaceae genus by Golovin (1958). Given as alga by Lemmerman (1910). Now in Capnodiaceae.	R
Dimerina	(Thiessen, 1912)	Septate ascospores. Given as synonym of Erysiphaceae genus by Golovin (1958). Now in Capnodiaceae.	R
Dimerisporium	(Jaczewski, 1896)	Septate ascospores.	R
Erysibe	(Tozzetti, 1767)	Given as synonym of *Podosphaera* by Salmon (1900). Has also been used for rusts and smuts.	R
Erysiphe	(Linnaeus, 1753)	Hypha-like appendages, several asci. Used as species name by Linnaeus. Used as genus name for all Erysiphaceae by De Candolle (1805). First used in its present meaning by Leveille (1851).	A
Erysiphella	(Peck, 1876)	Several asci, no appendages. Usually incorporated in *Erysiphe*.	D
Erysiphites	(Pampaloni, 1902?)	Fossil genus, not accepted by Salmon (1903).	R
Erysiphopsis	(Halsted, 1899)	Characters intermediate between *Erysiphe* and *Phyllactinia*.	R
Erysiphosis	(Sathe, 1969)	Instead of *Salmonomyces*.	R
Euoidium	(Brundza, 1934)	Imperfect genus for Erysiphaceae, with conidia borne in chains.	R
Eurotium	(Greville, 1825)	*E. roseum* is included in Salmon's *Sphaerotheca*. *E. herbariorum* is not in Erysiphaceae.	R
Ischnochaete	(Golovin, 1958)	Listed in Erysiphales by Ainsworth (1971).	R

TABLE 2 (*contd.*)

Generic name	Author	Justification	Recom- mendation
Kokkalera	(Ponnappa, 1970)	Like *Sphaerotheca* but without appendages.	R
Lanomyces	(Gaumann, 1922)	Included in Erysiphaceae by Clements and Shear, 1931. Usually in Perisporiaceae. See Blumer, 1967.	R
Lasiobotrys	(Schroeter?)	According to Pollacci (1911) Schroeter includes *Lasiobotrys* in Erysiphaceae.	R
Leucoconis	(Thiessen and Sydow, 1917)	Septate ascospores. Included in Erysiphaceae by Clements and Shear, 1931.	R
Leveillula	(Arnaud, 1921)	Created to accommodate truly endophytic mycelium.	A
Linkomyces	(Golovin, 1958)	A genus for present *Erysiphe communis*. probably justified, but tentatively rejected here in the cause of preservation of established names. Used by Blumer (1967) for species of *Erysiphe* with conidia borne singly.	D
Medusosphaera	(Golovin and Gamalitskya, 1962)	Based on serpentine appendates. Listed as valid genus by Blumer, 1967.	R
Meliola (*Meliopsis?*)	(Fries, 1825)	Given as synonym of *Sphaerotheca* by Salmon (1900). Now in Perisporiales.	R
Microsphaera	(Leveille, 1851)	Several asci, dichotomous appendages.	A
Microthyrium	(Demazieres, 1841)	Septate ascospores. Included in Erysiphaceae by Jaczewski (1896).	R
Monilia	(Hill, 1751)	Given as synonym for *Acrosporium* by Sumstine (1913).	R
Mucor	(Micheli, 1729)	Used as genus name for Erysiphaceae by Linnaeus (1753).	R
Oidiopsis	(Scalia, 1902)	Imperfect state of *Leveillula*,	R
Oidium	(Link, 1809)	A saprophytic member of Hyphomycetes. Widely and incorrectly used for imperfect stage of all Erysiphaceae (see Linder, 1942). Sometimes restricted to forms of Erysiphaceae with conidia formed in chains (Arnaud, 1921).	R
Oospora	(Wallroth, 1833?) = *Oidium.*	Listed as synonym of *Sphaerotheca* by Blumer (1933). Now in Moniliaceae.	R
Ovularia	(Saccardo, 1880?)	Given as synonym of *Erysiphe* by Salmon, 1905. Now in Moniliaceae	R

TABLE 2 (*contd.*)

Generic name	Author	Justification	Recommendation
Ovulariopsis	(Patouillard and Hariot, 1900)	Imperfect state of *Phyllactinia*. See Kendrick and Carmichael, 1973.	R
Perisporium	(Fries, 1829)	Given as synonym of *Erysiphe* by Salmon, 1900. Now in Perisporiaceae.	R
Phyllactinia	(Leveille, 1851)	Several asci, major appendages with bulbous base. The best defined genus in the Erysiphaceae.	A
Pleochaeta	(Saccardo, 1881 ?)	Several asci, straight appendages, internal mycelium. Usually included in *Uncinula*. See Kimbrough, 1963.	D
Podosphaera	(Kunze, 1823 ?)	Single ascus, dichotomous appendages.	A
Pseudoidium	(Brundza, 1934)	Imperfect genus for Erysiphaceae with conidia borne singly.	R
Queirozia	(Viegas, 1944)	Septate ascospores. Accepted by Hirata (1966).	R
Rhizocladia	(De Bary, 1871 ?)	Given as rejected genus by Golovin (1958). See Blumer (1967).	R
Saccardia	(Saccardo, 1882)	Septate ascospores.	R
Salmonia	(Blumer and Muller, 1964)	Small hyaline or pale brown perithecia. Close to *Erysiphe trina*. Blumer (1967) recommends incorporation into *Brasilomyces*.	D
Salmonomyces	(Chiddawar, 1959)	Undifferentiated appendages. Included in *Uncinula* by Ainsworth, 1971.	D
Sawadaea	Miyabe (Sawada, 1914)	Intermediate between *Podosphaera* and *Uncinula*. Recognized by Homma, 1937. *Incorporated* in *Uncinula* by Blumer, 1967.	R
Schinzia	(Nageli, 1842)	Nageli believed the penicillate appendages of *Phyllactinia* were a separate fungus.	R
Schistodes	(Thiessen and Sydow, 1917)	Septate ascospores. Included in Erysiphaceae by Clements and Shear, 1931.	R
Sclerotium	(Persoon, 1796)	Genus name for all Erysiphaceae. Now included in *Mycelia sterili*.	
Sphaeria	(Fries, 1823)	Listed as synonym of *Podosphaera* by Salmon (1900). See Wakefield, 1940.	R
Sphaerotheca	(Leveille, 1851)	Single ascus, hypha-like appendages.	A
Spolverinia	(Massalongo, 1855 ?)	Probably a *Phyllactinia* out of place. See Junell, 1964.	R
Sporotrichium	(Schweinitz, 1834)	Listed as synonym of *Oidium* by Anon. (1841).	R

TABLE 2 (contd.)

Genetic name	Author	Justification	Recom-mendation
Thelebolus	(Fries, 1823)	On dung. Perithecia similar to *Podosphaera*. See Cooke and Barr (1964).	R
Torula	(Persoon, 1796)	Imperfect genus (Tulasne and Tulasne, 1861). Listed as synonym of *Sphaerotheca* by Blumer (1933).	R
Trichocladia	(Neger, 1901)	Listed as synonym of *Microsphaera* by Blumer (1933).	R
Typhulochaeta	(Ito, 1915)	Clavate appendages, transparent perithecia, no conidiophores. Listed as valid genus by Blumer (1967).	D
Uncinula	(Leveille, 1851)	Several asci, uncinulate appendages.	A
Uncinulites	(Pampaloni, 1902?)	Fossil, not accepted by Salmon (1903).	R
Uncinulopsis	(Sawada, 1916)	Conidial stage of *Uncinula*. Accepted by Homma (1937).	R

V. Current Taxonomy

The number of species included in a taxonomic unit such as the genus or family depends on the differences used to separate them and on personal judgement. With the present status of powdery mildews a good case can be made for both the splitters (those who would create and/or maintain more genera and species) and the lumpers (those who would consolidate several species into one). The present number of named species of Erysiphaceae in the world is about 380 according to Hirata (1966) but this includes several synonyms. If the Erysiphaceae were separated on the basis of such small differences as have been used for the Perisporiaceae, the most closely related family, or the Uredinales, the most nearly comparable, heavily studied group, the number of species of Erysiphaceae would be greater than the some 1840 species of Perisporiaceae described by Hansford (1961) or the 530 species of Uredinales described by Arthur and Cummins (1934). For example, the conidiophores of the mildew on *Eschscholtzia californica* (California poppy) are morphologically distinct from most if not all other forms of powdery mildew I have examined though it is clearly an *Erysiphe polygoni* type (Yarwood, 1937). It is listed as *Erysiphe communis* (= *E. polygoni*) by Hirata, but whether this is on the basis of the conidial or perithecial stage is not stated.

In my opinion its inclusion as a new species would be as well justified as that of most species of *Erysiphe* or *Oidium* which have been described.

In addition to the 89 named species of *Oidium* (Hirata, 1966) there are some 1145 listings of *Oidium* spp. on some 1145 host species. It is likely that many, if not most, of these could be distinguished from each other morphologically and culturally and would on that basis merit ranking as a new species. If this logic were followed, we might have as many species of Erysiphaceae as there are host genera (1289 according to Hirata, 1966) or even host species (7606), or even more, since several genera or even species of higher plants are parasitized by more than one species of Erysiphaceae. I believe that in many cases the difference between those persons who give a powdery mildew a species name (e.g. *Oidium abelmoschi* on *Abelmoschus esculentus*) and those who do not (e.g. *Oidium* sp. on *Abutilon avicennae*) was merely the difference between a splitter and a lumper.

Physiologic specialization was reported in the Erysiphaceae by Neger (1902). As a result of this a popular and logical device of the splitters is to apply trinomials (e.g. *Erysiphe graminis tritici*) or even quadrinomials (e.g. *Erysiphe polygoni* f. *Robinae hespidae*, Blumer, 1933). These can be further subdivided and there are at least 72 races of *Erysiphe graminis hordei* (Anon., 1968) on the basis of host reaction. In this case I would prefer to designate them as strains of *E. graminis* from barley. New physiologic races may result from the fusion of genetically different mycelia (Siebs, 1958).

The arguments of the lumpers are just as logical and as unsatisfying as of the splitters. Salmon, who was the first to monograph the Erysiphaceae, was in my opinion a lumper. He combined 51 named species into one in the case of *E. polygoni* though some of these 51 were merely transfers from one genus to another and did not involve separate collections. With perhaps the greatest world collection of powdery mildews ever examined and reported on, he finished with only 49 species and other taxonomists examining the same material might well have finished with fewer. Hammarlund (1945) believed he showed experimentally that 12 *Oidium* species were different names for *Erysiphe polyphaga*.

More specifically, powdery mildew of sugar beet (*Beta vulgaris*) was first described by Vanha (1903) as *Microsphaera betae* n.sp.—the genus *Microsphaera* because of its dichotomously branched appendages and the species *beta* because it occurred on beet, did not infect clover and the strain on clover did not infect beet. This name seemed reasonably well justified though I guess Salmon would have included it in his *E. polygoni*, as would I. In 1933 Blumer called the fungus *Erysiphe communis*. In 1937 Yarwood found a conidial stage on beet corresponding to Vanha's *M. betae* but with no perfect stage and identified it as *E. polygoni* type which is morphologically similar to the *M. alni* type of conidiophore. In 1963 Weltzien reexamined the perfect and imperfect stages of this fungus and named it *Erysiphe betae* new comb., again with reasonable justification. In 1966 Hirata included the designation *Oidium* sp. for what is apparently the same fungus. In 1975 Coyier *et al.* reported the perfect stage for the first time in North America and, on the basis of data similar to that of Weltzien, decided it was *E. polygoni*. There are therefore five names applied with some justification to what is apparently the same fungus.

Since all powdery mildews are obligate parasites, we can usually only compare different collections on their natural hosts. Yet it is only when we compare collections on the same substrate and in the same environment that we can test if they are inherently different, other than in their host range. Many of the differences we see between species may be due to differences in hosts. Hammarlund (1945) found large differences in the conidial stage of a culture of *E. polyphaga* when it was produced on different hosts. Homma (1937) found the same for *E. graminis*. Yarwood (1957) found differences in the size of conidia of *E. polygoni* on *Trifolium pratense* due to environment. In this sense the validity of the difference between, say, *M. alni* and *S. lanestris* on the same tree of *Quercus agrifolia* is more certain than the difference between these same species on *Q. douglasii* and *Q. agrifolia* respectively. It is only when all Erysiphaceae can be cultured on the same artificial medium that we can be sure of the validity of the taxonomic units we are using but until that time, much caution should be used in creating new species and in recognizing old ones. Serology would be a good criterion for measuring relatedness of genera and species but apparently has not been used.

Some confusion with Salmon's taxonomy has arisen with *Erysiphe v. Microsphaera* and *Erysiphe v. Sphaerotheca*. Powdery mildews on lupine (Thompson, 1951; USDA, 1960), broad bean (McKenzie and Morral, 1975), red clover (Peterson, 1938), soybean (Paxton and Rogers, 1974), sweet pea (USDA, 1960) and sugar beet (Vanha, 1903; Coyier *et al.*, 1975) have been assigned to both *Erysiphe* and *Microsphaera*. That these collections on the same host species are really the same fungus species is indicated by the apparent identity of the conidiophores of *E. polygoni* and *M. alni*, by the apparent identity of the perithecia of *Erysiphe* and *Microsphaera* except for the ends of the appendages, by the overlap of dichotomous and hypha-like appendages (Yarwood, 1973) and by the otherwise rarity of *Microsphaera* on herbaceous plants. I do not accept the inference of Peterson (1938) that one reason for the apparent scarcity of *Erysiphe* and *Microsphaera* perithecia is because they can only be detected under magnification. I have looked for these perithecia on red clover on occasion for about 46 years and did not find them until August, 1975, at Kiev and Arkangelski, Russia and in both cases with no microscopic aid. I found some appendages with dichotomous branching but these were so hypha-like I readily assigned the specimens to *E. polygoni*. The confusion of *Erysiphe* and *Microsphaera* is also indicated by Salmon (1900), Neger (1901) and Homma (1937). Neger even set up the genus

Trichocladia as intermediate between *Erysiphe* and *Microsphaera*, and which might accommodate the above examples, but *Trichocladia* has not come into general use.

The confusion of *Erysiphe v. Sphaerotheca* has been mainly confined to *E. cichoracearum v. S. fuliginea* on cucurbits and has been partially reviewed by Nagy (1970). These two species have apparently identical conidiophores, except for the presence of fibrosin bodies in *Sphaerotheca*. Up to about 1960 (USDA, 1960) cucurbit powdery mildew, which rarely has perithecia, was usually listed as *E. cichoracearum*. With the delayed recognition of the diagnostic value of fibrosin bodies (Sawada, 1927), cucurbit mildew is now usually assigned to *S. fuliginea*.

Leveille's (1851), Salmon's (1900) and Blumer's (1967) taxonomy are remarkably similar. The main fault I find with all of these is the inclusion of the *E. cichoracearum* type, the *E. polygoni* type and the *E. graminis* type conidiophores in the same genus. Golovin (1958) has corrected this by the erection of the genus *Linkomyces* for *E. polygoni* (*E. communis*) and *Blumeria* for *E. graminis*. However, *E. polygoni* and *E. graminis* are so well established in the literature that it would be difficult and perhaps inadvisable to change them, at least until another comprehensive treatment of the Erysiphaceae of the world is undertaken.

I believe naming of Erysiphaceae by generic names, applied to the imperfect stages (*Acrosporium, Oidium, Oidiopsis, Ovulariopsis*, etc.) can and should be avoided if possible. Just as the apple scab fungus is named *Venturia inaequalis* when only the imperfect (*Fusicladium*) stage is present and the grape powdery mildew fungus is named *U. necator* when only the conidial (*Acrosporium* or *Oidium*) stage is present, so can other powdery mildew fungi usually be ascribed to a perfect stage genus when only the imperfect stage is present.

Keys for genera of Erysiphaceae based on the mycelia and perithecia, and of some genera and species based on the conidial stage are as follows:

Key to genera of Erysiphaceae based on mycelia and perithecia

1.	Mycelium usually superficial	2
	Mycelium partly internal	5
2.	Perithecia with one ascus	3
	Perithecia with two or more asci	4
3.	Appendages hypha-like or absent	*Sphaerotheca*
	Appendages dichotomously branched	*Podosphaera*
4.	Appendages hypha-like or absent	*Erysiphe*
	Appendages dichotomously branched	*Microsphaera*
	Appendages coiled at tips	*Uncinula*

5. Appendages hypha-like or absent   *Leveillula*
   Appendages with basal swelling   *Phyllactinia*
   Appendages coiled at tips   *Uncinula*

Key to some Erysiphaceae based on conidial stages

1. Conidiophores usually abundant   2
   Conidiophores rare   12
   Conidiophores unknown   13

2. Conidia borne singly   3
   Conidia borne in chains   8

3. Conidiophores emerging through stomata   *Leveillula*
   Conidiophores on superficial hyphae   4

4. Base of conidiophore twisted   5
   Base of conidiophore not twisted   6

5. On *Dalbergia*   *Phyllactinia*
   On *Celtis*   *Uncinula*

6. Conidia ellipsoidal   7
   Conidia pointed   *Phyllactinia*

7. On *Vitis, Salix* and other   *Uncinula*
   On *Quercus, Syringa* and other   *Microsphaera*
   On herbaceous plants   *Erysiphe polygoni*

8. Base of conidiophore swollen   *Erysiphe graminis*
   Base of conidiophore not swollen   9

9. Conidia barrel shaped   *Sphareotheca lanestris*
   Conidia ellipsoidal   10

10. With fibrosin bodies   11
    Without fibrosin bodies   *Erysiphe cichoracearum*

11. On herbaceous plants   *Sphaerotheca*
    On woody plants   *Podosphaera*

12. On *Quercus*   *Erysiphe trina*
    On *Cerocarpus*   *Sphaerotheca*
    On *Viburnum*   *Microsphaera*

13. On *Alnus*   *Erysiphe aggregata*
    On *Vaccinium* and *Oxycoccus*   *Podosphaera*
    On *Quercus*   *Typhulochaeta*

VI. Landmarks in Non-Taxonomic History

The three ways in which the pathology of powdery mildews differs most from that of other plant pathogens are probably the speed of spread of the diseases, the tolerance of the fungi to dryness and the ease of control with sulphur. Observations on each of these aspects extends back well over 100 years. Only sample data will be presented.

A. Rate of spread

The first well-recorded epidemic of a powdery mildew and probably the second well-publicized and well-documented extension of any plant disease was with *U. necator* in Europe from 1847–1851. Although the fungus had been known in America since 1834 (Schweinitz) it was apparently rather a mycological curiosity since most American varieties were rather resistant. Presumably the fungus was carried in grape cuttings to England, where it first appeared in destructive form on European varieties in greenhouses (Berkeley, 1847). By 1850 it had spread to field plants throughout the Mediterranean region and in 1854 it was believed that the French harvest of grapes was reduced to one fourth of normal as a result of mildew (Viala, 1886; Gaumann, 1945). Sulphur had been successfully used for the control of peach mildew (Robertson, 1821) and hop mildew and was applied, erratically at first, for the control of this new disease. By 1855 losses were greatly reduced as a result of control measures. Now grape powdery mildew is a chronic disease throughout the world but it is regularly and successfully controlled by sulphur dust.

Another example of the rapid spread of a powdery mildew is the movement of a strain of *E. polygoni* on red clover from New York in 1921 through several intermediate sites to Washington state in 1925 (Horsfall, 1930).

The most dramatic, rapid, and best documented spread of a powdery mildew, and probably of any disease, is the movement of *E. polygoni* on sugar beet from Imperial Valley, California, in April, 1974, to western Nebraska and central Washington in August, 1974 (Ruppel *et al.*, 1975). This approximately $300 \text{ km month}^{-1}$ movement of wind disseminated *Erysiphe* compares with the approximately 6 km month^{-1} for insect-disseminated *Ceratocystis ulmi* (data from Wysong and Willis, 1968). Of course the south to north seasonal movement of *Puccinia graminis* (Martin and Salmon, 1953) may be more rapid than that of *Erysiphe*, but the evidence for this is more indirect.

It is possible that this seasonal south to north movement of infections may also apply to clover powdery mildew, cucumber powdery mildew and other diseases in the United States.

B. Xerophytism

A unique but controversial aspect of the history of powdery mildews is their
relation to moisture. This discussion will be subdivided into moisture requirements for germination, change in volume of spores during germination,
moisture requirements for spore production, water and lipid content of
spores, soil moisture affecting disease, weather favouring disease and control
by spraying with water. The controversy concerns the conidial (pathogenic)
stage, since the necessity of free moisture for the release and germination of
ascospores (Cherewick, 1944) seems to be generally accepted.

It has been an almost universal belief that all fungus spores require external
free water for germination. For example Dugger (1909) wrote,

> Moisture generally augments the production of spores and is, of course, essential
> to their germination.

Brown and Wood (1953) wrote,

> . . . spore germination and penetration of the host tissue takes place only under
> moist conditions (as with fungal parasites generally) but the notable features of
> the powdery mildew type of parasite is the speed with which the successive
> operations of germination and entry into the host are brought about. The
> necessary humid conditions are provided by the dew period . . .

In contrast to this, Brodie (1945), Clayton (1942), Delp (1954), Yarwood
(1936c) and several others found that conidia of several powdery mildews
germinate at the lowest humidities which could be experimentally provided,
including atmospheres over concentrated sulphuric acid or dry calcium
chloride.

Three explanations have been offered for the germination of conidia at
low atmospheric humidities. Brodie (1945) believed that conidia of *E.
polygoni* and *E. graminis* had high osmotic pressures (and therefore low water
content) and suggested that these high osmotic pressures might aid the spores
in absorbing water from dry atmospheres. Foex (1926), Corner (1935), Jhooty
and McKeen (1965) and several others have reported that conidia have a high
water content; Hess *et al.* (1974), McKeen (1970) and Yarwood (1950) indicate
that conidia have a high water and lipid content. Kenaga (1970) gives high
moisture content and electric charge of powdery mildew conidia as a principle
of plant pathology. This high water content in the spores could supply the
water necessary for germination and the high lipid content of the wall area
could protect them from rapid desiccation in a dry environment. The third
interpretation is that of McKeen (1970) who suggests that respiration of oils
could yield some of the water needed for spore germination.

Spores of most airborne fungi have a relatively low water content (Cochrane, 1958), take up water during germination and increase in volume
(Barnes and Parker, 1966; Yarwood, 1936c). Yarwood (1952) believed that

conidia of powdery mildews have a high water content, do not take up water during germination and decrease in volume during germination. Brodie (1945) believes their volume remains unchanged.

Most fungi sporulate better at high than at low relative humidity (Cochrane, 1958; Yarwood 1956), though good quantitative data are rare. Powdery mildews are exceptional in that they sporulate luxuriantly under dry conditions. In the glasshouses at Berkeley, which are usually at about 50% relative humidity, the downy mildews of cucumber, onion, hop and lettuce do not normally sporulate unless placed in a special moist chamber, whereas the powdery mildews of aster, bean, cucumber, barley and sunflower sporulate luxuriantly in the normal glasshouse environment. Powdery mildew conidia produced in dry environments show higher germination than those produced in humid environments (Hammarlund, 1925).

The favourable effect of low soil moisture for powdery mildew has been noted by Knight (1818), Anon. (1859), Cook (1931), Yarwood (1949) and Drandarevski (1969). Several investigators have also reported that powdery mildews are favoured by high soil moisture. Most other aspects of moisture are probably direct effects on the pathogen but the effect of soil moisture must probably be a predisposition effect, since the fungus is always on tissues well removed from the soil.

Investigators and observers probably disagree more on the effect of weather on powdery mildews than on any other aspect. Among those indicating they are favoured by dry weather and/or are injured by water, rain or sprinkling are Leveille (1851), Graham (1852), Garman (1899), Corner (1935), Yarwood (1936c) and Boughey (1949). I believe that rain *per se* is injurious to powdery mildews because it inhibits the germination of conidia (Perera and Wheeler 1975; Corner, 1935), because it mechanically removes or injures mycelium and spores (Yarwood, 1939) and because it favours *Cicinnobolus*, the common parasite of powdery mildews (Yarwood, 1936c).

Part of the controversy over moisture and powdery mildews may result from the interaction of moisture, temperature, light and the host plant. Dry weather is commonly hot weather and the inhibitory effect of hot, dry weather might be attributed to the dryness rather than more correctly to the high temperature (Yarwood et al., 1954). Dry weather is commonly bright weather and the slow development of mildew in this environment might be attributed to the dryness rather than more correctly to the high light intensity. While moisture *per se* is injurious to powdery mildews, rainy weather may in some cases indirectly favour powdery mildews by causing a cooling of the environment, by reducing the light intensity and by washing off protective deposits of sulphur dust and by favouring the growth and inherent susceptibility of the host.

The San Francisco Bay area of California in summer is apparently a favourable site and time for powdery mildews (Gardner and Yarwood, 1974).

I believe this is probably because the rain-free summer provides the dry environment and the common high fog produces the cool, slightly shady environment.

If powdery mildew development in nature is inhibited by rain, it is logical to expect that spraying with water would control the disease. Evidence in support of this is given by Daniels (1860), Anon. (1861), Sheppard (1884), Yarwood (1939), McClellan (1942) and Ruppel *et al.* (1975) but contrary evidence is given by Crawford (1927), Sprague (1955) and many others. If syringing with water is used it must be frequent and vigorous, and this is usually considered less practical than occasional application of sulphur or one of the several other highly effective fungicides (Yarwood, 1957; Ruppel *et al.*, 1975; Kontaxis, 1976).

C. Control

The application of sulphur to control mildew was recommended in 1802, if not much earlier (Forsyth, 1824), it had become a standard control by 1855 (Viala, 1886) and is still the most effective control known for many powdery mildews (Kontaxis, 1976). Since 1802 when sulphur was first used many alternative chemicals have been developed for sulphur-sensitive crops; there have been improvements in the form of sulphur which is used and in methods of application. Systemic fungicides have been developed and many mildew-resistant varieties have been created. The most surprising aspect of this history is that, while mildews have acquired tolerance to several recently developed fungicides (Dekker, 1969), they have apparently not acquired resistance to sulphur during more than 150 years.

References

Agarah, J. G. (1852). "Species Genera et Ordines Floridiarum". Masson, Paris.
Ainsworth, G. C. (1970). "Dictionary of the Fungi". Commonwealth Mycological Institute, Kew.
Alcorn, J. L. (1969). *Aust. J. Sci.* **31**, 296–297.
Alexopoulos, C. J. (1952). "Introductory Mycology". Wiley, New York.
Allison, C. C. (1931). *Phytopathology* **24**, 305–307.
Anon. (1841). *Gdnrs. Chron.* **1**, 517.
Anon. (1848). *Gdnrs. Chron.* **8**, 523.
Anon. (1859). *Gdnrs. Chron.* **19**, 755–756.
Anon. (1861). *Calif. Farmer* **15**, 148.
Anon. (1968). *Rep. Pl. Breed. Inst.* **1966–1967**. 125.
Arnaud, G. (1921). *Annls Epiphyt.* **1**, 1–115.
Arthur, J. C. and Cummins, G. B. (1934). "Manual of the Rusts". Harper and Row, New York.

Ballantyne, B. (1963). *Aust. J. Sci.* **25**, 360–361.
Barnes, M. and Parker, M. S. (1966). *Trans. Br. mycol. Soc.* **49**, 487–494.
Berkeley, M. J. (1847). *Gdnrs. Chron.* **1847**, 779.
Berkeley, M. J. (1848). *J. R. hort. Soc.* **3**, 91–98.
Berkeley, M. J. (1855). *J. R. hort. Soc.* **9**, 61–70.
Berkeley, M. J. (1857). "Introduction to Cryptogamic Botany". Balliere Tindall, London.
Berkeley, M. J. and Curtis, M. A. (1858). *Proc. Am. Acad. Arts Sci.* **4**, 111–130.
Bessey, E. A. (1961). "Morphology and Taxonomy of Fungi". Hafner, New York.
Blumer, S. (1922). Das problem der "Bridging Species" bei den parasitichen Pilze. *Mitt. Naturf. Ges. Bern* **1922**, XLV–XLVI.
Blumer, S. (1933). "Die Erysiphacee Mitteleuropas". Fritz, Zurich.
Blumer, S. (1967). "Echte Mehltaupilze (Erysiphaceae)". Fischer, Jena.
Blumer, S. and Muller, E. (1964). *Phytopath. Z.* **50**, 379–385.
Bonorden, H. F. (1861). *Bot. Ztg.* **19**, 201–204.
Bos, L. (1970). "Symptoms of Virus Diseases in Plants". Institute of Phytopathological Research, Wageningen.
Boughey, A. S. (1949). *Trans. Br. mycol. Soc.* **32**, 179–189.
Bouwens, H. (1924). *Meded. phytopath. Lab. Willie Commelin Scholten* **8**, 3–28.
Bouwens, H. (1927). *Meded. phytopath. Lab. Willie Commelin Scholten* **10**, 3–31.
Brodie, H. J. (1945). *Can. J. Res.* **C20**, 41–62.
Brodie, H. J. and Neufeld, C. C. (1942). *Can. J. Res.* **C20**, 41–62.
Brown, W. and Wood, R. K. S. (1953). Ecological adaptations in fungi. *In* "Adaptation in Microorganisms". (R. Davies and E. F. Gale, Eds), pp. 326–339. Cambridge University Press, Cambridge.
Brundza, K. (1934). *Zemes Ukio Akad. Metr.* **1933**, 107–197.
Candolle, A. P. de (1805). *Flore Francaise* **2**, 272–275.
Candolle, A. P. de (1815). *Flore Francaise* **7**, 104–109.
Castagne, L. (1851). "Supplement au Catalogue des Plantes Geni Croissent Naturellement aux Environs de Marseille". pp. 187–192. Nicot et Pardigo, Aix.
Cauquil, J. and Sement, G. (1973). **28**, 279–286. Abstr. 1974 *Rev. Pl. Path.* **53**, 113, 1974.
Cherewick, W. J. (1944). *Can. J. Res. C*, **22**, 52–86.
Chiddawar, P. P. (1959). *Sydowia* **13**, 55–56.
Childs, J. F. L. (1940). *Phytopathology* **30**, 65–73.
Clare, B. G. (1964). *Pap. Dep. Bot. univ. Qd* **4**, 114–144.
Clayton, C. N. (1942). *Phytopathology* **32**, 921–943.
Clements, F. C., and Shear, C. L. (1931). "The Genera of Fungi". H. W. Wilson and Co., New York.
Clerk, G. C. and Ankara, J. K. (1969). *Can. J. Bot.* **47**, 1289–1290.
Cochrane, V. W. (1958). "Physiology of Fungi". Wiley, New York.
Cook, H. T. (1931). *Bull. Va. agric. Exp. Stn.* **74**, 931–940.
Cook, J. C. and Barr, M. E. (1964). *Mycologia* **56**, 763–769.
Cooke, M. C. (1878). "Microscopic Fungi". Hardwicke, London.
Corner, E. J. H. (1935). *New Phytol.* **34**, 180–200.
Couderc, G. (1893). *C. R. hebd. Séanc. Paris.* **116**, 210–211.
Coyier, D. L., Malvy, O. C. and Zalewski, J. C. (1975). *Proc. Am. Phytopath. Soc.* **2**, 112.
Crawford, R. F. (1927). *New Mexico agric. Exp. Stn. Bull.* **163**, 1–13.
Crepin, C. (1922). *Bull. Soc. Path. Veg. Fr.* **9**, 118–119.
D'Angremond, A. (1924). *Meded. Proefst. Tabak. Klaten.* **49**, 7–25. Abstr. (1924) *Rev. Appl. Mycol.* **3**, 434–435.
Daniels, W. (1860). *Calif. Culturist* **3**, 146–151.
Dash, J. S. (1913). *Rep. Queb. Soc. Prot. Pl.* **5**, 32–38.
Dekker, J. (1969). *Wld. Rev. Pest Control* **8**, 79–95.

Delp, C. J. (1954). *Phytopathology* **44**, 615.

Desmazieres, J. B. (1841). *Annl. Sci. nat.* 2, **5**, 129–146.

Drandarevski, C. A. (1969). *Phytopath. Z.* **65**, 124–154; 201–218.

Duby, J. E. (1830). "Botanicum Gallicum". Desray, Paris.

Duggar, B. M. (1909). "Fungus Diseases of Plants". Ginn, Boston.

DuToit, J. J. (1948). *Fmg. S. Afr.* **23**, 815–816.

Fischer, R. (1957). *Sydowia* **1**, 203–209.

Foex, E. (1912). *Annls Ec. natn. Agric. de Montpellier* **11**, 246–264.

Foex, E. (1924). Abstr *Rev. appl. Mycol.* **3**, 487.

Foex, E. (1926). *Bull. Trimest. Socl Mycol. Fr.* **41**, 417–438,

Forsyth, W. (1824). "A Treatise on the Culture and Management of Fruit Trees". 524 p. Longmans, London.

Fries, E. (1823). "Systema Mycologicum". Vol. 2. Ernest Mauirtius, Griefswald.

Fries, E. (1825). "Systema Orbis Vegetabilis". Typographia Academica, Lund.

Fries, E. (1829). "Systema Mycologicum". Vol. 3.

Gardner, M. W. and Yarwood, C. E. (1974). *Calif. Agric. Exp. Stn. Leaflet* 217.

Garman, H. (1899). *Bull. K. Agric. Exp. Stn.* **80**, 201–265.

Gaumann, E. (1922). *Annls. Jard. bot. Buitenz.* **32**, 43–63.

Gaumann, E. (1945). *Experimentia* **1**, 1–12.

Girard, A. (1889). *C. R. hebd. Acad. Sci. Séanc. Paris* **109**, 324–327.

Golovin, P. N. (1958). (in Russian) *Sborn. Rab. Inst. Prikl. Zool. Fitopat.* **5**, 101–139.

Golovin, P. N. and Gamalitskya, N. A. (1962). *Bot. Mater (Not. Syst. Sect. Crypt.) Inst. Bot. Acad. Sci. USSR* **15**, 91–93.

Graham, E. J. (1852). *Gdnrs Chron.* **12**, 453.

Greville, R. K. (1825). "Scottish Cryptogamic Flora III". Baldwin, London.

Hales, S. (1727). "Vegetable Statiks". Innys, London.

Halsted, B. D. (1884). *Proc. Am. pomol. Soc.* **1883**, 87–89.

Halsted, B. D. (1899). *Bull. Torrey bot. Club* **26**, 594–595.

Hammarlund, C. (1925). *Hereditas* **6**, 1–126.

Hammarlund, C. (1945). *Bot. Notiser* **1945**, 101–108.

Hansford, C. G. (1961). *Sydowia* **2**, 1–806.

Hermansen, J. E. (1972). *Friesia* **10**, 86–88.

Hess, W. M., Weber, D. J. and Johnson, D. (1974). *Am. J. Bot.* **61**, supp. p. 23.

Hill, J. (1951). "A General Natural History". Vol. 2. London. Printed for Thomas Osborne, in Gray's Inn, Holborn, London.

Hirata, K. (1966). "Host Range and Geographical Distribution of the Powdery Mildews". Mimeo. Niigata, Japan.

Hiura, U. (1962). *Phytopathology* **52**, 664–666.

Homma, Y. (1933). *Proc. imp. Acad. Japan* **9**, 186–187.

Homma, Y. (1937). *J. Fac. Agric. Hokkaido (Imp) Univ.* **38**, 186–461.

Horsfall, J. G. (1930). *Mem. Cornell Univ. agric. Exp. Stn* **130**, 139 p.

Hughes, S. J. (1958). *Can. J. Bot.* **36**, 727–836.

Ito, S. (1915). *Bot. Mag. Tokyo* **29**, 15–22.

Jaczewski, A. M. (1896). *Bull. Herb. Boissier* **4**, 722–755.

Jarvis, W. R. (1964). *Nature Lond.* **203**, 895.

Jhooty, J. S. (1971). *Indian Phytopath.* **24**, 67–73.

Jhooty, J. S. and McKeen, W. E. (1965). *Can. J. Microbiol.* **11**, 531–545.

Jorstad, I. (1925). *Nor. Vidensk. Akad. Oslo* **10**, 116 p.

Junell, L. (1964). *Svensk. bot. Tidskr* **58**, 55–61.

Karsten, P. A. (1885). *Middel. Soc. Fauna Flora Fenn. (Hilsingfors)* **11**, 21–28.

Kenaga, C. B. (1970). "Principles of Phytopathology". Balt, Lafayette.

Kendrick, W. B. and Carmichael, J. W. (1973). Hyphomyctes. "The Fungi. 4 A" (Ainsworth and Sussman, Eds), pp. 323–509. Academic Press, New York.
Kimbrough, J. W. (1963). *Mycologia* 55, 608–618.
Kimbrough, J. W. and Korf, R. P. (1963). *Mycologia* 55, 619–620.
Knight, T. A. (1818). *Trans. hort. Soc.* 2, 82–90.
Koltin, Y. and Kenneth, R. (1970). *Ann. appl. Biol.* 65, 263–268.
Kontaxis, D. G. (1976). *Calif. Agric.* 30, 13–14.
Kuntze, D. (1892). *Reviso Genera Plantarum* 3, 442.
Kunze, G. (1823). *Mykologische Hefte* 2, 111–113.
Large, E. C. (1940). "The Advance of the Fungi". Holt, New York.
Leach, C. M. (1975). *Phytopathology* 65, 1303–1312.
Lemmermann, E. (1910). "Algen l". Gebruder Borntraeger, Leipzig.
Leveille, J. A. (1851). *Ann. Sci. Nat. III* 15, 109–179.
Liabach, F. (1930). *Jb. wiss. Bot.* 72, 106–136.
Linder, D. H. (1942). *Lloydia* 5, 165–207.
Link, H. F. (1809). *Mag. Ges. Naturf. Freunde Z. Berlin* 3, 3–42.
Link, H. F. (1824). *Willdenow, Species Plantarum* 6, 100–117.
Linneas, C. von (1753). *Species Plantarum* 2, 1186.
Mains, E. B. (1923). *Proc. Indiana Acad. Sci.* 1922, 307–313.
Mamluk, O. F. and Weltzien, H. C. (1973). *Phytopath. Z.* 76, 285–302.
Martin, J. A. and Salmon, S. C. (1953). *Yb. US Dep. Agric.* 1953, 329–343.
McClellan, W. D. (1942). *Bull. Cornell Univ. agric. Exp. Stn.* 785, 1–39.
McKeen, W. E. (1970). *Phytopathology* 60, 1303.
McKenzie, D. L., and Morrall, R. A. A. (1975). *Can. Pl. Dis. Surv.* 55, 1–7.
Mendonca, A. L. D. V. E. (1965). *Agronomia Lusit.* 27, 123–126.
Menzies, J. R. (1950). *Pl. Dis. Rept* 34, 140–141.
Merat, F. V. (1831). "Nouvelle Flore des Environs de Paris". 2nd ed. Vol. 1, Paris.
Micheli, P. A. (1729). "Nova Plantarum Genera".
Migula, W. (1913). *Kryptogamen Flora Deutsch.* 4, 66–80.
Moyed, H. S (1964). *A. Rev. Microbiol.* 18, 347–366.
Murray, J. A. A. (Ed.) (1908). "A New English Dictionary". Clarendon, Oxford.
Nageli, K. (1842). *Linnaea* 16, 237–285.
Nagy, G. S. (1970). *Acta Phytopath. Hung.* 5, 231–248.
Nees von Esenbeck, C. G. (1817). "Das System der Pilze und Schwamme". Warzburg.
Neger, F. W. (1901) *Flora Jena* 88, 333–370.
Neger, F. W. (1902). *Flora Jena* 90, 221–272.
Neger, F. W. (1923). *Flora Jena* 116, 331–335.
Ogilvie, L. (1942). *Rep. agric. hort. Res. Stn. Univ. Bristol* 1942, 83–88.
Pady, S. M. Kramer, C. L. and Clary, R. (1969). *Phytopathology* 59, 844–848.
Patouillard, N. and Hariot, P. (1900). *J. Bot. Paris* 14, 245–246.
Paxton, J. D. and Rogers, D. P. (1974). *Mycologia* 66, 894–896.
Peck, C. A. (1876). *Rep. N.Y. St. Mus. nat. Hist.* 28, 31–88.
Perera, R. G. and Wheeler, B. E. J. (1975). *Trans. Br. mycol. Soc.* 64, 313–319.
Peries, O. S. (1966). *Nature Lond.* 212, 540–541.
Persoon, C. A. (1796). "Observationes Mycologique". Wolf, Lipsiea.
Peterson, G. A. (1938). *Mycologia* 30, 299–301.
Pollacci, G. (1911). *Atti Ist. Bot. Univ. Pavia* 9, 151–180.
Ponnappa, K. M. (1970). *Sydowia* 23, 4–7.
Price, T. V. (1970). *Ann. appl. Biol.* 65, 231–248.
Randall, T. E. and Menzies, J. D. (1956). *Pl. Dis. Reptr.* 40, 255.
Riley, C. V. (1886). *Proc. Am. pomol. Soc.* 1885, 49–54.
Robertson, J. (1821). *Trans. Hort. Soc.* 5, 175–185.

36   C. E. YARWOOD

Rogers, D. P. (1959). *Mycologia* **51**, 96–98.
Ruppel, E. G., Hills, F. J. and Mumford, D. L. (1975). *Pl. Dis. Reptr.* **59**, 283–286.
Saccardo, P. A. (1882). *Sylloge Fungorum* **1**, 24.
Salmon, E. S. (1900). *Bull. Torrey bot. Club* **9**, 1–292.
Salmon, E. S. (1903). *J. Bot. Lond.* **41**, 127–130.
Salmon, E. S. (1904). *Annls Mycol.* **2**, 255–266.
Salmon, E. S. (1905). *J. Bot. Lond.* **43**, 41–44.
Salmon, E. S. (1906a). *Phil. Trans. R. Soc. B.* **198**, 87–95.
Salmon, E. S. (1906b). *Ann. Bot.* **20**, 187–200.
Sathe, A. V. (1969). Erysiphosis replaces Salmonomyces. *Bull. Torrey bot. Club* **96**, 101–102.
Sawada, K. (1914). *Bull. Formosa agric. Exp. Stn.* **9**, 1–102.
Sawada, K. (1916). In Japanese. *Trans. Formosan Nat. Hist. Soc.* **6**, 27–35.
Sawada, K. (1927). *Rep. Dep. Agric. Res. Inst. Formosa* **24**.
Scalia, G. (1902). *L'Agric. Calabro Sicula* **27**, 393–397.
Schlectendal, D. F. L. von (1829). *Verh. Berl. Ges. Nat. Freund* **1**, 46–51.
Schweinitz, L. D. de (1834). *Trans. Am. philos. Soc.* **4**, 141–317.
Shepard, H. H. (1939). "The Chemistry and Toxicity of Insecticides". Burgess, Minneapolis
Sheppard, J. (1884). *Gdnrs'. Chron.* **1884**, 53.
Siebs, E. (1958). *Phytopath. Z.* **34**, 86–102.
Smith, C. G. (1970). *Trans. Br. mycol. Soc.* **55**, 355–365.
Somers, E., and Horsfall, J. G. (1966). *Phytopathology* **56**, 1031–1035.
Spegazzini, C. (1910). *Rev. Fac Agron. Vet.* **6**, 1–205.
Sprague, R. (1958). *West Fruit Grow.* **9**, 17.
Sprengel, C. (1806). "Flora Halensis". Kummelii, Halae.
Stakman, E. C. and Harrar, J. G. (1957). "Principles of Plant Pathology". Ronald, New York.
Steiner, J. A. (1908). *Cent. fur Bakt. (Parasit. und Krank.)* **21**, 677–736.
Stone, O. M. (1962). *Ann. appl. Biol.* **50**, 203–210.
Sumstine, D. R. (1913). *Mycologia.* **5**, 45–61.
Thiessen, F. (1912). *Beih. Bot. Cent.* **29**, 45–73.
Thiessen, F. and Sydow, H. (1917). *Annls. mycol.* **15**, 389–491.
Thirumalachar, M. J. (1947). *New Phytologist* **46**, 269–273.
Thompson, G. E. (1951). *Pl. Dis. Reptr* **35**, 221.
Tozzetti, G. I. (1767). "Altimurgia". Moucke, Florence. Phytopathological Classic 9.
Tulasne, L. R. (1853). *C. R. hebd. Séanc. Acad. Sci. Paris* **37**, 605–609.
Tulasne, L. R. and Tulasne, C. (1861). "Selecta Fungorum Carpologia". Imprimerie, Imperiale, Paris. English translation by W. B. Grove, Clarendon Press, Oxford. 1931.
United States Department of Agriculture (1960). *Handbk US Dep. Agric.* **165**.
Vanha, J. (1903). *Zuck. Ind. Böhm.* **27**, 180–186.
Van Hook, J. M. (1906). *Bull. Ohio agric. Exp. Stn.* **173**, 231–249.
Viala, P. (1886). "Les Maladies de la Vigne". Coulet, Montpellier.
Viegas, J. A. (1944). *Bragantia* **4**, 5–392.
Wakefield, E. M. (1940). *Trans. Br. Mycol. Soc.* **24**, 282–293.
Wallroth, F. W. (1819). *Ann. Witter Ges.* **4**, 226–247.
Weltzien, H. C. (1963). *Phytopath. Z.* **47**, 123–128.
Weltzien, H. C. and Yarwood, C. E. (1963). *Mycologia* **55**, 342–351.
Whetzel, H. H. (1918). "An Outline of the History of Phytopathology". Saunders, Philadelphia.
Wysong, D. S. and Willis, W. G. (1968). *Pl. Dis. Reptr.* **52**, 652–653.
Yarwood, C. E. (1935). *Science* **82**, 417–418.
Yarwood, C. E. (1936a). *J. agric. Res.* **52**, 645–657.
Yarwood, C. E. (1936b). *J. agric. Res.* **52**, 659–665.

Yarwood, C. E. (1936c). *Phytopathology* **26**, 845–859.
Yarwood, C. E. (1937). *Pl. Dis. Reptr* **21**, 180–182.
Yarwood, C. E. (1939). *Phytopathology* **29**, 288–290.
Yarwood, C. E. (1944). *Phytopathology* **34**, 937.
Yarwood, C. E. (1949). *Phytopathology* **39**, 780–788.
Yarwood, C. E. (1950). *Am. J. Bot.* **37**, 636–639.
Yarwood, C. E. (1952). *Mycologia* **44**, 506–522.
Yarwood, C. E. (1956). *Pl. Dis. Reptr* **40**, 318–321.
Yarwood, C. E. (1957). *Bot. Rev.* **23**, 235–500.
Yarwood, C. E. (1971). *Pl. Dis. Reptr.* **55**, 342–344.
Yarwood, C. E. (1973). Erysiphales. *In* "The Fungi IV A" (G. C. Ainsworth, F. K. Sparrow and A. S. Sussman, Eds). pp. 71–86. Academic Press, New York.
Yarwood, C. E. and Gardner, M. W. (1970). *Mycologia* **62**, 707–713.
Yarwood, C. E., and Gardner, M. W. (1972). *Mycologia* **64**, 799–805.
Yarwood, C. E., Gardner, M. W. and Duafala, T. (1972). *Pl. Dis. Reptr* **56**, 313–317.
Yarwood, C. E., Sidky, S., Cohen, M. and Santilli, V. (1954). *Hilgardia* **44**, 506–522.
Zaracovitis, C. (1965). *In* "The Fungus Spore" (M. F. Madelin, Ed.). pp. 273–286. Butterworth, London.
Zwirn, H. E. (1943). *Palest. J. Bot.* **3**, 52–53.
Zopf, W. (1887). *Ber. dt. bot.* **5**, 275–280.

Chapter 2

Geographical Distribution of Powdery Mildews

H. C. WELTZIEN

Institut für Pflanzenkrankheiten, Universität, Bonn, Germany

I. Introduction

The amount of effort which goes into the search for plant pathogenic organisms in any particular area is directly related to crop production and crop protection in that area. Most check lists of plant pathogens are based on local studies which are usually at a later date compiled into national lists but an international or regional approach is rarely tried. In the past therefore the problem of pathogen distribution was generally related to the occurrence of susceptible host plants, whilst the study of distribution patterns as independent characteristics of plant pathogenic organisms was usually neglected.

Biogeographical studies have always revealed that an unequal distribution of organisms based on endogenic and exogenic factors, is the rule rather than the exception. Although many cases are known where transport and spread leads to the colonization of new habitats, most organisms are restricted to specific areas, defined by ecological boundaries.

The application of these principles to plant pathology has led to the concept of Pathogeography by Reichert (1953, 1958). This concept was developed and systematically expanded by Weltzien (1967, 1972) as Geophyto-pathology. When it became apparent that plant protection work should be embedded in the wider concept of Phytiatry or Phytomedicine, the concept was extended as Geophytomedicine. Geophytomedicine is an integral part of Geomedicine and Medical Geography, which includes Human-, Veterinary- and Phytomedicine (Weltzien, 1973).

II. Occurrence of Powdery Mildews in Different Geographical Regions

Powdery mildew fungi and their respective host plants do not always have the same areas of distribution. Especially in high latitudes and alpine regions some host plants through better adaptation to more extreme climatic con-ditions escape damage. Blumer (1967) reported for example, that some hosts escape powdery mildew damage, when planted above the polar circle in Norway. The species involved were *Phyllactinia guttata* on hazelnut (*Corylus* spp.), willow (*Salix* spp.) and birch (*Betula* spp.), *Phyllactinia fraxini* on ash (*Fraxinus* spp.), *Microsphaera betulae* on birch, *Uncinula adunca* and *Podosphaera schlechtendalii*, both on willow. Similarly *Sphaerotheca volkartii* does not follow its host *Drya octopetala* above a 200 m altitude in the Swiss alps.

The most complete study on the world wide distribution of powdery mildews was compiled by Hirata (1966). His lists cover the 12 recognized powdery mildew genera on more than 7000 host plant species in 162 different geographic areas. If one tries to condense this vast work on a world map, the picture in Fig. 1 results. It has been supplemented for California (Gardner and Yarwood, 1974) and the Arab Republic of Yemen (unpublished data collected by Hindorf and Weltzien). A number of areas with an extremely high disease frequency and more than 500 known hosts per country become obvious. In Europe these centres are found in France, Germany, Italy, Romania and the European part of the USSR. In Asia there are two centres of high disease frequency, one in Kazakstan and one in Japan, whereas in the whole of North and South America the only area with high disease frequency is in California. Similar centres are lacking in Africa and Australia. On these latter two continents and in Central and South America most countries or states have between 10 and 99 reported hosts, whilst in Africa there is a remarkably large area without any recorded powdery mildews. In the USA, where a total of 1797 hosts are known, the "non diseased areas" only signalize an absence of state surveys or lists.

Groups of countries with similar disease densities are listed in Table 1. It is quite apparent that these figures do not in all cases represent the true global

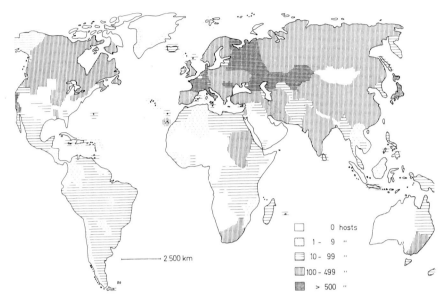

Fig. 1. World map of the number of reported hosts of powdery mildews in different countries and geographic regions, based on data by Hirata (1966).

distribution of powdery mildews. The high numbers of records from well surveyed areas such as Japan, France, Germany and Switzerland suggest that some of the figures represent the quantity and/or the quality of survey work performed. In those countries where the disease surveys can be assumed to have been of similar quality, the distribution pattern may reflect the true picture. A number of countries, where survey work should be intensified are listed in Table 2. In all of them, numerous new host records are to be expected. This is especially true for the North African and Near Eastern countries on the list. On the other hand, comparatively low figures for Iceland (21), Belgium (51) and Holland (110) suggest that maritime climates are less favourable for powdery mildews than are continental climates. The Scandi-navian countries show comparatively high figures: Denmark (263), Finland (256), Norway (251) and Sweden (190). One may conclude that this is due to the continental summer climate and the extended day length. It is also quite obvious that powdery mildews are not very frequent in the tropics. The only tropical countries where the number of hosts exceeds 100 are India and Sudan. Both of these are large countries with a number of different climatic zones, some of which are not tropical.

In general the map makes it clear that new surveys are needed for many areas where no figures at present exist and that elsewhere the addition of information from hitherto unpublished or difficult to obtain literature sources, would help to complete the picture.

TABLE 1

Groups of countries, states or islands with similar numbers of host species affected by powdery mildew

Number of host species	countries, states or islands
0 —no records according to Hirata (1966)—	Albania, Andorra, Angola, Bhutan, Cameroon, Chad, Cuba, Dahomey, Ecuador, Falkland I., Gambia, Indonesia, Inner Mongolia, Kuweit, Laos, Liberia, Lichtenstein, Luxemburg, Nepal, Niger, Nigeria, Oman, Saudi Arabia, Senegal, Somalia, Tibet, Togo
1–9	Algeria, Azores, Barbados, Borneo, Brit. Guinea, Chile, Columbia, Costa Rica, Faeroes, Fiji, Greenland, Guinea, Haiti, Hawaiian I., Honduras, Ivory Coast, Jordan, Mozambique, New Caledonia, Nicaragua, Norfolk I., Panama, Paraguay, Rio do Oro, Sardinia, Svalbard, Syria, Thailand, Tunisia, Vietnam
10–99	Afganistan, Alaska, Argentinia, Azerbaidzhan (USSR), Balearic I., Belgium, Bermuda, Bolivia, Brazil, Burma, Cambodia, Cyprus, Egypt, Eire, El Salvador, Estonia (USSR), Ethiopia, Georgia (USA), Ghana, Guatemala, Iceland, Iraq, Jamaica, Java, Kamtschatka (USSR), Kenya, Korea, Latvia (USSR), Lebanon, Lybia, Madagaskar, Madeira I., Malawi, Malaya, Malta, Marokko, Mauritius, Mexico, New Guinea, New Zealand, Oklahoma (USA), Oregon (USA), Pakistan-West, Peru, Philippines, Puerto Rico, Queensland (Austr.), Rep. Dominic., Rhodesia N., Rhodesia S., Sakhalin, Shetland, Sierra Leone, Sri Lanka, Tanzania, Tasmania (Austr.), Trinidad, Uganda, Uruguay, Uzbek (USSR), Venezuela, Virgin I., Yemen, Zaire
100–499	Austria, Armenia (USSR), Bulgaria, Canada, Canary I., Central Asia (USSR), China, Corsica, Crimea (USSR), Czechoslovakia, Denmark, Finland, Formosa, Greece, Hungary, Idaho (USA), Illinois (USA), India, Iowa (USA), Iran, Israel, Yugoslavia, Kirgiz (USSR), Lithuania (USSR), Minnesota (USA), Mississippi (USA), Missouri (USA), Montana (USA), Netherlands, New South Wales (Austr.), Norway, Ohio (USA), Pennsylvania (USA), Poland, Portugal, Siberia (USSR), South Africa, Spain, Sudan, Sweden, Turkey, Turkmen (USSR), Ukraine (USSR), United Kingdom, Washington (USA), Wisconsin (USA)
>500	California (USA), European Part of USSR, France, Germany, Italy, Japan, Kazak (USSR), Romania, Switzerland

TABLE 2

Some countries or islands with unusually low numbers of powdery mildews and number of diseased host species as reported by Hirata (1966)

Algeria	(8)	Sardinia	(3)
Brazil	(15)	Syria	(8)
Chile	(3)	Tunisia	(2)
Jordan	(8)	Vietnam	(1)

III. Epidemic Spread of Powdery Mildews

The static picture in Fig. 1 represents largely the host frequency as reported by Hirata (1966), which is the result of large and numerous movements of hosts and parasitic organisms from areas of origin to new habitats. The introduction into other parts of the world of solanaceous crop plants such as potato and tomato from the American continent during the past 300 years is a case in point. This resulted in a considerable enlargement of the habitat of such powdery mildew fungi as *Leveillula taurica, Erysiphe cichoracearum, Erysiphe communis* and *Erysiphe polyphaga*. Any other natural or anthropogenous spread of host plants would probably have similar effects. This aspect, however will not be analysed in this article.

The spread of powdery mildew fungi into new, previously uninhabited areas has, in some cases, been observed and documented. High economic losses resulting from new outbreaks usually arouse wide attention. The introduction of *Uncinula necator* from North America (or Japan) into England in 1845 and its subsequent spread throughout Europe and the neighbouring countries of Asia and Africa is probably the most well known and important

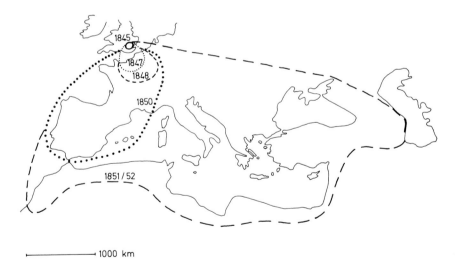

Fig. 2. The epidemic spread of the powdery mildew of grapes, *Uncinula necator*, between 1845 and 1852, according to Blumer (1933).

example. Figure 2 shows the progress of the epidemic based on data collected by Blumer (1933). One may note, that a distance of about 4300 km was

covered within eight years; an average annual distance of *c.* 540 km. Today
the parasite is found in 76 countries and on all continents and has probably
reached all grape cultivation areas.

At the beginning of the twentieth century the fruit and shoot powdery
mildew epidemic on gooseberries, known in Europe as "American gooseberry
mildew", seems to have started from infection loci in North-West Ireland and
Western Russia. Whether or not the fungus, *Sphaerotheca mors-uvae*, was
introduced from North America, where it had been known since 1834,
remains uncertain. Figure 3 shows the spread of the epidemic, as far as it can

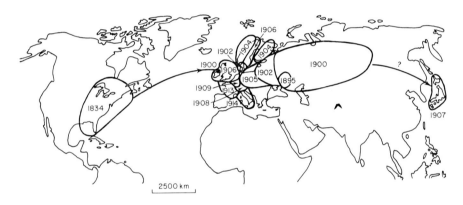

Fig. 3.   Epidemic spread of the gooseberry fruit and shoot powdery mildew, *Sphaerotheca*
mors uvae, compiled from various sources.

be reconstructed from the information now available (Eriksson, 1906;
Herter, 1907; CMI, 1964; Noack, 1928).

Whether the oak mildew epidemic which has caused severe and widespread
damage in Europe, can be explained only by the introduction and spread of
its causal fungi is doubtful. *Quercus* is the host genus with the greatest number
of powdery mildew species. Hirata (1966) lists the following 13 species from
seven genera: *Cystotheca lanestris*, *Erysiphe aquilegiae*, *E. communis*,
Microsphaera hypophylla, *Microsphaera alphitoides*, *Microsphaera alni*,
Phyllactinia suffulta, *Phyllactinia quercus*, *Phyllactinia roboris*, *Sphaerotheca*
phytoptophila, *Typhulochaeta japonis*, *Uncinula bifurcata* and *Uncinula septata*.
Since the fungus is often found only in the conidial stage, its identity in many
cases is uncertain and it is difficult therefore to reconstruct a picture of its
epidemic spread.

IV. Incomplete Distribution of Powdery Mildews

If epidemic spread of powdery mildews has caused such problems in the past, one may well ask if similarly serious outbreaks could occur in the future. When scanning reports for specific fungus species in various regions one notes that such a possibility exists. One of the most interesting cases concerns the genus *Leveillula* with its 709 registered host plants (Hirata, 1966). It occurs on all continents and in 68 countries with a preference for warm, dry summer climates such as those in Central Asia and the Mediterranean. Surprisingly it has never been found in California (Gardner and Yarwood, 1974) though the climate there is certainly favourable for *Leveillula* and numerous host plants do exist. One could therefore predict that the introduction of *Leveillula* spores into California would launch an epidemic on suitable hosts for example among the Solanaceae or the Leguminosae.

Brasiliomyces, a rather rare genus reported only from Peru, Brazil and Puerto Rico in Latin America and in one case from South Africa is another noteworthy case. The species *Brasiliomyces malachrae* parasitizes three species of cotton: *Gossypium barbadense*, *G. bissectum* and *G. peruvianum*. An unidentified species of *Brasiliomyces* from South Africa was found on a member of the Leguminosae (*Eutada spiccata*). Theoretically, therefore, the large cotton growing areas in Africa and Asia would be considered threatened by *Brasiliomyces*, if the fungus is accidentally introduced from Latin America.

V. Areas of Different Disease Intensity

The economic importance of pathogens is related more to the disease intensity than to the disease frequency. Records of plant diseases and pathogens from countries or territories must, therefore, be supplemented by disease intensity studies. The result is frequently a more differentiated picture of the geographic distribution with more economic significance given to plant production.

Within the total pathogen distribution area one can frequently recognize centres of high disease intensity, where the pathogen develops within or close to its ecological optimum. At the edge of its area of distribution, the pathogen becomes widely dispersed and survives only in suitable ecological niches. Between both extremes all levels of transition may be found. For practical reasons Weltzien (1972) delineated disease intensity into the following three zones: (1) areas of main damage, (2) areas of marginal damage and (3) areas of sporadic attack. They correspond with the three zones of different population density described for insect pests by Bremer (1929) and by Schwerdtfeger (1968).

Based on this system, Drandarevski (1969) analysed the ecology and distribution pattern of the powdery mildew of sugar beet (*E. betae*). By correlating experimental data and field studies on disease intensity with macro-climatic characteristics, guide lines for a world wide disease intensity map were calculated. The climatic diagrams of Walter and Lieth (1967) were used for climatic analysis. They consist of the long-term monthly means for temperature and rainfall, at a ratio of 1:2. Thus 10°C corresponds with 20 mm. Periods with the temperature line above the rainfall generally represent humid conditions, whereas if the temperature line is below the rainfall line, conditions are considered as arid. Thus macroclimatic conditions of any meteorological station can easily be classified for ecological significance.

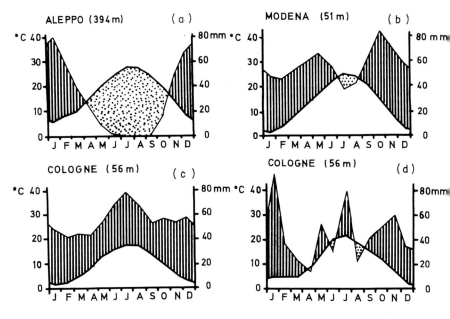

Fig. 4. Climatic diagrams representing conditions typical for areas of main damage, occasional damage and of sporadic attack by the powdery mildew of sugar beet, *Erysiphe betae*. (a) Areas of main disease; (b) areas of marginal damage; (c) areas of sporadic attack; (d) Cologne, 1976.

The four diagrams in Fig. 4 represent typical conditions characterizing the three disease intensity zones for sugarbeet powdery mildew. Aleppo (Fig. 4a) represents those areas with extended summer dryness and continuous aridity during the vegetative period for sugar beet. The average temperature during

this time is mostly above 18°C. This climate clearly favours fungus sporulation and conidia germination, thus allowing the development of regular
epidemics.

In contrast Cologne (Fig. 4c) represents a climate with continuous humidity
and an average temperature below 18°C. Here sporulation and germination
are limited and infection processes are slow. Primary infections can only start
late and thus heavy epidemics can not be expected.

Modena (Fig. 4b) is given as an example of one of the many possible
intermediate climates. The short arid period favours the disease to only a
limited extent. Here the influence of annual changes of climatic characteristics
may be very pronounced and years with heavy epidemics may be followed by
others with very weak infections. In rare cases this may even occur in places
with generally unsuitable climates such as Cologne. In Fig. 4d the 1976 data
for Cologne show three separate arid periods, two of them with temperatures
favourable for infection. It is not surprising that under these conditions a
rather severe epidemic resulted even in this area where the disease normally
occurs only sporadically.

Analysis of the sugar beet growing areas of the world according to the
macroclimatic characteristics resulted in the map given in Fig. 5. Infection

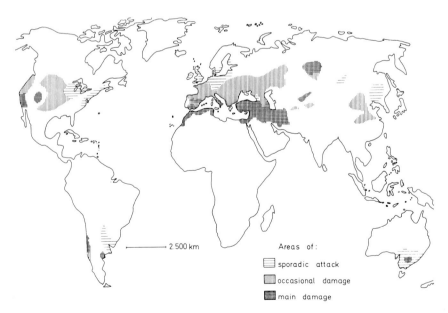

Fig. 5.   World map of observed or expected disease intensity of powdery mildew of sugar
beet, *Erysiphe betae*.

centres of high intensity in Armenia, Azerbaidzhan and Kazakstan in southern USSR and in Turkey, Lebanon, Syria, Iran and Iraq stand out. Subsequent field survey work in the Near and Middle East countries confirmed the
earlier findings and postulates (Weltzien, Mamluk, unpublished data).
Infection centres placed in China, Australia and the USA represented potential disease areas, since no records existed of detection of sugar beet powdery
mildew in these countries. However, a severe outbreak was recorded in 1974 in
Southern California (Kontaxis *et al.*, 1974) covering exactly the predicted
main damage area. The overall spread of the disease in the Western USA was
surveyed by Ruppel *et al.* (1975) who confirmed the predicted areas of
infection. One may therefore assume, that the single observation of a powdery
mildew on sugar beet in California 1934 by Yarwood (Gardner and Yarwood,
1974) was an accidental infection, possibly by a different species. If the
identity of the California *Erysiphe* with *E. betae* (Vanha) Weltzien (Weltzien,
1963) can be confirmed, one may assume its recent introduction into the
Californian sugar beet growing areas.

VI. Outlook

The question must be asked whether or not the study of geographic distribution patterns can be considered to be a useful research tool for future use in
plant pathology. There is no doubt that the reports on the occurrence of
powdery mildew diseases as summarized in Fig. 1 are very incomplete. Likewise, it can be seen from Table II that there are some extreme cases of
underrepresented countries but many others may also need more attention.
In general, disease survey work needs continued and more intensified attention, before a true picture can be drawn of powdery mildew occurrence in all
geographic areas. The future epidemic spread of powdery mildews cannot,
with certainty be anticipated though the examples of *Leveillula* and *Brasiliomyces* have been mentioned. Others may remain unnoticed until a sudden
disease outbreak demands attention. The epidemic spread of new physiological races of powdery mildews on crop plants will probably raise much interest
in the future, especially in the case of *E. graminis* on cereal crops.
    In relation to crop damage, powdery mildew research must supply con-
siderably more data on disease intensity through standardized disease assessment methods. If geographic zones of similar disease intensity can be
recognized and ecologically analysed, long term prognosis and advance
planning of control activities could become feasible. Thus the successful
prediction of the powdery mildew epidemic on sugar beet in California may
serve as encouragement for future work.

References

Blumer, S. (1933). Beiträge zur Kryptogamenflora der Schweiz, Bd. VII, H. 1 "Erysiphaceen Mitteleuropas". Fretz AG, Zürich.
Blumer, S. (1967). "Echte Mehltaupilze (Erysiphaceen)". G. Fischer, Jena.
Bremer, H. (1929). Z. angew. Ent. 14, 254–272.
Commonwealth Mycological Institute (1964). "Distribution Maps of Plant Diseases", Map No. 16, 2nd ed.
Drandarevski, C. A. (1969). Phytopath. Z. 65, 201–218.
Eriksson, J. (1906). Z. Pflkrankh. 16, 83–90.
Gardner, M. W. and Yarwood, C. E. (1974). "A List of Powdery Mildews of California" Calif. Agric. Exper. Stat. Ext. Serv.
Herter, W. (1907). Zentbl. Bakt. Parasitkde II., 18, 828–830.
Hirata, K. (1966). "Host range and Geographical Distribution of the Powdery Mildews". Niigata Univ., Japan.
Kontaxis, D. G., Meister, H. and Sharma R. K. (1974). Pl. Dis. Reptr 58, 904–905.
Noack, M. (1928). In "Handbuch der Pflanzenkrankheiten" (P. Sorauer Ed.,) Bd. II, 5. Aufl., 515, Paul Parey, Berlin.
Reichert, I. (1953). Proc. 12. Int. Bot. Cong. Stockholm 1950, 730–731.
Reichert, I. (1958). Trans. N.Y. Acad. Sci. II 20, 333–339.
Ruppel, E. G., Hills, F. J. and Mumford, D. L. (1975). Pl. Dis. Reptr 59, 283–286.
Schwerdtfeger, F. (1968). "Demökologie", Paul Parey, Berlin.
Walter, H. and Lieth, H. (1967). "Klimadiagramm-Weltat-las". G. Fischer, Jena.
Weltzien, H. C. (1963). Phytopath. Z. 47, 123–128.
Weltzien, H. C. (1967). Z. Pflkrankh. 74, 176–189.
Weltzien, H. C. (1972). A. Rev. Phytopath. 10, 277–298.
Weltzien, H. C. (1973). Geogr. Z. Bei. Forts. Geomed. Forsch. 110–114.

Chapter 3

Epidemiology of Powdery Mildews

D. J. BUTT

East Malling Research Station, Maidstone, Kent, England

I. Introduction

Epidemiology—the science of disease in populations—has made rapid advances in the twenty years since Yarwood (1957) reviewed the powdery mildews—a unique family of plant pathogens. Van der Plank's (1963) analysis of the growth of epidemics is a landmark in plant pathology which stimulated research into population dynamics, mathematical analysis and modelling (Kranz, 1974). Meanwhile, the widespread use of spore traps has added a new dimension to field investigations and the discovery of many fungicides active against powdery mildews has motivated epidemiological studies and awakened interest in forecasting systems.

Epidemiology is a branch of ecology concerned with coexisting populations of plant parasites and their hosts in natural and agricultural ecosystems (Zadoks, 1974) and this chapter is mainly autecological in approach, examining powdery mildews in relation to their physical and biotic environment. Particular emphasis is given to primary inoculum which initiates epidemics, to the biology of these fungi and their multiplication and to features which characterize and regulate epidemics. Some of the information is presented elsewhere in this book but no apology is needed because epidemiology integrates experience at the population, individual and cellular levels in the quest for a better understanding of temporal and spatial aspects of disease and a better formulation of control strategy and tactics.

II. Perennation

Powdery mildews cause diseases of shoots, flowers and fruits and so annual epidemics, whether on herbaceous or woody hosts, recur during seasons of vegetative growth. Between epidemics, the fungi are less active and survive (perennate) in the perfect state as cleistothecia, or remain in the conidial (imperfect) state on leaves or within dormant buds.

Three regions in the temperate zone of the northern hemisphere serve to illustrate the diversity of conditions which powdery mildews endure in the inter-epidemic period, although the geographical range of the family includes the tropics and sub-tropics. In the Mediterranean climate field crops are cultivated in the moist and relatively cool winter months and so the fungi must endure (oversummer) the long, dry, hot summer. Further north, winter is the season which threatens the survival of inoculum between epidemics; compare, however, conditions in snow-covered regions of North America and eastern Europe, where the winter of the continental climate is cold and dry, with the milder and wetter weather of North-west Europe where the oceanic climate allows the growing of winter cereals.

A. The perfect state

There seems no reason to dispute Yarwood's (1957) opinion that

the role of perithecia in the life history of powdery mildews has probably been overemphasised;

there are many examples of successful mildews which do not produce fertile fruit bodies or which discharge apparently functionless ascospores.

Cleistothecia mature on senescent leaves and woody stems towards the end of an epidemic but not all varieties of a host (e.g. rose) necessarily support their production (Price, 1970). The number of fruit bodies may vary from year to year. On oak and apple, for example, there are many reports of their abundance after exceptionally hot, dry summers, and it has been shown that high temperature and low humidity encourage their formation on sugar beet (Mamluk and Weltzien, 1973). Some species are heterothallic (Coyier, 1974; Jackson and Wheeler, 1975) and so an absence of cleistothecia may be explained by the absence of compatible strains; indeed, Smith (1970) suggested that the late formation of fruit bodies in a season may be due to the delayed arrival of appropriate mating types, and a similar suggestion has been made to explain the finding of the perfect state on sugar beet in England in recent years (Byford and Bentley, 1976).

B. The imperfect state

In view of the obligate parasitism of the Erysiphaceae, perennation in the conidial (asexual) state depends upon the presence of living host tissue at a time of the year when perennial hosts are dormant or quiescent and herbaceous hosts may be scarce and not actively growing.

1. Hibernation in dormant buds

Powdery mildews of many deciduous woody plants such as peach, apple, rose, oak, gooseberry and hawthorn invade newly formed buds during the vegetative season and hibernate in dormant buds during winter. Details of the cycle for apple buds are given by Burchill (1958) and Kosswig (1958). This type of overwintering also occurs in herbaceous plants such as hop, on which the fungus resides in the crown buds of the rootstock (Liyanage and Royle, 1976).

2. Perpetuation on leaves

Some powdery mildews bridge annual epidemics as mycelial colonies on leaves which persist between seasons of active vegetative growth. In the tropics, for instance, rubber trees remain in leaf throughout the year except for a short season of "wintering", and between epidemics of *Oidium heveae* which develop on young foliage at the times of refoliation the fungus subsists in a state of relative inactivity on the mature canopy (Populer, 1972). Likewise, in the Mediterranean region, mildew on apple and perhaps other deciduous fruit trees survives as mycelium on living leaves which persist throughout the mild winter (Palti, 1972), and this form of perennation is also described for *Microsphaera* on the deciduous oak *Quercus pubescens* and several evergreen species (Grasso, 1973).

Mildew colonies are capable of withstanding winter temperatures if the leaves of the host are not killed; thus, in England, mycelia of *Sphaerotheca* are able to survive on leaves of strawberry and rose (Peries, 1961; Howden, 1968). Cereal mildews provide the best examples of winter tolerance for on winter crops colonies of *Erysiphe graminis* continue to grow slowly (Turner, 1956; Bainbridge, 1972; Jenkyn, 1973, 1974a). Furthermore, Powers (1957) observed sporulation of this species after a night when the temperature fell below $-6°C$, and viable conidia have been collected in months when the mean minimum air temperature was $-13°C$ (Johnston, 1974). The conidia of powdery mildews are short-lived in summer but have an average longevity of 40 days at 0 to $-10°C$ (Yarwood *et al.*, 1954); strawberry mildew conidia, for example, can survive for more than 40 days at $0°C$ (Jhooty and McKeen, 1965a) and lettuce mildew conidia (*Erysiphe cichoracearum*) have survived a week at $-5°C$ (Schnathorst, 1960a).

III. Primary Inoculum

The target population upon which an epidemic is initiated by primary inoculum may be a stand of cereals, the foliage of an orchard, a field of sugar beet, or wild vegetation. The source of primary inoculum may reside in the target population, as with mildewed buds in a peach orchard, or may be associated with nearby or distant wild or cultivated plants. The primary inoculum may be ascospores (with the advantage of genetic novelty) or conidia generated by a remanant population of the parasite which has survived from an earlier epidemic (see Perennation), or conidia transported from an existing epidemic separated from the target by distance alone.

A. Ascospores

1. Ascospores as inoculum in semi-arid climates

According to Palti (1971, 1972), cleistothecia in Israel form more abundantly on weeds and wild vegetation than on crop plants and appear at the end of the vegetative season in late winter and early spring. The role of wild graminaceous hosts in the oversummering of cereal mildew is clear from the work of Koltin and Kenneth (1970) and Eshed and Wahl (1975) who concluded that this source of ascosporic inoculum is important for the infection of cereals sown at the end of the five to six hot, rainless, summer months. The importance of ascospores as primary inoculum is undoubtedly enhanced in the case of non-specific species of powdery mildew because of their wide host range; in this respect Dinoor (1974) discussed *Xanthium*, *Senecio* and *Papaver* as sources of epidemics of *Sphaerotheca fuliginea* on cucurbits. Schnathorst (1959) reasoned that ascospores probably initiate epidemics on field lettuce in California, because the earliest disease appeared on sites where cleistothecia had been plentiful on the previous crop.

2. Ascospores as autumn inoculum in cold temperate climates

Evidence of an autumn role for cleistothecia seems limited to the cereal mildews. Cherewick (1944) was doubtful of ascospores infecting spring-sown cereals and his opinion that their discharge takes place in autumn has been verified in several countries (Smith and Blair, 1950; Turner, 1956). In England, Turner (1956) showed that ascospores of *E. graminis* are discharged in late summer and early autumn and she drew attention to the importance of post-harvest stubble as a source of this inoculum. In Denmark, most of the cleistothecia have dehisced by the end of the year and the remainder are sterile (Smedégard-Petersen, 1967).

3. Ascospores as spring inoculum in cold temperate climates

Smith and Wheeler (1969) wrote

> there is still relatively little information about the role of cleistocarps in the overwintering of powdery mildews . . . overwintering . . . under *natural* conditions, and discharge of infective ascospores from *these* cleistocarps . . . (my italics).

Studies by Wheeler and his colleagues (see Section IIID) lead to the conclusion that cleistothecia are not equipped to survive winters which are neither sufficiently dry nor cold. Field evidence of ascospores as primary inoculum in spring has been obtained by Jordan (1966) on blackcurrant, by Liyanage and Royle (1976) on hop, and by Itoi *et al.* (1962) on mulberry.

B. Conidia from sources residential in the target population

Populations of powdery mildews which have hibernated in buds are ideally positioned to initiate epidemics immediately vegetative growth recommences and as soon as conditions allow these foci to sporulate. The term *primary mildew* is often used to describe the totally mildewed growths which emerge from diseased buds in spring and distinguishes this phase of a disease from *secondary mildew* which is the subsequent epidemic of conidial infections.

Buds are strong sources of primary inoculum because of their location within the target population and because vigorous buds (e.g. apical vegetative buds) produce mildewed shoots which release an enormous quantity of conidia before their growth stops prematurely. Consequently, the concentration of conidia at the start of the vegetative season is often high and is one reason epidemics of diseases such as apple mildew begin early and prove difficult to control.

Mycelial colonies which have persisted on leaves since the previous epidemic likewise threaten new growth on the parent plants but these foci are individually less productive of conidia than are foci of primary mildew. Epidemics on rubber are initiated in this way (Populer, 1972). In the case of winter wheat in Maryland, USA, Powers (1957) described how a mildew population was static during the cold months of January and February, but new lesions which appeared in March were the start of an epidemic which spread rapidly in the same crop.

C. Conidia from sources outside the target population

1. Primary inoculum from weeds and volunteer plants

Rains which trigger the dehiscence of cleistothecia also stimulate the growth of volunteer seedlings and the mycelial colonies which arise from ascospores which infect these plants produce conidial inoculum for crops sown soon afterwards. Volunteer plants are known to play this bridging role in the autumn infection of winter cereals (Turner, 1956) and in Israel this same function is ascribed to wild grasses (Koltin and Kenneth, 1970; Eshed and Wahl, 1975). Weeds can be important "carriers" of species which, like Leveillula taurica, have a wide host range. Nour (1958) commented upon the rarity of cleistothecia in the tropics and sub-tropics and the difficulty of identifying sources of primary inoculum; he showed that in the dry season in the Sudan the common weed Euphorbia heterophylla is an alternative host for L. taurica and a likely source of primary inoculum attacking broad beans. Dinoor (1974) mentioned the probable role of wild plants in initiating severe epidemics of Oidiopsis on peppers, eggplants and other crops at isolated desert sites.

2. Primary inoculum transferred from neighbouring crops

The spread of conidia from autumn- to spring-sown cereals is a serious problem in Europe (Turner, 1956; Stephan, 1957) and North America (Cherewick, 1944; Johnston, 1974) and in this context the winter crops are described as "green bridges" which enable E. graminis easily to overwinter after the discharge of ascospores in late summer and autumn. Smith and

Davies (1973) found that proximity to winter barley increased the conidia in the vicinity of spring-sown barley five- to six-fold early in the season and the existence of disease gradients provide additional and convincing evidence that winter cereals are a source of primary inoculum (see Section IIID). The spread of conidia from mature to young crops of herbaceous plants is considered a hazard in Israel (Palti, 1972, 1973). Two factors which aggravate this problem are an increasing proneness to infection as the crops age and the wide host range of several important mildews. In consequence, *Sphaerotheca fuliginea* moves freely between all cucurbit crops and *L. taurica* cross-infects some important solanaceous hosts. The problem is even more serious where, with the aid of plastic covers and irrigation, vegetables are cultivated throughout the year.

3. Primary inoculum from distant sources

Mehta (1930) suggested that conidia blown from uplands might be responsible for the annual appearance of *E. graminis* on cereals in certain foothills in India. The convincing evidence of remote sources of primary inoculum has been provided by Hermansen (1968) to explain widespread outbreaks of mildew in early summer on spring-sown barley in Denmark. He has reasoned that in the absence of local overwintering sources spore showers originate in distant source areas of winter barley, such as North Germany (see Section IIID). Likewise, oat mildew is considered to have its source in more distant parts of Europe. His theory has been strengthened by the interception at the Danish coast of infectious barley mildew conidia in onshore winds traced back to likely source areas in Europe, and on the basis of this and other evidence the inoculum which initiated a severe epidemic on Emir barley in 1971 is believed to have originated in Britain (Hermansen *et al.*, 1975).

Barley mildew in the Faroes may owe its origin to conidia blown from western Europe (Hermansen and Wiberg, 1972), and a northward dispersal of conidia from continental Europe may account for the earliness of infection on cereals in eastern England (Yarham and Pye, 1969; Doling *et al.*, 1971). Smith (1969) suggests a continental source to explain the absence of *Erysiphe polygoni* on peas in England until late summer.

D. Factors determining the quantity of primary inoculum

1. Source strength

(a) Ascospores.

The cleistothecia of barley mildew have been known to discharge infective ascospores after thirteen years storage at 10°C (Moseman and Powers, 1957), but more relevant is the effect of normal weathering on longevity. Merriman and Wheeler (1968) found that the ascospore potential of mildewed blackcurrant leaves in England fell rapidly in late winter, especially if leaf debris was lying on soil. This is important because discharge precedes the start of the vegetative season. Wheeler *et al.* (1973) suggested that weakening of the wall of the fruit body by microbial action and physical weathering might explain an increased dehiscence of *E. cichoracearum* between December and mid-February; later work provided evidence that this weakening can culminate in degeneration, especially if host residues are in an environment which favours degradation (Jackson and Wheeler, 1974). For example, cleistothecia on blackcurrant leaves suspended in air survived the winter better than those on leaves on soil.

The significance of winter conditions is evident if these results of Wheeler and his colleagues in England are compared with those of Smith (1971) in Canada. In Alberta, cleistothecia on leaf debris on the ground were preserved by snow and functioned as effectively as those exposed to air which was sometimes below −30°C. Even after the air warmed to above 0°C in spring, the soil remained frozen and ascospores were eventually discharged when the snow melted. Smith has emphasized the difference between conditions such as those in Canada, which favour survival of the fruit bodies, and the milder and wetter conditions in England which lead to degeneration and premature dehiscence.

The dehiscence mechanism is activated by physical factors which therefore regulate the source strength. Cutter and Wheeler (1968) observed that the number of discharged ascospores of *E. cichoracearum* in any period, increased with the length of time above 5°C but Moseman and Powers (1957) found moisture to be more important than temperature as factors controlling the dehiscence of *E. graminis*. The need for external water has frequently been noted by laboratory and field workers and it is interesting that this is the only event in the life history of powdery mildews which requires liquid water. Cherewick (1944) and Turner (1956) remarked on the need for rains to activate fruit bodies of cereal mildews, and Padalino *et al.* (1970) stated that although ascospores of wheat mildew first appear in southern Italy after summer rains, most are discharged from September to November when rains follow the summer drought. Likewise, in Israel the oversummering of barley mildew is terminated by cool, rainy weather.

(b) Conidia from buds.

The amount of primary mildew depends upon the progress and severity of the previous epidemic (Butt, 1971a, 1972a). The sporulation potential of each infected bud can depend upon when it was invaded in the previous season;

early infection often produces greater damage and so less sporing tissue (Kosswig, 1958; Mygind, 1965). There is evidence that the number of buds with viable mildew declines during winter; Weinhold (1961a) recorded 24% mildewed peach buds in January but only 9% in March. Cold weather reduces the survival of hibernating mildew to the extent that a few hours at -20 to $-25°C$ will virtually eradicate overwintering populations of apple mildew (Roosje and Besemer, 1965; Cimanowski, 1969) and even cold spells in the moderate winters of North-west Europe can have this sanitary effect, as Preece (1965) observed on gooseberry.

(c) Conidia from leaves.

The quantity of primary inoculum depends upon the number of mycelial colonies which persist through the inter-epidemic period and where this entails overwintering, a crucial factor is the survival of living host tissues. Cherewick (1944) suggested that the relative freedom from mildew of cereals on the plains of Saskatchewan, Canada, might be due to the lack of protective snow cover and hence the winter-kill of potential carrier hosts. According to Mygind (1970), volunteer cereal plants in Denmark possibly survive as carriers only in exceptionally mild winters. Yarham et al. (1971) associated the disappointing control of mildew on winter barley in a fungicide trial in England with the many volunteer plants present in stubble fields in a mild autumn. The large acreage of winter cereals grown in recent years has undoubtedly increased the strength of this source of primary inoculum which annually threatens the spring-sown crop.

Where primary inoculum arrives from an active epidemic outside the target the quantity clearly depends upon the severity of the epidemic in the source area. A cultural factor which affects the severity of E. graminis on winter cereals and so determines the quantity of inoculum to threaten spring-sown cereals is the nutritional status of the crop. Last (1953, 1954) has shown that the susceptibility of winter wheat increases with growth rate, and he has stressed the importance of applying nitrogenous fertilizer early in winter so that crop responses occur before conditions favour the rapid multiplication of the mildew (see also Section V).

2. Distance of source from target

The best indications of the effect of distance of source from target are disease gradients measured on spring-sown cereals. Gradients over a long distance have been detected on spring barley in Denmark (Hermansen, 1968), increasing from north to south towards the suspected source area of winter barley in North Germany. Over short distances there is clear evidence of dilution of inoculum; Yarham and Pye (1969) and Yarham et al. (1971) measured gradients extending 183 metres from winter barley, and Mygind

(1970) detected the influence of source at a distance of 1000 metres. (See also Section IIIC.)

IV. Effect of Physical Environment on the Disease Cycle

This section examines the effects of physical factors upon events in each generation of the asexual state of the parasite. These successive events in the disease cycle, sometimes called the infection chain, are, for example, infection, colonization, sporulation and dispersal. The activity of the fungus at each stage of the cycle is determined not only by the physical environment but also by attributes of the parasite (e.g. aggressiveness) and the host (e.g. susceptibility) and by the interaction of all three. An understanding of each phase of the cycle is fundamental to the analysis of the polycyclic epidemics which develop after initial infections are established and to the synthesis of conceptual models of the disease system.

A. The infection phase

1. Water relationships

(a) The conidium.

A unique feature of these large spores is their ability to germinate and infect in the absence of external liquid water. Yarwood (1950) suggested that this is made possible by their relatively large water content, and his concept of a self-sufficient conidium is supported by the generous volume of vacuoles in which water appears to be stored (Mitchell and McKeen, 1970) and by the gelatinous sheath which makes most of the cell wall impervious (McKeen et al., 1967). Somers and Horsfall (1966) considered that the water-retaining ability of these spores may be more important than the water content but whatever the physico-chemical reasons, the powdery mildew conidium is a powerfully independent propagule.

(b) Atmospheric moisture.

(i) Germination of conidia. Yarwood (1936a) showed that conidia of E. polygoni not only germinate without liquid water but can do so in atmospheres approaching complete dryness. Not all powdery mildews exhibit this degree of xerophytism for species differ in their tolerance of atmospheric moisture stress. Two workers have separately classified these fungi into three groups using as their criteria response to moisture stress at germination (Schnathorst, 1965), or rate of germination (Zaracovitis, 1965).

Like *E. polygoni*, *Uncinula necator* germinates over a wide range of moisture stress (Delp, 1954) and both species germinate rapidly ($>$ 70% germinated in 5 h in saturated air at 21°C) which suggests that fast germination is associated with tolerance of moisture stress (vapour pressure deficit (VPD)). In contrast, *Sphaerotheca pannosa* and *Podosphaera leucotricha* are in the group least tolerant of moisture stress with a humidity requirement at or near saturation (Berwith, 1936; Weinhold, 1961b), and both species germinate slowly ($<$ 70% in 10 h). In both classifications *E. graminis* and *E. cichoracearum* are placed in the middle group of species which, according to Schnathorst, germinate best at low moisture stress but have a proportion of conidia capable of germinating under high stress.

The response of an organism to moisture stress is, of course, a function of ambient temperature, and humidity should not be divorced from air temperature in any study of germination. Tolerance to low relative humidity diminishes, for example, as temperature rises and VPD increases (Peries, 1962; Manners and Hossain, 1963), and the conidia of even *E. polygoni*, a very xerophytic species, shrivel as temperature rises from 13–22°C (Yarwood, 1936a). Shrivelling, accompanied by an irreversible loss of turgidity and viability, is often suffered by conidia exposed to low humidity (Nour, 1958; Jhooty and McKeen, 1965a; Zaracovitis, 1966; Clerk and Ankora, 1971).

(*ii*) *Growth of germ tubes.* An early caution by Brodie and Neufeld (1942) that conditions which favour germination may not be optimal for other stages of infection proved to be wise counselling for it seems that germ tubes are more sensitive to moisture stress than are their parent conidia. On glass, conidia of *E. graminis* germinate at all humidities, including zero, but germ-tube growth is poor below 98% r.h. (Manners and Hossain, 1963). Data presented by Nour (1958) for *E. cichoracearum* give an average germ-tube length of 32 μm at saturation but only 3–8 μm at 85% r.h. Even with *Phyllactinia corylea* and *Erysiphe betae*, which germinate well at all humidities, tubes increase in length with rising humidity (Drandarevski, 1969; Clerk and Ankora, 1971).

(c) Liquid water.

It has long been known that the internal structure of powdery mildew conidia collapses when the spores are in water (Corner, 1935; Zaracovitis, 1966), and immersion for as brief as 3 min can kill 50% of the conidia (Peries, 1962). Only in a few cases has it been possible to use suspensions in water for experimental inoculations (Reuveni and Rotem, 1973; Aldwinckle *et al.*, 1975). Condensation water reduces germination (Delp, 1954; Manners and

Hossain, 1963) and produces a typical growth of germ tubes (Populer, 1972). The position of spores in relation to the water is important. Nour (1958), for example, noted that although conidia of *L. taurica* are damaged when submerged they produce appressoria when in a thin film of water. Germination of *S. pannosa* on peach leaves improved in the presence of condensation droplets, a finding discussed by Weinhold (1961b) in the context of the harmful effect of water films. Water may interfere with penetration of the host and Perera and Wheeler (1975) postulated that a fine spray applied to rose leaves soon after inoculation floated the conidia and prevented contact with the leaf surface. However, in studies of *L. taurica* on *Euphorbia heterophylla* and *E. graminis* on barley, simulated rain applied three hours after inoculation had little effect on the development of disease, a result interpreted by Nour (1958) as evidence of the adhesion of conidia and germ tubes to leaves.

2. Temperature relationships

Every epidemic has a dimension in time, and germination, like every other event in the disease cycle, is a function of time. Current interest in systems analysis is directing attention to rates of change. The significance of rates in powdery mildew germination studies was stressed by Weinhold (1961b), and Manners and Hossain (1963) measured rate of development as the best criterion of the response of germination to temperature. Weinhold obtained maximum germination of *S. pannosa* in the temperature range 18–24°C after 6 hours, whereas after 12 hours the optimum range extended to 27°C, a temperature which had earlier supported rather slow growth. Another example of dependence upon time is the finding by Schnathorst (1960a) that the optimum temperature for germination of *E. cichoracearum* decreased with time. Although optima sometimes extend over a broad temperature range the action of a limiting factor such as time can sharpen the response in the manner described by Colhoun (1973), and this is nicely shown in germination data for *P. leucotricha* after 6 and 24 hours (Coyier, 1968).

Time is not the only factor which affects the optimum temperature and it is not surprising that for spores which germinate in air, atmospheric humidity is a limiting variable in temperature-germination relationships. In several studies the optimum temperature falls as moisture stress increases (Manners and Hossain, 1963; Hewitt, 1974). The magnitude of this interaction probably varies between species and might be expected to be smallest with extreme xerophytes such as *U. necator* (Delp, 1954).

In general, germination declines sharply at or just above 30°C, and although Jhooty and McKeen (1965a) obtained some germination of *Sphaerotheca macularis* at 35°C in saturated air, the conidia and germ tubes shrivelled. Germination at low temperature is, of course, less affected by moisture stress;

reported minima are often around 2°C, and with *P. leucotricha* Coyier (1968) obtained 44% germination at 4°C on young apple leaves.

Schnathorst (1965) noted that in general the maximum and minimum temperatures for infection are lower and higher respectively than the cardinal values for germination; there is some evidence that in comparison with germination, germ tubes prefer slightly higher temperatures (Manners and Hossain, 1963; Hewitt, 1974).

3. Light

The author is not aware of any failures to infect plants in darkness and in the field, part of the infection phase almost certainly occurs at night following the diurnal liberation of spores. Yarwood (1936b) improved the germination of *E. polygoni* conidia collected at night by exposing them to light, but this treatment did not affect spores collected in the daytime. Drandarevski (1969) found that light stimulated the germination of *E. betae*.

Ultraviolet light is harmful to the germinability of *E. graminis* conidia (Martin *et al.*, 1975) and their infectivity is even more sensitive (Buxton *et al.*, 1957; Moseman and Greeley, 1966). The sporeling is less vulnerable to ultraviolet after the haustoria have formed so that more radiation is required to kill colonies than to prevent infection (Hey and Carter, 1931; Mount and Ellingboe, 1968).

4. The quality of conidia

The environment can affect the quality of inoculum before conidia are liberated and researchers have observed variations in germinability and infectivity according to the day, place and time of season (Nour, 1958; Peries, 1962; Benada, 1970). Cherewick (1944) obtained high germination rates with spores collected on bright days and a favourable effect of light on inoculum quality has also been observed by Yarwood (1936b), Aust (1973) and Moseman and Greeley (1966). Stavely and Hanson (1966) found no effect of sporulation temperature on subsequent germination (although spore size increased with temperature) but a high humidity during sporulation can improve germinability (Ward and Manners, 1974). Yarwood *et al.* (1954) described powdery mildew conidia as among the shortest lived of airborne fungal spores and time may therefore be a factor contributing to certain of these observations. The quality of inoculum is known to depend upon the age of the conidia and upon the age of the parent colonies (Nair and Ellingboe, 1962) and it is nowadays common practice to shake stock plants a day before collecting spores in order to avoid using stale inoculum (Peries, 1962; Morrison, 1964).

B. The colonization phase

In this post-infection phase of the disease cycle the sporeling develops and establishes a superficial thallus which is seen as a colony or pustule on the surface of the host.

Colonization does not end with the start of sporulation (the end of the latent period) or with the appearance of visible symptoms (the end of incubation) but continues until radial growth of the thallus ceases.

1. Water relationships

A consequence of ectophytism is the exposure of the thallus to rain and dew and the hyphae of powdery mildews are, like the conidia, easily damaged by contact with liquid water. Yarwood (1936a) observed an improvement in mycelial growth when mildewed plants were protected from rain and young colonies grew abnormally when sprayed with water (Cherewick, 1944). By this latter method it is possible to obtain complete control of mildew (Yarwood, 1939) but the duration of wetting is important (Rogers, 1959).

The thallus seems to be less sensitive than conidia and germ tubes to atmospheric moisture stress (Rogers, 1959). Thus in the case of *S. fuliginea* on squash a low humidity (55–60% r.h.) favours colonization and sporulation whereas a high humidity favours infection (Reuveni and Rotem, 1974). Mildew colonies can be damaged by high moisture stress, however, and although the radial growth of *S. macularis* is not affected by humidity in the range 12–100% r.h. (Peries, 1962), its hyphae have been observed to shrivel under extreme dryness (Jhooty and McKeen, 1965a). Even colonies of *E. polygoni*, a species which exhibits considerable xerophytism at germination, die if the moisture stress is too high (Yarwood, 1936a).

2. Temperature relationships

Although certain environmental races are adapted to high temperatures (Newton and Cherewick, 1947; Schnathorst *et al.*, 1958), Yarwood *et al.* (1954) considered that among plant pathogens the powdery mildews have low optima, averaging 21°C.

The mycelia of some species grow at 5°C (Bainbridge, 1972) but slow development at low temperature lengthens the latent and incubation periods and in the case of strawberry mildew Peries (1962) obtained no sporulation after three weeks in the range. 5–13°C Barley mildew in Denmark had a latent period of 56–87 days at 2–4°C, reducing to seven days in summer (Hermansen, 1968). Jenkyn (1973, 1974a) measured incubation periods on barley seedlings exposed to field conditions in England. He found that in winter the period ranged from 16 to > 35 days, but was only 5–8 days (also the length of latency) in summer, and there was an inverse relationship between incubation time and infection severity.

Bainbridge (1972) reported that *E. graminis* on wheat and barley grows slowly above 25°C and Newton and Cherewick (1947) described the pustules of this species as thin mycelial mats at this temperature. However, just as time is a factor in the response of germination to temperature so temperature-colonization relationships are time dependent. Perhaps the best illustration was given by Stavely and Hanson (1966). The development of *E. polygoni* two days after inoculation was similar at 24° and 28°C but after four days, growth was better and sporulation had begun at the lower temperature; even after 14 days mycelial growth and sporulation remained markedly inhibited at 28°C. A reduction of sporulation at temperatures above the optimum is obviously of epidemiological importance.

C. The sporulation phase

1. The periodicity of spore production

Some species produce a single conidium on each erect conidiophore every 24 hours (Yarwood, 1936b; Drandarevski, 1969), whilst others are more fruitful and produce a daily chain of conidia. *E. polygoni, E. betae, U. necator, P. corylea* and *O. heveae* are monoconidial types whilst *S. pannosa, E. cichoracearum* and *P. leucotricha* are typical chain-forming mildews.

With the notable exception of *E. graminis* (Hammett and Manners, 1973), the activity of the spore-producing apparatus is diurnal in both groups of mildews (Yarwood, 1936b; Childs, 1940; Pady *et al.*, 1969). In the chain formers this periodicity is determined not by the initiation of spores by divisions of the generative cell in the stalk of the conidiophore but by the final stage of sporulation, when the conidia abstrict and are ready for liberation. This maturation stage is photosensitive and proceeds faster in light than in dark (Cole and Fernandes, 1970) with the result that most of the conidia initiated at night remain immature until the following daytime.

2. Quantitative aspects of spore production

Ward and Manners (1974) showed a large effect of temperature on the productivity of a colony. Working with barley mildew they obtained distinct optimum at 20°C in the number of conidia produced per unit area of colony; sensitivity to temperature is evident from the large difference in sporulation between 20° and 25°C. At the higher temperature productivity was one quarter and infectivity one tenth that at 20°C so that effective inoculum was reduced to 2·4%. This result agrees with that of Newton and Cherewick (1947) who found that colonies of *E. graminis* at 25°C ceased sporulating 2–3 days after reaching maximum development. As mentioned above, the sporulation of *E. polygoni* is markedly inhibited at 28°C.

Ward and Manners (1974) obtained maximum sporulation of barley mildew at 100% r.h. and Butt (1972b, 1975) found that sporulation of apple mildew is enhanced by high humidity at night. Reuveni and Rotem (1974) reported, however, that a relatively low humidity (55–60% r.h.) improves sporulation of *S. fuliginea* on squash, as Hammarlund (1925) found for *S. pannosa*.

3. Spore liberation

Apart from the dubious advantage of the conidiophore bending in *P. corylea* (Clerk and Ankora, 1969), spore liberation in powdery mildews is a passive event, as is evident from the accumulation of dense mats of spore aggregates and the development of spore chains by monoconidial species when in still air (Clerk and Ankora, 1969; Hammett and Manners, 1973). Yarwood (1936b) suggested that wind speed and atmospheric humidity control spore liberation, and Hammett and Manners (1971) showed that the liberation of *E. graminis* from field barley increases with increasing wind speed, low humidity, dry leaves and high temperature.

Wind tunnels have been used to measure the interaction of air speed and humidity on spore release in *E. graminis* and *P. leucotricha* (Butt, 1972b; Hammett and Manners, 1974). Release increases with air speed until the supply of conidia is depleted, the majority being liberated as air accelerates (gusts); relatively few conidia are detached in steady-state conditions. Spores of apple mildew are most easily liberated in air below 90% r.h., which is in accord with a report of increased "stickiness" of conidia at high humidity (Schnathorst, 1959): in contrast, *E. graminis* release does not respond to humidity changes and in the case of *E. cichoracearum* on tobacco the response is very slight (Cole and Fernandes, 1970). Perhaps this variation in sensitivity to humidity is due to differences in the hygroscopic properties of the spore wall. Another factor in spore detachment is the shaking of leaves by wind (Bainbridge and Legg, 1976).

Clearly, the regulation of spore production by photoperiodicity ensures that the majority of conidia are mature when meteorological conditions are most favourable for liberation and dispersal. Cole (1976) discusses this aspect of the synchronization of conidiophores.

The phenomenon of "tap-and-puff" by which raindrops cause the dry liberation of spores (Hirst and Stedman, 1963) effectively detaches powdery mildew conidia at the onset of rain (Hammett and Manners, 1971; Jenkyn, 1974b), although in some weather conditions this temporary increase in airborne conidia is associated with turbulence immediately before a storm (Fernando, 1971a; Eversmeyer *et al.*, 1973).

D. The spore dispersal phase

1. Spore clouds

The propagules in spore clouds are not necessarily single conidia because the spore chains of chain-forming species fragment into short lengths which, with clumps of spores, disperse as aggregates (Peries, 1962; Carter, 1972; Cole, 1976). Schnathorst (1959) found that more than 70% of the total catch of spores of E. *cichoracearum* in fields of lettuce were in clumps, a state of flight which possibly has some survival value.

The conidia of powdery mildews belong to the dry-air spora, and in view of the normal daily pattern of the release agencies and the diurnal spore maturation of most species it is not surprising that the concentration of spore clouds is highest during daytime; numbers peak around midday in and above field crops (Peries, 1962; Sreeramulu, 1964; Cole, 1966a) and orchards (Burchill, 1965). Minor deviations from this periodicity may be caused by local atmospheric disturbances, in the way that rising convective air probably explains the brief fall in spore concentration around midday in rubber plantations in equatorial Africa (Populer, 1972). The data collected by Smith and Davies (1973) in their study of cereal mildew indicate that as wind speed increases there is an eventual decline in spore concentration, presumably due to dispersal and dilution of the cloud.

An onset of rain temporarily increases the concentration of conidia at any time of the day or night. After heavy rain spore numbers (and new infections) are low for several days due to the damage to the superficial colonies (Butt, 1970). After a thunderstorm the reduction in airborne conidia of E. *graminis* lasted for three days (Hirst, 1953) and Peries (1962) recorded a three-day lag before strawberry mildew conidia recovered to their pre-rains concentration.

A steep concentration gradient has been recorded in the proximity of diseased plants, which suggests rapid rall-out. Peries (1962) trapped 90% of airborne conidia of strawberry mildew within a radius of 1·5 metres from the plants and fewer than 10% of the propagules were caught in flight above the crop. The proportion of spores above a crop depends upon whether their source is endo- or exophytotic and upon meteorological conditions, in particular turbulence. With cereals, most endogenous conidia remain within the crop and tend to concentrate below the level of production (Last, 1955; Stedman and Hirst, 1973).

2. Long-distance dispersal

In Section IIIC, D examples were given of the movement of inoculum from distant sources.

Aeroplanes have trapped *Erysiphe* conidia over Denmark (Hermansen *et al.*, 1965) and above the English Channel (Hirst *et al.*, 1967). Viable propagules of this mildew have been intercepted at altitudes up to 1500 metres and there is evidence that conidia of *E. graminis* are still infectious after crossing the North Sea from Britain to Denmark (Hermansen, 1968; Hermansen *et al.*, 1975). If mildew conidia have survived in air masses arriving at the Danish coast the popular view of these spores as short-lived must be revised. In this respect conidia of barley mildew have withstood storage at −60°C for more than a year (Hermansen, 1972).

V. Aspects of Host Resistance

Disease resistance is a subject which is dealt with in several other chapters in this book but certain aspects are mentioned here, in particular mature leaf resistance, adult plant resistance, and changes in disease proneness as a crop matures.

Leaves of trees and shrubs are very susceptible to powdery mildew when they emerge but rapidly acquire resistance as they unfold and expand (Mence and Hildebrandt, 1966; Butt, 1971b). On rose, for instance, a leaf begins to acquire resistance as early as four days after unfolding and is almost fully resistant after two weeks (Rogers, 1959). Populer (1972) described in detail the changes in resistance which accompany the growth of individual rubber leaves; they remain susceptible for about two weeks but infections towards the end of this period result in small colonies.

It is well known that the resistance of cereals to *E. graminis* increases with plant age so that infection rate and ultimate disease severity are greatest on the lower leaves (Large and Doling, 1962; Shaner, 1973a). The expression of adult plant resistance (apr) is very complex and Jones and Hayes (1971) and Jones (1975) have revealed the complicated interrelationships between onset of apr in individual leaves and physiologic age of plants, date of sowing and rate of growth. Furthermore, Tapke (1953) showed the extent to which environment can modify apr. The upper leaves of cereals support fewer and smaller colonies and less sporulation than the lower leaves (Hyde, 1976; Russell *et al.*, 1976) and Shaner (1973b) has calculated that sporulation potential (product of colony size and conidiophore density) accounts for the low infection rate on "slow-mildewing" wheat.

Crops of some herbaceous plants behave differently, becoming more prone to powdery mildews as they mature (Chorin and Palti, 1962). (See also Section X.) Thus in the 1974 sugar beet epidemic in the USA mildew did not become serious until after the crop canopy formed (Ruppel *et al.*, 1975). Such behaviour is perhaps attributable to changes in microclimate, although it is to be noted that the leaves of some herbaceous hosts do increase in susceptibility after they emerge (Wark, 1950; Palti, 1961; Cole, 1966b).

Bainbridge (1974) described how nitrogenous fertilizer increases infection, colonization and the sporulation potential of *E. graminis* on barley. Other aspects of the effect of nitrogen on resistance and disease progress are mentioned under Sections IIID and VIII.

VI. Weather

Yarwood (1957) wrote

the greater incidence of *Sphaerotheca pannosa* on roses under dry . . . conditions is one of the first observations on the epidemiology of specific plant diseases by a competent observer,

and the generalization that powdery mildews are more severe in dry weather is basically true. Boughey (1949) recognized the association of these fungi with regions of low rainfall and they certainly thrive in Mediterranean and semi-arid climates (Palti, 1972).

It is important to realize that rain is not always harmful, for whilst free water causes poor hyphal growth and impacting raindrops damage conidiophores, showers can stimulate powdery mildews by raising the atmospheric humidity. Wastie (1969, 1972) distinguished between these opposing effects of rain in his study of rubber mildew in Malaysia; rain for 1–2 days at the start of refoliation is important for "triggering" an epidemic, which develops rapidly when the following weather is fine with occasional short showers. If there is frequent rain after the initial "trigger" the epidemic is unlikely to be severe.

Although these fungi do not prosper in wet weather (*Erysiphe trina* may be an exception—see Yarwood and Gardner, 1972a), there has been too much emphasis in the past on the dry-loving nature of the family. There can be no denying that powdery mildews are adapted to dry environments but species vary in their degree of xerophytism and many respond favourably to high humidity, mist and dew, especially in the infection phase. In Ceylon it is not hot, sunny weather which encourages rubber mildew but humid conditions accompanied by night mists (Fernando, 1971b). Likewise, strawberry mildew in Finland is less severe inland than 20 km away near the coast, where the night humidity is higher (Tapio, 1972), and in western Europe apple mildew becomes serious in fine, dry weather when there is dew at night (Roosje and Besemer, 1965; Butt, 1975). The positive correlation between dew and mildew severity has been observed by several workers (Sprague, 1953; Peries, 1973; Royle, 1975).

In these circumstances it is not surprising that the descriptions of weather associated with certain mildews seem to be self-contradictory. Consider, for example, the following extracts from Butler and Jones (1955).

Clover mildew

any correlation between the incidence and severity of mildew on clover and the relative humidity of the atmosphere is difficult to understand, as infections may start and continue during seasons of light rainfall and be equally severe during periods of comparative drought.

Cucurbit mildew

the disease is liable to develop during wet periods . . .. But even during comparatively dry periods the fungus may still be active provided heavy dews occur at night, but long periods of dry weather definitely arrest the trouble.

Rose mildew

mildew appears to break out under most diverse conditions of the environment. If warm muggy weather occurs in late spring or early summer, the disease may develop rapidly, but it varies considerably in intensity from season to season, and is often severe during a dry summer.

Apple mildew

there is no definite evidence that outbreaks of mildew are favoured by one type of climate more than another, and while it is stated that the trouble in some localities is worse in hot and dry seasons, in others it is said to be encouraged by warm, moist summers, or by cool, damp sunless weather.

The apparent anomalies in these descriptions reflect the enigmatic relationship between powdery mildews and moisture, and the sensitivity of atmospheric humidity to location, topography and crop environment.

Atmospheric moisture stress is a major habitat factor which controls not only the severity of powdery mildews but also their distribution. Chorin and Palti (1962) classified the powdery mildews in Israel in three groups: the largest tolerates a wide range of moisture stress and includes species like *U. necator* on vine, *S. fuliginea* on cucurbits and *L. taurica* on artichokes, with distributions largely independent of humidity. In the second group, *E. graminis* on cereals and *E. cichoracearum* on tobacco prefer regions with a low moisture stress. Finally, there are species like *Oidium* on potato which need a high moisture stress. It is interesting that the host can interact with the pathogen–humidity relationship and as an example Palti (1972) cited *L. taurica* which on tomato develops well at a relatively low daytime humidity (50% r.h.) but on pepper grows better at 85% r.h. On tomato, this species extended its habitat to the humid coastal plain in Israel when susceptible cultivars were cultivated.

The influence of temperature is perhaps most conspicuous when epidemics are suppressed in hot weather. The curtailment of the more xerophytic species is likely to be a direct effect of high temperature rather than low humidity. In California, for example, the mid-summer arrest of *U. necator* on vines is probably a direct effect of temperature (Delp, 1954), and in Wisconsin, *E. polygoni* does not thrive on red clover in early summer because the high temperature inhibits sporulation (Stavely and Hanson, 1966).

Just as the habitats of powdery mildews range from stands of lettuce and cereal to the canopies of plantation trees, and from field to glasshouse, so the microclimate varies with corresponding diversity. Schnathorst (1960b) concluded that the microclimate around field lettuce in California is of no relevance to the development of mildew but in the case of herbaceous crops, which form a closed canopy, the microclimate may be important and Palti (1973) suggested that the lower radiation and greater humidity under a closed canopy can partly account for increased disease proneness as such crops age. A microclimate is established during the refoliation of rubber trees and meteorological parameters recorded within the canopy are used to forecast the first critical period of infection by *O. heveae* (Lim, 1972).

VII. Temporal and Spatial Aspects of Epidemics

The severity of an epidemic is a function of the quantity of primary inoculum, so that on apple the amount of primary mildew in spring is correlated with the severity of secondary mildew in summer (Sprague, 1955; Cimanowski, 1969). In some years, however, primary inoculum is not a reliable index of epidemic severity, as when high humidity in summer supported an unexpectedly serious epidemic of apple mildew even though overwintering of the parasite had been greatly reduced by a cold winter (Powell and Doll, 1970). Similarly, a high rate of infection on cereals can result in an epidemic which exceeds expectations based upon primary inoculum (Johnson *et al.*, 1971).

Mildews which perennate residentially in the target population can be expected to multiply rapidly in the first phase of an epidemic, especially those species which emerge as primary mildew from infected buds to produce conidial inoculum for a protracted period in spring and early summer. Diseases with this high initial momentum are often difficult to control, as on apple, whereas on hop the carry-over of hibernating mildew is less and the early progress of the epidemic is much slower than on apple (Royle, 1975). It should be noted that primary mildew generates a "simple interest" component of disease increase (*sensu* Van der Plank, 1963) which augments the "compound" multiplication of secondary colonies already present.

Although an early appearance of cereal mildew is often followed by a severe epidemic there are years when the disease does not reach its potential (Saunders and Doodson, 1970). In cold temperate climates low temperatures and/or rain in spring retard mildew epidemics; Price (1970) showed that cold weather was the reason for the long delay between the appearance of primary mildew on rose (22 March) and his first observation of secondary mildew (17 April).

The middle phase of an epidemic often coincides with the period when vegetative growth is most rapid and this is certainly true for susceptible trees and shrubs. Annual epidemics of rubber mildew are compressed into the relatively brief season of refoliation, the only time when there is a plentiful supply of susceptible leaves (Populer, 1972); if "wintering" is early and rapid, refoliation is also early and rapid leaving insufficient time for the disease to become serious (Peries, 1973; Lim and Rao, 1975). It is interesting that on outdoor roses in England there are two periods of vegetative growth and the annual epidemic of leaf infections occurs mainly during the second flush (Price, 1970). With cereals, the seasonal peak in the number of airborne conidia occurs when stems are extending rapidly (Sreeramulu, 1964).

In the final phase of an epidemic the rate of infection declines. Jenkyn (1974b) suggested that the mid-season reduction in the number of airborne conidia of cereal mildew may be caused by adult plant resistance and by the greater retention of liberated spores within the crop; with the onset of flowering, cereal mildew epidemics wane and terminate when straw forms. The decline of epidemics on rubber is accelerated by the abscisson of mildewed leaves (Populer, 1972), a condition called secondary leaf-fall.

Evidence of long-distance dispersal has been presented in Sections IIIC, D and IVD and two examples of the spread of powdery mildew diseases are the advance of vine mildew in Europe from southern England to the Mediterranean in six years after its first appearance in 1845 (Zadoks, 1967) and the spread of clover mildew in the USA from New York State to Washington State between 1921 and 1924 (Horsfall, 1930). More recently, an outbreak of sugar beet mildew in California in 1974 spread rapidly in the same season (Ruppel et al., 1975). In Denmark and England cereal mildew has been shown to appear each year on successively later dates as distance increases from suspected source areas of overwintering inoculum in the south and south-west respectively (Hermansen, 1968; Doling et al., 1971) but it is possible that this apparent advance of the disease front is in fact the progressive delay in the seasonal onset of favourable conditions.

VIII. Crop Management Practices

Some examples are given by Yarwood and Gardner (1972b) of an increased proneness to mildew of plants growing in or near tilled soil.

The revolution in ground management which has accompanied the introduction of herbicides has not taken place without some effect on powdery mildews. Minimum cultivation techniques in cereal farming encourage the survival of self-sown hosts on unploughed land and hence the overwintering of obligate foliar pathogens (Doling, 1966; Yarham, 1975). The abandonment of soil cultivation in hop gardens has increased the number of infected dormant buds (Liyanage and Royle, 1976). Simazine might have contributed to the increase of American Gooseberry Mildew in blackcurrant plantations by altering the nutritional status of treated plants (Upstone and Davies, 1967).

Overhead (sprinkler) irrigation has reduced the severity of *L. taurica* on tomato (Rotem and Cohen, 1966) and *E. betae* on sugar beet (Ruppel *et al.*, 1975), but on apple *P. leucotricha* was increased by sprinkling (Sprague, 1955), a difference in response possibly related to the high moisture requirement of the latter species. Palti (1971) mentioned that the degree to which sprinkling suppresses *L. taurica* depends upon the crop and atmospheric conditions; on very susceptible tomato cultivars and certain other hosts this species is not markedly inhibited, and the method is less effective when the humidity is very low.

It is sometimes possible to manipulate the timing of the vegetative season in order to impede an epidemic. This technique has been explored on rubber by using defoliants to ensure early leaf-fall. After such an induced "wintering" refoliation is rapid and a serious epidemic is avoided. This interesting development was described by Lim and Rao (1975), who also discussed the application of nitrogenous fertilizer as an alternative way of achieving rapid refoliation. In cereal farming it is well known that disease potential is reduced by sowing winter crops late (Jenkyn, 1970; 1974c) and spring crops early (Last, 1957; Marshall *et al.*, 1971).

The expanding area of winter cereals has aroused anxiety because of the importance of these crops as "green bridges" for the overwintering of *E. graminis* and Johnston (1974) has urged the breeding of resistant cultivars to reduce the strength of this source of spring inoculum. A ban on winter barley in Denmark did not prevent the widespread infection of spring-sown barley in the summer and explanations for this failure have been mentioned in Sections IIIC, D and IVD. In England, the sowing of fungicide-treated seed of winter barley in a large-scale trial was not successful as a method of preventing contamination of the spring-sown crop (Yarham *et al.*, 1971).

Van der Plank (1963) has described why sanitation treatments which reduce primary inoculum are unlikely to be sufficient for the control of pathogens which, like powdery mildews, have a potentially high infection rate. The removal of mildewed shoots in winter and early spring is, however, widely practised as a supplement to chemical control on hosts such as rose and apple, although experience with apple indicates that a very high sanitation ratio is needed before a worthwhile delay of the epidemic is achieved. Excessive cutting of trees and shrubs can exacerbate certain powdery mildews because the new shoots which proliferate are often very susceptible. Indeed, oak mildew becomes a serious problem on pollarded trees.

IX. Experimentation

By their nature, powdery mildews create difficulties in laboratory and field studies. Major problems arise in investigations of the infection phase and in extrapolating the results of controlled experiments to the field.

Yarwood (1936a) warned of the differential behaviour of conidia on glass and leaf surfaces and reported how clover mildew conidia were killed on glass but remained infectious on leaves exposed to the same conditions. Schnathorst (1965) also stressed the importance of working on leaves but despite these warnings germination studies are still undertaken on glass. The "stimulatory" effect of leaves, where conidia are buffered against high temperature and moisture stress (Jhooty and McKeen, 1965a,b), is well known and extends to non-hosts (Peries, 1962). The cause of this effect is still not clear and although Yarwood and Hazen (1944) suggested chemical stimulation, Jhooty and McKeen (1965b) concluded that the relative humidity at the leaf surface can be advantageously high. Whatever the explanation, glass is an inappropriate substitute. Furthermore, it would seem advisable to conduct infection-phase studies in environments which simulate field conditions in order to avoid abnormal microclimates in the phyllosphere and in this respect air movement is important.

The production of plants free of mildew is often made difficult by the ease with which contaminations occur. An apparatus designed by Jenkyn et al. (1973) has solved this problem for seedlings, for each plant in the propagator is supplied with filtered air. For some hosts didecyldimethylammonium, bromide is a useful spray or dip, being a powerful mildew eradicant with the advantage of weak protective action (Kirby et al., 1963).

A use for mildew-free plants is to monitor infections in field conditions (Polley and King, 1973) but a problem which arises with powdery mildews is the difficulty of confidently ascribing disease to infection activity which occurs in the period when the plants are exposed as traps. With most leaf-infecting fungi the experimenter terminates infection by drying the leaves but this simple expedient is not possible with powdery mildews. In their work with barley mildew Polley and King (1973) attempted to resolve this problem by dipping the seedlings in 50% alcohol, after exposure but prior to incubation, in order to kill ungerminated conidia. An alternative technique might be to use a controlled dose of ultraviolet light before incubation.

Inter-plot interference due to inoculum transfer in field experiments is not peculiar to powdery mildews but these fungi spread rapidly and Bainbridge and Jenkyn (1976) have shown how in cereal mildew trials this source of experimental error leads to erroneous interpretations.

Two types of spore trap deserve mention because of their value in epidemiological studies. The first filters air through a sheath of detached host leaves and so detects infectious conidia of known pathogenicity (Hermansen *et al.*, 1967). The second is a stationary rod which has proved adequate for measuring the cumulative number of airborne conidia, a parameter of disease severity convenient for plotting epidemic progress curves (Jenkyn, 1974b).

X. Forecasting Powdery Mildews

In recent years there has been a growing interest in forecasting the critical date for a single application of fungicide (e.g. on barley), the starting date for a fungicide programme (e.g. on rubber), or the timing of applications in intensive spray programmes (e.g. on apple).

For the majority of leaf-infecting fungi infection is limited to periods when surfaces are wet and the infection phase of the disease cycle is therefore an easily monitored event with a recognizable beginning and end. Short-range predictions of such diseases are assured of reasonable accuracy when wetness duration is used as a key parameter. The irrelevance of wetness as a determinant of powdery mildew infection has a profound consequence upon the philosophy of forecasting these diseases and upon the predictive systems used, because there is no meteorological parameter in the infection phase with the dominating role that wetness plays for the majority of leaf fungi.

The philosophy is affected because infections occur almost daily during the growth of an epidemic; there is continuous infection pressure rather than the stop-go pattern dictated by the infection requirements of most fungal pathogens. The main objective for the forecaster of powdery mildews is the prediction of "critical periods" when there are peaks of infection but the control achieved by the timing of fungicide sprays according to these critical periods cannot be as effective as when curative sprays are timed to negate the distinct infection periods of other foliar fungi.

As has been discussed above, atmospheric moisture has a strong influence in the infection phase of some species and relative humidity is included as a component in the predictive systems which have been developed for mildew on rubber (Peries, 1973; Lim and Rao, 1975) and apple (Aerts and Soenen, 1962). For species like *P. leucotricha*, with a strong preference for low moisture stress in the infection phase, there is undoubtedly some value in monitoring atmospheric humidity but this is likely to be less rewarding with more xerophytic species. Butt (1975) has shown that for apple mildew a low moisture stress is important in the night and several other workers have also referred to the importance of moisture at night, which therefore seems to be a time worthy of special consideration in mildew prediction systems (Manners and Hossain, 1963; Clerk and Ankora, 1971; Palti, 1973; Reuveni and Rotem, 1973).

Analyses of field data have revealed that in the absence of the wetness requirement, inoculum assumes the role of key determinant of infection severity (Weinhold, 1961a; Polley and King, 1973; Butt, 1975; Royle, 1975). Whilst there have been many studies of the circadian rhythm of conidial production and of the conditions affecting spore liberation, relatively little is known about the effects of environment on the capacity to sporulate. In this context the ectophytic habit of most species is important because the thallus is superficial and directly exposed to the environment; there is no sheltered incubation period. The capacity of a colony to sporulate is therefore strongly influenced by physical factors which control mycelial growth during colonization. Furthermore, any damage suffered by a thallus during its infectious period results not only in the loss of the immediate crop of spores but also in successive crops until the colony recovers. Clearly, any conceptual model of the disease cycle must incorporate the responses of the thallus to physical factors during and after the latent period.

It is not surprising that recent empirical predictive systems developed in England for barley mildew are based upon the identification of dates when airborne conidia reach high concentrations and meteorological parameters used in these systems have been selected for their influence on spore production, liberation and dispersal (Polley and King, 1973; Smith and Davies, 1973; Smith, 1975). In view of the key role of inoculum it is especially important that models incorporate estimates of current mildew status in addition to meteorological parameters. Thus the correct interpretation of predictions of optimum spray dates on spring barley depends upon the distance of the target from the source crop of winter barley (Polley and Smith, 1973). In models it may be important to include primary inoculum in terms of its strength or its distance from target, or alternatively, to include the level already reached by the epidemic on the day of the forecast. Additional crop-based parameters may be needed; in the case of woody hosts Wheeler (1975) emphasized the importance of the rate of shoot growth, since this determines the availability of susceptible leaves.

The predictive system developed by Palti (1961, 1971, 1972) for herbaceous crops in Israel is an example of negative forecasting and is based upon increasing proneness to powdery mildews as the crops age. When, for example, cucurbits are in their early phase of growth, mildews seldom appear and in the intermediate phase these diseases only develop if the environment is favourable. On crops in the mature phase, however, epidemics develop even in unfavourable weather. "Infection calendars" indicate the length of time crops are expected to remain free of disease after sowing, taking into account the effect of temperature on the rate of growth of each host and the limitiation imposed by humidity on the distribution of certain mildews. Klose (1975) has developed a system of negative forecasting for spring barley.

The powdery mildews are not represented among the first generation of conceptual models of disease systems perhaps this is partly explained by the challenge of disease cycles in which infection is not dominated by a critical environmental factor such as wetness. There is, however, a wealth of quantitative data waiting to be synthetized, and the success of predictive models may be greatest for species with a high optimal moisture requirement at germination if it is easier to characterize the infection phase of such species.

Acknowledgement

I am pleased to record my appreciation of the excellent help given by Mrs Gillian M. Tardivel with the survey and collation of the literature, and with the preparation of the text.

References

Aerts, R. and Soenen, A. (1962). *Agricultura, Louvain*, 2nd Series **10**, 513–547.
Aldwinckle, H. S., Watson, J. P. and Gustafson, H. L. (1975). *Pl. Dis. Reptr* **59**, 185–188.
Aust, H. J. (1973). *Phytopath. Z.* **76**, 179–181.
Bainbridge, A. (1972). *Rep. Rothamsted exp. Stn.* **1971**, part 1, 137.
Bainbridge, A. (1974). *Pl. Path.* **23**, 160–161.
Bainbridge, A. and Jenkyn, J. F. (1976). *Ann. appl. Biol.* **82**, 477–484.
Bainbridge, A. and Legg, B. J. (1976). *Trans. Br. mycol. Soc.* **66**, 495–498.
Benada, J. (1970). *Phytopath. Z.* **69**, 273–276.
Berwith, C. E. (1936). *Phytopathology* **26**, 1071–1073.
Boughey, A. S. (1949). *Trans. Br. mycol. Soc.* **32**, 179–189.
Brodie, H. J. and Neufeld, C. C. (1942). *Can. J. Res.* C **20**, 41–61.
Burchill, R. T. (1958). *Rep. agric. hort. Res. Stn Univ. Bristol* **1957**, 114–123.
Burchill, R. T. (1965). *Ann. appl. Biol.* **55**, 409–415.
Butler, E. J. and Jones, S. G. (1955). "Plant Pathology". Macmillan, London.
Butt, D. J. (1970). *Rep. E. Malling Res. Stn* **1969**, 48.
Butt, D. J. (1971a). *Ann. appl. Biol.* **68**, 149–157.
Butt, D. J. (1971b). *Rep. E. Malling Res. Stn* **1970**, 108.
Butt, D. J. (1972a). *Ann. appl. Biol.* **72**, 239–248.
Butt, D. J. (1972b). *Rep. E. Malling Res. Stn* **1971**, 116.
Butt, D. J. (1975). *In* "Climate and the Orchard" (H. C. Pereira, Ed.), 125–126. *Res. Rev. Commonw. Bur. Hort. Plantn Crops*, No. 5.
Buxton, E. W., Last, F. T. and Nour, M. A. (1957). *J. gen. Microbiol.* **16**, 764–773.
Byford, W. J. and Bentley, K. (1976). *Trans. Br. mycol. Soc.* **67**, 544–545.
Carter, M. V. (1972). *Aust. Pl. Path. Soc. Newsletter* **1**, 2.
Cherewick, W. J. (1944). *Can. J. Res.* C **22**, 52–86.
Childs, J. F. L. (1940). *Phytopathology* **30**, 65–73.
Chorin, M. and Palti, J. (1962). *Israel J. agric. Res.* **12**, 153–166.
Cimanowski, J. (1969). *Acta agrobot.* **22**, 253–263.
Clerk, G. C. and Ankora, J. K. (1969). *Can. J. Bot.* **47**, 1289–1290.
Clerk, G. C. and Ankora, J. K. (1971). *Trans. Br. mycol. Soc.* **57**, 162–164.
Cole, J. S. (1966a). *Ann. appl. Biol.* **57**, 445–450.
Cole, J. S. (1966b). *Ann. appl. Biol.* **57**, 435–444.

Cole, J. S. (1976). *In* "Microbiology of Aerial Plant Surfaces". (C. H. Dickinson and T. F. Preece, Eds) 627–636.. Academic Press, London, New York, San Francisco.
Cole, J. S. and Fernandes, D. L. (1970). *Trans. Br. mycol. Soc.* **55**, 345–353.
Colhoun, J. (1973). *A. Rev. Phytopath.* **11**, 343–364.
Corner, E. J. H. (1935). *New Phytol.* **34**, 180–200.
Coyier, D. L. (1968). *Phytopathology* **58**, 1047–1048.
Coyier, D. L. (1974). *Phytopathology* **64**, 246–248.
Cutter, E. C. and Wheeler, B. E. J. (1968). *Trans. Br. mycol. Soc.* **51**, 791–795.
Delp, C. J. (1954). *Phytopathology* **44**, 615–626.
Dinoor, A. (1974). *A. Rev. Phytopath.* **12**, 413–436.
Doling, D. A. (1966). *J. natn. Inst. agric. Bot.* **10**, (Suppl.), 12–15.
Doling, D. A., Saunders, P. J. W. and Doodson, J. K. (1971). *Ann. appl. Biol.* **67**, 1–11.
Drandarevski, C. A. (1969). *Phytopath. Z.* **65**, 124–154.
Eshed, N. and Wahl, I. (1975). *Phytopathology* **65**, 57–63.
Eversmeyer, M. G., Kramer, C. L. and Burleigh, J. R. (1973). *Phytopathology* **63**, 211–218.
Fernando, T. M. (1971a). *Q. Jl Rubb. Res. Inst. Ceylon* **48**, 100–111.
Fernando, T. M. (1971b). *A. Rev. Rubb. Res. Inst. Ceylon* **1970**, 48–49.
Grasso, V. (1973). *Relatos III Congr. Un. fitop. medit.* Oeiras **1972**, 23–53.
Hammarlund, C. (1925). *Hereditas* **6**, 1–126.
Hammett, K. R. W. and Manners, J. G. (1971). *Trans. Br. mycol. Soc.* **56**, 387–401.
Hammett, K. R. W. and Manners, J. G. (1973). *Trans. Br. mycol. Soc.* **61**, 121–133.
Hammett, K. R. W. and Manners, J. G. (1974). *Trans. Br. mycol. Soc.* **62**, 267–282.
Hermansen, J. E. (1968). *Friesia* **8**, 161–359.
Hermansen, J. E. (1972). *Friesia* **10**, 86–88.
Hermansen, J. E. and Wiberg, A. (1972). *Friesia* **10**, 30–34.
Hermansen, J. E., Johansen, H. B., Hansen, H. W. and Carstensen, P. (1965). *Åsskr. K. Vet.-Landbohøjsk.* **1965**, 121–129.
Hermansen, J. E., Johansen, H. B. and Hansen, H. W. (1967). *Åsskr. K. Vet.-Landbohøjsk.* **1967**, 77–81.
Hermansen, J. E., Torp, U. and Prahm, L. (1975). *Åsskr. K. Vet.-Landbohøjsk.* **1975**, 17–30.
Hewitt, H. G. (1974). *Trans. Br. mycol. Soc.* **63**, 587–589.
Hey, G. L. and Carter, J. E. (1931). *Phytopathology* **21**, 695–699.
Hirst, J. M. (1953). *Trans. Br. mycol. Soc.* **36**, 375–393.
Hirst, J. M. and Stedman, O. J. (1963). *J. gen. Microbiol.* **33**, 335–344.
Hirst, J. M., Stedman, O. J. and Hogg, W. H. (1967). *J. gen. Microbiol.* **48**, 329–355.
Horsfall, J. G. (1930). *Mem. Cornell Univ. agric. Exp. Stn* No. 130, 139 pp.
Howden, J. C. W. (1968). *Natn. Rose Soc. Rose A 1968*, 131–136.
Hyde, P. M. (1976). *Phytopath. Z.* **85**, 289–297.
Itoi, S., Nakayama, K. and Kubomura, Y. (1962). *Bull. seric. Exp. Stn Japan* **17**, 321–445.
Jackson, G. V. H. and Wheeler, B. E. J. (1974). *Trans. Br. mycol. Soc.* **62**, 73–87.
Jackson, G. V. H. and Wheeler, B. E. J. (1975). *Trans. Br. mycol. Soc.* **65**, 491–496.
Jenkyn, J. F. (1970). *Rep. Rothamsted exp. Stn 1969* part I, 151.
Jenkyn, J. F. (1973). *Ann. appl. Biol.* **73**, 15–18.
Jenkyn, J. F. (1974a). *Ann. appl. Biol.* **78**, 289–293.
Jenkyn, J. F. (1974b). *Ann. appl. Biol.* **76**, 257–267.
Jenkyn, J. F. (1974c). *Rep. Rothamsted exp. Stn 1973* part I, 125.
Jenkyn, J. F., Hirst, J. M. and King G. (1973). *Ann. appl. Biol.* **73**, 9–13.
Jhooty, J. S. and McKeen, W. E. (1965a). *Phytopathology* **55**, 281–285.
Jhooty, J. S. and McKeen, W. E. (1965b). *Can. J. Microbiol.* **11**, 539–545.
Johnson, R., Scott, P. R. and Wolfe, M. S. (1971). *Rep. Pl. Breed. Inst.* **1970**. 108.
Johnston, H. W. (1974). *Can. Pl. Dis. Surv.* **54**, 71–73.
Jones, I. T. (1975). *Ann. appl. Biol.* **80**, 301–309.

Jones, I. T. and Hayes, J. D. (1971). *Ann. appl. Biol.* **68**, 31–39.
Jordan, V. W. L. (1966). *Rep. agric. hort. Res. Stn Univ. Bristol* **1965**, 178–183.
Kirby, A. H. M., Frick, E. L., Burchill, R. T. and Moore, M. H. (1963). *Nature, Lond.* **197**, 514.
Klose, A. (1975). *Z. PflKrank. PflSchutz* **82**, 13–21.
Koltin, Y. and Kenneth, R. (1970). *Ann. appl. Biol.* **65**, 263–268.
Kosswig, W. (1958). *Höfchenbr. Bayer PflSchutz-Nachr.* **11**, 14–24.
Kranz, J. (1974). "Epidemics of Plant Diseases: Mathematical Analysis and Modeling". Ecological Studies, 13 Springer-Verlag, Berlin, Heidelberg, New York.
Large, E. C. and Doling, D. A. (1962). *Pl. Path.* **11**, 47–57.
Last, F. T. (1953). *Ann. appl. Biol.* **40**, 312–322.
Last, F. T. (1954). *Ann. appl. Biol.* **41**, 381–392.
Last, F. T. (1955). *Trans. Br. mycol. Soc.* **38**, 453–464.
Last, F. T. (1957). *Ann. appl. Biol.* **45**, 1–10.
Lim, T. M. (1972). *Proc. Rubb. Res. Inst. Malaysia Planters' Conf. Kuala Lumpur,* 169–179.
Lim, T. M. and Rao, B. S. (1975). *International Rubber Conference 1975, Kuala Lumpur,* 1–19.
Liyanage, A. de S. and Royle, D. J. (1976). *Ann. appl. Biol.* **83**, 381–394.
Mamluk, O. F. and Weltzien, H. C. (1973). *Phytopath. Z.* **76**, 221–252.
Manners, J. G. and Hossain, S. M. M. (1963). *Trans. Br. mycol. Soc.* **46**, 225–234.
Marshall, R., Doodson, J. K. and Jemmett, J. L. (1971). *J. natn. Inst. agric. Bot.* **12**, 286–298.
Martin, T. J., Stuckey, R. E., Safir, G. R. and Ellingboe, A. H. (1975). *Physiol. Plant Pathol.* **7**, 71–77.
McKeen, W. E., Mitchell, N. and Smith, R. (1967). *Can. J. Bot.* **45**, 1489–1496.
Mehta, K. C. (1930). *Agric. J. India* **25**, 283–285.
Mence, M. J. and Hildebrandt, A. C. (1966). *Ann. appl. Biol.* **58**, 309–320.
Merriman, P. R. and Wheeler, B. E. J. (1968). *Ann. appl. Biol.* **61**, 387–397.
Mitchell, N. L. and McKeen, W. E. (1970). *Can. J. Microbiol.* **16**, 273–280.
Morrison, R. M. (1964). *Mycologia* **56**, 232–236.
Moseman, J. G. and Greeley, L. W. (1966). *Phytopathology* **56**, 1357–1360.
Moseman, J. G. and Powers, H. R. (1957). *Phytopathology* **47**, 53–56.
Mount, M. S. and Ellingboe, A. H. (1968). *Phytopathology* **58**, 1171–1175.
Mygind, H. (1965). *Tidsskr. PlAvl* **69**, 216–239.
Mygind, H. (1970). *Tidsskr. PlAvl* **74**, 177–195.
Nair, K. R. S. and Ellingboe, A. H. (1962). *Phytopathology* **52**, 714.
Newton, M. and Cherewick, W. J. (1947). *Can. J. Res. C* **25**. 73–93.
Nour, M. A. (1958). *Trans. Br. mycol. Soc.* **41**, 17–38.
Padalino, O., Antonicelli, M. and Grasso, V. (1970). *Phytopath. medit.* **9**, 122–135.
Pady, S. M., Kramer, C. L. and Clary, R. (1969). *Phytopathology* **59**, 844–848.
Palti, J. (1961). *Bull. Res. Coun. Israel D*, **10**, 236–249.
Palti, J. (1971). *Phytopath. medit.* **10**, 139–153.
Palti, J. (1972). *Actas III Congr. Un. fitop. medit., Oeiras* **1972**, 177–183.
Palti, J. (1973). *Relatos III Congr. Un. fitop. medit., Oeiras* **1972**, 55–62.
Perera, R. G. and Wheeler, B. E. J. (1975). *Trans. Br. mycol. Soc.* **64**, 313–319.
Peries, O. S. (1961). *Pl. Path.* **10**, 65–66.
Peries, O. S. (1962). *Ann. appl. Biol.* **50**, 211–224.
Peries, O. S. (1973). *Q. Jl Rubb. Res. Inst. Sri Lanka* **50**, 208–217.
Polley, R. W. and King, G. (1973). *Pl. Path.* **22**, 11–16.
Polley, R. W. and Smith, L. P. (1973). *Proc. 7th Brit. Insectic. Fungic. Conf. 1973* **2**, 373–378.
Populer, C. (1972). *Publs Inst. natn. Étude agron. Congo belge. Série scientifique* No. 115, pp. 368.
Powell, D. and Doll, C. C. (1970). *Trans. Ill. St. hort. Soc. for 1969,* 127–130.

80 D. J. BUTT

Powers, H. R. (1957). Pl. Dis. Reptr 41, 845–847.
Preece, T. F. (1965). Pl. Path. 14, 83–86.
Price, T. V. (1970). Ann. appl. Biol. 65, 231–248.
Reuveni, R. and Rotem, J. (1973). Phytopath. Z. 76, 153–157.
Reuveni, R. and Rotem, J. (1974). Phytoparasitica 2, 25–33.
Rogers, M. N. (1959). Mem. Cornell Univ. agric. Exp. Stn No. 363, 37 pp.
Roosje, G. S. and Besemer, A. F. H. (1965). Meded. Inst. plziektenk. Onderz. Wageningen, No. 369, pp. 154.
Rotem, J. and Cohen, Y. (1966). Pl. Dis. Reptr 50, 635–639.
Royle, D. J. (1975). Rep. Dep. Hop Res. Wye Coll. 1974, 24.
Ruppel, E. G., Hills, F. J. and Mumford, D. L. (1975). Pl. Dis. Reptr 59, 283–286.
Russell, G. E., Andrews, C. R. and Bishop, C. D. (1976). Ann. appl. Biol. 82, 467–476.
Saunders, P. J. W. and Doodson, J. K. (1970). Trans. Br. mycol. Soc. 55, 318–321.
Schnathorst, W. C. (1959). Phytopathology 49, 464–468.
Schnathorst, W. C. (1960a). Phytopathology 50, 304–308.
Schnathorst, W. C. (1960b). Phytopathology 50, 450–454.
Schnathorst, W. C. (1965). A. Rev. Phytopath. 3, 343–366.
Schnathorst, W. C., Grogan, R. G. and Bardin, R. (1958). Phytopathology 48, 538–543.
Shaner, G. (1973a). Phytopathology 63, 867–872.
Shaner, G. (1973b). Phytopathology 63, 1307–1311.
Smedêgard-Petersen, V. (1967). Åsskr. K. Vet.-Landbohøjsk. 1967, 1–28.
Smith, C. G. (1969). Trans. Br. mycol. Soc. 53, 69–76.
Smith, C. G. (1970). Trans. Br. mycol. Soc. 55, 355–365.
Smith, C. G. (1971). Trans. Br. mycol. Soc. 56, 275–279.
Smith, C. G. and Wheeler, B. E. J. (1969). Trans. Br. mycol. Soc. 52, 437–445.
Smith, H. C. and Blair, I. D. (1950). Ann. appl. Biol. 37, 570–583.
Smith, L. P. (1975). "Methods in Agricultural Meteorology". Developments in Atmospheric Science, Vol. 3. Elsevier, Amsterdam.
Smith, L. P. and Davies, R. R. (1973). Pl. Path. 22, 1–10.
Somers, E. and Horsfall, J. G. (1966). Phytopathology 56, 1031–1035.
Sprague, R. (1953). In "Plant Diseases: the Yearbook of Agriculture 1953" (A. Stefferud, Ed.), 667–670. United States Department of Agriculture, Washington.
Sprague, R. (1955). Bull. Wash. agric. Exp. Stn No. 560, pp. 36.
Sreeramulu, T. (1964). Trans. Br. mycol. Soc. 47, 31–38.
Stavely, J. R. and Hanson, E. W. (1966). Phytopathology 56, 940–943.
Stedman, O. J. and Hirst, J. M. (1973). Rep. Rothamsted exp. Stn 1972 part 1, 128.
Stephan, S. (1957). NachrBl. dt. PflSchutzdienst., Berl. 11, 169–177.
Tapio, E. (1972). Ann. Agric. Fenn. 11, 79–84.
Tapke, V. F. (1953). Phytopathology 43, 162–166.
Turner, D. M. (1956). Trans. Br. mycol. Soc. 39, 495–506.
Upstone, M. E. and Davies, J. C. (1967). Pl. Path. 16, 68–69.
Van der Plank, J. E. (1963). "Plant diseases: Epidemics and control". Academic Press, New York, London.
Ward, S. V. and Manners, J. G. (1974). Trans. Br. mycol. Soc. 62, 119–128.
Wark, D. C. (1950). J. Aust. Inst. agric. Sci. 16, 32–33.
Wastie, R. L. (1969). Planter, Kuala Lumpur 45, 587–591.
Wastie, R. L. (1972). J. Rubb. Res. Inst. Malaya 23, 232–247.
Weinhold, A. R. (1961a). Phytopathology 51, 478–481.
Weinhold, A. R. (1961b). Phytopathology 51, 699–703.
Wheeler, B. E. J. (1975). Ann. appl. Biol. 79, 177–188.
Wheeler, B. E. J., Cook, R. T. A., Fagan, H. J., Chandarasrivongs, C., Khan, A. N. and Achavasmit, P. (1973). Trans. Br. mycol. Soc. 60, 177–186.

Yarham, D. J. (1975). *Outl. Agric.* **8**, 245–247.
Yarham, D. J. and Pye, D. (1969). *Proc. 5th Br. Insectic. Fungic. Conf. 1969* **1**, 25–33.
Yarham, D. J., Bacon, E. T. G. and Hayward, C. F. (1971). *Proc. 6th Br. Insectic. Fungic. Conf. 1971.* **1**, 15–25.
Yarwood, C. E. (1936a). *Phytopathology* **26**, 845–859.
Yarwood, C. E. (1936b). *J. agric. Res.* **52**, 645–657.
Yarwood, C. E. (1939). *Phytopathology* **29**, 288–290.
Yarwood, C. E. (1950). *Am. J. Bot.* **37**, 636–639.
Yarwood, C. E. (1957). *Bot. Rev.* **23**, 235–301.
Yarwood, C. E. and Gardner, M. W. (1972a). *Mycologia* **64**, 799–805.
Yarwood, C. E. and Gardner, M. W. (1972b). *Pl. Dis. Reptr* **56**, 852–855.
Yarwood, C. E. and Hazen, W. E. (1944). *Am. J. Bot.* **31**, 129–135.
Yarwood, C. E., Sidky, S., Cohen, M. and Santilli, V. (1954). *Hilgardia* **22**, 603–622.
Zadoks, J. C. (1967). *Nethl J. Pl. Path.* **73**, Suppl. 1, 61–80.
Zadoks, J. C. (1974). *Phytopathology* **64**, 918–923.
Zaracovitis, C. (1965). *Trans. Br. mycol. Soc.* **48**, 553–558.
Zaracovitis, C. (1966). *Annls Inst. phytopath. Benaki*, N.S. **7**, 219–226.

Chapter 4

Significance of Wild Relatives of Small Grains and Other Wild Grasses in Cereal Powdery Mildews

I. WAHL, NAVA ESHED*, A. SEGAL and Z. SOBEL

Faculty of Life Sciences, the Tel-Aviv University, Tel-Aviv, Israel
*Department of Plant Pathology and Microbiology, the Faculty of Agriculture, the Hebrew University of Jerusalem, Rehovot, Israel.

I. Introduction

Powdery mildew diseases of barley, wheat and oats, caused, respectively, by *Erysiphe graminis* f. sp. *hordei*, *E. graminis* f. sp. *tritici* and *E. graminis* f. sp. *avenae*, are widely distributed in the world and are of great economic importance (Blumer, 1967; Yarwood, 1973). Powdery mildew is one of the most important diseases of barley all over the world (Brooks, 1970; Hoffmann and Kuckuck, 1938; Schaller, 1951; Slootmaker, 1974), reducing its malting quality, kernel weight and total yield. Powdery mildew of wheat, though less hazardous than that on barley (Yarwood, 1973), consistently attacks the crop in the USA (Anon., 1965), some countries of South America (W. F. Kugler, personal communication). Europe (Leijerstam, 1972a; Mraz, 1971; Smiljaković, 1966), Australia (Anon., 1970), Africa and Asia (Saari and Wilcoxson, 1974).

Oat powdery mildew is insignificant in the USA (Sherwood et al., 1977), but is the most important disease of this crop in Great Britain (Lawes and Hayes, 1965; Aung and Thomas, 1976).

Breeding for resistance is considered to be the most effective and economically feasible means of powdery mildew control in cereals. Ordinarily, breeding programmes have involved incorporation of major genes that condition resistance associated with hypersensitivity (Brückner, 1976; Leijerstam, 1972a; Slootmaker, 1971; Wolfe, 1972a, b). Resistance of this type has often proven to be ephemeral. The major genes have often been overcome by virulent races of the pathogen even before extensive use of the cultivars containing them (Doling, 1969; Hayes, 1973). This phenomenon known as "boom and bust" (Wolfe, 1973), or "vicious circle" (Browning and Frey, 1969) is an outcome of preferential selection pressure exerted by the resistant variety on the pathogen populations. The man-guided evolution (Johnson, 1961) of the pathogen has particularly serious consequences in powdery mildews of cereals because of their variability and long-distance dissemination. These fungi hybridize readily and give rise to ascospores presenting a wide range of pathogenic differentiation (Brückner, 1976; Eyal et al., 1973; Koltin et al., 1964). The pathogenic diversity of E. graminis is further enhanced by fusion of hyphae (Naito and Hirata, 1969) and mutation. Leijerstam (1972a) reported that for certain virulence loci in E. graminis f. sp. tritici the mutation frequency could be as high as 2000 $\text{locus}^{-1}\,\text{ha}^{-1}\,\text{day}^{-1}$.

The build-up and spread of new forms of virulence is particularly rapid and dangerous in windborne fungi like powdery mildews (Leijerstam, 1972a).

The disappointments of the past and the threat of depletion of the known reservoirs of race-specific resistance (Lehmann et al., 1976) have prompted development of new concepts of resistance and new ideas concerning gene management. The form of resistance expressed by low infection severity and slow progress of the disease has gained wide interest. Cultivars characterized by resistance of this type and referred to as "slow mildewers", are known in wheat (cv. Knox), barley (cv. Minerva and Vada), and oats (cv. Maldwyn). Some researchers (Slootmaker, 1971) consider slow mildewing as non-race specific resistance. This type of resistance is regarded as durable because it does not exert selective effect on the pathogen and does not offset the balance in the pathogen populations. Riley (1973) has anticipated that

for the future . . . our major concern in breeding is going to be with non-specific resistance.

Leijerstam (1972a) concluded that

an attempt must be made to obtain a high degree of race non-specific resistance.

The new concepts of gene management emphasize the need for diversifying, genetically, the crop either by combining various genes of resistance and types of resistance into the same plant or by producing multiline crops with each of the components endowed with a different gene for resistance (Browning, 1974). Resolution No. 1 of the Third International Barley Genetics Symposium, 1975, concluded that to increase the durability of resistance to powdery mildew

it is essential to have as wide a range as possible of different resistances in a commercial crop, both at the national and farm level.

All programmes and concepts of breeding for stabilized resistance stress the need for genetic diversity. Regrettably, resistance resources erode rapidly and become increasingly scarce.

Wild relatives of crops have always been successfully exploited as sources of disease resistance. "They have been used for this purpose in almost every crop grown by man" (Harlan, 1976). They have coexisted with some of their parasites for long periods of time and various types and levels of protection against diseases have evolved as a result of natural selection (Wahl, 1970).

Our publication deals with the following problems:

1. wild relatives of small grains as sources of resistance to powdery mildews of barley, wheat and oats;
2. parasitic specialization of *E. graminis* on wild grasses;
3. role of wild grasses in epidemics of powdery mildew on small grains. The paper is confined to wild grasses and does not include primitive "land varieties".

II. Resistance of Wild Barleys

A. Studies on *Hordeum spontaneum* C. Koch originated outside Israel

Biffen (1907) was probably the first to study the resistance of *H. spontaneum*, the putative ancestor of cultivated barleys and readily crossable with them. He stated that plants of that species are "as a rule completely free from mildew". In crosses with the susceptible *Hordeum hexasticofurcatum* (K.H.) the inheritance of *H. spontaneum* resistance was conditioned by a single recessive gene. Honecker (1937) used *H. spontaneum nigrum* in studies of the physiologic specialization of *E. graminis* f. sp. *hordei* and stressed its resistance. Hoffmann and Kuckuck (1938) reported resistance in *H. spontaneum nigrum*, line H 204—from Russia, to practically all races of barley mildew. They demonstrated that the resistance of H 204 is dominant and basically monogenic. Complementary genes modify the effect of the major gene. In the ensuing years the resistance of H 204 gained in importance in Germany because of the breakdown of resistance controlled by the gene *Mlg* (from the cultivar Pflugs Intensiv). Rudorf and Wienhues (1951) incorporated resistance of H 204 to winter and summer barleys and produced hybrids combining the good agronomic characteristics of commercial cultivars such as malting quality and high yielding with a broad spectrum of resistance to powdery mildew. They, too, concluded that the resistance of H 204 is dominant and monogenic but it is modified by complementary genes and therefore, crosses involving H 204 and susceptible cultivars did not always segregate in a 3:1 ratio in the F_2 generation. Moseman *et al.* (1965) designated the resistance gene of H 204 and its derivatives as *Mla6* located on chromosome 5. They assumed that the gene *Mla6* is the same as gene *JMlsn* identified in *H. spontaneum nigrum* by Hiura (1960).

A number of barley cultivars possess gene *Mla6*. They include Akme, Allasch, Maris Badger C.I. 11803, Maris Concord C.I. 12114, Hispont C.I. 8828, HB 279/5/1/3, Voldagsen 8141/44 and Vogelsanger Gold. The following barley cultivars contain gene *Mla6* in combination with the gene *Mlg*, Gerda, Inis, Matura, Impala, and Carina. The cultivar Maris Canon comprises resistance of *Mla6* "*H. laevigatum*", and *Mlg*. Over the years the gene *Mla6* has become less important owing to evolution and build-up of new virulent cultures which have rendered it ineffective. Nevertheless, the cultivars Gerda, Carina and Vogelsanger Gold continue to be prominent in barley production in Germany. The cultivar Vogelsanger Gold is rated above average among winter barleys resistant to powdery mildew.

In Wiberg's studies (1974), seedlings of *H. spontaneum nigrum* were resistant to 30 cultures and susceptible to eight cultures of *E. graminis* f. sp. *hordei* isolated in Belgium, Czechoslovakia, Great Britain and Sweden. Besides *H. spontaneum nigrum*, several other accessions of *H. spontaneum* were proven to be resistant, namely (i) *H.* 6586 from Afghanistan which offers protection to all cultures tested by Wiberg, as well as to cultures from Japan, and the German race Amsel C2 (Wiberg, 1974), (ii) *H. spontaneum* C. Koch var. *euspontaneum* Åberg (Wiberg, 1974).

Nover and Lehmann (1973) reported uniform, moderate resistance in *H. spontaneum* HOR 2826 from Iran to all races tested by them. The resistance has not appeared to be conditioned by genes in loci *Mla*, *Mlg* and *mlo*. Another seven accessions of *H. spontaneum* and two accessions of *H. agriocrithon* Åberg were resistant to some of the investigated races.

B. Mildew resistance in *H. spontaneum* of Israeli origin

Israel is located in one of the centres of origin and diversification of *H. spontaneum*. Heterogeneic populations of the species are of country-wide distribution (Harlan and Zohary, 1966) and are annually attacked by *E. graminis* f. sp. *hordei*. The concept of host-parasite coevolution implies that the genetic variation of the host is matched by corresponding diversity of the fungus which is also reflected in its virulence. The mechanism of evolution of pathogen virulence was discussed by Moseman (1971). The prevalence throughout the country, of fertile cleistothecia of *E. graminis* that liberate infectious ascospores concurrently with the development in nature of congenial hosts, enhances pathogenic diversity of the fungus. Some of the virulent strains occurring in Israel were unknown elsewhere (Moseman, 1971; Eyal *et al.*, 1973; Wiberg, 1974).

The host-parasite coevolution lasting from remote antiquity has resulted in the development of various types and levels of protection from the disease due to natural selection. Joint studies conducted by Moseman* and researchers of the Tel-Aviv University have revealed that *H. spontaneum* in Israel is a vast reservoir of mildew resistance of the conventional type. Studies of Fischbeck† with the Tel-Aviv University group have demonstrated that the same is true also of other types of resistance expressed by slow mildewing.

Moseman and Craddock (1976, and personal communication) in the USA tested 277 accessions of *H. spontaneum* from Israel and reported that over 75% had some resistant plants. Many of the plants were derived from the more humid areas where the pathogen is abundant. Their preliminary results suggest that there are more genes for resistance to *E. graminis* f. sp. *hordei* in barley from Israel than are present in barleys from Turkey and Ethiopia. It is noteworthy that over 40% of the accessions were resistant to both the powdery mildew and leaf rust cultures.

Preliminary results published by Fischbeck *et al.* (1976a) indicate that derivatives of about 50% of 2700 unselected populations of *H. spontaneum* collected at 69 locations in Israel were attacked only slightly or not at all by European field-races of mildew. "Some lines showed resistance to all mildew cultures thus far used for infection which covers a very broad spectrum of virulence." Most of the investigated accessions of *H. spontaneum* were randomly sampled, regardless of their reaction to the disease, but some were selected on the basis of the reaction to mildew in natural habitats. In these

* Research supported by PL 480 grant, FG-Is-260, A10-CR-96, of the United States Department of Agriculture, and a grant of the United States-Israel Binational Science Foundation, Jerusalem, Israel.
† Research supported by a grant of the Deutsche Forschungsgemeinschaft (DFG), the Federal Republic of Germany.

accessions seed was harvested from plants which showed low infection severity or small pustules or symptoms of hypersensitivity at sites where heavy mildew incidence prevailed. Progenies of the accessions were tested partly in the greenhouse and mainly in shaded field nurseries. In all trials prolific development of mildew was secured. The nurseries were repeatedly inoculated with cultures of E. graminis f. sp. hordei of country-wide origin. Field nurseries were maintained at several locations and for several years in succession and there was always good agreement in results obtained in different trials. The recorded mildew reactions included immunity, heavy necrosis, pustules surrounded by necrotic lesions, small and appressed pustules with scanty mycelial cast, low incidence of pustules of susceptible type and full susceptibility. Some plants displayed more than one type of reaction (Fischbeck et al., 1976b). The map showing the geographic distribution of resistance to powdery mildew in H. spontaneum in Israel demonstrates its scarcity in regions with low rainfall (Fischbeck et al., 1976b). Similar results were obtained with Avena sterilis for resistance for crown rust (Wahl, 1970) and H. spontaneum for resistance to leaf rust (Anikster et al., 1976). The overall resistance of various types to powdery mildew was more prevalent in samples collected at the Golan Heights, the Galilee areas and in the Samarian and Judean Highlands (Fischbeck, Schwarzbach, Segal, Wahl, unpublished). However, distinct fluctuations in the frequency of the occurrence of resistance could be observed among sites of each region.

Plants of some accessions have consistently behaved as slow mildewers in Israel and in Germany both in greenhouse and field studies (Fischbeck et al., 1976b). For example, the accessions 01-B-84 and 01-B-87 that had been selected in 1972 because of their slow mildewing performance in nature, retained this behaviour in greenhouse and field trials in Israel in 1973–74, and 1974–75, and were virtually immune in Germany. About 84% of samples preselected in Israel for low mildew infectibility displayed low to moderate infection in tests at Weihenstephan, proving thus the reliability of preselection tests (Fischbeck et al., 1976b). Accessions of H. spontaneum selected in Israel because of their slow mildewing performance in nature have supported low fungus sporulation when infected at Weihenstephan with cultures representing a wide range of virulence (Chaudhary et al., 1976).

Stands of H. spontaneum that had been sampled across the country contained, in varying proportions, different types of resistance to powdery mildew. At certain sites resistance was manifested by infection types 1–2, while at others there was a distinct prevalence of resistance of the slow mildewing type-associated with low to moderate infectibility. This is shown in Figs 1 and 2 (Fischbeck, Schwarzbach, Segal, Wahl, unpublished). The results were attained with the aid of the transect sampling procedure. The method involves harvesting single heads from 80–100 plants per site at intervals of 1 m, regardless of the host reaction. The seed was planted in

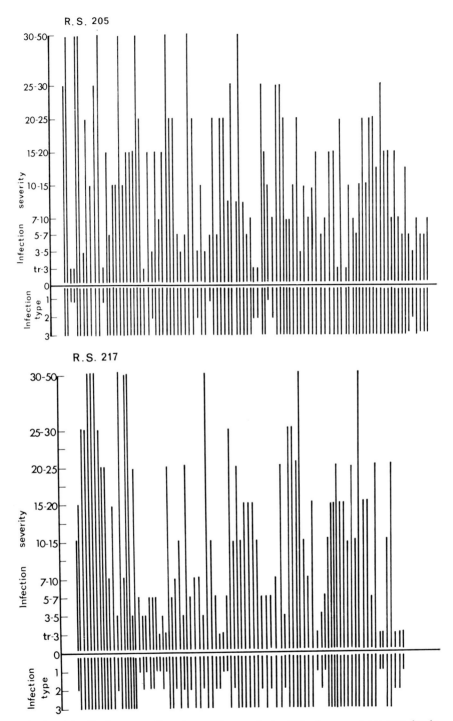

Figs 1 and 2.   Performance of powdery mildew on plants of *Hordeum spontaneum* randomly sampled in two transects, R.S. 205 (Fig. 1), and R.S. 217 (Fig. 2). Infection severity denotes the percentage of leaf area (mainly of penultimate leaf F-1, and leaf F-2) covered with pustules. Designations of infection types are those of Moseman (1968). Note in transect R.S. 217 the relatively frequent occurrence of plants with infection types 1–2.

field nurseries exposed to severe mildew incidence. A total of 17 transects from various regions was studied in several nurseries. Of the approximately 1700 plants assayed, 17·5% showed hypersensitive reaction, 27% displayed slow mildew development with normal pustules and 17·2% harboured small, appressed and prematurely senescing pustules. It should be noted that the 17 transects were made at locations where resistance to powdery mildew had been observed in preceding years. This may explain the high percentage of resistance recorded in the trials. In a field nursery trial in Israel, 27% of about 4000 plants of *H. spontaneum* randomly sampled across the country from 81 locations showed resistance of various types. Heavy infection in the nursery was obtained by using inoculum of country-wide origin.

The differences in the disease progress on slow mildewing accessions of *H. spontaneum* and on the slow mildewing cultivar Vada as compared with fast mildewers represented by Proctor and Manchuria, are shown in Fig. 3. The

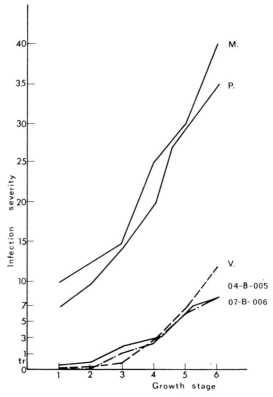

Fig. 3. Development of powdery mildew at different stages of growth on the two slow mildewing accessions of *Hordeum spontaneum* 04–B–005 and 07–B–006, on the slow mildewing cultivar Vada (V.), and on the two fast mildewing cultivars Proctor (P.) and Manchuria (M.). Designations of growth stages are those of Large (1954).

plants were maintained under conditions that assured heavy disease incidence. The inoculum used in the test does not induce hypersensitive reactions in the plants involved.

C. Resistance of "*Hordeum laevigatum*"*

The Dutch barley cultivars Minerva and Vada derive their resistance to mildew from the botanical species "*H. laevigatum*" (Dros, 1957). The resistance is referred to as "moderate resistance" (Brückner, 1968), "middle resistance" (Brückner, 1976), "incomplete resistance" (Wolfe, 1972a), "intermediate resistance" (Wolfe, 1972b) or "partial resistance" (Slootmaker, 1974). Some researchers maintain that the resistance is conditioned by a single gene (Brückner, 1968) while others consider it to be more complex (Wolfe, 1972a). Seedlings of Minerva have shown susceptible reactions when inoculated with various mildew cultures, whereas adult plants have produced similar reactions but have given relatively high yields under severe epidemics (Slootmaker, 1971). The incomplete resistance is reported to be of long-term stability (Brückner, 1976). Slootmaker (1971) suggested that the resistance of Vada and Minerva "is an example of possibly non-race specific resistance". However, according to Hayes (1973), mildew resistance of Vada, previously accepted to be race-non-specific, is now known to be race-specific.

In Israel seedlings of Vada and Minerva infected with mildew cultures representing a wide range of pathogenicity, exhibited susceptible reactions but showed distinctly lower rates of infectability, in that fewer pustules developed on them than on check plants. Similar observations were made at the more advanced stages of plant growth (Fig. 3). In field tests, adult plants of Minerva and Vada showed susceptible reactions but behaved as slow mildewers. The level of infection was relatively low and the disease progressed slower than on fast mildewers. The results were obtained in repeated studies, at different locations and under severe epidemics that had been incited by inoculation with cultures of country-wide origin.

D. Resistance of *Hordeum bulbosum*.

H. bulbosum L. of the tetraploid type ($2n = 28$), is very common in Israel and numerous accessions of the species have displayed resistance to *E. graminis* f. sp. *hordei* (Wahl, unpublished). Their utilization in barley breeding is envisaged with the progress of hybridization *H. bulbosum* with *H. vulgare* (Fukuyama and Takahashi, 1976).

---

* "The name *Hordeum laevigatum* has never been validly published", F. A. Stafleu (personal communication).

III. Resistance of Wild Wheats

Effective sources of resistance to *E. graminis* f. sp. *tritici* were selected in
Triticum timopheevi Zhuk. (Allard and Shands, 1954; Jørgensen and Jensen,
1972), its spontaneous mutant *T. militinae* Zhuk. and Migusch. (Miguschova
and Yamaleev, 1974). *T. monococcum* L., *T. dicoccum* Schübl. var. *georgicum*
Dekap and Menab (Lupton, 1956), *T. persicum* Vav. = *T. carthlicum* Nevsky
(Jørgensen and Jensen, 1972; Lupton and Macer, 1958). They are not dis-
cussed by us since the listed species are cultivated in some parts of the world
(Zeven and Zhukovsky, 1975; Zhukovsky, 1971).

Accessions resistant to powdery mildew have been described in the wild
emmer wheat, *T. dicoccoides* Körn. (Filatenko, 1969; Leijerstam, 1972b;
Lehmann, 1968). This species, crossable with cultivated wheat, is indigenous
in Israel, Jordan and Syria.

Plants resistant to *E. graminis* f. sp. *tritici* were found in *Aegilops* species,
such as *A. caudata* L. var. *typica* Eig, *A. speltoides* Tausch. var. *Aucheri*
(Boiss.) Born., *A. ovata* L. var. *vulgaris* Eig (Lupton, 1956). Amphiploids
involving *A. caudata* or *A. ovata* and *Triticum* plants were resistant to powdery
mildew (Lupton, 1956).

Amphiploids produced by crossing susceptible wheat with *Agropyron
intermedium* (Host) Beauv., which is resistant to wheat mildew, were resistant
to *E. graminis* f. sp. *tritici*. The resistance is conditioned by two *Agropyron*
genes (Sinigovets and Lapchenko, 1975). *Agropyron repens* (L.) Beauv. is the
source of mildew resistance in the wheat cultivar Mildress (Macer, 1968).

IV. Resistance of Wild Oats

The hexaploid species *Avena sterilis* L.—the putative progenitor of cultivated
oats and readily crossable with them, is an important source of resistance to
E. graminis f. sp. *avenae*. In field tests in Israel, numerous accessions of *A.
sterilis* and of *A. barbata* Pott. have shown mildew resistance (Wahl, unpublished). Zillinsky and Murphy (1967) found resistance to mildew in *A.
sterilis* in diverse areas of the Mediterranean region. Their results suggest
"that genes for resistance to powdery mildew occur with greater frequency
among wild species than among cultivated varieties". Mildew resistance in *A.
sterilis* has been reported in the USSR by Mordvinkina (1969). Resistant
strains of *A. sterilis* and *A. fatua* L. have been studied in Great Britain
(*Cambridge Plant Breeding Institute, Annual Report*, 1962–63).

The powdery mildew resistance of *A. ludoviciana* Dur. has been investigated
in Great Britain and utilized for breeding purposes (*Cambridge Plant Breeding*

Institute, Annual Report, 1958–59, 1959–60, 1960–61, 1961–62, 1963–64, 1965–66).

The location of the gene controlling mildew resistance in *A. barbata* was determined and the gene was transferred to cultivated oats (Aung and Thomas, 1976). Hayes and Jones (1966) related mildew resistance in *A. strigosa* Schreb. var. *hirtula* and var. *glabrota*. Ladizinsky (personal communication) recorded mildew resistance in tests in Israel on accessions of two species from Spain, *A. prostrata* Ladiz. ($2n = 14$), and *A. murphyi* Ladiz. ($2n = 28$).

V. Parasitic Specialization of *Erysiphe graminis* on Wild Grasses

A. Parasitic Specialization of *E. graminis* isolated from wild grasses

According to Stebbins (1956), "most of the common species of grasses . . . contain in varying proportions gene combinations derived from two, three, four or more separate and sometimes widely divergent ancestors". Our study was based on the hypothesis that wild grasses, owing to their genetic interrelationships, are congenial hosts for powdery mildew cultures isolated from a wide range of plants. We suggest that the results of the study may contribute to the better understanding of the taxonomy and phylogeny of Gramineae (Savile, 1954, 1971; Watson, 1972).

In studies of Eshed and Wahl (unpublished),* seedlings of wild grasses were inoculated separately with four cultures of *E. graminis*;

1. composite culture of ascospore origin isolated from several plants of *Hordeum murinum* L. (Hordeae tribe);
2. culture of ascospore origin secured from several plants of *Bromus rigens* L. (Festuceae tribe);
3. composite culture of ascospore origin that were produced on several plants of *Phalaris paradoxa* L. (Phalarideae tribe);
4. composite culture of conidial origin derived from several plants of *Alopecurus myosuroides* Huds. (Agrostideae tribe).

The cultures will be referred to according to their source host. Only cultures capable of infecting seedlings of the source host were considered as compatible with the artificially inoculated grass host from which they were reisolated.

The *H. murinum* culture was compatible with seedlings of the following 13 genera (in fractions in brackets, the denominator indicates the number of

* Research conducted at the Faculty of Agriculture of the Hebrew University of Jerusalem, Rehovat, Israel, and supported under P.L.480 by the US Department of Agriculture, Grant No. FG-Is-109, A10-CR-1.

species inoculated, and the numerator—the number of species successfully infected in the genus): *Elymus* (1/2); *Eremopyrum* (1/1); *Hordeum* (3/4); *Psilurus* (1/1)—of the Hordeae tribe; *Bromus* (3/15); *Cutandia* (1/3); *Echinaria* (1/1); *Lamarckia* (1/1); *Lolium* (1/5); *Sphenopus* (1/1); *Vulpia* (1/5)—of the Festuceae tribe; *Avena* (1/4)—of the Aveneae tribe; *Stipa* (1/1)—of the Stipeae tribe.

The tested culture was incompatible with plants of *H. spontaneum* but was capable of parasitizing seedlings of *H. bulbosum* and *H. murinum*.

The culture from *Bromus rigens* successfully infected seedlings in 11 genera: *Elymus* (1/2); *Pholiurus* (1/2)—of the Hordeae tribe; *Boissiera* (1/1); *Briza* (1/1); *Bromus* (8/13); *Cutandia* (1/3); *Echinaria* (1/1); *Lamarckia* (1/1); *Sphenopus* (1/1)—of the Festuceae tribe; *Corynephorus* (1/1); *Trisetum* (1/4)—of the Aveneae tribe.

The culture from *Phalaris paradoxa* had congenial seedling hosts in species of the 20 genera: *Elymus* (1/2); *Lepturus* (1/1); *Pholiurus* (2/2); *Psilurus* (1/1)—of the Hordeae tribe; *Ammochloa* (1/1); *Briza* (2/2); *Cutandia* (3/3); *Cynosurus* (2/3); *Dactylis* (1/1); *Echinaria* (1/1); *Lamarckia* (1/1); *Sclerochloa* (1/1), *Sphenopus* (1/1); *Vulpia* (4/5)—of the Festuceae tribe; *Phalaris* (4/5)—of the Phalarideae tribe; *Trisetum* (3/3)—of the Aveneae tribe; *Alopecurus* (1/3); *Gastridium* (1/1)—of the Agrostideae tribe; *Oryzopsis* (1/1)—of the Stipeae tribe; *Schismus* (1/2)—of the Danthonieae tribe.

The culture from *Alopecurus myosuroides* was compatible with seedlings in the following 30 genera: *Elymus* (2/2); *Eremopyrum* (1/1), *Hordeum* (2/4); *Lepturus* (1/1), *Pholiurus* (2/2); *Psilurus* (1/1)—of the Hordeae tribe; *Briza* (2/2); *Bromus* (4/13), *Cutandia* (3/3), *Cynosurus* (1/3); *Dactylis* (1/1); *Echinaria* (1/1); *Lamarckia* (1/1); *Lolium* (4/5); *Poa* (3/3); *Sphenopus* (1/1); *Vulpia* (5/5)—of the Festuceae tribe; *Phalaris* (1/4)—of the Phalarideae tribe; *Avena* (1/4); *Gaudinia* (1/1); *Holcus* (1/1); *Koeleria* (1/1); *Pilgerochloa* (1/1); *Trisetum* (3/3)—of the Aveneae tribe; *Alopecurus* (3/3), *Cornucopiae* (1/1); *Gastridium* (1/1); *Lagurus* (1/1); *Polypogon* (1/2)—of the Agrostideae tribe; *Schismus* (2/2)—of the Danthonieae tribe.

A list of the tested species will be published. The results presented above demonstrate the wide host ranges of each of the investigated cultures. They embrace a number of genera and even tribes. The *H. murinum* culture has proven infectious on plants of 13 genera belonging to four tribes. The *Phalaris paradoxa* culture has parasitized seedlings of 20 genera from seven tribes and the *Alopecurus myosuroides* has shown virulence on plants belonging to a larger number of genera but fewer tribes. According to Hardison (1944, 1945), all mildew cultures isolated from grasses outside the tribe Hordeae, infected only plants outside that tribe. Conversely, cultures that had originated on plants in the tribe Hordeae infected grasses only of that tribe. Cultures in Israel have behaved differently. For example, the *H. murinum* culture attacked plants belonging to the tribe Hordeae and to three other tribes. The remaining three cultures, each isolated from plants outside the tribe Hordeae, had congenial hosts also in the Hordeae tribe. Also Hardison (1945) indicated the contrast between the narrow specialization of powdery mildew from *Bromus* and the parasitic versatility of the fungus from plants of the tribe Agrostideae.

These results support the previous conclusions regarding the duality of the parasitic nature of *E. graminis* (Eshed and Wahl, 1970). The fungus has proved to be incompatible with some plants of the preferred genus, preferred species or even of its source collection on one hand, yet has parasitized seedlings of remote tribes on the other. Pleophagy of *E. graminis* has been demonstrated by Hardison (1944) and confirmed by Mühle and Frauenstein (1962, 1963) and others (Eshed and Wahl, 1970). In contrast, in the studies of Dodoff and Kunovsky (1971), the host ranges of *E. graminis* f. sp. *tritici*, *E. graminis* f. sp. *avenae*, *E. graminis* f. sp. *agropyri* and *E. graminis* f. sp. *bromi*, were confined to plants of the respective source genus. It is noteworthy that those cultures which had been isolated, respectively, from *H. murinum*, *B. rigens*, *P. paradoxa*, and *A. myosuroides*, were non-infectious on cultivated barley, wheat and oats, except for weak virulence of the *H. murinum* cultures on some seedlings of *H. vulgare*.

Plant collections in some grass species, such as *Sphenopus divaricatus* collection no. 534, contained components compatible with at least one of each of seven varieties of *E. graminis*. The seven varieties included cultures of the four varieties isolated from grasses and described above, and the three varieties, *E. graminis* f. sp. *hordei*, *E. graminis* f. sp. *tritici*, and *E. graminis* f. sp. *avenae*, obtained from cultivated crops. Eshed and Wahl (1970) found that in a number of grass species single plants served as common host to two varieties, viz. *E. graminis* f. sp. *tritici* and *E. graminis* f. sp. *hordei*, or *E. graminis* f. sp. *tritici* and *E. graminis* f. sp. *avenae*, or *E. graminis* f. sp. *hordei* and *E. graminis* f. sp. *avenae*.

Sexual and somatic hybridization between different varieties of *E. graminis* has been experimentally demonstrated (Hiura, 1964; Naito and Hirata, 1969) and hybridization would be expected to occur more often on a common host (Hardison, 1945; Eshed and Wahl, 1970). Luig and Watson (1972) demonstrated that *Agropyron scabrum* and *Hordeum leporinum* are substrates of somatic hybridization between *Puccinia graminis* f. sp. *tritici* and *P. graminis* f. sp. secalis. The evolutionary implications of hybridization between varieties of *P. graminis* and their significance were discussed by Green (1971). The principles and implications involved in hybridization of varieties of *P. graminis* are most likely applicable to the corresponding processes in *E. graminis*. According to Moseman (1966), crossing varieties of *E. graminis* can be useful in studies of their taxonomy and other relationships, as well as in elucidation of "cytologic, histologic, physiologic, biochemical and genetic relationships between host and pathogen".

B. Parasitic specialization of *E. graminis* isolated from small grain crops

Hardison (1944, 1945) found that in the United States the host ranges of powdery mildew cultures isolated, respectively, from cultivated barley, wheat and oats, were restricted to plants belonging to their source tribes. According to Mühle and Frauenstein (1970), none of the mildew cultures derived from barley, wheat and rye was infectious on even one of the 27 fodder grass varieties of the genera *Agrostis*, *Alopecurus*, *Arrhenatherum*, *Dactylis*, *Festuca*, *Lolium*, *Phalaris*, *Phleum* and *Poa*. Each of the varieties, *E. graminis* f. sp. *hordei*, *E. graminis* f. sp. *tritici* and *E. graminis* f. sp. *avenae*, possesses a wider host range in Israel than elsewhere (Eshed and Wahl, 1970). *E. graminis hordei* attacked accessions of 37 species from 18 genera belonging to the tribes Hordeae, Aveneae, Festuceae and Stipeae. Isolates of *E. graminis* f. sp. *tritici* were compatible with grasses of 47 species belonging to 16 genera of the above four tribes. Accessions of 31 species of 20 genera, representing the mentioned tribes, were congenial hosts of *E. graminis* f. sp. *avenae*.

The wide host ranges of powdery mildew cultures isolated from small grain crops contrast with the finding that mildew cultures infectious on cultivated barley, wheat and oats originated almost exclusively from intra-generic grasses, mainly from allied species (Eshed and Wahl, 1975). One exception was that conidia and ascospores from *Triticum dicoccoides* were compatible with cultivated wheat and barley (Eshed and Wahl, 1975). It was postulated (Eshed and Wahl, 1970) that the wide host ranges are at least partly accountable by the major trends in the phylogeny of grasses, that

> were accompanied by a series of cycles of divergent and convergent or reticulate evolution, the latter being brought about by hybridization . . . or amphiploidy (Stebbins, 1956).

Conceivably, wild grasses, owing to their genetic interrelationships, are less specialized hosts than cultivated cereals, which are man-produced in an effort to obtain distinct, well-delimited entities.

VI. Role of Wild Grasses in Epidemics of Powdery Mildew on Small Grains

Hardison's (1944) studies brought out that,

> wild grasses must now be considered in the epiphytology of the (powdery mildew) disease on cereals as sources of primary infection and perennial stations for the fungus.

Hardison (1945) concluded that

> wild grasses can harbor wheat and barley mildews and supply the inoculum for infection of these cereals.

According to Frauenstein (1970), cultivated grasses in Germany are of no practical importance in the distribution of powdery mildews in cereals.

> It was found that neither the spread of grass mildew to cereals nor that of cereal mildew to grasses are likely to occur.

Studies in Israel however (Eshed and Wahl, 1975) have proved that cultivated barley, wheat and oats become infected with mildew cultures from intrageneric wild grasses and mainly from closely related species. Despite the limited number of wild species that harbour pathogens infectious on small grains, they are important in the epidemics of powdery mildew because of their wide distribution. These grasses become infected in the autumn by primary ascosporic inoculum produced in cleistothecia which had oversummered on their stubble. After infection, the conidia multiply and disseminate to cultivated small grain crops throughout the growing season. Cleistothecia preserved on the debris of wild grasses enable the oversummering of *E. graminis* f. sp. *hordei*, *E. graminis* f. sp. *tritici*, *E. graminis* f. sp. *avenae* and other varieties of *E. graminis* (Eshed and Wahl, 1975).

Race populations of *E. graminis* f. sp. *hordei* on *H. spontaneum* contain very virulent components rendering ineffective, important sources of resistance to the pathogen (Eyal *et al.*, 1973).

VII. Summary

Wild relatives of cultivated barley, wheat and oats are rich and diversified resources of resistance to powdery mildew diseases on the respective crops.

Populations of *Hordeum spontaneum* are abundant in Israel and provide very variable gene pools of resistance to *Erysiphe graminis* f. sp. *hordei*. The resistance is of different types and may be associated with a hypersensitive reaction or with symptoms of slow mildewing protection (low infectibility of the host, small pustules, reduced longevity of pustules). All types of resistance are effective against a wide range of pathogen virulence and are heritable.

The geographic distribution of resistance to powdery mildew in populations of *H. spontaneum* in Israel shows distinct patterns, and is scarce in regions of low rainfall.

Even in small stands of *H. spontaneum* the structures of mildew resistance are complex. They combine, in varying proportions, resistance associated with hypersensitivity and other forms of protection against the disease.

The knowledge of patterns of integration of various types of resistance in natural ecosystems may be relevant to the development of gene management programmes for stabilizing disease resistance in agroecosystems (Browning, 1974).

In Macer's (1968) opinion, resistance derived from grasses may not "differ qualitatively from more readily accessible forms for resistance". Green (1971) postulated that the most effective genes of resistance "to counteract the greater aggressiveness and specialization of the changing pathogen", may be those from other species or genera, especially when used in combinations. This important problem needs to be thoroughly studied.

Cultures of *Erysiphe graminis* isolated from small grains or wild grasses in Israel have wider host ranges than their counterparts elsewhere. Some of the indigenous cultures represent new and very virulent strains. Both phenomena are attributable to the abundance of fertile cleistothecia in nature. The presence of cultures infectious to both, cultivated wheat and barley, is of special interest.

The frequent occurrence of hosts common to different combinations of varieties of *E. graminis*, presumably enhances sexual and somatic hybridization.

The heterogenic patchwork of genotypes in wild grasses in Israel offers many selective ecological niches enabling the survival of diverse fungus strains.

Grasses in Israel are important in the epidemics of powdery mildews on small grains. They become infected in the autumn, with the onset of the rainy season, by primary ascosporic inoculum liberated from cleistothecia which have oversummered on the stubble of grasses. After infection, the conidia multiply and disseminate to cultivated barley, wheat and oats throughout the growing season.

Acknowledgements

This publication is dedicated to J. G. Moseman for his inspiring leadership of research on the powdery mildews of wild grasses in Israel over two decades.

Sincerest thanks are due to G. Fischbeck who initiated our joint study on slow mildewing of barley and has acted as its leader.

We are grateful to R. C. F. Macer and M. S. Wolfe for providing important information for this publication.

References

Allard, R. W. and Shands, R. G. (1954). *Phytopathology* **44**, 266–274.
Anikster, Y., Moseman, J. G. and Wahl, I. (1976). *In* "Barley Genetics" Vol. III, pp. 468–469. Proc. 3rd Int. Barley Genet. Symp. Garching 1975.

Anonymous (1965). "Losses in agriculture". Agriculture Handbook No. 291, ARS, US Department of Agriculture, Washington, D.C. 120 pp.

Anonymous (1970). "Powdery mildew of wheat". N.S.W. Dept. Agric., Biol. Branch, Plant Disease Bull. 32.

Aung, T. and Thomas, H. (1976). *Nature* **26** (5552), 603–604.

Biffen, R. H. (1907). *J. agric. Sci.* **2**, 109–128.

Blumer, S. (1967). "Echte Mehltaupilze (Erysiphaceae)". 436 pp. G. Fischer Verlag, Jena.

Brooks, D. H. (1970). *Outl. Agric.* **6**, 122–127.

Browning, J. A. (1974). *Proc. Am. Phytopath. Soc.* **1**, 191–199.

Browning, J. A. and Frey, K. J. (1969). *A. Rev. Phytopath.* **7**, 355–382.

Brückner, F. (1968). *Genetika a šlechtěni* **4**, 99–104.

Brückner, F. (1976). *In* "Barley Genetics". Vol. III, 418–420. Proc. 3rd Int. Barley Genet. Symp. Garching 1975.

Chaudhary, R. C., Schwarzbach, E. and Fischbeck, G. (1976). *In* "Barley Genetics". Vol. III, 456–463. Proc. 3rd In. Barley Genet. Symp. Garching 1975.

Dodoff, D. N. and Kunovsky, Zh. (1971). *Rasteniovedni nauki* **8**, 147–160.

Doling, D. A. (1969). *Pl. Path.* **18**, 56–61.

Dros, J. (1957). *Euphytica* **6**, 45–48.

Eshed, N. and Wahl, I. (1970). *Phytopathology* **60**, 628–634.

Eshed, N. and Wahl, I. (1970). *Phytopathology* **65**, 57–63.

Eyal, Z., Yurman, R. Moseman, J. G. and Wahl, I. (1973). *Phytopathology* **63**, 1330–1334.

Filatenko, A. A. (1969). *Trudy prikl. Bot., Genet. Selek.* **39**, 29–51.

Fischbeck, G., Schwarzbach, E., Sobel, Z. and Wahl, I. (1976a). *Z. Pflanzenzuchtg.* **76**, 163–166.

Fischbeck, G., Schwarzbach, E., Sobel, S. and Wahl, I. (1976b). *In* "Barley Genetics". Vol. III, 412–417. Proc. 3rd Int. Barley Genet. Symp. Garching 1975.

Frauenstein, K. (1970). *NachrBl dt. PflSchutzdienst, Berl.* **24**, 47–51.

Fukuyama, T. and Takahashi, R. (1976). *In* "Barley Genetics". Vol. III, 351–360. Proc. 3rd Int. Barley Genet. Symp. Garching 1975.

Green, J. G. (1971). *Can. J. Bot.* **49**, 2089–2095.

Hardison, J. R. (1944). *Phytopathology* **34**, 1–20.

Hardison, J. R. (1945). *Phytopathology* **35**, 394–405.

Harlan, J. R. (1976). *Crop Sci.* **16**, 329–333.

Harlan, J. R. and Zohart, D. (1966). *Science* **153**, 1074–1080.

Hayes, J. D. (1973). *Ann. appl. Biol.* **75**, 140–144.

Hayes, J. D. and Jones, I. T. (1966). *Euphytica* **15**, 80–86.

Hiura, U. (1960). *Ber. Ohara Inst. Landw. Biol.* **11**, 235–300.

Hiura, U. (1964). *Ber. Ohara Inst. Landw. Biol.* **12**, 131–132.

Hoffmann, W. and Kuckuck, H. (1938). *Z. Zucht.* **22**, 271–302.

Honecker, L. (1937). *Phytopath. Z.* **10**, 197–222.

Johnson, T. (1961). *Science* **133**, 357–362.

Jørgensen, J. H. and Jensen, C. J. (1972). *Euphytica* **21**, 121–128.

Koltin, Y., Kenneth, R. and Wahi, I. (1964). *In* "Barley Genetics". Vol. I, 228–235. Proc. 1st Int. Barley Genet. Symp., Wageningen 1963.

Large, E. C. (1954). *Pl. Path.* **3**, 128–129.

Lawes, D. A. and Hayes, J. D. (1965). *Pl. Path.* **14**, 125–128.

Lehmann, C. O. (1968). *Genetika a šlechtěni* **4**, 69–80.

Lehmann, C. O., Nover, I. and Scholz, F. (1976). *In* "Barley Genetics". Vol. III, 64–69. Proc. 3rd Int. Barley Genet. Symp. Garching 1975.

Leijerstam, B. (1972a). *Statens Växtskyddsanstalt. Meddelanden* **15**, 231–248.

Leijerstam, B. (1972b). *Statens Växtskyddsanstalt. Meddelanden* **15**, 251–270.

Luig, N. H. and Watson, I. A. (1972). *Aust. J. biol. Sci.* **25**, 335–342.
Lupton, F. G. H. (1956). *Trans. Br. mycol. Soc.* **39**, 51–59.
Lupton, F. G. H. and Macer, R. C. F. (1958). *Agriculture, Lond.* **64**, 540–544.
Macer, R. C. F. (1968). *In* Cereal Rusts Conference, Oeiras—Portugal. 47–51.
Miguschova, E. F. and Yamaleev, A. M. (1974). *Trudy prikl. Bot. Genet. Selek.* **53**, 66–69.
Mordvinkina, A. I. (1969). *Trudy prikl. Bot. Genet. Selek.* **39**, 233–242.
Moseman, J. G. (1966). *A. Rev. Phytopathol.* **4**, 269–290.
Moseman, J. G. (1968). *Pl. Dis. Reptr* **52**, 463–467.
Moseman, J. G. (1971). *In* "Barley Genetics". Vol. II, 450–456. Proc. 2nd Int Barley Genet. Symp. Pullman, Washington, Washington State University Press.
Moseman, J. G. and Craddock, J. C. (1976). *In* "Barley Genetics". Vol. III, 51–57. Proc. 3rd. Int. Barley Genet. Symp. Garching 1975.
Moseman, J. G., Macer, R. C. F. and Greeley, L. W. (1965). *Trans. Br. mycol. Soc.* **48**, 479–489.
Mraz, F. (1971). *Rostlinna vyroba. Praha* **17**, 399–404.
Mühle, E. and Frauenstein, K. (1962). *Der Züchter* **32**, 345–352.
Mühle, E. and Frauenstein, K. (1963). *Der Züchter* **33**, 124–131.
Mühle, E. and Frauenstein, K. (1970). *Theoret. appl. Genet.* **40**, 56–58.
Naito, H. and Hirata, K. (1969). *Niigata Agric. Sci.*, **21**, 29–36.
Nover, I. and Lehmann, C. O. (1973). *Kulturpflanze* **21**, 275–294.
Riley, R. (1973). *Ann. appl. Bio.* **75**, 128–132.
Rudorf, W. and Wienhues, F. (1951). *Z. Pflanzenzücht.* **30**, 445–463.
Saari, E. E. and Wilcoxson, R. D. (1974). *A. Rev. Phytopathol.* **12**, 49–68.
Savile, D. B. O. (1954). *Science* **120**, 583–585.
Savile, D. B. O. (1971). *Naturaliste can.* **98**, 535–552.
Schaller, C. W. (1951). *Agron. J.* **43**, 183–188.
Sherwood, R. T., Hite, R. E. and Marshall, H. G. (1977). *Pl. Dis. Reptr* **61**, 37–41.
Sinigovets, M. E. and Lapchenko, G. D. (1975). *Cytologia i Genetika* **9**, 439–442.
Slootmaker, L. A. J. (1971). Non-specific versus race-specific resistance in barley mildew (*Erysiphe graminis* f. sp. *hordei*). Eucarpia, Proc. Dijon.
Slootmaker, L. A. J. (1974). *Outl. Agric.* **8**, 133–140.
Smiljakovíc, H. (1966). *Savremena poljoprivreda* **14**, 357–364.
Stebbins, G. L. (1956). *Am. J. Bot.* **43**, 890–905.
Wahl, I. (1970). *Phytopathology* **60**, 746–749.
Watson, L. (1972). *Q. Rev. Biol.* **47**, 46–62.
Wiberg, A. (1974). *Hereditas* **78**, 1–40.
Wolfe, M. S. (1972a). 1972. *Rev. Pl. Path.* **51**, 507–522.
Wolfe, M. S. (1972b). *Outl. Agric.* **7**, 27–31.
Wolfe, M. S. (1973). *Ann. Appl. Biol.* **75**, 132–136.
Yarwood, C. E. (1973). *In* "The Fungi" (G. C. Ainsworth, Sparrow, F. K., Sussman, A. S. Eds) Vol. 4, 71–86. Academic Press, New York and London, 621 pp.
Zeven, A. C. and Zhukovsky, P. M. (1975). "Dictionary of cultivated plants and their centers of diversity". Pudoc. Centre for Agricultural Publishing and Documentation, Wageningen, 219 pp.
Zhukovsky, P. M. (1971). "Kulturnye rastennia i ikh sorodichi (Cultivated plants and their wild relatives)". Publ. "Kolos", Leningrad, 751 pp.
Zillinsky, F. J. and Murphy, H. C. (1967). *Pl. Dis. Reptr* **51**, 391–395.

Chapter 5

Genetic Basis of Formae Speciales in *Erysiphe graminis* DC.

U. HIURA

Institute for Agricultural and Biological Sciences, Okayama University, Kurashiki, Japan

I. Introduction

On the basis of host specialization, Marchal (1902) distinguished seven formae speciales within *Erysiphe graminis*. Mains (1933) reported that a number of grasses other than *Triticum* were very resistant to *E. graminis* f. sp. *tritici*. Cherewick (1944) considered that formae speciales of *E. graminis* were restricted to their own host genera, but in contrast to the previous concept of restriction of infection to a single host genus, Hardison (1944) found that all the cultures of *E. graminis* which he studied produced infection on species of two or more genera. Mühle and Frauenstein (1962a, 1962b, 1963) considered that the boundaries of formae speciales were not distinct, though they suggested that under natural conditions powdery mildews of barley, wheat and rye did not infect the varieties of forage grasses which they examined (Mühle and Frauenstein, 1970). Likewise Eshed and Wahl (1970) found that formae speciales of the cereal mildews possessed a wide host range, but that seedlings of barley, wheat, and oats were congenial hosts almost exclusively for isolates derived from intrageneric grasses, and mainly related species (Eshed and Wahl, 1975). Thus it is clear that the parasitism of a forma specialis is not limited to a single genus. However, as Johnson (1968) stated, each specialized form has its characteristic adaptation. For example, wheat stem rust, *Puccinia graminis* f. sp. *tritici*, is an assemblage of individuals that resemble one another more in pathogenicity than they resemble the individuals of other specialized forms. They all have in common the capacity to attack one or more species of *Triticum*. What then is the genetic basis of the common capacity for pathogenicity?

The gene–for–gene concept has been either demonstrated or suggested for a number of host–parasite systems (Flor, 1971; Day, 1974). The question arises whether the differential interactions between a forma specialis and its host genus are analogous to the differential interactions between a physiologic race and its host cultivar.

In *Puccinia graminis*, formae speciales have the common alternate host, *Berberis vulgaris* L. As early as 1932 Johnson *et al.* showed that *P. graminis* f. sp. *tritici* and *P. graminis* f. sp. *secalis* are interfertile and Green (1971) has reported the pathogenicity of hybrid cultures derived from artificial crosses between *P. graminis* f. sp. *tritici* and *P. graminis* f. sp. *secalis* on wheat and rye. Although a wide host range for formae speciales of cereal mildews was reported in Israel (Eshed and Wahl, 1970), mildews on barley, wheat, rye and oats rarely attack any crop plant other than their own hosts. It seems therefore, that there is no suitable host here, on which hybridization between formae speciales can occur. The genetics of interactions between formae speciales of *Erysiphe graminis* and their host species, therefore, had not been studied until Hiura (1962) first found that different formae speciales of *E. graminis* could be hybridized.

II. Hybridization Between Formae Speciales

Hiura (1962) first found that hybridization occurred between formae speciales of *E. graminis* when conidia of an incongenial forma specialis (i.e., one which would not normally attack that host) were placed on the pustules of the other forma specialis growing on a congenial host. Moseman *et al.* (1965) found that when a barley cultivar, infected with *E. graminis* f. sp. *hordei*, was inoculated with *E. graminis* f. sp. *tritici*, a few conidia of *E. graminis* f. sp.

tritici were produced and when a wheat cultivar, infected with *E. graminis* f. sp. tritici, was inoculated with *E. graminis* f. sp. *hordei*, a few conidia of *E. graminis* f. sp. *hordei* were produced. Tsuchiya and Hirata (1973) demonstrated that when the wheat or wheatgrass powdery mildew was inoculated onto barley leaves already bearing young colonies of the barley fungus, the first haustoria of *E. graminis* f. sp. *tritici* or *E. graminis* f. sp. *agropyri* were formed in the epidermal cells harbouring the haustoria of the barley fungus and in their neighbouring epidermal cells. Thus, preliminary infection with a congenial culture of *E. graminis* rendered the host plant accessible to an incongenial forma specialis (Ouchi *et al.*, 1974). By this procedure, Hiura (1962, 1965) has studied cross compatibility between formae speciales, *hordei, tritici, agropyri, secalis, avenae* and *poae* of *E. graminis*.

E. graminis is heterothallic (Powers and Moseman, 1956; Hiura, *et al.*, 1961). In the matings within and between the three formae speciales, *agropyri, tritici* and *secalis*, cleistothecia and ascospores were formed only when opposite mating types were paired (Table 1). Apparently the cross compatibility between the three formae speciales is controlled by one pair of genes.

TABLE 1

Heterothallism and variation in cross-compatibility between formae speciales in *Erysiphe graminis* D.C.

Culture[a]	h–4	a–2	t–2	H–1	A–1	T–1
h–9	—[b]	—*	—*	+	±	±
a–3	—	—	—	±	+	+
t–3	—*	—	—	±	+	+
H–14	+	±	±	—	—*	—*
A–4	±	+	+	—	—	—
T–4	±	+	+	—*	—	—
S–1	±	+	+	—*	—	—

[a] H, h = *hordei*, A, a = *agropyri*, T, t = *tritici*, S = *secalis*. Capital denotes mating type *Mt*, small denotes mating type *mt*.
[b] + = Formed cleistothecia with ascospores.
± = Formed cleistothecia but no ascospores.
— = Neither cleistothecium nor gametangium and white mycelial mat.
—* = No cleistothecium but formed gametangia and white mycelial mat.

In the matings of *E. graminis* f. sp. *hordei* with the three formae speciales, *agropyri, tritici* and *secalis*, regardless of mating types, white mycelial mats developed which indicated that hybridization had occurred. When opposite mating types were paired, normal cleistothecia (150–200 μm) were formed in the white mycelial mats and asci developed in the cleistothecia but produced very few ascospores. Occasionally a few ascospore-like cells developed in some asci but they did not germinate. When cultures of the same mating type were paired, sclerotium-like bodies (15–80 μm) were formed in the white mycelial mats, but they did not develop further (Fig. 1 b–2). Under a microscope, three days after mating, characteristic gametangia were observed at the point of mycelial contact of the two cultures (Fig. 1 b–1).

A gametangium never develops on mycelium derived from a single spore of *E. graminis*. In ordinary mating, (other than with *E. graminis* f. sp. *hordei*) gametangia develop only at the point of mycelial contact of the two cultures with opposite mating types (Fig. 1 a–1). About 30 days after gametangial contact a mature cleistothecium is formed (Fig. 1 a–2). At the same time as gametangial contact, seta-like hyphae develop to surround the gametangia. The seta-like hyphae have a determinate growth and do not form haustoria nor become conidiophores. Seven to ten days after gamentangial contact many seta-like hyphae develop and appear to the naked eye like a white mycelial mat. This is the first visible evidence of hybridization.

In the matings of *E. graminis* f. sp. *hordei* with the other three formae speciales, however, gametangia and white mycelial mats developed even when the same mating type of cultures were paired. These results indicate that in the matings with *E. graminis* f. sp. *hordei* some complementary factor or factors for gametangial development are involved. Hybridization did not occur when *E. graminis* f. sp. *avenae* and *E. graminis* f. sp. *poae* were mated or when paired with the cultures of *E. graminis* f. sp. *hordei*, *E. graminis* f. sp. *agropyri* and *E. graminis* f. sp. *tritici* (Hiura, 1962). The data for the two formae speciales, therefore, were omitted from Table 1.

Only one culture of each of the *E. graminis* formae speciales *avenae* and *poae* was tested. It is not clear whether the cultures themselves do not have sexual stages or whether there is cross-incompatibility between formae speciales *avenae* and *poae* and between these two formae speciales and other formae speciales tested. It appears that there are various levels of reproductive isolation among formae speciales of *E. graminis*. Nevertheless, hybridization occurred without any barriers between the three *E. graminis* formae speciales, *agropyri*, *tritici* and *secalis* and produced viable ascospores. It is perhaps significant that the host species of these three formae speciales, namely *Agropyron*, *Triticum* and *Secale*, are also capable of hybridization.

III. Inheritance of Pathogenicity in Formae Speciales

The parent powdery mildew cultures of *E. graminis* f. sp. *tritici* and of *E. graminis* f.sp. *agropyri* were t–2 and A–1 respectively. Culture t–2 was

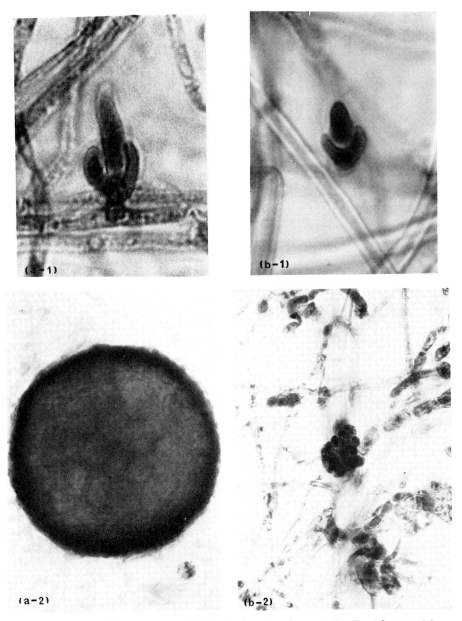

Fig. 1. Gametangial contact and cleistothecium development in *Erysiphe graminis*. a—Mating between cultures T-1 and t-2 of forma specialis *tritici* with opposite mating types. a-1 shows gametangial contact 3 days after mating (×1650), a-2 shows mature cleistothecium 30 days after mating (×330). b—Mating between culture t-2 of f. sp. *tritici* and culture h-9 of f. sp. *hordei* with the same mating type. b-1 shows gametangial contact 3 days after mating (×1650), b-2 shows sclerotium-like body 30 days after mating (×330). (Photo. taken H. Heta.)

virulent on the cultivars of *Triticum* spp. and avirulent on the strains of *Agropyron* spp. used in this study. The virulence of culture A–1 on the *Triticum* spp. and *Agropyron* spp. was the reverse of that of t–2. They were also of opposite mating types. The cultures of these two formae speciales hybridized with each other as easily as they crossed within each forma specialis on either wheat or wheatgrass. The crossing of cultures t–2 and A–1 produced a number of normal cleistothecia with abundant viable ascospores.

E. *graminis* is an obligate parasite which can be cultured only on living plants. The pathogenicity of hybrid cultures is selected by the genes which condition the reaction of the plant on which those cultures are maintained. Hybrid cultures from the cross of cultures t–2 and A–1 were isolated and maintained on two different host plants, cultivar Norin 4 of *Triticum aestivum* L. and strain Mishima of *Agropyron tsukushiense* var. *transiens* Ohwi. The progeny cultures were isolated by removing single conidia from individual pustules produced by ascospores discharged from the cleistothecia which resulted from the hybridization. The cultures were maintained on the same type of host plant as that on which the original pustule was formed. By this procedure 94 cultures were isolated and maintained on wheat cultivar Norin 4 and 96 cultures on wheatgrass strain Mishima. Hereafter, the progeny cultures maintained on wheat cultivar Norin 4 are called W-isolate and cultures on wheatgrass strain Mishima are called A-isolate. The plants used for isolating and maintaining a pure progeny culture, such as wheat cultivar Norin 4 and wheatgrass strain Mishima, are called maintaining host. The wheat cultivars and wheatgrass strains used for testing pathogenicity of the hybrid cultures are called test plant.

Sixty five wheat cultivars and 25 wheatgrass strains were inoculated with the 94 cultures of W-isolate and 96 cultures of A-isolate, respectively. Notes were taken according to the five classes of infection types used on barley mildew (Hiura, 1960). Of the 65 wheat cultivars all except two were susceptible to the parent culture t–2, producing infection types 3–4 or 4, and they were resistant to culture A–1, producing infection types 0 or 0–1. The two exceptions gave infection type 3 (Hiura and Heta, 1973). By contrast, 15 of the 25 wheatgrass strains were susceptible, producing infection types 3–4 or 4 and 10 produced infection type 3 to the culture A–1 but all of the 25 strains were resistant to the culture t–2 ,producing infection types 0 or 0–1, (Hiura and Heta, 1977).

Though there was similarity among the cultivars in their reactions to the parent mildew cultures, very different segregation occurred on each of the test plants for pathogenicity of the hybrid cultures (Table 2). When the 94 cultures of W-isolate were inoculated onto 65 wheat cultivars 6110 infections were obtained and of these about 10% were types 3–4 or 4, and about 20% were type 0. The remaining 70% of the infections produced intermediate reactions, giving infection types 0–1, 1, 2, or 3. Minor factors conditioning

TABLE 2

Segregation of 94 cultures of W-isolate and 96 cultures of A-isolate from a cross of culture t-2 of *Erysiphe graminis* f.sp. *tritici* and culture A-1 of *E. graminis* f. sp. *agropyri* for pathogenicity on seedlings of wheat cultivars and wheatgrass strains

Test plant[a]	Hybrid culture inoculated	Number of cultivars or strains on which avirulent[b] and virulent segregated with the ratio indicated					Total number of cultivars or strains tested
		1:3	1:1	3:1	7:1	more than 15:1	
Wheat	W-isolate	4	12	18	19	12	65
	A-isolate		1	13	26	25	65
Wheatgrass	W-isolate	3	3	14	1	7	25
	A-isolate	9	9	8			17

[a] Wheat included 58 *Triticum aestivum*, 4 *T. durum*, 2 *T. spelta* and 1 *T. dicoccum*, Wheatgrass included 21 *Agropyron tsukushiense* var. *transiens* and 4 *A. ciliare*.

[b] Avirulent = infection types 0 and 0–1 (sub-infection), Virulent = infection types 1–4.

intermediate reactions are likely to be involved in the hybrid cultures and if
these minor factors were to be taken into consideration, the host–parasite
interaction between the test plants and the hybrid cultures would be too
complicated to allow for reasonable discussion. Furthermore, finer infection
type differences of *E. graminis* f. sp. *tritici* on wheat cultivars are dependent
upon the environmental conditions at the time of testing as has been pointed
out by Wolfe (1965). The presence of intermediate reactions was more
marked on wheatgrass than on wheat. For these reasons the pathogenicity of
the progeny cultures was divided into those which were avirulent and those
which were virulent on the following basis.

avirulent = infection types 0 and 0–1

virulent = infection types 1 to 4

The sense in which the terms, virulence, aggressiveness and pathogenicity
are used here is slightly different from that used by Green (1971, 1975).
Here, virulence and avirulence are intended to describe a specific relationship
between a host cultivar and a mildew culture. Virulence is expressed when
host resistance genes are matched by virulence genes in the pathogen.
Aggressiveness is the ability of a pathogen to increase to epidemic proportions
in a host population. Pathogenicity is used in a broader sense as the ability
of a pathogen to cause disease on its host. It includes virulence, avirulence
and aggressiveness.

A. Genes for pathogenicity on wheat

Since a culture of *Erysiphe graminis* is haploid a 1:1 ratio is expected for the
frequency of avirulence and virulence in the hybrid progeny cultures, if the
parent cultures differed by one major gene for avirulence on a host. If one
parent culture differed from the other by two major genes for avirulence then
a 3:1 ratio would be expected for avirulent versus virulent cultures.

As shown in Table 2, when 65 wheat cultivars were inoculated with 96
cultures of A-isolate, three, or more than four-factorial segregation occurred
for pathogenicity on 51 cultivars. This suggests that a considerable number of
genes for avirulence on the cultivars are involved in the hybrid cultures. When
65 wheat cultivars were inoculated with 94 cultures of W-isolate, three or more
than four-factorial segregation for pathogenicity was observed on 31 cultivars.
It would thus appear that on most cultivars smaller numbers of genes for
avirulence were involved in the W-isolate than in the A-isolate. The W-isolate
maintained on wheat cv. Norin 4 comprises a group of cultures which do not
involve genes for avirulence on Norin 4. It is to be expected therefore that
each cultivar would have a number of genes for mildew resistance and that
some of these genes would be common to those of Norin 4. For example, as

TABLE 3

Relationships of 96 cultures of A-isolate for pathogenicity on some wheat cultivars on which frequency of avirulent and virulent cultures were 3 :1

Wheat cultivar		Number of cultures with pathogenic types				Value of P for 9:3:3:1	Value of P for 5:1:1:1
x	y	$A_xA_y{}^a$	A_xV_y	V_xA_y	$V_xV_y{}^a$		
Norin 4	Chancellor	75	2	3	16	small	small
	Seneca	65	12	8	11	small	0·7–0·5
	Turkey Red	66	11	7	12	small	0·5–0·3
Seneca	Chancellor	64	9	14	9	small	0·7–0·5
	Turkey Red	71	2	2	21	small	small

[a] A_x and V_x indicate the pathogenicity of cultures on wheat cv. x,
A_y and V_y indicate the pathogenicity of cultures on wheat cv. y.
A = avirulent, V = virulent

shown in Table 3, the observed number of cultures of A-isolate avirulent and virulent on each of cv. Norin 4 and cv. Seneca fits the 3:1 ratio which would be expected if two major genes for avirulence on each cultivar were involved in the A-isolate cultures. If the two genes for avirulence on each of Norin 4 and Seneca were different and inherited independently a 9:3:3:1 segregation ratio of the four pathogenically different cultures would be expected and these would be respectively, avirulent on both Norin 4 and Seneca, avirulent on Norin 4 only, avirulent on Seneca only, and virulent on both Norin 4 and Seneca. The observed number of the four pathogenically different cultures on the two cultivars does not fit the 9:3:3:1 ratio, but fits the 5:1:1:1 ratio. A 5:1:1:1 ratio would be expected if one of the two genes conditioning the avirulence of the hybrid cultures on the two cultivars was the same but the other gene conditioning avirulence on each cultivar was different. The data in Table 3 show that the frequency of the four pathogenically different cultures on either of the two pairs of cultivars, Norin 4 and Turkey Red, or Seneca and Chancellor, fits the 5:1:1:1 ratio. These results therefore suggest that one of two genes conditioning the avirulence of the hybrid cultures of A-isolate on Norin 4, Seneca, Turkey Red and Chancellor is the same in each culture, and that the second gene for avirulence on each cultivar is different. Only a smaller number of the hybrid cultures differed from their parent cultures in pathogenicity on Norin 4 and Chancellor, and on Seneca and Turkey Red (Table 3). These results indicate that some of the genes for pathogenicity on these two pairs of cultivars may be linked.

On nine cultivars, as shown in Table 4, there was a monofactorial segregation for pathogenicity of the W-isolate. The linkage relationships of the genes conditioning pathogenicity on the nine cultivars were determined by comparing the pathogenicity of the 94 cultures of W-isolate on those cultivars. The probability values from the chi-square test for goodness of fit to a 1:1:1:1 ratio for the four pathogenically different cultures were used to show the relationship of the genes conditioning pathogenicity on the two cultivars (Table 4). These probability values indicate that the gene conditioning pathogenicity of Seneca is linked with the gene conditioning pathogenicity on each of the cultivars Mayo 64, Ramona 50, Taichu 2 and Turkey Red and that the gene conditioning pathogenicity on Chancellor is linked with the gene conditioning pathogenicity on each of the cultivars Swan 86, isogenic lines Chul × Cc8 and Sonora × Cc8 (Briggle, 1969).

It is concluded that a number of genes condition avirulence of the A-isolate cultures to each wheat cultivar and that some of these genes are common to many cultivars. It is also apparent that many of the genes which condition avirulence on wheat cultivars are inherited by the coupling of two linkage groups. This might explain why the W-isolate maintained on wheat cv. Norin 4 involves a smaller number of avirulence genes on wheat cultivars than are involved in conditioning avirulence of the A-isolate.

TABLE 4

Relationship of 94 cultures of W-isolate from cultures t-2 × A-1 for pathogenicity on 9 wheat cultivars on which frequency of avirulent and virulent were 1 : 1

Wheat cultivar		Number of cultures with pathogenic type				Value of P for
x	y	A_xA_y	A_xV_y	V_xA_y	V_xV_y	1:1:1:1
Seneca	Mayo 64	44	10	5	35	small
	Ramona 50	45	9	6	34	small
	Taichu 2	47	7	8	32	small
	Turkey Red	50	4	4	36	small
	Chancellor	28	26	26	14	0·2 − 0·1
	Swan 86	28	26	26	14	0·2 − 0·1
	Chul × Cc8	26	28	30	10	0·02 − 0·01
	Sonora × Cc8	28	26	27	13	0·1 − 0·05
Chancellor	Mayo 64	25	29	26	16	0·3 − 0·2
	Ramona 50	27	27	24	16	0·5 − 0·3
	Taichu 2	29	25	26	14	0·2 − 0·1
	Turkey Red	27	27	27	13	0·1 − 0·05
	Swan 86	50	4	4	36	small
	Chul × Cc8	48	6	8	32	small
	Sonora × Cc8	50	4	5	35	small

B. Genes for pathogenicity on wheatgrass

As shown in Table 2, when 25 wheatgrass strains were inoculated with 94 cultures of W-isolate, monofactorial segregation for pathogenicity occurred on three strains and bifactorial segregation occurred on 14 strains. When the 14 wheatgrass strains, on which mono- and bifactorial segregation were observed with the W-isolate, were inoculated with 96 cultures of A-isolate, on nine strains monofactorial segregation occurred and on eight strains bifactorial segregation occurred for pathogenicity. It is evident that a smaller number of genes which condition avirulence on most wheatgrass strains are involved in the A-isolate than in the W-isolate. The frequency of avirulent and virulent cultures of the W-isolate on each of seven strains of wheatgrass, Hashima, Noken 4, Yura, Ashiya, Gifu 1, Hikimi and Okayama 1 was 3 : 1 (Table 5) whereas the corresponding ratio for the A-isolate was 1 : 1 (Table 6). From these data it was expected that two genes would condition avirulence of the W-isolate cultures on each of the seven strains and that one of the two genes would be the same as the gene conditioning avirulence of the W-isolate cultures on Mishima, which was the maintaining host of the A-isolate. The relationships of 94 cultures of the W-isolate for pathogenicity on Mishima and the seven wheatgrass strains are shown in Table 5. The frequency of avirulent

TABLE 5

Relationships of 94 cultures of W-isolate for pathogenicity on wheatgrass strain Mishima and other 7 strains

Wheatgrass strain		Number of cultures with pathogenic type				Value of P for
x^a	y^a	A_xA_y	A_xV_y	V_xA_y	V_xV_y	3:1:3:1
Agt. Mishima	*Agt.* Hashima[b]	45	4	25	20	small
	Agt. Noken 4	45	4	26	19	small
	Agt. Yura	45	4	24	21	small
	Agc. Ashiya[b]	41	8	30	15	0·3–0·2
	Agc. Gifu 1	39	10	29	16	0·5–0·3
	Agc. Hikimi	35	14	28	17	0·3–0·2
	Agc. Okayama 1	42	7	33	12	0·5–0·3

[a] Frequency of avirulent and virulent culture on x strain and y strain were 1:1 and 3:1, respectively.

[b] *Agt.* = *Agropyron tsukushiense* var. *transiens* Ohwi, *Agc.* = *A. ciliare* Franchet.

TABLE 6

Relationships of 96 cultures of A-isolate from cultures t-2 × A-1 for pathogenicity on the wheatgrass strains on which frequency of avirulent and virulent were 1 : 1

Wheatgrass strain		Number of cultures with pathogenic type				Value of P for
x	y	A_xA_y	A_xV_y	V_xA_y	V_xV_y	1:1:1:1
Ashiya	Hifu 1	43	1	6	46	small
	Gikimi	44	0	5	47	small
	Okayama 1	34	10	10	42	small
	Hashima	22	22	28	24	0·9 –0·8
	Noken 3	20	24	25	27	0·8 –0·7
	Noken 4	26	28	31	21	0·3 –0·2
	Niigata 1	24	20	25	27	[0·8 –0·7
	Yura	24	20	25	27	0·8 –0·7
Noken 3	Gifu 1	22	23	27	24	0·9 –0·8
	Hikimi	23	22	26	25	0·99 –0·9
	Okayama 1	16	29	28	23	0·3 –0·2
	Hashima	28	17	22	29	0·3 –0·2
	Noken 4	41	4	16	35	small
	Niigata 1	26	19	23	28	0·7 –0·5
	Yura	25	20	24	27	0·8 –0·7

and virulent cultures of W-isolate on Mishima was 1:1 suggesting that one gene conditioned avirulence of the cultures on Mishima. Therefore, if the gene for avirulence on Mishima and two genes for avirulence on Hashima were different from each other and inherited independently, then a 3:1:3:1 ratio of the four pathogenically different cultures would be expected and these would be respectively, avirulent on both Mishima and Hashima, avirulent on Mishima only, avirulent on Hashima only and virulent on both Mishima and Hashima. However, the observed number of the four pathogenically different cultures on the two strains did not fit this ratio. If one of the two genes on Hashima was the same as the gene on Mishima, then a culture avirulent on Mishima only would not be expected, but four cultures were found to be avirulent on Mishima only. It was concluded therefore that the gene for avirulence on Mishima was different from the two genes for avirulence on Hashima and that one of the two genes on Hashima might be linked with the gene on Mishima. Similar results were obtained on strains Noken 4 and Yura.

On Mishima and on each of strains Ashiya, Gifu 1, Hikimi and Okayama 1, the observed number of the four pathogenically different cultures fits the 3:1:3:1 ratio. Contrary to expectation, two genes in each of the four strains and the gene in Mishima were inherited independently. It is not clear, therefore, why bifactorial segregation occurred for pathogenicity of the W-isolate and monofactorial segregation occurred for pathogenicity of the A-isolate on the four strains. A minor gene which has the effect of lowering the infection type on wheatgrass might possibly be involved in the W-isolate.

On nine strains, as shown in Table 6, there was a monofactorial segregation for pathogenicity of the A-isolate for which the linkage relationships of the genes were determined. The observed number of the four pathogenically different cultures on Ashiya and on Gifu 1, Hikimi and Okayama 1, does not fit the 1:1:1:1 ratio. The gene conditioning pathogenicity on Ashiya is linked with the gene conditioning pathogenicity on each of the other three strains and the genes conditioning pathogenicity on these three strains are also linked to each other. The gene conditioning pathogenicity on Noken 3 is linked with the gene conditioning pathogenicity on Noken 4. Thus two linkage groups of genes which condition pathogenicity were found on wheatgrass.

Person and Sidhu (1971) pointed out that linkage of genes for virulence is reported only occasionally and no report of allelism has been found in the literature. Kimber and Wolfe (1966) found *E. graminis* f. spp. *hordei*, *tritici* and *avenae* to have only two chromosomes each. Despite this small number of chromosomes, Moseman (1963) and Hiura (1964) reported that the linkage of genes for virulence on barley cultivars was found only occasionally. By contrast many linkages were found between genes for pathogenicity on both wheat and wheatgrass in the hybrid cultures from the cross of cultures t–2 and A–1. Supposing that there are three genes, R_a, R_b and R_c, for mildew resistance in wheat cultivars and that R_a and R_b are common to most cultivars but R_c is rare. Then probable genotypes for resistance of cultivars will be $R_a R_a R_b R_b r_c r_c$ and $R_a R_a R_b R_b R_c R_c$. Suppose also that in mildew cultures there are three genes, V_a, V_b and V_c, for virulence which correspond to the resistance genes in wheat cultivars and that V_a is closely linked with V_b but V_c is independent. Then postulated genotypes of mildew cultures are $V_a V_b A_c$ and $V_a V_b V_c$ (A = avirulent). The cultures having these two genotypes will be recognized as physiologic races. But R_a and R_b genes for resistance and V_a and V_b genes for virulence and their linkage relationship would never be recognized unless cultures possessing unusual genes for avirulence such as hybrids between formae speciales are used.

It will be shown later that most wheat cultivars have common genes for mildew resistance. Mildew cultures on wheat, therefore, must have common genes for virulence which correspond to the genes for resistance. These common genes for virulence in wheat mildew might be the genetic background of the forma specialis *tritici* and it is interesting, therefore in relation to host specialization that frequently these genes are linked.

C. Relationships between genes for pathogenicity on wheat and wheatgrass

Two linkage groups of genes which condition the pathogenicity of progeny cultures from the crossing of cultures t–2 and A–1 were found on each of the wheat cultivars and wheatgrass strains. Since there are only two chromosomes in *Erysiphe graminis* f. sp. *tritici* (Kimber and Wolfe, 1966), it might be expected that a gene for avirulence on wheat would be linked with a gene for avirulence on wheatgrass but as yet such linkage relationships have not been found.

Monofactorial segregations for pathogenicity were observed only on one wheat cultivar and on three wheatgrass strains when the test plants were inoculated with A-isolate cultures and W-isolate cultures, respectively (Table 2). Very few data, therefore, are available to determine the linkage relationships and additional testing will be necessary before a definite conclusion can be reached.

IV. Survival of the Progeny Cultures

A. Pathogenicity of survivors of progeny cultures

As mentioned above, the genes which condition avirulence on wheat and wheatgrass are inherited independently. Therefore, as expected, a culture virulent on both wheat and wheatgrass was produced by crossing the formae speciales *tritici* and *agropyri*. In nature, however, specialized pathogenic forms such as culture t–2 or culture A–1 are prevalent and a culture which is virulent on both wheat and wheatgrass rarely survives. This phenomenon can be demonstrated experimentally in a greenhouse. Thus when the progeny cultures from a cross of culture t–2 and culture A–1 were grown in a mixture on wheat for several consecutive conidial generations the percentage of cultures which were virulent on wheatgrass gradually decreased and the cultures virulent on wheat only became predominant. Conversely when the same mixture was grown on wheatgrass, the proportion of cultures virulent on wheat gradually decreased and the cultures virulent on wheatgrass only became predominant. (Heta and Hiura, 1972, 1973). Fig. 2 shows one of the experimental results on survival of the progeny cultures from a cross of cultures t–2 × A–1 (Hiura, 1973). In this experiment the progeny cultures were repeatedly grown in a mixture on wheat Norin 4. The pathogenicity of cultures which survived on Norin 4 were tested at 0, 5, 10 and 15 conidial generations. Six wheat cultivars Chancellor, Durum, Hati, Seneca, Sinvalucho and Titan, and one wheatgrass strain Yamanashi were used as test plants for pathogenicity. Virulence 0 means the culture is avirulent on the six test cultivars. Virulence 1, 2, 3, 4, 5 or 6 indicates that the culture is virulent on 1, 2, 3, 4, 5 or 6 of the six cultivars. As shown in Fig. 2, at the

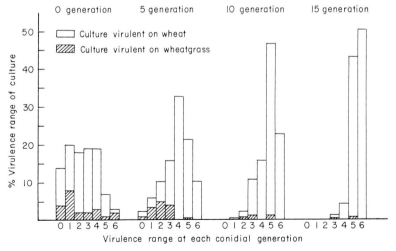

Fig. 2. Frequency of virulence range of the progeny cultures from cultures t-2 × A-1 on six test wheat cultivars and one wheatgrass strain when the progeny cultures were grown in mixture on Norin 4 for several consecutive conidial generations. Virulence ranges, 0, 1, 2, 3, 4, 5 and 6 indicate virulent on any 0, 1, 2, 3, 4, 5 and 6 of 6 test cultivars, respectively. The 6 test cultivars includes Chancellor, Durum, Hati, Seneca, Sinvalucho and Titan.

first generation there were similar percentages of cultures in the virulence range from 0 to 4 (approximately 15–20%) but only 7 and 3% of cultures were virulent with grades 5 and 6 respectively of the six test cultivars. The frequency of cultures with virulence grades 5 and 6, however, gradually increased from generation to generation and at the fifteenth generation more than 90% of cultures were in virulence grades 5 and 6. It was also found that at the first generation a gene conditioning virulence on the single wheatgrass, strain Yamanashi, was randomly distributed over the entire virulence range. The frequency of cultures virulent on wheatgrass gradually decreased from generation to generation and at the fifteenth generation no cultures were virulent on the wheatgrass at grade 6. It was clear that when progeny cultures from a cross of *E. graminis* f. sp. *tritici* with *E. graminis* f. sp. *agropyri* were grown repeatedly on wheat cultivars those cultures with a wide virulence range on wheat came out predominant and that these cultures were scarcely virulent on wheatgrass.

B. Relationships between aggressiveness and virulence

In the experiment shown in Fig. 2, the hybrid cultures were grown repeatedly in a mixture on Norin 4. It was to be expected, therefore, that cultures which were aggressive on Norin 4 should come out predominant and that virulence on wheatgrass would decrease. Therefore, aggressiveness on Norin 4 might be related to virulence on a wide range of wheat cultivars and to lesser aggressiveness on wheatgrass. The relationship between aggressiveness on Norin 4 and the virulence range on 30 wheat cultivars or virulence range on seventeen wheatgrass strains in the 94 cultures of W-isolate are shown in Fig. 3.

Infection type on the host may be an index of aggressiveness on that host. In Fig. 3 the 94 cultures were divided into three classes of aggressiveness by the infection types produced on Norin 4. Twenty three showed lesser aggressiveness and produced infection types 1 and 2, thirty nine were intermediate in aggressiveness producing infection types 2–3 and 3, and thirty two were graded as more aggressive producing infection types 3–4 and 4. The virulence range of a culture on the test plants was shown by the number of cultivars or strains infected with the culture. In Fig. 3 the data are presented for 30 wheat cultivars and 17 wheatgrass strains where the frequency of avirulent and virulent cultures of the W-isolate was either 1:1 or 3:1 (Table 2). Among the selected wheat cultivars were 29 of *Triticum aestivum* L. and one of *T. durum* Desf. and in the wheatgrass strains were 13 of *Agropyron tsukushiense* var. *transiens* Ohwi and four of *A. ciliare* Franchet.

As shown in Fig. 3, the virulence range on the wheat cultivars becomes wider and virulence range on the wheatgrass strains becomes narrower as aggressiveness on Norin 4 increases. These results support and confirm the results shown in Fig. 2. Figure 3 is concerned with aggressiveness on Norin 4 only, but it is desirable to demonstrate a general relationship between aggressiveness and virulence on many test plants. For this purpose the data for 94

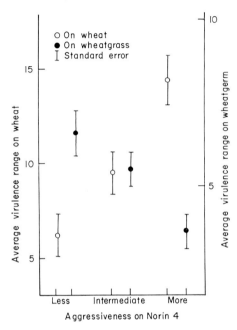

Fig. 3. Relationships between aggressiveness on wheat cv. Norin 4 and virulence range on 30 wheat cultivars or 17 wheatgrass strains in 94 cultures of W-isolate from cultures t-2 × A-1. Less aggressiveness included 23 cultures producing infection types 1 and 2, intermediate aggressiveness included 39 cultures producing infection types 2–3 and 3, more aggressiveness included 32 cultures producing infection types 3–4 and 4. Virulence range of a culture was shown by the number of virulent infections caused by the culture on 30 wheat cultivars or 17 wheatgrass strains.

cultures of W-isolate on the 30 wheat cultivars were used again and are presented in Table 7.

The 94 cultures were divided into four classes of virulence range, viz. avirulent, narrow, intermediate and wide which indicate virulence on none of the cultivars, on 1–6 cultivars, on 7–14 cultivars and on 15–30 of the 30 test cultivars, respectively. As mentioned above, where an obligate parasite is concerned, the infection type on the host can usually be used as an index of aggressiveness on that host and thus, in our experiments, the frequency of each infection type on the 30 test cultivars might be regarded as a relative index of the general aggressiveness of the cultures on those cultivars. The

TABLE 7

Relationships between virulence range and ratio of infection type of 94 cultures
of W-isolate on 30 wheat cultivars

Virulence[a] range	No. of culture observed	Total No. of infection	Percent of 3 classes of infection type[b]		
			Low	Intermediate	High
Avirulent	5	0			
Narrow	29	107	62·6**	27·1	10·3**
Intermediate	31	325	43·7**	34·8	21·5**
Wide	29	544	33·1**	34·0	32·9**

[a] Avirulent = no infection, Narrow = virulent on 1 to 6 cvs., Intermediate = virulent on 7 to 14 cvs., Wide = virulent on 15 to 30 cvs.

[b] Low = infection types 1 and 1–2, Intermediate = infection types 2 and 3, High = infection types 3–4 and 4.

** indicate a significant difference between virulence ranges at the 0·01 level.

ratios of the infection types in the four classes of virulence range are shown in Table 7, where infection types were grouped into three classes; low, intermediate and high, to include infection types 1 and 1–2, 2 and 3, and 3–4 and 4, respectively. Clear relationships between infection type and virulence range are apparent. The percentage of the high infection type was high in the wide virulence range and low in the narrow virulence range and, conversely, the percentage of low infection type was high in the narrow virulence range and low in the wide virulence range. From these results, it may be said that a wide virulence range is related to more aggressiveness and that a narrow virulence range is related to less aggressiveness.

C. Negative correlation between virulence on wheat and on wheatgrass

There was an indication (Fig. 3) that where the virulence range of cultures was wide on wheat it was narrow on wheatgrass and where it was narrow on wheat it was wide on wheatgrass. The correlation of the 94 cultures of W-isolate for virulence range on the 30 wheat cultivars and 17 wheatgrass strains are shown in Fig. 4 where a moderately negative correlation is evident. The calculated correlation coefficient was −0·3978. It was concluded that cultures which were more aggressive on wheat were less aggressive on wheatgrass and vice versa and that cultures which were virulent on both wheat and wheatgrass were intermediate in aggressiveness or were less aggressive on both hosts. This conclusion was also supported by the results of a serial inoculation experiment on wheat with mixtures of culture t–2 and hybrid cultures which were virulent on both wheat and wheatgrass. The hybrid cultures virulent on both wheat

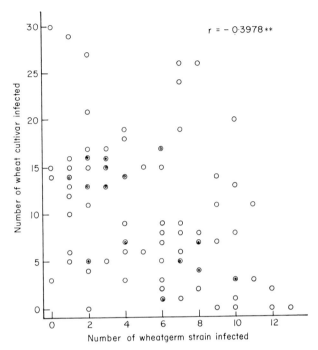

Fig. 4.   Correlation of 94 cultures of W-isolate for virulence ranges on 30 wheat cultivars and 17 wheatgrass strains. ○, ◎ and ⊙ indicate one, two and three cultures, respectively. ** indicates significance at the 0·01 level.

and wheatgrass were obtained by the following procedures. Wheatgrass strain Mishima was inoculated with ascospores produced by a crossing of cultures t–2 and A–1. The hybrid conidia produced on thes train Mishima were again inoculated onto wheat cv. Norin 4. The conidia then produced on Norin 4 were found to be virulent on both wheat and wheatgrass. Various batches of these conidia, virulent on both hosts, were mixed with conidia of culture t–2 which was virulent on wheat only. The mixtures were grown repeatedly on Norin 4 and the frequency of t–2 cultures was checked at each conidial generation. The results obtained are shown in Table 8 (Hiura and Heta, 1972). The frequency of culture t–2 in the mixture increased rapidly and in some mixtures reached 100% after three conidial generations. This clearly indicates that those hybrid cultures which are virulent on both wheat and wheatgrass are less aggressive than culture t–2 on wheat.

As mentioned previously, the genes which govern pathogenicity on wheat and wheatgrass are inherited independently. If these genes recombine freely, a culture which is as aggressive as both parent cultures on both the hosts should be expected, but as shown in Figs 2 and 4 hybrid cultures virulent on

TABLE 8

Rapid increasing of culture t-2 in a mixture with hybrid cultures[a] which were virulent on wheat and wheatgrass when the mixture was grown repeatedly on wheat cv. Norin 4

Conidial generation	Percent of culture t-2 in mixture					
	Mixture 1	Mixture 2	Mixture 3	Mixture 4	Mixture 5	Mixture 6
1	12·5	12·5	12·5	27·5	20·0	15·0
2	72·5	57·5	67·5	87·5	95·0	85·0
3	95·0	92·3	90·0	100·0	95·0	100·0
4	90·0	90·0	70·0	100·0	100·0	100·0
5	95·0	92·5	70·0	100·0	97·5	100·0
6				100·0	100·0	100·0
7				100·0	97·5	100·0

[a] Hybrid culture from a cross of cultures t–2 × A–1

both wheat and wheatgrass are intermediate in aggressiveness or are less aggressive on the respective hosts.

In studies on the hybrid from a cross of *Nicotiana langsdorffii* and *N. alata*, Anderson (1949) stated

the recombination of the F_2, however manifold they may seem, are in reality but a narrow segment of the total imaginable recombinations of the parental species.

He designated this narrow segment of recombinations as a "recombination spindle" and interpreted it in the following manner:

If each gene were on a tiny separate chromosome and the germplasm was composed of hundreds or thousands of such units, then we might get complete recombination of specific differences in hybrid populations. Germplasms, however, are not constructed in this way or in anything like this way. The genes are carried in long, protein, threadlike units, the chromosomes. Within each differing chromosome pair in a hybrid nucleus, only a very limited amount of exchange is possible.

The recombinations shown in Fig. 4 resemble Anderson's recombination spindle. The extreme condition of high aggressiveness on both wheat and wheatgrass was not obtained. This is attributable to linkage, as proposed by Anderson, in this case among the genes which condition pathogenicity on wheat (Table 4) and also among the genes which condition pathogenicity on wheatgrass (Table 6).

V. Mechanism of Host Specialization

A. Specific relationship between host genotype and hybrid genotype

Cotter and Roberts (1963) recognized resistance factors in many oat varieties previously considered susceptible, by inoculating them with a hybrid between *Puccinia graminis* var. *avenae* and *P. graminis* var. *agrostidis*. Sanghi and Luig (1971) also isolated genes conditioning rust resistance from wheat cultivars Mentana and Yalta by the use of a culture of *P. graminis tritici* possessing unusual genes for avirulence. Hiura and Heta (1973) studied the inheritance of pathogenicity on wheat cultivars in a cross between culture t–2 of *E. graminis* forma specialis *tritici* and culture A–1 of forma specialis *agropyri*, but they did not analyse the genetics of resistance to mildew in wheat by crossing resistant and susceptible wheat cultivars.

Flor (1955) has shown that for each gene conditioning the reaction of the host there is a corresponding gene conditioning the pathogenicity of the pathogen. If Flor's gene–for–gene hypothesis is valid between the genes conditioning the reaction of wheat and genes conditioning the pathogenicity of the hybrid cultures, genetic studies on the pathogenicity of mildew can be used to obtain information regarding the genetics of the wheat cultivars (Moseman, 1964).

It was recognized that one or two genes conditioned avirulence of the 94 cultures of W-isolate on each of the 30 wheat cultivars. These results were based on the assumption that infection types 0 and 0–1 were avirulent and all infection types 1 through 4 were virulent. As mentioned previously, about 70% of the infections obtained by inoculating with 94 cultures of W-isolate onto 65 wheat cultivars were sub-infection or intermediate infection types. It is reasonable to assume that a number of genes conditioning the intermediate reaction might be present in the cultures and that a number of corresponding genes for intermediate resistance might be expected to occur in the wheat cultivars. From the information already presented on the specific relationship between wheat cultivars and hybrid cultures from a cross of two formae speciales the following working hypothesis was suggested:

A wheat cultivar, even though previously considered to be susceptible to mildew, has a number of genes for resistance to the pathogen. Each of these genes is specific against a gene for pathogenicity in the mildew culture and generally shows as an intermediate mildew reaction. A set of complementary genes for resistance in a host and avirulence in a parasite (abbreviated as *R–A* set) induces a lowering of the infection type. The more the *R–A* set is involved the narrower is the virulence range and less aggressiveness is the result. To produce infection type 4 on a host a mildew culture must have as many virulence genes as there are corresponding genes for resistance in the host. Many wheat cultivars have common genes for resistance to mildew.

Mildew cultures, therefore, have common genes for pathogenicity on wheat.

On the basis of this working hypothesis a model for a specific relationship between host cultivars and hybrid cultures from a cross of two formae speciales has been proposed (see Table 9). According to Flor's gene-for-gene hypothesis, when resistance in the host and avirulence of the parasite are dominant, the resistance reaction occurs only when complementary genes in both the host and the parasite are dominant. If either or both of each pair of complementary genes are recessive, then susceptibility results. This indicates that in a case where the parasite has a gene for avirulence but its host does not have the complementary gene for resistance to the avirulence gene, susceptibility results. In relation to evolution, it is improbable that a parasite with the avirulence gene has survived because its host happened to be susceptible. It is more reasonable to consider that the plant always has genes for resistance and that the parasite has evolved to adapt itself for survival even on the host with genes for resistance. Favret (1971) has proposed a new adaptation of Flor's theory, namely that susceptibility is always the result of the action of a gene for virulence in the pathogen. In making the model we accepted Favret's new hypothesis.

The model presented in Table 9 is for the simplest imaginable case. Supposing that there are three genes for resistance, R_1, R_2, and R_3, in hosts, and that each cultivar has at least two of the three genes, then four host genotypes for resistance are expected. The postulated genotype of one parent culture which is virulent on a host having the three genes for resistance is $V_1V_2V_3$ and that of the other parent culture which is avirulent on the host is $A_1A_2A_3$. Then, eight genotypes for pathogenicity are expected in the hybrid cultures from a cross of $V_1V_2V_3 \times A_1A_2A_3$. The phenotypes resulting from interaction between host genotypes and parasite genotypes are shown by V, I, sA, and A in Table 9. V denotes virulence indicating no $R–A$ set. I denotes intermediate reaction with one $R–A$ set. sA denotes sub-avirulence with two $R–A$ sets and A denotes avirulence with three $R–A$ sets.

Table 10 is a comparison between the expected ratio of the four phenotypes, V, I, sA, and A, calculated from the model and the actual ratio obtained by inoculating with 94 cultures of W-isolate onto the 30 wheat cultivars. The percentages of the four phenotypes obtained experimentally were based on the assumption that V includes infection types 3–4 and 4, I includes infection types 1, 2 and 3, sA is infection type 0–1, and A is infection type 0. Thus, when five genes for resistance are postulated, the ratio of the four phenotypes calculated from the model approximates to the actual ratio obtained by inoculating with 94 cultures of the W-isolate onto 30 wheat cultivars. However, in the case of the A-isolate the calculated ratio is very different from the actual ratio obtained by inoculating with 96 cultures, suggesting that far more genes are involved in the interaction between wheat cultivars and A-isolate

that is avirulent on wheatgrass is by chance more aggressive on wheat than cultures with genes, V_{a-1} or V_{a-2} or both V_{a-1} and V_{a-2}, then it will become prevalent on wheat.

Mildew on wheat has continuously evolved to adapt itself to wheat by mutation, recombination, transgressive segregation and by the selection brought about by genes for mildew resistance in wheat. In the process of adaptation to wheat, genes for virulence on other plants if they were independent of aggressiveness on wheat, might be left behind by the wheat mildew population. In this manner a forma specialis *tritici* that is specific on wheat has developed and the mildew on wheatgrass has evolved towards the forma specialis *agropyri*, which is specific on wheatgrass.

C. Existence of formae speciales

As stated above, a culture of forma specialis *tritici* would have a considerable number of genes for virulence necessary for its survival on wheat. Many of these genes are linked and make up two linkage groups. Likewise a culture of forma specialis *agropyri* would have a number of linked genes for virulence on wheatgrass. When cultures of these two different formae speciales are hybridized, the sets of genes which fit the pathogen for survival on one or other of wheat and wheatgrass will be broken up in the recombination process. Because, virulence genes are often linked to each other, the exchange of genes would occur in groups resulting in the loss of a number of genes from the original set. This would produce a lowering effect on the aggressiveness of the hybrid progeny cultures on either wheat and wheatgrass. Johnson *et al.* (1932) and Johnson (1949) reported that hybrids between *Puccinia graminis* f. sp. *tritici* and *P. graminis* f. sp. *secalis* were low in virulence on wheat and rye. Green (1971) also stated that the progeny of crosses between wheat stem rust and rye stem rust have less virulence on rye than the rye stem rust parent and less virulence on wheat than the wheat stem rust parent. These cultures appear to have a broader host range than their parents but apparently they achieved it at the expense of vigour on one or other of the parental hosts. This lowered aggressiveness of a hybrid progeny has been observed not only in a obligate parasite but also in a facultative parasite. *Cochliobolus carborum* produces corn-specific toxin but does not produce oat-specific toxin. *Cochliobolus victoriae* produces oat-specific toxin but does not produce corn-specific toxin. When these two were hybridized, strains of the fungus that produced both corn- and oat-specific toxins and were pathogenic to both host species were synthesized. To date, no such isolates have been found in nature (Scheffer *et al.*, 1967). From this it may be inferred that the synthetic strains lack fitness for survival on one or other of the parental hosts. Since the classic studies of the ecological control of hybridization in *Quercus* by Muller

(1952), it has been a matter of common knowledge that when two parents require very different environmental conditions for growth their hybrid progenies will not be able to survive under both environmental conditions if they are fitted to growth in conditions suitable for one or the other of the parents. *Erysiphe graminis* f. sp. *tritici* and *E. graminis* f. sp. *agropyri* are very different in requirement for growth because they grow on hosts belonging to different genera. It is, therefore, inevitable that their hybrids will be less aggressive on both wheat and wheatgrass. This lesser aggressiveness in hybrid progeny leads to reproductive isolation because the progeny may not survive. The wild type of *Erysiphe graminis* for example might have been originally a Mendelian population and the wild type mildew on different hosts might have hybridized without any barrier but it is inferred that mildew on a particular host gradually became specialized and even though the specialization might not have been complete, reversion to a wild type would have been prevented by reproductive isolation in the hybridization. Thus host specialization would have developed more and more and formae speciales have thus become established.

The breakdown of the forma specialis is also prevented by cross-incompatibility in hybridization between formae speciales as we have seen previously in *Erysiphe graminis* (Hiura, 1965) and *Puccinia graminis* (Johnson et al., 1932). These cross-incompatibilities might have occurred through mutation and have been maintained in the formae speciales, because, cross-incompatibility in hybridization between formae speciales is advantageous to retain the fitness for survival of the specialized pathogen on a specific host.

On the other hand hybridization between formae speciales is not always harmful to a parasite. When a new gene for resistance is introduced into a plant from a different species, the most effective method to adapt to the new resistant plant may be to hybridize with the forma specialis that is virulent on the donor plant of the new resistance gene. For example, a new physiologic race of *Puccinia graminis* which attacks rye, wheat and *Triticale* has been reported (Lopez et al., 1974). This new race might have been derived from introgressive hybridization between *Puccinia graminis* f. sp. *tritici* and *P. graminis* f. sp. *secalis*.

Host specialization occurs not only on hosts which belong to different genera but also on different varieties of the same species. Varieties of rice, *Oryza sativa*, comprise two types, Indica and Japonica which have been grown in the different areas in the world. There is reproductive isolation between the two types and according to Morishima (1969), races of the rice blast fungus, *Pyricularia oryzae* can be classified into two groups, one pathogenic mainly on Indica and the other mainly on Japonica. From this, she inferred that the Indica and Japonica types have different genes controlling blast resistance and that the blast races are differentiated accordingly. Difference between formae speciales can be attributed to general pathogenicity conditioned by many genes and from this point of view, the two race groups of rice blast fungus may be comparable to formae speciales.

The pathogenicity of a parasite will evolve as its host evolves. This is especially true of obligate parasites, or near obligate parasites, where pathogenic specialization will reach the extreme of fitness to the host genotype. The relationship between the genotype for pathogenicity of an obligate parasite and the genotype for resistance of the host can be likened to the relationship between a solid object and its shadow; as one moves so does the other follow,

References

Anderson, E. (1949). "Introgressive hybridization". John Wiley and Sons Inc.
Briggle, L. W. (1969). *Crop Sci.* **9**, 70–72.
Cherewick, W. J. (1944). *Can. J. Res.* **C22**, 52–86.
Cotter, R. U. and Roberts, B. J. (1963). *Phytopathology* **53**, 344–346.
Day, P. R. (1974). "Genetics of host-parasite interaction". Freeman and Company, San Francisco.
Eshed, Nava and Wahl, I. (1970). *Phytopathology* **60**, 628–634.
Eshed, Nava and Wahl, I. (1975). *Phytopathology* **65**, 57–63.
Favret, E. A. (1971). "Barley Genetics". Vol. Proc. Second Int. Barley Genet. Sympos. Pullman, Washington, Washington State University Press. II, 457–471.
Flor, H. H. (1955). *Phytopathology* **45**, 680–685.
Flor, H. H. (1971). *Rev. Phytopathol.* **9**, 275–296.
Green, G. J. (1971). *Can. J. Bot.* **49**, 2089–2905.
Green, G. J. (1975). *Can. J. Bot.* **53**, 1377–1386.
Hardison, J. R. (1944). *Phytopathology* **34**, 1–20.
Heta, H. and Hiura, U. (1972). *Ann. Phytopath. Soc. Japan* **38**, 179. (in Japanese).
Heta, H. and Hiura, U. (1973). *Ann. Phytopath. Soc. Japan* **39**, 157. (in Japanese).
Hiura, U. (1960). *Ber. Ohara Inst. landw. Biol. Okayama Univ.* **11**, 235–300.
Hiura, U. (1962). *Phytopathology* **52**, 664–666.
Hiura, U. (1964). *Ber. Ohara Inst. landw. Biol. Okayama Univ.* **12**, 121–129.
Hiura, U. (1965). *Nogaku Kenkyu* **51**, 67–74. (in Japanese).
Hiura, U. (1973). *2nd. Inter. Cong. Pl. Path. Abst.* 0421.
Hiura, U. and Heta, H. (1969). *Ber. Ohara Inst. landw. Biol. Okayama Univ.* **14**, 203–209.
Hiura, U. and Heta, H. (1972). *Ann. Phytopath. Soc. Japan* **38**, 179. (in Japanese).
Hiura, U. and Heta, H. (1973). *Rep. Tottori Mycol. Inst.* **10**, 505–510. (in Japanese).
Hiura, U. and Heta, H. (1977). *Nogaku Kenkyu* **56**, 239–247. (in Japanese).
Hiura, U., Heta, H. and Tsushima, T. (1961). *Nogaku Kenkyu* **48**, 49–54. (in Japanese).
Johnson, T. (1949). *Can. J. Res.* **C27**, 45–65.
Johnson, T. (1968). *In* "The Fungi". (G. C. Ainsworth and A. Sussman Eds) Vol. 3, 543–556. Academic Press, New York and London.
Johnson, T., Newton, M. and Brown, A. M. (1932). *Sci. Agric.* **13**, 141–153.
Kimber, G. and Wolfe, M. S. (1966). *Nature* **212**, 318–319.
Lopez, A., Rajaram, S. and Bauer, L. I. de (1974). *Phytopathology* **64**, 266–267.
Mains, E. B. (1933). *Proc. Nat. Acad. Sci.* **19**, 49–53.
Marchal, E. (1902). *C. R. hebd. Séanc. Acad. Sci. Paris* **135**, 210–212.
Morishima, Hiroko (1969). *Sabrao Newsletter* **1**, 81–94.
Moseman, J. G. (1963). *Phytopathology* **53**, 1326–1330.

Moseman, J. G. (1964). "Barley Genetics", Vol. I, 215–221. Proc. 1st. Int. Barley Genet. Symp., Wangeningen 1963.
Moseman, J. G., Scharen, A. L. and Greeley, L. W. (1965). *Phytopathology* **55**, 92–96.
Mühle, E. und Frauenstein, K. (1962a). *Züchter* **32**, 324–327.
Mühle, E. und Frauenstein, K. (1962b). *Züchter* **32**, 345–352.
Mühle, E. und Frauenstein, K. (1963). *Züchter* **33**, 124–131.
Mühle, E. und Frauenstein, K. (1970). *Theoret. App. Genet.* **40**, 56–58.
Muller, C. H. (1952). *Evolution* **6**, 147–161.
Ouchi, S., Oku, H., Hibino, C. and Akiyama, I. (1974). *Phytopath. Z.* **79**, 24–34.
Person, C. and Sidhu, G. (1971). *IAEA-PL*-412/4, 31–38.
Powers, H. R., Jr. and Moseman, J. G. (1956). *Phytopathology* **46**, 23.
Sanghi, A. K. and Luig, N. H. (1971). *Can. J. Genet. Cytol.* **13**, 119–127.
Scheffer, R. P., Nelson, R. R. and Ullstrup, A. J. (1967). *Phytopathology* **57**, 1288–1291.
Tsuchiya, K. and Hirata, K. (1973). *Ann. Phytopath. Soc. Japan* **39**, 396–403.
Wolfe, M. S. (1965). *Trans. Br. mycol. Soc.* **48**, 315–326.

Chapter 6

The Recent History of the Evolution of
Barley Powdery Mildew in Europe

M. S. WOLFE [1] and E. SCHWARZBACH [2]

[1]*Plant Breeding Institute, Trumpington, Cambridge, England*
[2]*Lehrstuhl für Pflanzenbau und Pflanzenzüchtung Freising-Weihenstephan, FRG*

I. Introduction

The insignificance of cereal mildew during the nineteenth century was
summarized by Graham (1852):

. . . I only mention the occurrence (of mildew) for the sake of recording its existence, as it is possible, at some future time, we may hear more about it than may be desired.

Indeed, Sprengel (1847) reported only two diseases of importance on barley, namely rust (unspecified) and loose smut. From other reports, it seems that mildew on barley was generally unimportant in Europe, although there were numerous reports of wheat mildew. As late as 1896, Frank reported that, although mildew was often found on wheat and grasses in Germany, England and North America, barley was apparently free from the disease. By 1903, the situation had changed dramatically. Systematic field disease reports of the plant protection committee of the German Society of Agronomy (DLG), started in 1893, revealed in 1901 the first observation of severe mildew on winter barley, and in 1903, the first severe epidemic in which it was considered that the spring barley crop was being damaged by this disease. Since that time barley mildew has remained a constant problem in Europe.

Continued European studies of the interaction of cultivated barley with the pathogen have led to an improved understanding of the evolution of this major pathogen under agricultural conditions. The model of the interaction which is currently emerging has further helped in understanding data which have been accumulating over the past half century or so. Our attempt to interpret the recent history of barley mildew in Europe may therefore be clarified by first describing our present view of the way in which host and pathogen interact and then showing in detail how this model may be used in the interpretation of the facts which are currently available. It is hoped that the development of such studies will help to rationalize attempts to develop more effective and durable disease control in the future.

II. A Logical Analysis of the Dynamics of Pathogen Response to the Use of Resistant Hosts

A. The occurrence of pathogen variation

Because of the large size of the pathogen populations, mutations for virulence against any source of resistance might not be easily detectable, either because of their rarity, or because they have only a small effect. Indeed, with one exception, virulence against all sources of resistance used so far in breeding has been detected during exploitation of the resistance. The exception is virulence against the $ml\text{-}o$ gene (Jorgensen, 1971), but pathogen populations from areas where the $ml\text{-}o$ gene is common (Ethiopia) have not yet been examined, and no commercial variety carrying $ml\text{-}o$ has yet been released.

The absolute frequency of unselected virulence mutants in the field may be considerably greater than is commonly supposed. First, mutation rates of fungi, commonly of the order of 10^{-6} to 10^{-8}, have to be considered in relation to the massive spore production of a field epidemic. For example, Leijerstam (1972) calculated the rate of production of mildew mutants in moderately infected Swedish spring wheat fields to be as high as 1000 or 2000 $locus^{-1} ha^{-1} day^{-1}$. Second, if the mutations are not lethal, they will accumulate in the population to a frequency dependent on the rates of reproduction and back-mutation. If the reproductive rate is the same as that of the remaining pathogen population and the forward exceeds the backward mutation rate, the mutation will eventually predominate in the population. Since the growth rate of particular mutants may differ on different host varieties, the overall frequency of the mutants in a pathogen population will be a function of the structure of the current and past host population. Consequently, the initial response of the pathogen population to a newly introduced resistant variety will be influenced by the present and previous use of other varieties. Another source of variation may lie in mildew populations from wild barleys. For example, high frequencies of virulence for Algerian (CI 1179) and other barleys, resistant in Europe, have been reported from wild stands of *Hordeum spontaneum* in Israel (Koltin *et al.*, 1963; Fischbeck *et al.*, 1976). High frequencies of virulence for Algerian have also been found in pathogen populations elsewhere in the Mediterranean region (Caddel, 1976) even though this resistance did not occur in any European commercial variety. It can be seen from the table in the Appendix that occasional isolates with virulence for Algerian have been identified in Europe, but these are relatively rare.

B. Importance of airborne inoculum

It is generally accepted that primary infection of a barley crop in the spring is from airborne conidia. For winter barley crops, ascospore sources may be of considerably greater significance: there is no information available on the relative importance of the two sources. An unknown proportion of the spores produced in a source is carried into the lower layers of the atmosphere and becomes rapidly diluted: mathematical analyses of these processes have been developed elsewhere (see Kranz, 1974; Jenkyn and Bainbridge, in this volume). The direct effect of this airborne release is most evident from the increased disease levels in spring barley crops adjacent to winter barley sources (Pape and Rademacher, 1934). The effect has been recognized for distances up to approximately 200 m from the source (Benada, 1962), although there may be considerable local variability (Yarham and Pye, 1969). At a greater distance from the source, the spore concentration eventually becomes too low for detection by conventional spore-trapping techniques. For example, the spore concentration above newly emerged spring crops is often less than one spore per cubic metre: assuming a constant concentration of one spore per

10 cubic metres in the air, a standard spore trap would have to run for approximately 20 h to catch a single spore. If it is assumed that the spore-trapping efficiency of a crop is equal only to that of a flat horizontal surface, and that the terminal velocity of spores is 1 cm sec^{-1}, then one square metre of crop surface would catch a single spore from the same constant spore concentration in only 17 min (for details on trapping efficiency, see Gregory, 1973). This consideration partly explains the crucial lack of information on the point at which the spore concentration in the atmosphere reaches a level significant for the development of an epidemic.

Because of the effect of dilution of the spore cloud, infection in a particular field will develop from foci and these will tend to vary both in the degree of virulence for the particular variety in the field and for the frequency of other virulences. Infection foci are usually difficult to find, partly because there may be many of them and partly because they rapidly join up through local spore dispersals. Nevertheless the effects of focal infection may persist for a long period, as illustrated in the survey in Table 1.

TABLE 1

Relative numbers of colonies produced on test leaves of Hassan, Lofa Abed, Midas and Tern, by infections from 24 points on a grid in a field of Maris Otter. The grid points were about 50 m apart

13	3[a]	4	2	13	4	3	2
2	4	0	0	2	5	4	1
0	0	2	3	5	9	6	3
2	0	0	1	0	6	1	3
5	1	8	1	3	16	10	2
2	6	3	12	5	6	2	—
5	11	12	4	23	8	5	4
4	8	2	2	0	7	1	2
13	11	15	17	21	9	23	0
4	4	9	3	1	19	7	11
10	17	11	5	9	14	7	15
9	4	6	8	8	15	5	6

[a] Layout at each point: Hassan   Lofa Abed
                             Midas   Tern

In this survey of an apparently uniform heavy infection in a crop of cv. Maris Otter, analyses of the pathogen population structure were made at 24 points on a grid, separated by about 50 m in each direction. Isolates from each point were assessed for the ability to produce colonies on the test varieties Hassan, Lofa Abed, Midas and Tern. From Table 1 it can be seen that there was considerable variation in virulence frequency at the different points, with a tendency towards higher frequencies in the more exposed area of the grid. The survey was made at the late tillering stage after several selection cycles on the Maris Otter host. This suggest that heterogeneity in the original inoculum source may have been greater than that indicated by the survey.

Because of the effects of mixing of spore clouds in the atmosphere, differences in frequencies of the common virulences will tend to be relatively consistent in the primary inoculum over large areas, distant from local winter barley sources. For example, seedlings of a variety susceptible to the commonly occurring pathogen virulences were exposed at eight locations more than 100 km distant from each other for one week in the spring of 1976 in south Germany. Mildew bulks isolated from the seedlings were not significantly different in composition and revealed a high frequency of virulence for varieties with *Mlg* and *Mla6* resistance, a low frequency of virulence for varieties with *Mla4/7** (5%) and only traces of virulence for varieties with *Mlas* resistance (Schwarzbach, unpublished. For the key to the gene symbols, please see Table 2a).

C. Selective forces acting on the pathogen

Following the introduction of a variety with a new resistance, it is possible that there will be some infection from a range of appropriate pathogen genotypes, even though it may be difficult to observe. Spores produced from such infections will contribute to the total pathogen population, so that their frequency will increase with time until an equilibrium is reached at which the rate of accumulation is equalled by the rate of selection against those genotypes (e.g., due to a lower reproduction rate). If the newly selected pathogen genotypes reproduce at least at the same rate on the original varieties as does the rest of the pathogen population, then they will finally predominate, even if the proportion of the new variety remains small.

For spores from distant sources that land on crop plants, there is a simple linear relationship between the probability of a spore hitting a susceptible host and the proportion of the crop area occupied by that host. Thus, if the host population contains varieties with different resistances, spores of pathogen genotypes with complex virulence corresponding to the range of resistant varieties, will have a high probability of survival. In other words, there is selection for flexibility in the spore population landing on newly emerged crops at the beginning of the season. As the epidemic becomes established, new infections originate increasingly from plants within the crop so that the contribution to the epidemic of spores from distant sources quickly becomes

---

* This is an approximate symbol since there is confusion in the literature over the inheritance of resistance derived from the Lyallpur 3645 source.

negligible; the main impact of selection for flexibility in pure stands is thus limited to a few cycles during the establishment of epidemics.

When the proportion of the crop area occupied by a resistant variety increases, the magnitude of the selective advantage of flexibility becomes high. For example, if a resistant variety is grown on 60% of the crop area, then it would require only five selection cycles for a pathogen genotype, able to grow on all hosts in the region, to increase from an initial frequency of 1%, up to 50% of the pathogen population. This assumes that all pathogen genotypes are able to grow equally well on their corresponding hosts. This change in frequency is derived from a simple mathematical function, i.e.

$$p_n = \frac{p_0}{p_0 + q_0 f^n},$$

where p_n is the final frequency and p_0, q_0, the initial frequencies respectively of the flexible and non-flexible pathogen genotypes: f represents the proportion of the crop area susceptible to the original pathogen population and n is the number of pathogen generations under selection.

As the epidemic develops within the crop, infection from distant sources and thus flexibility becomes unimportant. However, amongst the pathogen genotypes within the crop epidemic, there will be variation in the rate of reproduction and selection will favour those genotypes with the highest. Because there is a large number of selection cycles, ten or more within the season, then small differences in reproduction rate may lead to considerable changes in the pathogen population. For example, a pathogen genotype with a rate of reproduction 20% higher than that of the remainder of the population, will increase in frequency from an initial value of 20% to 62% in ten mildew generations. For this calculation, the above formula can be used by substituting relative reproduction for flexibility. Summarizing, it is apparent that the selection which acts on spores in the aerial pool differs in kind and timing from that which acts on spores retained within the crop stand. Selection from the aerial pool is for complex virulence and is most important in the early stages of the epidemic. Selection within the crop stand is for increased virulence and proceeds throughout the growing season. Consequently, variation in complexity of virulence within the crop may be considerably reduced by the end of the growing season. A more comprehensive analysis of the selection response of the pathogen will be published elsewhere (Barrett et al., in preparation).

Since an arbitrary and limited set of test varieties cannot be used to detect the real number of genes for virulence in a pathogen genotype, it is not logical to expect any relationship between the number of detectable genes for virulence and the reproductive ability of the pathogen on a particular host. For example, if pathogen genotypes selected for the most complex virulence are compared with those selected for the highest level of reproduction on a particular host, then a negative correlation can be expected for purely statistical reasons. Such a correlation may be spurious, as, for example, that described by Van der Plank (1968).

Although a specific gene or genes may be crucial to enable a pathogen to reproduce on a particular host, many other genes may influence the degree of reproduction on that host (see also Clarke, 1976). Selection for maximum reproduction may thus favour a particular combination of many genes in the pathogen, which may be unique with respect to that host. The particular combination of pathogen genes favoured on one host may affect in different ways the ability of the pathogen to grow on other hosts. For example, we have consistently found an unexpectedly low frequency of virulence for the $Mla4/7$ resistance in pathogen populations on $Mlas$ varieties and vice versa. There are several possible explanations, but it seems most likely that the same virulence gene can simultaneously be essential for the pathogen to grow on one variety, but reduce its ability to grow on the other. Other gene combinations may, of course, have positive interactions.

If new resistances are introduced into the host population, selection in the pathogen can, of course, proceed simultaneously for increased flexibility and for increased ability to grow on each host especially in multiline cultivars or mixed crops. Because of the complexity of the pathogen population, initial progress may be rapid. For example, when three different resistances became widely used in Europe (Mlg, Mlh and $Mla6$), pathogen genotypes with a high level of virulence for all three resistances soon predominated. If the host population becomes more complex, however, the rate of pathogen adaptation slows down, essentially for two reasons.

First, the addition of a virulence gene to a pathogen genotype, represents only a small gain in flexibility if there are many virulence genes already present. This results in less selection for flexibility in complex than in simpler host systems. Second, the frequency of pathogen genotypes which possess the maximum number of virulence genes is an inverse exponential function of the complexity of the system. Thus, more selection cycles are required in a complex than in a simple system to raise the frequency of the most flexible pathogen genotypes to an appreciable level. Currently in Europe, there are seven resistances and two fungicides which have been used on a wide scale for several years: pathogen races able to overcome all or most of these factors are yet unknown.

D. Classification of genetic variability in the pathogen

Following the early description of *Erysiphe graminis* DC. (De Lamarck and De Candolle, 1815), Marchal (1902) discovered genetic diversity within the species and described seven formae speciales, including f. *sp hordei*. Further genetic diversity, within the formae specialis, was noted by Mains and Dietz (1930) who proposed sub-categories of "physiologic forms" based on their interactions with an arbitrary set of host varieties. Honecker (1934, 1936) confirmed the existence of physiologic forms or races in Europe but found that European cultures were not so clearly differentiated by the original test varieties; he therefore recommended a modified set of differential hosts. Nover (1957) was able to reveal further genetic diversity within the earlier defined physiologic races, so that numerical sub-divisions were given to Honecker's original alphabetic classification. Newly evident variation (e.g. Nover, 1968; Mayer, 1974) necessitated further sub-division of the accepted physiologic races (Nover *et al.*, 1968) into "sub-races". For convenience of reference, the current status of the physiologic race classification is given in the appendix. The ten standard varieties should theoretically differentiate a maximum of $2^{10} = 1024$ physiologic races, using a resistant versus susceptible classification: approximately 10% of this total has been identified.

Since populations of the mildew pathogen are large, it is likely that for any arbitrary set of host varieties, all combinations of the corresponding virulences will occur but at frequencies which may or may not be detectable by standard sampling techniques. Consequently, there is no justification for the erection of a hierarchical system of classification. Obviously, however, a practical classification is necessary for the use of breeders and farmers, strictly relevant to the resistance of current and potential commercial varieties. A system of non-hierarchical classification based on the study of frequencies of individual virulences was proposed by Wolfe and Schwarzbach (1975); the practical application of the system was further discussed by Wolfe *et al.* (1975).

III. Early History of Breeding for Mildew Resistance

Parallel with the apparently abrupt increase in disease severity at the beginning of the century were two major developments in barley cultivation. First, there was general recognition of the value of mineral fertilizer treatment for crop production. This stimulated the development of comparative variety trials which in turn led to the second major development, the replacement of heterogeneous land varieties by homogeneous higher yielding selections. For example, from a variety report for north-east Germany, it appears that by 1900 most of the barley area was already covered by homogeneous varieties such as Chevalier, Golden Melon, Hanna and Imperial (Gisevius, 1901). It thus seems that the increased severity of barley mildew was a product of improved agriculture.

The period from the turn of the century to about 1930 was characterized by efforts to improve the general health of breeding lines through continuous visual selection. As a consequence, Hülsenberg (1930) reported a considerable difference in the degree of mildew attack between susceptible varieties in field trials. Outstanding at that time was cv. Isaria, a variety produced from a cross between lines selected from the same land variety (Bavaria × Danubia), which had consistently low levels of disease. Most commercial varieties were susceptible and it was evident in the severe mildew epidemic in Germany in 1929 that there was a need for improved levels of resistance. In a field test of 51 commercial varieties in that year (Honecker, 1931) only one was completely free from mildew and four others relatively resistant; the remainder were highly susceptible. The mildew-free variety, Pflugs Intensiv, was a selection out of a European land variety which subsequently formed a major source of mildew resistance for commercial barley varieties in Europe.

To try to identify further sources of mildew resistance, Honecker (1931) tested several hundred barley entries from different parts of the world for their mildew response. Although a wide range of resistance was found, it was concluded that the exotic sources such as *H. spontaneum*, Nigrate, (CI 2444) and others, could not be usefully exploited in breeding since they carried many adverse agronomic characters in addition to the resistance (Honecker, 1943). From the winter barleys, the only source considered useful was a line isolated from a Balkan land variety, Ragusa B, which provided the resistance for many later European winter varieties such as Dea and Senta. Among the spring entries, Honecker considered that Pflugs Intensiv provided the only agronomically useful source of resistance. The whole of the resistance breeding programme thus became based on only two selections from European land varieties.

The possibility of incorporating exotic sources of resistance into cultivated varieties had been recognized much earlier. In 1907, Biffen reported that a mildew resistant line of *H. spontaneum nigrum* could be crossed with cultivated barley; this gave a Mendelian segregation between resistant and susceptible plants indicating that the resistance was controlled by a single recessive gene. Fruwirth (1910) considered this as a possible source for breeding for mildew resistance. However, neither this material nor crosses made by Honecker from exotic sources were used successfully in breeding work. In 1928, a similar attempt with an Ethiopian line, later known as *H. distichon* var. *laevigatum*, also failed (Dros, 1957).

Exotic resistance sources were exploited commercially, but considerably later. A line of *H. spontaneum nigrum* from southern Russia and kept in the Müncheberg collection as line H. 204, was found to be resistant (Honecker, 1936). After an initial failure (Honecker, 1943) the resistance was introduced into breeding programmes in Germany, following a series of backcrosses (Hoffmann and Kuckuck, 1938; Rudorf and Wienhues, 1951). Later it was introduced into England and the Netherlands. Similarly, successful breeding with *H. dist. laevigatum* was not begun until 1947–49 in the Netherlands, resulting in the varieties Minerva and Vada which were included on the Dutch variety list respectively in 1955 and 1956 (Dros, 1957). The so-called "Lyallpur" resistance source, which is incorporated into a number of modern European varieties, was originally obtained more than 40 years ago, in the German Hindu Kush expedition of 1935 (Freisleben, 1940). An important derivative from this material was line 6831 (Hoffman and Nover, 1959) which combined the resistance of Lyallpur 3645 with that of Heines Pirol (Pflugs Intensiv resistance) and together with Heines 4808, became the parent of many of the modern commercial varieties. More recently, three other sources of resistance have been exploited commercially, namely, "Arabische", Monte Cristo (CI 1017) and Rupee. The sources of resistance and some examples of varieties carrying the resistances are listed in Table 2a; combinations of the resistances which have been developed are shown in Table 2b.

The introduction of exotic sources of mildew resistance into commercial barley varieties in Europe can be summarized by three characteristics. First, many years have usually elapsed between the first recognition of the value of a particular source of resistance and its commercial use, usually because it was difficult to select acceptable host genotypes from the first attempts at hybridization with adapted varieties. Second, it has been observed generally that the derivative lines had less complex resistance than was available in the original resistant parent because of loss of integrity of the donor genome. Third, because of the difficulties of exploitation of the resistances, the small group of resistant lines which was developed was used simultaneously in different breeding programmes throughout Europe.

IV. The Range of Host Resistance and the Pathogen Response

In the nineteenth century it seems likely that the population of the mildew pathogen was to some extent buffered by variability in the population of land varieties. It is impossible to determine the extent of that variation in host resistance but it is evident that numerous genes must have been involved. For example, at least two resistance genes ($Mla8$, Mlh; Moseman and Jorgensen, 1971, and Wiberg, 1974) have been identified in resistant selections from the Hanna region of Moravia. Several selections with considerable levels of mildew resistance have been derived from Bavarian land varieties, e.g. "Isaria" (Honecker, 1942). From tests with non-adapted pathogen populations, Nishikado et al. (1951) were able to detect the occurrence of resistance genes in varieties such as Chevalier, Golden Melon, Moravia, Binder and in barley from Russia. Specific resistance was also evident in the old varieties White Pearl and Zeiners Immune (Nover and Mansfeld, 1956) and Hoffmann and Kuckuck (1938) reported lines with high levels of resistance from Spain

TABLE 2a

Major resistance sources, gene symbols and examples of derivative varieties in
breeding for resistance to barley mildew in Europe since *c.* 1930

Source	Gene symbol[a]	Variety examples
Pflugs Intensiv	*MLg*	Weih.MRII; Haisa II; Union; Johanna, Zephyr; Julia
Ragusa B	*Mlh*	Senta; Malta; Astrix
H. spontaneum nigrum	*Mla6*	Vold.8141; Allasch; Maris Concord; Midas
H. distichon laevigatum	*Mlv*	Minerva; Vada; Lofa Abed
Arabische	*Mlas*	Emir; Sultan; Hassan
Lyallpur 3645	*Mla4/7*	Amsel; Tern; Wing
Monte Cristo	*Mla9*	Mona
Rupee	*Mlar*	Rupal

[a] The gene symbols used approximate to the actual genes; see Wiberg (1974) for more
detailed information. "a" in the symbol indicates a gene at or near the *Mla* locus on
chromosome 5.

TABLE 2b

Existing and possible examples of combinations of the different mildew resistance
sources used in European barley breeding

	Mlar	*Mla9*	*Mla4/7*	*Mlas*	*Mlv*	*Mla6*	*Mlh*
Mlg	—	Akka	Mazurka Ortolan	Aramir Maris Mink	Sundance Universe	Impala Inis	Astrix Malta
Mlh	—	—	—	—	—	—	—
Mla6	X[a]	X	X	X	M. Canon M. Jupiter		
Mlv	—	—	Belfor	—			
Mlas	X	X	X				
Mla4/7	X	X					
Mla9	X						

[a]X = combination unlikely because of allelism or close linkage. Thus, of 18 likely
combinations, 8 have so far been produced as commercial varieties.

and Switzerland. We have also found that modern susceptible varieties, such
as Diamant and Golden Promise, which have no known resistant parents
exhibit considerable resistance when tested with mildew cultures of Mediterranean origin.

With the exception of Pflugs Intensiv and Ragusa B, the effects of most of
these resistances are of low magnitude against European mildew populations
but in all cases so far tested they show differential interactions with the pathogen. The size of such quantitative genetic effects may be judged from a field

trial involving the varieties Pirkka and Union. Although both varieties are currently regarded as susceptible, we found from measurement of spore production that Pirkka produced about ten times more mildew spores per unit area of leaf than did Union. The specificity of the quantitative effects can be observed from comparisons of the performance of mildew isolates collected in England in the early 1970s on the varieties Julia and Zephyr, both of which possess the qualitative resistance, *Mlg*. The isolates could be grouped into those that produced the same number of colonies on leaves of both varieties, those that produced significantly more on Zephyr and those that produced significantly more on Julia (own data). Although this result was produced with seedling leaves of the two varieties, the effects of quantitative resistance in modern varieties is generally most evident at the adult plant stage. Indeed, mildew infection of the awns and ears of European varieties is rarely if every observed, even in so-called susceptible varieties: this contrasts with observations made on some exotic barleys (Chaudhary *et al.*, 1976).

V. History of the Individual Major Resistance Sources

A. Pflugs Intensiv: *Mlg*

Pflugs Intensiv was first recognized officially as an original variety in Germany in 1921. The area of cultivation is unknown but the variety did not become widely grown being only one among more than 50 officially recognized commercial varieties; because it was free of mildew in the field and was agronomically acceptable, it was introduced into breeding programmes. At the time pathogen genotypes with virulence for the variety must have been generally uncommon. However, Honecker (1934) reported a race (B), virulent on this source (*Mlg*), in 1931, and then another (C) from Bavaria in 1934 (Honecker, 1935) with combined virulence for *Mlg* and *Mlh*. Although uncommon, race B was found to be widespread; race C was more restricted. The variety CP 127422 and other lines derived from crosses with Pflugs Intensiv were widely used in breeding programmes but varieties with the *Mlg* resistance did not increase greatly in popularity until the end of World War II: by 1948, its area in Germany had increased to more than 10% of the total and all of the *Mlg* varieties were classified as resistant. The subsequent increase in area up to about 70% in 1960 created an enormous selective advantage for pathogen genotypes with corresponding virulence, so that, by 1960, the *Mlg* varieties were fully susceptible. The progress curves for area of host cultivation, virulence frequency and infection levels in Germany are illustrated in Fig. 1. Figure 1 clearly shows the sequence of events. Following the increase in area

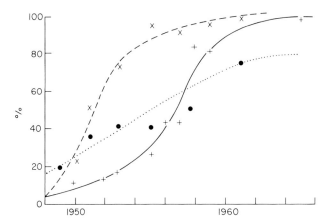

Fig. 1. Diagram showing the relative increase in frequency of *Mlg* virulence and infection of *Mlg* varieties consequent upon the increase in area of *Mlg* varieties, in Germany. .... = area, ——— = infection relative to Donaria and Isaria = 100, – – – – = frequency of *Mlg* virulence.

cultivated with the resistance source, creating a large selective advantage for virulent pathogen genotypes, the frequency of the virulence in the pathogen population rapidly increased, thus increasing the probability that airborne spores would land on a suitable target. Consequently, there was a rapid increase in the infection levels on the varieties involved.

The effect of this change on host yield was recorded by Lau (1962) who showed that varieties such as Pirol, Stanka and Frigga (Freja), all with *Mlg*, yielded approximately 12% more than susceptible controls (Haisa I, Donaria) in the period 1947–52. From 1953–58 however, the yield advantage was reduced to 4% because of the increase in the C races.

Spread of the C races into other parts of Europe tended to occur later, but they were reported from Yugoslavia around 1960 (Numic and Aganovic, 1961) and they were common in north-east Hungary in the early 1960s (Pollhammer, 1963) and in Poland in 1964–66 (Ralski and Mikolajewicz, 1969). In the latter areas, the frequency of the *Mlg* virulence had increased to more than 80% of all isolates tested at the time. In Russia, however, Union (*Mlg*) was still resistant in some areas during the 1960s (Starcenkova, 1969).

Haisa II was the first *Mlg* variety grown in the UK and during the early 1950s it was highly resistant to mildew (Eade, 1957); by the mid-1950s, however, Pirol (*Mlg*) was reported to be similar to the moderately susceptible variety Proctor in mildew reaction (Eade and Elliott, 1958), which indicates an increase in *Mlg* virulence. The rapid appearance of infection on this source in the UK compared with the slow initial increase in central Europe suggests that some of the early inoculum may have been windborne from the continent. The subsequent increase in frequency of the *Mlg* virulence was

142   M. S. WOLFE AND E. SCHWARZBACH

described by Wolfe (1968). From 1968 to the present time, the overall frequency of this virulence in the UK has always been more than 50%, whilst remaining close to 100% for the whole of this period in central Europe.

Amongst the many varieties produced throughout Europe with *Mlg* resistance there has been a wide range of susceptibility to virulent populations of the pathogen. The variation is presumably due to differences in the "background" genes of the varieties, against which, more or less virulence was available in the pathogen population at the time. As an example, during their introductory years in the UK, the varieties Zephyr, Julia and Deba Abed respectively, showed relatively high, moderate and low levels of the disease caused by the current populations of *Mlg* virulent pathogen strains. At the time of its introduction, the level of infection on Julia was about the same or a little less than the average infection on all other varieties: at the same time, Zephyr was considerably more susceptible than average. During the period 1970–72 however, there was a rapid increase in the area of Julia grown, thus producing a large selective advantage for pathogen genotypes with a high rate of reproduction on Julia. Consequently, there was a considerable increase in infection of Julia, above the mean of all other varieties, and indeed above that of Zephyr (Fig. 2).

As the area of Julia subsequently declined, so did the selective advantage for Julia-specific isolates, so that by 1976 the infection level had returned to a value close to that of the mean of all other varieties. This suggests that pathogen genotypes with maximum virulence on Julia did not gain sufficient flexibility during the period of selection to allow their persistence in the pathogen population when the selective host was no longer common. During that period, it was possible to differentiate isolates, described above, with or without specific virulence for either Julia or Zephyr.

Deba Abed was considerably more resistant in the field than either Julia or Zephyr and remained so during the period 1967–70 but it was not grown on a large scale. From Fig. 2 it can be seen that in 1973, however, after several years of cultivation and before its popularity had completely declined, the infection level was considerably increased. This is consistent with an increase in pathogen genotypes with specific virulence for the variety, but their occurrence in the pathogen population is not proven.

B. Ragusa B: *Mlh*

The history of the use of the winter barley resistance, *Mlh*, and the pathogen response to it is not well documented but appears similar to that of *Mlg*. The *Mlh* resistance was introduced widely into winter barley varieties after World War II and the consequent upsurge of C races in central Europe reflects the rapid increase of combined virulence in the pathogen population for both

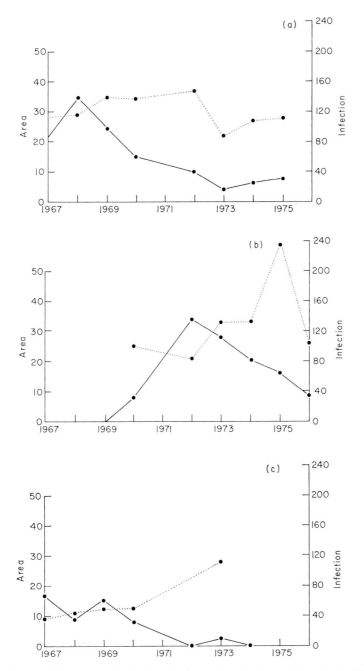

Fig. 2. Area and infection relative to all other spring barley varieties in the UK, of (a) Zephyr, (b) Julia and (c) Deba Abed, 1967–76 (calculated from ADAS cereal foliar disease surveys). ——— = area, ········ = infection.

resistance sources. Clearly, there was a selective advantage in the pathogen population for flexibility with respect to virulence for the major winter and spring varieties, allowing pathogen growth and reproduction throughout the year. Virulence for *Mlh* is now common throughout European mildew populations on all varieties, winter and spring. It is not clear to what extent this increase has been due to improved winter survival, or to the relatively high fitness of *Mlh* virulence on non-*Mlh* hosts.

C. *Hordeum spontaneum nigrum*: *Mla6*

The second source of resistance widely used in European spring barley breeding programmes was derived from H.204, a line of *H. spontaneum nigrum* first mentioned by Honecker (1936). The common pathogen cultures originally tested on H.204 gave no visible signs of infection but Honecker (1936) noted a mildew culture from the field which did produce a visible though slight reaction: this isolate was also virulent on *Mlg*. In 1937, Honecker isolated a single culture from a large pathogen collection which had a high level of virulence for both *Mla6* and *Mlg*. Because of the rarity of such isolates they were regarded as being of no practical importance and not representing a danger. Incorporation of this source of resistance into agronomically acceptable breeding lines was first reported by Rudorf and Wienhues (1951). However, before the first commercial release of *Mla6* varieties, Nover and Mansfeld (1956) further detected the presence of mildew genotypes with *Mlg* and *Mla6* virulence, both alone and in combination; indeed in the period 1953–57, the percentage of isolates virulent on H.204 increased from 6 to 14% (Hoffmann and Nover, 1959). The first commercial variety derived from this source was Akme, which had *Mlg* and *Mla6* combined and which was released in Germany in 1961. Development of the disease on these varieties in field trials was observed in 1964 and they became fully susceptible in about 1969. A different situation developed in the UK because the introduction of this resistance source into the breeding programmes led to the commercial release in the early 1960s of Maris Badger and Maris Concord, both of which possessed *Mla6* alone. Using the conventional techniques, it was not possible to identify isolates with combined virulence for *Mlg* and *Mla6* from infections on these or other varieties. However, it is likely that during the early 1960s, pathogen genotypes with the combined virulence did increase in frequency, simply because the pathogen populations on *Mlg* and *Mla6* varieties increased in size. The frequency of the combined virulence rose dramatically in the late 1960s following the rapid increase in the area occupied by Impala (*Mlg* + *Mla6*). This development was reported by Wolfe (1968) and analysed further by Wolfe and Barrett (1976). The widescale use of the variety with the combined resistance probably caused rapid selection for increased virulence of the pathogen genotypes, which already had the advantage of flexibility, which could explain why they persisted at a relatively high frequency after the rapid decline in area of the host varieties with the corresponding resistance. An important effect of this selection was that, when Impala became heavily infected, infection of the varieties carrying the two resistances separately also increased, because of the overall increase in frequency of the appropriate virulences.

D. Arabische: *Mlas*

Varieties with the "Arabische" source of resistance became generally available in Europe in the late 1960s from the Cebeco breeding programme. One of the varieties, Sultan, quickly became popular in the UK and by 1969 occupied almost a quarter of the spring barley area; in the following year it was heavily infected in most parts of the country. Pathogen cultures had been isolated from test plots of Sultan and Emir in 1967 but none of these had a high level of *Mlas* virulence. Isolates which did have a high degree of this virulence were first confirmed in 1968 when their frequency was about 29% of the pathogen population on Sultan. The subsequent increase was very rapid because of the varietal popularity and the large selective advantage for *Mlas* virulence.

Differences in the degree of virulence against Sultan and Emir were observed to be similar amongst isolates obtained in Germany in the later 1960s. However, the *Mlas* varieties have been grown on only a small scale in Germany until recently, occupying less than 1% of the barley area up to 1974. Following the current rapid increase in popularity of Aramir (*Mlg + Mlas*), it can be expected that there will be corresponding dramatic changes in the pathogen population in the near future. The occurrence of *Mlas* virulence in the pathogen populations in northern Europe has been considered to be due principally to conidiospore import in southerly winds from the UK (Hermansen *et al.*, 1974).

E. Lyallpur resistance: *Mla4/7*

Much of the European breeding material with *Mla4/7* resistance has been derived from a cross involving Lyallpur 3645 (*Mla4/7*) and Pirol (*Mlg*) with other varieties which gave rise to parent lines such as line 6831 and Ho 4595 (Hoffman and Nover, 1959). Consequently a number of the commercial varieties produced from this source have the *Mla4/7* resistance combined with *Mlg* (e.g. Mazurka), so that most of the *Mla4/7* virulent isolates of the pathogen which have been examined also possess *Mlg* virulence.

Virulence for *Mla4/7* was first observed in Europe generally in 1968–70 (Nover, 1968; Plate and Fischbeck, 1969; Slootmaker, 1970), several years after commercial varieties with this resistance were first grown in Europe. The increase in frequency of *Mla4/7* virulence and infection of the corresponding hosts followed the increase in area cultivated of the relevant varieties, in a pattern similar to that of the previous examples. Indeed, selection for *Mla4/7* virulence in the UK was closely analogous to that for *Mlas*. In 1970, only one in three of the isolates obtained from *Mla4/7* varieties had a high level of the corresponding virulence; by 1972 the figure was 82%, reaching 100% by 1974. Combinations of the two virulences, *Mla4/7* and *Mlas*, are still rare in Europe (see below).

F. *Hordeum laevigatum* resistance: *Mlv*

Unlike varieties with the previously described sources, which were more or less completely resistant at the time of their introduction, *Mlv* varieties, such as Vada, Minerva and Lofa Abed, have been only partially resistant showing low rates of sporulation with necrotic reactions. This characteristic and the erroneous impression that quantitative resistance is equivalent to "horizontal resistance", sensu Van der Plank (1968), led to the suggestion that this type of resistance would be more difficult for the pathogen to overcome than the others described. Different degrees of virulence against *Mlv* have been recorded among pathogen isolates (own data) but *Mlv* varieties have not been used extensively in Europe, so that the frequency of high virulence against them has remained low. There was evidence from the ADAS Foliar Disease Surveys (1968–76) that, when the area cultivated with *Mlv* varieties in the UK did increase appreciably, there was a concomitant though delayed increase in infection (Fig. 3) in the same way as with varieties carrying other sources of resistance.

More recently, the area occupied by Vada and Lofa Abed (*Mlv*) has decreased simultaneously with an increase in area of a range of newer resistant varieties. From isolates which we identified in the UK in 1976 the frequency of *Mlv* virulence on the newer varieties was low, indicating that it was not widespread in the source of the newly selected pathogen populations and that there was less selection for this virulence.

G. Monte Cristo resistance: *Mla9*

The mildew resistance of Monte Cristo barley was first recognized by Mains and Martini (1932). Prior to its exploitation, Monte Cristo had been highly resistant in extensive laboratory and field nursery tests in Europe and both

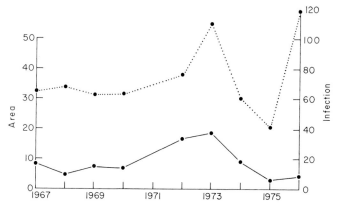

Fig. 3.   Area and infection relative to all other spring barley varieties in the UK, of *Mlv* varieties, i.e. Vada and Lofa Abed, 1967–76 (calculated from ADAS cereal foliar disease surveys) ——— = area, ······· = infection.

Americas although virulence against it had been reported in wild stands of *H. spontaneum* (Koltin *et al.*, 1964). The first commercial varieties with *Mla9* resistance were produced in Sweden, e.g. Akka and Mona but they have not yet been extensively grown in Europe. Subsequently, *Mla9* virulence increased rapidly following the initial cultivation of Akka (Wolfe and Wright, unpublished). All of the first found isolates had a combination of virulence for *Mla9* and *Mla4/7*; indeed, the only varieties other than those with *Mla9* from which *Mla9* virulence was isolated were those with *Mla4/7*. It appears that the majority of current varieties with these two resistances, have one factor in common, which is probably the gene identified alone in HOR 1063 (see Wolfe, 1972; Wiberg, 1974). It is therefore likely that the increasing infection of the more popular *Mla4/7* varieties provided a stepping-stone for the rapid emergence of *Mla9* virulent genotypes. Currently virulence for *Mla9* is widespread but infrequent in the UK and Scandinavia but not yet reported from Germany. Since varieties with *Mla9* are becoming popular in Germany, however, the corresponding virulence can be expected to increase rapidly in the near future.

H. Fungicides: ethirimol and tridemorph

In order to combat the continued high level of mildew on barley in Europe, there has been a considerable increase in recent years of the use of fungicides. In areas where these materials have been used extensively, there is evidence of the selection and increase of pathogen genotypes showing different degrees of reduced sensitivity to ethirimol (Wolfe and Dinoor, 1973) and to tridemorph

M. S. WOLFE AND E. SCHWARZBACH

(Walmsley-Woodward, unpublished data). From various sources of evidence, Wolfe and Barrett (to be published) concluded that there has been a slow but general decline in effectiveness of both fungicides, particularly ethirimol. Although the response of the pathogen population to fungicides and host resistance is essentially similar, the effects with fungicides have been less dramatic and more difficult to measure than those observed as a consequence of the cultivation of initially resistant varieties. It is proposed that this may be due mainly to the shorter persistence of the fungicidal action relative to that of host resistance. In other words, there is a shorter period in the growing season of the host during which pathogen genotypes insensitive to fungicides have a selective advantage. Such a consideration may also explain why pathogen response to ethirimol appears to be greater than to tridemorph, even though both fungicides have been used to a similar extent. With the seed-applied, more persistent ethirimol, there are more cycles of selection for fungicide-insensitivity, at a time when the pathogen population is small in size, than with the foliar-applied, less persistent tridemorph.

VI. Complexity of the Host Population and Selection for Flexibility

During the first commercial utilization of high levels of mildew resistance, only two resistance sources were used, *Mlg* and *Mlh*. In this simple system, the selective advantage of pathogen genotypes with virulence for both components was very high particularly after World War II, when the area occupied by the resistant varieties rapidly increased. Within a few years the pathogen population responded so well that pathogen genotypes with combined virulence became predominant. The subsequent addition of *Mla6* resistance did not greatly increase the complexity of the system on the continent because of the already high frequency of pathogen genotypes with combined virulence for *Mlg* and *Mlh*. Pathogen genotypes with the triple combination of virulence were rapidly selected and there were increasing reports of the occurrence of appropriate "C" races, e.g. C4, C5, C10 and C26 (see Appendix, and Wolfe, 1968).

In the UK, *Mlg* varieties remained common but there was a rapid decline in the popularity of *Mla6* varieties, thus reducing the selective advantage of the combined virulence. As a consequence, pathogen genotypes with combined *Mlg* + *Mla6* virulence persisted on varieties with *Mla6* only but their frequency declined, albeit slowly, on *Mlg* varieties (Table 3a).

Following the extensive use of *Mlas* varieties in the UK, a different situation developed (Table 3). Pathogen genotypes with combined virulence for *Mlg* and *Mlas* increased to only a limited extent on *Mlg* varieties and to a

TABLE 3

Relative frequencies of pairs of virulences on their corresponding hosts, in the UK, 1968–75, from the Physiologic Race Surveys (Wolfe and Wright, 1968–76 unpublished).

(a) from *Mlg* and *Mla6* vars.

Host	Path.	'68	'69	'70	'72	'72	'73	'74	'75
Mlg	*Mlg*-virul.	100	97	98	100	100	100	94	100
	Mla6-virul.	100	87	78	65	43	62	56	35
Mla6	*Mlg*-virul.	100	100	100	100	100	100	100	100
	Mla6-virul.	100	80	100	100	100	100	100	100

(b) from *Mlg* and *Mlas* vars.

Host	Path.	'68	'69	'70	'72	'72	'73	'74	'75
Mlg	*Mlg*-virul.	100	97	98	100	100	100	94	100
	Mlas-virul.	0	3	15	14	43	19	19	16
Mlas	*Mlg*-virul.	82	27	35	20	67	70	38	63
	Mlas-virul.	29	88	96	97	100	100	100	100

(c) from *Mlas* and *Mla4/7* vars.

Host	Path.	'68	'69	'70	'72	'72	'73	'74	'75
Mlas	*Mlas* virul.	29	88	96	97	100	100	100	100
	Mla4/7 virul.	0	0	0	0	0	0	0	2
Mla4/7	*Mlas* virul.	0	0	67	33	24	9	21	4
	Mla4/7 virul.	0	0	33	67	82	93	100	100

higher level on *Mlas* varieties. The high relative frequency of *Mlg* virulence on *Mlas* varieties in 1968 (Table 3b) was due to the occurrence of *Mlg* virulent genotypes with low levels of Sultan virulence in the small pathogen population on Sultan before it became severely infected by highly virulent pathogen genotypes. The relatively poor performance of the pathogen genotypes with combined virulence appears to be in contrast with their performance in continental Europe, where combined virulence for *Mlg* and *Mlas* is relatively common among the considerably smaller population of *Mlas* virulent genotypes. This presumably reflects the historical difference in the use of the two groups of resistant varieties: the more extensive use of *Mlg* on the continent probably ensured that *Mlas* virulent pathogen genotypes would increase only in *Mlg* virulent populations of the pathogen.

There is no doubt, however, that the frequency of this virulence combination is currently rising in many parts of Europe because of the increased cultivation of varieties such as Maris Mink and Aramir, which possess the combination *Mlg* + *Mlas* and which are therefore susceptible only to the corresponding combined virulence.

In contrast with the previous examples, extensive cultivation of varieties with *Mlas* or *Mla4/7*, particularly in the UK but also on the continent, has not led to a rapid increase in pathogen genotypes with combined virulence (Table 3). The few isolates with apparent combined virulence have often been found to differ considerably in the level of virulence for the two varieties, such that only a low level of infection was produced on one component. The low frequency of this particular virulence combination is particularly surprising in the UK where, in the last two years, half of the total spring barley area has been more or less equally divided between *Mlas* and *Mla4/7* varieties, ensuring a large selective advantage for the combined virulence. Clearly there has been a severe limitation of unknown cause on the production of pathogen genotypes with these particular virulences in common. A lower than expected frequency of combined virulence has been recognized in several other cases (Wolfe and Barrett, 1977; Wolfe, unpublished).

The limitations on the emergence of pathogen genotypes with particular virulence combinations led to the construction of a variety diversification scheme in the UK. This took the form of a set of recommendations for the choice of varieties to grow on the farm which would exploit the advantages of limited flexibility. For example, dividing the farm area between *Mlas* and *Mla4/7* varieties, rather than concentrating on one or other, would be more likely to restrict epidemic development. That there is now a possibility for such recommendations is a reflection of the present complexity of European variety lists. Schwarzbach and Wolfe (1976) illustrated the way in which the range of available resistant varieties has increased in Germany and the UK over the last twenty years, which, together with the increased use of fungicides, has multiplied the range of problems for the pathogen and thus delayed the rate of progress towards overall flexibility. Unfortunately, much of this potential advantage is likely to be lost in the near future because of the tendency in current breeding programmes to concentrate on varieties which have combinations of resistances (see Table 2b). Extensive use of such varieties will, of course, accelerate selection of virulent pathogen genotypes which possess a number of virulence factors. As can be seen from Table 2b, this development is limited only by the fact that a number of the resistances are conditioned by alleles or closely linked genes at the *Mla* locus, which thus limits the selection of resistance combinations.

VII. Conclusions

Prior to the beginning of conscious breeding for mildew resistance in Europe, there existed greater genetic variation both in the host and in the pathogen for mildew reaction. The effect of breeding for resistance and using fungicides, is to impose two major forces of selection on the pathogen. First, at the beginning of the season, the probability of a spore from the primary airborne inoculum landing on a variety on which it can reproduce, is dependent on the proportion of the crop area occupied by such varieties. Selection will thus favour mildew genotypes able to grow on the greatest range of the widely grown varieties. This selection towards flexibility is limited to the early part of the season.

Second, during the growing season within a stand of a pure variety or a fungicide-treated variety, there is selection towards increased virulence or insensitivity, respectively; mildew genotypes which reproduce more rapidly than others will increase in proportion, dependent on the differences in reproductive ability and on the number of pathogen generations during the season. The final composition of the mildew population is the result of selection for both flexibility and high virulence. The observed changes in barley mildew in Europe in recent years can be explained in these terms.

Changes in the pathogen population have been consistently related to the introduction and subsequent large increase in area of varieties with new resistance sources, the most obvious being the levels of infection of specific varieties. Such changes have so far appeared to depend more on the area of use of a resistance source or fungicide rather than on the kind of resistance or chemical.

Changes in the pathogen population have been expressed in the occurrence and spread of physiologic races of the pathogen. The changes can be more effectively monitored, however, by estimation of changes in virulence frequencies relevant to important varieties carrying particular resistances (Wolfe and Schwarzbach, 1975).

The interpretation of the dynamics of pathogen response to control measures is limited by lack of knowledge of essential parts of the process, for example, the density and spread of the primary airborne inoculum, the reproductive rate of the pathogen in the early stages of disease development, the importance of the cleistothecia in pathogen survival and in the composition of the airborne inoculum.

The effective control of a disease such as barley mildew in Europe depends on the availability of resistance sources or fungicides, and on the strategy of their use. During the early part of the century, breeding for general health of the host was common and relatively little attention was paid either to the sources of resistance or to the pathogen involved. Although the many varieties produced were often highly susceptible there were some which afforded a considerable level of disease protection, generally based on a quantitative rather than on a qualitative improvement in disease reaction. Subsequently, breeding for resistance concentrated on the high levels of resistance derived from few sources and exhibited in a relatively small number of varieties, which were then grown on a large area. It seems timely therefore, to consider the kinds of resistance which may be expected to be inherently more durable and the methods of using such resistance so as to extend their durability. It can be expected that in the future, new sources of resistance will cause further shifts in the pathogen population towards loss of their effectiveness. It seems necessary therefore to provide a steady supply of new resistance sources for the plant breeder. Examination of populations of wild barley (Fischbeck et al., 1976; Moseman and Smith 1976) and of existing collections suggests that there is a large pool of sources of resistance available. The major difficulty at present is the evaluation and incorporation of these sources.

Development of strategies for the best use of host resistance and fungicides has been considered elsewhere (Wolfe and Barrett, 1977), but consists essentially in exploiting the control measures in a way which is both agronomically acceptable and which reduces the capacity of the pathogen to respond to the changing circumstances. We need to set more complex problems for the pathogen than we have been able to do in the past.

Acknowledgements

We wish to thank Dr J. A. Barrett, Genetics Department, Cambridge University, for valuable discussions and helpful criticism, and Dr J. E. King, ADAS Plant Pathology Laboratory, Harpenden, for access to original data in the annual cereal foliar disease surveys.

References

ADAS Pl. Path. Lab., Surveys of Foliar Diseases of Cereals, 1968–1976.
Anon. Jahresbericht des Sonderausschusses für Pflanzenschutz der DLG (Ann. Reps, 1895–1903).
Benada, J. (1962). *Rostl. výroba* **8**, 1287–1304.
Biffen, R. H. (1907). *J. agric. Sci.* **2**, 109–128.
Caddel, J. L. (1976). *Pl. Dis. Reptr* **60**, 65–68.
Chaudhary, R. C., Schwarzbach, E. and Fischbeck, G. (1976). *In* "Barley Genetics". Vol. III 456–463. Proc. 3rd Int. Barley Genet. Symp., Garching 1975.
Clarke, B. (1976). *Symp. Br. Soc. Parasitol.* **14**, 87–103.
De Lamarck, J. B. M. C. and De Candolle, A. P. L. P. (1815). *Flore Francaise,* **6**.
Dros J. (1957). *Euphytica* **6**, 45–48.
Eade, A. J. (1957). *J. nat. Inst. agric. Bot.* **8**, 36–49.
Eade, A. J. and Elliott, C. S. (1958). *J. nat. Inst. agric. Bot.* **8**, 342–350.
Fischbeck, G., Schwarzbach, E., Sobel, Z. and Wahl, I. (1976). *In* "Barley Genetics". Vol. III 412–417. Proc. 3rd Int. Barley Genet. Symp., Garching 1975.
Frank, A. B. (1897). "Kampfbuch gegen die Schädlinge unserer Feldfrüchte". Paul Parey, Berlin.
Freisleben, R. (1940). *Kühn-Archiv* **54**, 295–368.

Fruwirth, C. (1910). "Die Züchtung der landwirtschaftlichen Kulturpflanzen". Paul Parey, Berlin.
Gisevius (1901). "Die Sortenfrage in den Nordost-Provinzen". Paul Parey, Berlin.
Graham, F. J. (1852). *Gdnrs. Chron.* **12**, 501.
Gregory, P. H. (1973). "Microbiology of the Atmosphere". 2nd ed. Leonard Hill, Aylesbury.
Hermansen, J. E., Torp, U. and Prahm, L. (1974). Yearbook, Royal Vet. and Agric. University, Copenhagen, Denmark, 17–30.
Hoffmann, W. and Kuckuck, H. (1938). *Z. Zücht.* **A22**, 271–302.
Hoffmann, W. and Nover, I. (1959). *Z. Pflanzenz.* **42**, 68–78.
Honecker, L. (1931). *Pfl Bau. Pfl Schutz Pfll Zücht.* **8**, 78–84; 89–106.
Honecker, L. (1934). *Z. Zücht.* **A19**, 577–603.
Honecker, L. (1935). *Züchter* **7**, 113–119.
Honecker, L. (1936). *Prakt. Bl. Pfl Bau Pfl Schutz* **13**, 309–320.
Honecker, L. (1937). *Phytopath. Z.* **10**, 197–222.
Honecker, L. (1942). *Z. PflZücht.* **24**, 429–506.
Honecker, L. (1943). *Z. PflZücht.* **25**, 209–234.
Hülsenberg, H. (1930). *Forsch. der Landw.* **5**, 233–235.
Jørgensen, J. H. (1971). In "Mutation breeding for disease resistance". IAEA, Vienna.
Klein, M., Meyer, H. and Wallstabe, E. (1976). *Arch. Phytopathol. Pfl Schutz, Berlin* **12**, 169–184.
Koltin, Y., Kenneth, R. and Wahl, I. (1963). In "Barley Genetics". Vol. I. 228–235. Proc. 1st Int. Barley Genet. Symp. Wageningen (1963).
Kranz, J. (1974). "Epidemics of Plant Diseases". (J. Kranz, Ed.) Springer Verlag, Berlin, Heidelberg, New York.
Lau, D. (1962). *Z. PflZücht.* **48**, 80–90.
Leijerstam, B. (1972). *Meddn. Stat. Växtskyddsanstalt* **15**, 231–248.
Mains, E. B. and Dietz, S. M. (1930). *Phytopathology* **20**, 229–239.
Mains, E. B. and Martini, M. L. (1932). *USDA Tech. Bull.* no. 295.
Marchal, E. (1902). *C. r. hebd. Séanc. Acad. Sci., Paris* **135**, 210–212.
Meyer, H. (1974). Symp. zur Schadderregerüberwachung in der industriemässigen Getreide produktion. M-Luther Universität, Halle-Wittenberg.
Moseman, J. G. and Jørgensen, J. H. (1971). *Crop. Sci.* **11**, 547–550.
Moseman, J. G. and Smith, R. T. (1976). In "Barley Genetics" Vol. III 421–425. Proc. 3rd Int. Barley Genet. Symp. Garching 1975.
Nishikado, Y. Takahashi, R. and Huira, U. (1951). *Ber. Ohara Inst. landw. Biol.* **9**, 411–423.
Nover, I. (1957). *Phytopath. Z.* **31**, 85–107.
Nover, I. (1968). *Phytopath. Z.* **62**, 199–201.
Nover, I. and Mansfield, R. (1956). *Kulturpflanze* **3**, 105–113.
Nover, I., Brückner, F., Wiberg, A. and Wolfe, M. S. (1968). *Z. PflKranth PflPath PflSchutz* **75**, 350–353.
Numić, R. and Aganović, Z. (1961). *Zast. Bilja* 1961 (63–64), 81–87.
Pape, H. and Rademacher, B. (1934). *Angew. Bot.* **16**, 225–250.
Plate, D. and Fischbeck, G. (1969). *Z. PflZücht.* **61**, 225–231.
Pollhammer, E. (1963). *Növénytermelés* **12**, 231–240.
Ralski, E. and Mikolajewicz, T. (1969). *Hodowla Rodl. Aklim. Nasienn.* **13**, 359–373.
Rudorf, W. and Wienhues, F. (1951). *Z. PflZücht.* **30**, 445–463.
Schwarzbach, E. and Wolfe, M. S. (1976). In "Barley Genetics". Vol. III 426–432. Proc. 3rd Int. Barley Genet. Symp. Garching 1975.
Slootmaker, L. A. J. (1970). *Neth. J. Pl. Path.* **76**, 63–69.
Sprengel, C. (1847). "Meine Erfahrungen in Gebiete der allgemeinen und speciellen Pflanzen-Cultur". Baumgärtners Buchhandlung, Leipzig.

Starcenkova, M. V. (1969). *Trudy prikl. Bot. Genet. Selek.* **39**, 221–232.
Van der Plank, J. E. (1968). "Disease Resistance in Plants", Academic Press, New York and London.
Wiberg, A. (1970). *Phytopath. Z.* **69**, 344–365.
Wiberg, A. (1974. *Hereditas* **78**, 1–40.
Wolfe, M. S. (1968). *Pl. Path.* **17**, 82–87.
Wolfe, M. S. (1972). *Rev. Pl. Path.* **51**, 507–522.
Wolfe, M. S. and Barrett, J. A. (1976). In "Barley Genetics". Vol. III 433–439. Proc. 3rd Int. Barley Genet. Symp., Garching 1975.
Wolfe, M. S. and Barrett, J. A. (1977). *Ann. N.Y. Acad. Sci.* **287**, 151–163.
Wolfe, M. S., and Dinoor, A. (1974). *Proc. 7th Br. Insectic. Fungic. Conf.* 11–19.
Wolfe, M. S., and Schwarzbach, E. (1975). *Phytopath. Z.* **82**, 297–307.
Wolfe, M. S., Barrett, J. A. Shattock, R. C., Shaw, D.S. and Whitbread, R. (1975). *Ann. appl. Biol.* **82**, 369–374.
Yarham, D. J. and Pye, D. (1969). *Proc. 5th Br. Insectic. Fungic. Conf.* 25–33.

Appendix Races of barley powdery mildew isolated in north western Europe and their reaction types on the differential varieties. s = susceptible; no sign = resistant; — = reaction unknown; 1–2 = reaction on 0–4 scale, where 0 is fully resistant. Compiled after Nover, I., Bruckner, F., Wiberg, A. and Wolfe, M. S., 1968; Wiberg, A., 1970; Wiberg, A. 1974 and Klein, M., Meyer, H. and Wallstabe, E., 1976.

		Weih. CP 127 422	Weih. 37/136	Weih. 41/145	Vold. 8141/44	Gat. Mut. 511	Gat. Mut. 501	Ant. HOR 1063	Indien HOR 1657	Balk. HOR 1036	Algerian CI 1179	Amsel*	Emir*
A	0												s
	1					s			s				
	2	s							s				s
	3			s		s			s				s
	4	s				s			s				
	5						s		—				
	6								s		s		
	7			s					s				s
	8	s			s	s	s		s				
	9				s	s			s				
	10								—		1–2		
	11	s				s			s	s			
	14	s			s	s	s	s		s	s		
	15	s							s	s			
	16	s			s			s					
	17	s			s		s	s					
	18										s		
	19						s						
	20			s	s	s			s				
	21	s											
	22	s							s		1–2		
	23								s				s
	24			s	s	s	s		s				
	25				s	s							
	26			s		s			s		s		
	27			s	s	s		s	s				
	28			s					s		s		
	29			s									s
	30	s		s			s		s				s
D	1	s	s						s				s
	3	s	s			s			s				s
	5	s	s										s
	6	s	s			s							s
	7	s	s			s	s	s					
	8	s	s	s			s	s					
	9	s	s	s	s	s	s		s	s			
	10	s	s	s	s	s			s	s			

		Weih. CP 127 422	Weih. 37/136	Weih. 41/145	Vold. 8141/44	Gat. Mut. 511	Gat. Mut. 501	Anat. HOR 1063	Indien HOR 1657	Balk. HOR 1036	Algerian CI 1179	Amsel*	Emir*
D	11		s	s		s			s		s		
	12		s	s						s			
	13		s	s		s		s	s				
	14		s	s	s		s						
	15		s	s	s	s	s		s				
	16		s	s		s		s					
	17		s	s		s			s				
	18		s	s					s				
	19		s	s				s	s				
	20		s	s	s				s				
	21		s	s	s	s	s						s
	22		s	s					s	s			
	23		s	s		s			s	s			
	24		s	s				s					
	25		s	s		s		s					
	26		s	s		s			s				
	27												
	28												
B	0	s											
	3	s		s					s				s
	4	s				s			s				
	6	s	s		s	s	s						
	7	s	s						s				
	8	s	s			s			s	s			
	9	s							s		s		
	10	s		s		s			s		1–2		
	11	s		s		s			s	s			
	12	s		s		s			s				s
	13	s		s									
	14	s	s						s	s			
	15	s	s			s			s				s
	16	s		s	s	s	s		s				s
	17	s	s										s
	18	s	s					s					
	19	s	s			s							
	20	s		s			s					s	s
C	2	s	s	s					s			s	s
	3	s	s	s		s		s				s	
	4	s	s	s	s		s						s
	5	s	s	s	s	s	s		s			s	s
	6	s	s	s		s							s
	7	s	s	s		s			s			s	s
	8	s	s	s				s		s			

		Weih. CP 127 422	Weih. 37/136	Weigh. 41/145	Vold. 8141/44	Gat. Mut. 511	Gat. Mut. 501	Anat. HOR 1063	Indien HOR 1657	Balk. HOR 1036	Algerian CI 1179	Amsel*	Emir*
C	9	s	s	s									
	10	s	s	s	s		s		s			s	s
	11	s	s	s			s		s				
	12	s	s	s		s		s	s			s	
	13	s	s	s				s				s	s
	14	s	s	s		s			s	s			s
	15	s	s	s		s			s	s			
	16	s	s	s					s		s		
	17	s	s	s				s	s			s	s
	21	s	s	s					s		1–2		
	22	s	s	s							1–2		
	23	s	s	s						s			
	24	s	s	s	s		s	s				s	
	25	s	s	s				s			1–2		
	26	s	s	s	s	s	s					s	s
	27	s	s	s		s			s		1–2		
	28	s	s	s			s						
	29	s	s	s	s	s	s	s	s			s	
	30	s	s	s		s	s						
K	1	1–2		s		s			s				s
	2	1–2		s					s		1–2		

Chapter 7

A Genetic Analysis of Host-Parasite Interactions

ALBERT H. ELLINGBOE

Genetics Program and Department of Botany and Plant Pathology Michigan State University, USA

I. Introduction

The powdery mildews have some very excellent characteristics for studies on interactions between host and parasite. Probably one of the most important is that it is possible, with several of the mildew diseases, to do a genetic analysis, in both host and parasite, of the traits one wants to study. Other characteristics which make the mildews favourable for studies of host–parasite interactions include:

1. distinct morphological stages of differentiation in primary infection;
2. high infection efficiencies;
3. reasonably synchronous development of a parasite population during early interactions;
4. the parasite is primarily on the surface of the leaf with only the haustoria in the host cell and, therefore, the parasite (except the haustorium) can be removed from the host (Ellingboe, 1972).

The obligate parasitism of the powdery mildews does not hinder studies of interactions between host and parasite but it is a serious handicap for studies of the parasite independent of the host.

One of the main points I want to discuss in this paper is whether the genes which control the naturally-occurring variability in host–parasite interactions have a positive function for compatible host–parasite relationships or a positive function for incompatible host–parasite relationships. This has been a much debated issue in the past few years. An answer to this question is not only crucial for interpretations of published research in both basic and applied studies but is also crucial for the design of experiments to elucidate the interactions between host and parasite. Experiments would be designed quite differently if it was assumed that a gene associated with virulence in *Erysiphe graminis* f. sp. *hordei* had a positive function rather than assuming that it was associated with virulence because it was unable to confer avirulence.

In this chapter I will:

1. discuss the discovery and inheritance of the variability in interactions between host and parasite, including the inheritance in both host and parasite (the mildews have proved excellent material for these studies);
2. describe the morphological development of a parasite during primary infection of its host;
3. describe the expression of different incompatible parasite/host genotypes on the process of infection and on transfer of radioactive elements ^{35}S and ^{32}P from host to parasite;
4. discuss the phenomenon which has been termed "induced susceptibility";
5. attempt to summarize and present some uniform concepts of interactions between host and parasite with mildews.

II. Identification of Disease Resistance

One of the most effective ways to control diseases of plants, particularly of field crops, has been the selection of resistance, and the ensuing breeding of this resistance into commercially accepted varieties. Identifying techniques have affected, accordingly, the types of resistance covered. A popular procedure has been to collect seed from host species from as many parts of the world as possible, emphasizing their diverse origins, in the hopes that these collections, which sometimes contain thousands of entries, have not had recent common parentage. The collections were then grown in the field in very small plots or in glasshouses. Plants were bombarded with inoculum— in glasshouses by direct inoculation of the entries—and in the field either by direct inoculation or by infesting spreader rows. Resistance was expressed as the ability to survive high levels of inoculum in environments favourable for disease. Such procedures are useful to detect the presence of genes giving high levels of resistance. They do, however, have limited value in detecting genes whose effects are so small that they are detected primarily after several generations of infections or from observing the rate at which an epidemic develops in a commercial field. The techniques used to screen large numbers of host lines have, therefore, limited the types of resistance used in breeding for disease resistance. Plant breeders and pathologists have emphasized selection for immunity—or at least for high levels of resistance. Genes which give high levels of resistance are also easiest to handle in segregating generations and genes which give high levels of resistance, as contrasted to their alternate alleles, give discontinuous segregations in F_2 populations. If the differences in disease reaction of the two alternate alleles in a cross is small and the environmental variability is considerable, the F_2 may have continuous variation in disease reaction. The interpretation of such data may be that the variation is due to many gene differences between the parasite host lines. It is because of these considerations that more is known about the, inheritance of disease reaction expressed as infection type in plants, rather than genes whose effects are small. In this context it is interesting to note that the numbers of papers reporting qualitative, compared to quantitative, inheritance has increased (Person and Sidhu, 1971). A comparable argument can be made for the studies of genetic variability in pathogens. Small differences are generally not detected or dealt with. Only very rarely are data collected on the number of successful infections (i.e. number of pustules per unit leaf area or inoculation density), or the time necessary to develop a given infection type, or the relative number of viable spores produced in a pustule, etc. The reason for the usual approach is efficiency; thousands of host lines can be evaluated in a short period of time if data are collected on infection type only, at a single time after inoculation. Similarly it is possible to evaluate thousands of lines in the field in a short time if only immunity or high levels of resistance are selected under conditions of a man-created epidemic. The standard techniques allow for the easy detection of genes which confer infection type 0 or 1, more difficulty for detection of genes which confer infection type 2, great difficulty for detection of genes conferring infection type 3 and for all practical purposes do not allow the detection of genes which allow an infection type 4 but limit the number of successful infections, affect the generation time of the parasite and the degree of sporulation of the parasite, etc. There is no reason to suspect that the genes controlling very low infection type follow different inheritance patterns from genes controlling low, medium, medium-high infection types, or even genes which control infection efficiency, etc. (Ellingboe, 1975). Evidence is accumulating that essentially all variability follows the same patterns of inheritance of interactions (Ellingboe, 1975). When experiments are performed with more precise control of environment, more precise methods of production of uniform inoculum and quantification of inoculation techniques, we can see that the specificity of interactions appears to be the same as for parasite and host genes which give major effects. Evidence is also accumulating that the same host gene can have major effects with one strain of a parasite and very small effects with a second strain of the parasite (Martin and Ellingboe, 1976). With one strain of *Erysiphe graminis* f. sp. *tritici* an infection type 0 is obtained on wheat plants with *Pm4*. An infection type 4 is obtained with all cultures of *E. graminis* f. sp. *tritici* on plants with *pm4*. With a second culture the parasite has a reduction in growth rate, but still infection type 4 on plants with *Pm4* but normal growth on plants with *pm4*. Two additional cultures have reduced infection efficiencies on plants with *Pm4* but with infection type 4, and normal infection efficiencies on plants with *pm4*. Clearly the gene *Pm4* would never have been detected if the last three cultures had been used in the initial screening for mildew resistance in wheat. The usual methods used in the detection of genes affecting host–parasite interactions derermine which genes can be identified. If crude screening is performed, only loci at which the difference between alternate alleles is large will be detected. As the refinement in screening procedures is increased, our ability to detect smaller and smaller differences in the expression of alternate alleles is increased but the differences may still become so small that they are not detectable when the host is inundated with inoculum. Differences may be revealed, however, when inoculum is controlled, possibly over several generations of inoculations, so that small effects over each inoculation are magnified over several inoculation cycles. The idea that small effects are controlled by different gene systems is not substantiated. Recently evidence has been accumulating that essentially all variability follows the same pattern, the gene–for–gene relationship which is described in the following section.

The purpose of the above discussion is not to criticize past procedures but to understand how the inheritance, in both host and parasite, of the naturally-occurring variability in host–parasite interactions has been uncovered. The knowledge of the methods used to detect naturally-occurring genetic variability may give us a more accurate perspective of the detailed studies of inheritance of interactions and render us better able to interpret the results. We are also reminded that there is variability in host–parasite interactions, the inheritance of which has not been determined. The intent is not to imply that the inheritance of small differences will follow a different genetic pattern controlling interactions but rather to give proper perspective.

III. Inheritance of Genetic Variability Affecting Host-Parasite Interactions

The most comprehensive genetic analysis of host–parasite interactions where the inheritance was determined on both host and parasite was with flax rust disease (Flor, 1946, 1947, 1955). A comparable, though not as extensive, analysis has been performed for powdery mildew of barley (Moseman, 1959). The reactions of many cultivars of barley to many isolates of *E. graminis* f. sp. *hordei* gave evidence that the cultivars Goldfoil and Kwan had different genes for reaction to *E. graminis* f. sp. *hordei*. Inheritance studies showed that each variety had one gene for resistanse to culture CR3, and that the genes were at different loci and not alleles (Briggs and Stanford, 1938). The evidence for this conclusion was the appearance of susceptible progeny in a cross of Goldfoil × Kwan.

The cultivar Manchuria is susceptible to culture CR3 (Moseman, 1959). A cross of Goldfoil × Manchuria and inoculation of F_3 families with culture CR3 or culture 21·1 revealed segregation ratios of 1 homozygous resistant:2 segregating:1 homozygous susceptible. Goldfoil differs from Manchuria by one gene as detected by culture CR3 or culture 21·1. The F_3 families gave identical reactions to both culture CR3 and 21·1. A cross of Kwan × Manchuria also gave F_3 family segregation ratios of 1 homozygous resistant:2 segregating: 1 homozygous susceptible following inoculation with culture CR3 or culture 21·1. There was an absolute association of reactions to the two cultures. Therefore, both were able to detect only one gene segregation in each of the two crosses. The genotype of Goldfoil was designated *MlgMlg mlkmlk*; the genotype of Kwan was designated *mlgmlg MlkMlk*.

With a third culture of *E. graminis* f. sp. *hordei*, CAN12, high infection type was obtained on Manchuria, Goldfoil and Kwan (Moseman, 1959). The reactions of the three host lines to the three cultures of *E. graminis* f. sp. *hordei* is given in Table 1. Unfortunately, a host line of genotype *MlgMlg MlkMlk* was not tested.

Moseman then crossed culture 21·1 with CAN12. Of 49 progeny isolated, all gave a compatible reaction with Manchuria, 23 gave an incompatible reaction and 26 gave a compatible reaction with Goldfoil; 21 gave an incompatible reaction and 28 gave a compatible reaction with Kwan (Table 2). I have given the designation *P* to the segregating genes in the parasite. There were four kinds of progeny in the cross of cultures 21·1 and CAN12, each occurring with approximately equal frequency (see Table 2). A 1:1:1:1 segregation ratio in a haploid organism such as *E. graminis* f. sp. *hordei* suggests that two unlinked loci are involved. Manchuria is not a differential for the two parent cultures or the progeny of the cross 21·1 × CAN12. If the ability to give a compatible relationship was a positive function, one might

164 ALBERT H. ELLINGBOE

TABLE 1

Infection type[a] obtained on three barley cultivars with three cultivars of *E. graminis* f. sp. *hordei*

Host		parasite culture		
Cultivar	proposed genotype	CR3	21·1	CAN12
Manchuria	*mlgmlg mlkmlk*	H	H	H
Goldfoil	*MlgMlg mlkmlk*	L	L	H
Kwan	*mlgmlg MlkMlk*	L	L	H

[a] L = low infection type
H = high infection type
classification is based on Loegering (1966).

TABLE 2

The reaction[a] of four host lines upon inoculation with cultures 21.1 and CAN12 and 49 progeny of a cross 21.1 × CAN12

		host genotype		
		(Goldfoil)	(Kwan)	(Manchuria)
parasite genotype		*MlgMlg*	*mlgmlg*	*mlgmlg*
parent cultures		*mlkmlk*	*MlkMlk*	*mlkmlk*
(21.1) *PgPk*		—	—	+
(CAN12) *pgpk*		+	+	+
progeny	numbers			
1. *PgPk*	8	—	—	+
2. *Pgpk*	15	—	+	+
3. *pgPk*	13	+	—	+
4. *pgpk*	39	+	+	+
		(−):(+)	(−):(+)	(−):(+)
		23:26	21:28	0:49

[a] − = incompatible relationship
+ = compatible relationship

expect that some progeny would give an incompatible relationship with Manchuria. However, only genes affecting interactions with Goldfoil and Kwan were observed to segregate. Segregation in the parasite could be observed only in the presence of an *Ml* gene.

Let us begin the analysis of these data with the simplest types of comparisons. Culture 21·1 allows us to distinguish Goldfoil from Manchuria. Culture CAN12 does not allow for a distinction between Goldfoil and Manchuria. Goldfoil, but not Manchuria, allows us to distinguish between cultures 21·1 and CAN12 (see Table 3). If this analysis is correct, then crosses

TABLE 3

The reaction[a] of Goldfoil and Manchuria following inoculation with cultures 21.1 or CAN12 of *E. graminis* f. sp. *hordei*

	Host	
parasite culture	Goldfoil (*MlgMlg*)	Manchuria (*mlgmlg*)
culture 21·1 (*Pg*)	—	+
CAN12 (*pg*)	+	+

[a] — = incompatible relationship
+ = compatible relationship

between Goldfoil and Manchuria should show one gene segregation with culture 21·1 and crosses between 21·1 and CAN12 should show one gene segregation with Goldfoil. Segregation ratios consistent with this interpretation were found in the original crosses (see Tables 1 and 2). If the interpretation is correct, then one should be able to predict the results of test crosses with the progeny. Moseman (1959) did cross representatives of each of the four types of progeny with each of the two parent cultures. All crosses gave results consistent with the interpretation given in Table 2, i.e. 21·1 × progeny type 1 showed no segregation on any of the three hosts, 21·1 × progeny type 2 showed segregation only on Kwan, 21·1 × progeny type 3 showed segregation on Goldfoil, and 21·1 × progeny type 4 showed segregation on Goldfoil and Kwan but not Manchuria.

The simplest interpretation of the pattern in Table 3 is that the specific recognition is parasite/host genotype *Pg/MlgMlg*. To say that specific recognition is necessary for compatible relationships between host and parasite, one must assume that three different parasite/host genotypes, namely, *Pg/mlgmlg*, *pg/MlgMlg* and *pg/mlgmlg* all have the same phenotype. Only one cell in the quadratic check presented in Table 3 is unique. The simplest explanation of this pattern, especially when one considers possible molecular mechanisms, is that the product of *Pg* interacts specifically with the product of *Mlg* to give an incompatible relationship. The other three cells represent non-recognition.

Many of my colleagues have suggested that the quadratic check as presented in Table 3 is an artefact of experimental procedure. For example, if we take the first two types of progeny listed in Table 2 we obtain a pattern with Goldfoil and Kwan of the type presented in Table 4. Such a pattern also seems to fit

TABLE 4

The reaction[a] of two host lines Goldfoil and Kwan, upon inoculation with two parasite cultures of the types 1 and 2 listed in Table 2

	Host	
Parasite	Goldfoil ($r1r1$)	Kwan ($R1R1$)
progeny type 1 ($p1$)	—	—
progeny type 2 ($P1$)	—	+

[a] — = incompatible relationship
  + = compatible relationship

for *Helminthosporium victoriae* cultures which do, or do not, produce a host-specific toxin on oat cultivars with Vb or vb gene (Ellingboe, 1976). There is greater development of the parasite and effect on the host if the parasite produces a toxin and the host has the Vb gene. Toxin without the Vb gene or no toxin and either Vb or vb gene gives comparable results. The simplest explanation of Table 4 is that there is specific recognition with $P1/R1$ to give a compatible host–parasite relationship.

If, in Tables 2 and 4, progeny types 1 and 2 are crossed we would expect to see segregation only on Kwan, and that is what Moseman (1959) found. In a cross between Goldfoil and Kwan, no segregation would be expected with progeny of type 1 but segregation would be expected with progeny type 2. Thus far in the analysis the patterns of segregation suggest that the simplest explanation for the data in Table 4 is specific recognition for compatibility and that Table 4 should be interpreted as is given in the parentheses. The product of $P1$ interacts specifically with the product of $R1^{\times 2}$ to give a compatible relationship between host and parasite. The other three parasite/host genotypes, $P1/r1r1$, $p1/R1R1$, and $p1/r1r1$, represent no interaction on the part of the products of these genes. The test crosses that Moseman made show that the genetic interpretation given in Table 4 is not verified. When Moseman crossed a culture of progeny type 1 with the two parent cultures he was able to show that type 1 had two genes. Crosses between the host lines showed that they differed by two genes. When the individual genes are sorted out, the pattern given in Table 3 emerges again. The pattern given in Table 4 is misleading because it cannot be verified in test crosses.

Of the naturally occurring variability in host–parasite interactions with the mildews, the pattern given in Table 3 seems to explain essentially all data. Occasionally data are obtained which appear to be inconsistent with this pattern, but more precise investigations continuously show that this pattern seems to prevail (Ellingboe, 1975).

Incompatibility is usually, but not always, associated with the dominant allele. The fact that the alleles which given incompatibility are usually dominant is further evidence that there is a positive function for incompatibility. This argument is not sufficient for one to conclude that the specific interaction is for incompatibility but it is consistent with the evidence. Dominance is dependent on, among other things, the concentration of gene products relative to what is needed to accomplish its job, the relative concentration of the products of the alternate alleles and the obvious, our ability to recognize small changes in phenotype of the interaction.

The idea that specific interactions are for incompatibility is given additional support when one looks at many parasite/host gene pairs. The basic pattern for three gene pairs is given in Table 5. The first observation to make is that an incompatible relationship is achieved only when the corresponding P and R genes are together. For example, parasite culture 2 has two P genes but

TABLE 5

The interactions between host and parasite involving three parasite/host gene pairs with 2 alleles for each locus

	host lines							
	1 R1R1 R2R2 R3R3	2 R1R1 R2R2 r3r3	3 R1R1 r2r2 R3R3	4 R1R1 r2r2 r3r3	5 r1r1 R2R2 R3R3	6 r1r1 R2R2 r3r3	7 r1r1 r2r2 R3R3	8 r1r1 r2r2 r3r3
Parasite cultures								
1). P1 P2 P3	−	−	−	−	−	−	−	+
2). P1 P2 p3	−	−	−	−	−	−	+	+
3). P1 p2 P3	−	−	−	−	−	+	−	+
4). P1 p2 p3	−	−	−	−	+	+	+	+
5). p1 P2 P3	−	−	−	+	−	−	−	+
6). p1 P2 p3	−	−	+	+	−	−	+	+
7). p1 p2 P3	−	+	−	+	−	−	+	+
8). p1 p2 p3	+	+	+	+	+	+	+	+

− = incompatible relationship
+ = compatible relationship

168   ABLBERT H. ELLINGBOE

gives a compatible relationship with host line 7 which has one R gene. An incompatible relationship is obtained only when there is a P gene and its corresponding R gene, e.g., $P1/R1$ or $P2/R2$ or $P3/R3$. Let us compare this situation to the basic pattern of interactions given in Table 3. Parasite culture 2 with host line 7 is in the upper right hand quadrant for genes $P1/r1$ and $P2/r2$ but lower left hand quadrant for genes $p3/R3$. Parasite culture 2 with host line 6 gives an incompatible relationship. The parasite/host genotype is $P1P2p3/r1R2r3$. $P1/r1$ should specify compatibility (see Table 3); $P2/R2$ should specify incompatibility and $p3/r3$ should specify compatibility. The second observation is that if just one gene pair specifies incompatibility, that gene pair is epistatic over all other gene pairs which specify compatibility of the relationship. Again I believe the simplest interpretation of these data is that specific recognition of corresponding gene pairs occurs for incompatibility. For the data in Table 5 only three specific interactions ($P1/R1$, $P2/R2$ and $P3/R3$) would be necessary to explain all the data. If specific interaction was necessary for compatibility on the part of these genes in host and parasite it would be necessary to have interactions of $P1/r1$, $p1/R1$ or $P1/R1$ plus $P2/r2$, $p2/R2$ or $p2/r2$ plus $P3/r3$, $p3/R3$ or $p3/r3$. The idea that three interactions for each of three parasite/host gene pairs all give the same phenotype is difficult to explain at the molecular level. It is not impossible, but I consider it highly unlikely. I perfer the simpler hypothesis of specific recognition for incompatibility.

Multiple allelic series of R genes in the hosts are common. Barley, for example, appears to have an allelic series at the Mla locus. The type of data from which this conclusion is drawn is given in the example of crosses involving cultivars Ricardo, Algerian and Manchuria and cultures CR3 and 062 of E. graminis f. sp. hordei (Moseman and Schaller, 1960). Crosses of Algerian × Manchuria showed segregation ratios in the F_2, following inoculation with culture CR3, which were consistent with one locus distinguishing the two cultivars. The gene in Algerian was designated Mla. Manchuria was considered to have the allele mla. Inoculation with CR3 of crosses of Ricardo × Manchuria also showed segregation consistent with one locus distinguishing the two cultivars. When F_2 and F_3 progeny of the cross Ricardo × Algerian were inoculated with culture CR3, no plants were found which gave a compatible relationship with CR3. No segregation suggests that Ricardo and Algerian have the same Mla gene. However, Ricardo gave a compatible reaction with culture 062 and Algerian gave an incompatible reaction with culture 062. Segregation in the cross Ricardo × Algerian was evident when F_2 plants were inoculated with culture 062. Other test crosses involving Ricardo were consistent with the interpretation that Ricardo had an allele at the Mla locus which was different from the allele in Algerian. The allele in Ricardo was designated $Mla3$.

Crosses of cultures 062 and CR3 gave results which were consistent with one locus controlling interactions with Ricardo and another non-linked locus controlling interactions with Algerian.

On the basis of numerous such analyses it now appears that there are 12 "alleles" at the *Mla* locus and two alleles at each of three other loci, *Mlg*, *Mlk* and *Mlp* (Moseman and Jorgensen, 1971). There is no evidence that any of the corresponding *P* genes in the pathogen are allelic. To date, there is no hard evidence that there are more than two alleles at any *P* locus in a parasite. (Some evidence suggestive of an allelic series at a *P* locus will be presented later.) The genetic relationships between *P* genes in *E. graminis* f. sp. *hordei* and the *Ml* genes in barley is consistent with the data given in Table 6. There appear to be ten distinct loci in the parasite for recognition of ten of the eleven alleles at the *Mla* locus.

TABLE 6

The genetic relationships between *P* genes in *E. graminis* f. sp. *hordei* and the *Ml* genes in barley

Parasite	*Pa*	*Pa2*	*Pa3*	*Pa4*	*Pa5*	*Pa6*	*Pa7*	*Pa8*	*Pa9*	*Pa10*	*Pg*	*Pk*	*Pp*
	pa	*pa2*	*pa3*	*pa4*	*pa5*	*pa6*	*pa7*	*pa8*	*pa9*	*pa10*	*pg*	*pk*	*pp*

Host	*Mla*	*Mlg*	*Mlk*	*Mlp*
	Mla2	*mlg*	*mlk*	*mlp*
	Mla3			
	Mla4			
	Mla5			
	Mla6			
	Mla7			
	Mla8			
	Mla9			
	Mla10			
	Mla			

The pattern of interaction between the parasite gene *P* and its alternate allele *pa* with the genes at the *Mla* locus in barley is presented in Table 7. The figure is an expansion of the pattern in Table 3 to take into account the fact that there are several different alleles which can occupy the *Mla* locus. The simplest explanation of these data is that specific recognition occurs for incompatible relationships. To argue otherwise is to argue that there are 23 different parasite/host genotypes which have specific recognition for compatibility and only one parasite/host genotype which does not have specific recognition between gene products and, therefore, cannot function to give a

TABLE 7

The interaction between the *Pa* locus in *E. graminis* f. sp. *hordei* and the alleles at the *Mla* locus in barley

	Host genes										
parasite *Mla*	*Mla2*	*Mla3*	*Mla4*	*Mla5*	*Mla6*	*Mla7*	*Mla8*	*Mla9*	*Mla10*	*Mla11*	*ml*
Pa −	+	+	+	+	+	+	+	+	+	+	+
pa +	+	+	+	+	+	+	+	+	+	+	+

− = incompatible relationship
+ = compatible relationship

compatible relationship. Since the argument for specific recognition for 23 parasite/host genotypes, versus specific recognition for only one parasite/host genotype, becomes so unlikely, I use this illustration to strengthen the argument that specific recognition, within the naturally-occurring variability in host/parasite interactions, is for incompatible relationships. Parasite gene *Pa* treats all alleles, other than *Mla*, as though they were equal to the recessive allele *mla*.

Barley is a diploid plant. It will not have more than two alleles at any one locus in any one plant. As a self pollinated plant, it will usually be homozygous for one allele. A plant homozygous for *Mla3* will not have the other *Mla* alleles present. The parasite, however, has the corresponding gene at a different locus for each *Mla* allele in the host and will therefore, always have either *P* or *p* for each of the host alleles (see Table 8). What happens to the non-*Pa3* genes in a parasite culture inoculated onto a host of genotype

TABLE 8

The reaction[a] of barley with *Mla3Mla3* to inoculation with four cultures of *E. graminis* f. sp. *hordei*

	host genotype	
parasite genotypes	*Mla3 Mla3*	*Mla7 Mla7*
1). *Pa Pa2 Pa3 pa4 pa5 Pa7 pa7 Pa8 Pa9 pa10 pa11*	−	+
2). *Pa pa2 Pa3 Pa4 pa5 pa6 Pa7 Pa8 pa9 Pa10 Pa11*	−	−
3). *Pa pa2 pa3 pa4 Pa5 pa6 pa7 Pa8 pa9 pa10 Pa11*	+	+
4). *Pa Pa2 pa3 Pa4 pa5 pa6 Pa7 pa8 pa9 Pa10 pa11*	+	−

[a] − = incompatible relationship
+ = compatible relationship

Mla3 Mla3? It has been suggested that they are "unoccupied" (Person and Mayo, 1974)—a connotation which eludes me. There are several reasons, which will not be discussed here, for believing that the *P* genes have a function other than for interactions with the host. The *Pa, Pa2, Pa4, Pa5* . . . loci each treat a host of genotype *Mla3 Mla3* as though that host had the recessive *mla* gene. *Pa* is not any more "unoccupied" with a host of genotype *Mla3 Mla3* than is *Pg* in culture 21·1 on the cultivar Manchuria with genotype *mlg mlg* (see Table 3). There is no evidence that the *Pa* gene distinguishes between *Mla2, Mla3, Mla4* . . ., etc.; it treats them all as though they were the same allele (see Table 7). When all the combinations of parasite/host genotypes are considered for the *Mla* locus in barley and the corresponding *P* genes in *E. graminis* f. sp. *hordei*, it seems even more likely to me that the specific recognition of corresponding genes is for the incompatible relationship. To argue to the contrary it becomes necessary to postulate specific interactions for 23 parasite/host genotypes involving the *Pa* locus in the parasite (see Table 7), 23 parasite/host genotypes involving the *Pa2* locus in the parasite, and so on.

There is the possibility that the genes at the *Mla* locus are not true alleles but may be very closely linked genes. If they are true alleles and if specific recognition between host and parasite is for incompatibility, the expectation would be for recombinants to give a compatible reaction to the two cultures used to distinguish the alternate alleles. Only one true breeding recombinant was obtained in a cross involving *Mla* and *Mla3* and that gave a compatible relationship with both of the two cultures used to identify *Mla* and *Mla3* (Jorgensen and Moseman, 1972). It is not possible to distinguish between intercistronic and intracistronic recombination with the type of recombinant obtained. More detailed genetic analyses of the *L* locus in *Linum usitatissimum* demonstrated convincingly that recombination is intracistronic, that two types of recombinants are produced both of which give a compatible reaction with the cultures used to distinguish the parent alleles, *L2* and *L10*, that the recombinants can be distinguished by test crosses and that one of the specificities, *L10*, was recovered when some of the recombinants were test crossed (Shepherd and Mayo, 1972). Recombinants from intracistronic recombination would be expected to have altered specificities and, therefore, be unable to interact specifically to give an incompatible relationship with either culture of the parasite. The fact that both recombinants gave compatible reactions with the parasite cultures used to distinguish the parents is consistent with the argument that specific recognition is for incompatible relationships.

Temperature sensitivity has also been used in arguments to decide if specific interactions are for compatibility or incompatibility (Ellingboe, 1976). Temperature-sensitive mutations are generally considered to be missense mutations which do not affect the tertiary structure of the protein at normal temperatures but do affect the tertiary structure at a higher temperature (Edgar and Lielausis, 1964). The protein will have biological activity at the normal temperature but no or reduced biological activity at the higher temperature. If specific interactions are for incompatibility, a parasite/host genotype which was temperature-sensitive should give incompatibility at the normal temperature but compatibility at the higher temperature. If specific interactions are for compatibility, then compatibility would be expected at the normal temperature and incompatibility at the higher temperature. The known temperature-sensitive parasite/host genotypes give incompatibility at the normal temperature and compatibility at the higher temperature (Loegering, 1966). The data on temperature-sensitivity of host–parasite interactions are consistent with the concept that specific interactions between corresponding host and parasite genes are for incompatible relationships.

I have presented several different arguments which suggest that the naturally-occurring genetic variability which affects the interactions between host and parasite follows the pattern presented in Table 3. The simplest interpretation of these data is that the specific recognition occurs between the parasite gene P and the corresponding host gene R to give an incompatible relationship. The three other parasite/host genotypes, P/r, p/R and p/r represent no interaction between the product of the parasite gene and the product of the host gene and, therefore, a compatible interaction is allowed. An R gene in the host represents a positive function. A P gene in the parasite represents a positive function. The idea that a parasite has a positive function for incompatible relationships with its host is clearly not a dominant concept in plant pathology. The data from the genetic analyses are essentially impossible to explain except by concluding that the P allele is the allele whose product interacts with the host. There is no reason to expect that P interacts with r, or that p interacts with r but there are reasons to suspect that there may be an interaction between p and R (Martin and Ellingboe, 1976).

Person (1967) has argued that the evolution of the gene–for–gene interaction can be rationalized if we begin with the parasite/host genotype P/r. A mutation in the host, which gives an allele whose product interacts with P to give an incompatible relationship, should have a selective advantage in nature over the plants which have a different allele at that locus. When the host population primarily contains R, there will be a selective advantage for a mutation in the parasite to p. If P/R represents a specific interaction, a mutation $P \rightarrow p$ should be a mutation to the loss of recognition of the product of R but the mutation does not have to lead to a complete loss of recognition of the product of R. Many mutations which lead to a diminished recognition of R should have a selective advantage over the parent gene P. There is the expectation, therefore, that the parasite/host genotype p/R may not have the same degree of compatibility as P/r or p/r. Evidence that p/R may not give the same degree of compatibility as P/r or p/r has been presented with *E. graminis* f. sp. *tritici* and wheat with *Pm4* (Martin, 1974). Cultures of *E. graminis* f. sp. *tritici* have been obtained which give an infection type 4 on host plants with *Pm4* or *pm4*. One of the cultures is very slightly slower growing on hosts with *Pm4* but normal on plants with *pm4*; primary infection kinetics are delayed about 2 hours and the amount of mildew disease produced in 7 days is about what would be expected in 6·5 days but only on plants with *Pm4*. Therefore, this culture does not just have a gene for slow growth because it has normal development on host plants with *pm4*. Two other cultures of *E. graminis* f. sp. *tritici* gave reduced infection efficiencies on plants with *Pm4* but normal infection efficiencies on plants with *pm4*. The parasite units which do produce successful primary infections on plants with *Pm4* apparently develop completely normally and produce infection type 4 reactions 7 days after inoculation. I interpret these results to mean that these three cultures have mutations to increased compatibility with plants with *Pm4*. Whether the mutations are at the *P4* locus or modifiers of the *P4/Pm4* specificity is not known. The data again are consistent with the concept that, of the naturally-occurring variability, the specific interactions of host–parasite interactions are for incompatibility of the relationships.

The above discussion on genetics of interactions must be put into context. The results of analysing the naturally-occurring variability in hosts and parasites affect the observable interactions between host and parasite. The same genetic pattern of interactions is observed whether the differences between "+" and "−" (see Table 3) are great or small (Ellingboe, 1975). The only requirement is that the differences between "+" and "−" can be distinguished, even if environment and inoculation protocols must be controlled very precisely to minimize the within-genotype variability. The intent is not to argue the non-existence of genes whose function is necessary for successful parasitism since there are excellent reasons for postulating their existence. They would be expected to follow a pattern as presented in

TABLE 9

The pattern of interactions predicted if specific recognition between host and parasite is necessary for compatible interactions.

Parasite genotype	Host genotype	
	RxRx	*rxrx*
Px	+	−
px	−	−

+ = compatible relationships
− = incompatible relationships

Table 9. The product of the *Px* gene interacts with the product of the *Rx* gene to give a compatible relationship. Why does this pattern never emerge from the numerous genetic studies of the host–parasite interactions with the mildews? It is easy to see in Table 9 that a mutation from *Px* → *px* in the parasite would be lethal to the parasite because the mutant cannot survive on either host. The only examples of host–parasite interactions which appear to fit the pattern in Table 9 are the examples where the parasite produces a host-specific toxin. To my knowledge, there are no cases, with the mildews, which follow the pattern presented in Table 9. There are combinations of host and parasites which appear to follow this pattern but, as pointed out earlier, test crosses do not support this interpretation.

IV. Genetic Control of Morphological Development During Infection

Erysiphe graminis goes through several definite morphological stages in the process of infection of host plants (Ellingboe, 1972). Protocols have been developed whereby a high percentage of the spores applied to a host plant go through the necessary differentiation to establish successful primary infections with compatible parasite/host genotypes. Given the appropriate sequence of environmental conditions, the parasite units will go through these identifiable stages with a reasonable degree of synchrony (McCoy and Ellingboe, 1966). This background information has made it possible to determine when the expression of different parasite/host genotypes become manifested in the ontogeny of host–parasite interactions. An attempt has been made to establish the *Px/rx* parasite/host genotype as the basis with which all other genotypes are compared. Ideally the parasite would have only *P* genes and the host would have only *r* genes. Obviously one can never be certain that all parasite/host genes are in this combination (Person and Mayo, 1974). Host lines with known *r* genes and parasite cultures which have many *P* genes have been used as the standard for comparison. It is possible to compare the effect of one *P/R* gene pair on the ontogeny of the interaction between host and parasite with the genotype *P/r* by the systematic introduction of one *R* gene at a time in highly isogenic lines (Ellingboe, 1972). The pattern is given in Table 10. The standard for comparison has been culture MS–1 with the cultivar Chancellor. Culture MS–1 has *P* genes. Chancellor has *pm* genes. One host line with *Pm1Pm1* is incompatible with culture MS–1 on the basis of *P1/Pm1* interactions. Another host line with *Pm2Pm2* is incompatible with culture MS–1 on the basis of *P2/Pm2* interactions. *Pm3a* and *Pm3b* are considered as alleles in the host but the respective corresponding genes in the parasite, *P3a* and *P3b*, are at separate loci. The five incompatible interactions given in Table 10 are on the basis of five different parasite/host gene pairs. There are two possible interpretations of function of the incompatible parasite/host genotypes. One possibility is that each gene pair functions to

TABLE 10

The interactions[a] of the parasite and host genotypes used in studying genetic control of interactions with *Erysiphe graminis* f. sp. *tritici* and wheat

	designation and host genotype					
	(Chancellor)	(Pm1)	(Pm2)	(Pm3a)	(Pm3b)	(Pm4)
	pm1pm1	*Pm1Pm1*	*pm1pm1*	*pm1pm1*	*pm1pm1*	*pm1pm1*
	pm2pm2	*pm2pm2*	*Pm2Pm2*	*pm1pm1*	*pm1pm1*	*pm1pm1*
parasite	*pm3pm3*	*pm3pm3*	*pm3pm3*	*Pm3aPm3a*	*Pm3bPm3b*	*pm3pm3*
genotype	*pm4pm4*	*pm4pm4*	*pm4pm4*	*pm4pm4*	*pm4pm4*	*Pm4Pm4*
P1 P2						
P3a P3b						
P4 (MS-1)	+	−	−	−	−	−

[a] − = incompatible relationships
+ = compatible relationships

affect a unique system in the interactions. A second possibility is that each of these gene pairs represents an independent way to trigger a common mechanism for incompatibility. Each of these possibilities has its own predictions.

The evidence with *E. graminis* f. sp. *tritici* and wheat would suggests that the gene pairs *P1/Pm1, P2/PM2, P3a/Pm3a*, and *P4/Pm4* each affect unique stages in the ontogeny of the interaction (Ellingboe, 1972). None of the parasite/host gene pairs affected germination of the spores, formation of appressoria or attempted penetrations but three, *P1/pm1, P3a/Pm3a*, and *P4/Pm4*, had major effects on the proportion of parasite units which were successful in producing haustoria. Not all parasite units were prevented from producing haustoria, but those which did not produce haustoria did not produce very long secondary hyphae. The percentage of parasite units which produced haustoria and elongating secondary hyphae was different for each of the three parasite/host genotypes *P1/Pm1, P3a/Pm3a*, and *P4/Pm4* and their subsequent fate was also dependent on the parasite/host genotype. Some parasite units began to produce haustoria which did not attain normal size. Only those which formed normal-looking haustoria appeared to be capable of supporting the production of elongating secondary hyphae. The proportion of parasite units which had produced successful primary infections and were also capable of producing secondary haustoria was also affected by the different incompatible parasite/host genotypes. The proportion of infected host cells which take up aniline-blue dye is very low for a compatible parasite/host genotype and one of the first signs of a developing incompatibility is a capability on the part of the infected host cell to take up the aniline-blue dye (Haywood, 1975). This ability suggests that even though morphological development of the parasite proceeds quite normally with $P2/Pm2$, there is a developing incompatibility between host and parasite.

From observations with the microscope for the first 60 hours after inoculation and from macroscopic observations, one could conclude that there are several distinct stages of development at which the different parasite/host genes act. $P1/Pm1$ and $P4/Pm4$ appear to prevent the parasite from developing sufficiently to produce a macroscopically visible symptom on the plant. $P2/Pm2$ has its greatest effect on the parasite at the time of initiation of secondary and subsequent haustoria, followed by death of infected host cells and cells adjacent to the infected cells (what some call a hypersensitive response). $P3a/Pm3a$ has some easily observed effects early in primary infection but there are always a few large pustules formed within seven days after inoculation.

Some large pustules would have been observed with all incompatible parasite/host genotypes if enough plants were inoculated. The proportion is determined by each incompatible parasite/host genotype. Because there are several distinct stages at which it is possible to see the effects of each genotype, it appears that each genotype presents a series of hurdles which a certain portion of the parasites may overcome. The proportion that overcomes each hurdle determines the proportion of parasite units which reaches the final stage, that is, the production of a large pustule with abundant sporulation.

There are two tentative conclusions to draw from the studies of primary infection by $E.$ $graminis$. Each incompatible parasite/host genotype affects the ontogeny of interactions in a unique way and each has several different effects, on a time scale, during the ontogeny of interactions between host and parasite. The latter observation may be helpful to explain the so-called "X" reaction whereby a single plant has all infection types when inoculated with a genetically pure strain of the parasite. A parasite unit stopped early (an early hurdle) in its interaction with its host may have a final infection type 0 or 1; one which succeeds in overcoming all hurdles may have an infection type 4. One point is clear from the studies with $Erysiphe$ $graminis$, there is no evidence that a single incompatible parasite/host genotype can stop all parasite units in a single stage of ontogeny of the interaction between host and parasite. This observation is very crucial to the interpretation of data from several different laboratories. It is also very important to remember for the design of experiments on the comparative biochemistry, physiology or ultrastructure of compatible and incompatible host/parasite interactions.

V. Transfer of ^{35}S and ^{32}P from Host to Parasite During Primary Infection

Studies on the transfer of radioisotopes from host to parasite are easier with the powdery mildews than with other plant diseases because most of the parasite is on the surface of the host leaf; only the haustoria cannot be removed from the host with ease.

Host leaves inoculated with *Erysiphe graminis* were allowed to take up radioactive compounds for five hour periods starting at various hours after inoculation (Mount and Ellingboe, 1969). After the five hour labelling period, a solution of parlodion in ether:alcohol was placed on the inoculated leaf surface. The ether:alcohol mixture evaporated rapidly leaving a parlodion strip on the leaf in which the portion of the parasite on the surface of the leaf was embedded. The parlodion was removed from the leaf and placed in a scintillation vial, dried, scintillation fluid was added and radioactivity determined in a scintillation counter.

Plotting of the data gives a rate of transfer as a function of time after inoculation, a crude first derivative of transfer from host to parasite (Slesinski and Ellingboe, 1971). It was possible to base the data on counts min^{-1} unit number of spores^{-1} unit amount of radioactivity^{-1} in the epidermal cells available for transference to the parasite by considering several of the factors which affect the final results. The data suggest that there is no detectable transfer of ^{35}S from host to parasite up to 14 hours after inoculation. The most rapid rate of transfer during primary infection is 18–20 hours after inoculation (Stuckey and Ellingboe, 1975) for compatible parasite/host genotypes.

As stated earlier, one of the effects of the incompatible parasite/host genotypes was to reduce the percentage of spores which produced primary infections. The rates of transfer of ^{35}S from barley to *E. graminis* f. sp. *hordei* during primary infection with four incompatible parasite/host genotypes and one compatible genotype were consistent with the expectations based on infection efficiencies (Hsu and Ellingboe, 1972). 75–80% of parasite units produced successful primary infection with the compatible genotype (*Px/mlx*) and it had the highest rate of transfer. Genotypes with lower infection efficiencies had lower rates of transfer. The data were consistent with the interpretation that the rate of transfer per parasite unit which produced haustoria and elongating secondary hyphae was constant for compatible and incompatible parasite/host genotypes.

The relationship between infection efficiencies and rate of ^{35}S transfer from host to parasite for *E. graminis* f. sp. *tritici* and wheat (Stuckey and Ellingboe, 1975) was not as good as with *E. graminis* f. sp. *hordei* and barley (Hsu and Ellingboe, 1972). Nevertheless the calculations on label transfer per parasite unit which had produced a haustorium and an elongating secondary hypha was within a range which could have been due to experimental error. When the rates of ^{32}P transfer from host to parasite were examined, there were two differences from the data on ^{35}S transfer (Martin, 1974). It was possible to detect low levels of ^{32}P transfer as early as ten hours after inoculation. There were only very small differences between one compatible and four incompatible parasite/host genotypes in the amount of label transferred in the period of 20–26 hours after inoculation. The five parasite/host genotypes differed very significantly in terms of the percentage of parasite units which produce elongating secondary hyphae but differed only slightly in terms of ^{32}P transferred from host to parasite. When the calculations were made to determine the rate of ^{32}P transfer per parasite unit which produced an elongating secondary hypha, very great differences between genotypes were observed. With genotype $P4/Pm4$, few elongating secondary hyphae were produced; the calculated rate of ^{32}P transfer per parasite unit which produced an elongating secondary hypha was ten times the rate for the compatible genotype Px/pmx. Either the rate of transfer is much higher in an incompatible relationship or there is some inherent error in our perception of what is happening between host and parasite. It seems rational that an incompatible genotype, which is putting stress on the host–parasite relationship, may cause a greater mobilization of ^{32}P to the site of interactions than ^{35}S since phosphorus is an important element in the energy status of a cell.

If the transfer of both ^{32}P and ^{35}S from host to parasite were passive, one might expect similar kinetics of transfer of the two isotopes. The fact that the transfer kinetics of these two isotopes are not similar with the incompatible parasite/host genotypes is suggestive of an active transport for one of the isotopes and the evidence to date suggests that ^{32}P transfer is not just a passive function. There are many possible ways to explain the differences in transfer of ^{32}P and ^{35}S; at present we only know there is a difference in rates of transfer between the two with incompatible parasite/host genotypes.

VI. Induced Resistance or Susceptibility: Different Genes?

All the studies on the inheritance of host and parasite genetic variability which affects host–parasite interactions are most easily explained if we assume that specific interactions are for incompatibility. Hypotheses have been put forward which assume induced susceptibility (Favret, 1971). This seems reasonable when two hosts and one strain of the pathogen are used. When the quadratic check is completed for one parasite/host gene pair, the induced susceptibility hypothesis becomes less tenable because it is necessary to explain how three parasite/host genotypes can give a positive interaction and one alone cannot. When consideration is given to the allelic series in hosts, the induced susceptibility hypothesis becomes even more untenable. Fine structure mapping of the L series in *Linum ustitatissimum* (Shepherd and Mayo, 1972) essentially closes the argument as to whether the specific interaction is for incompatibility or compatibility. For all practical purposes we have to conclude that genes controlling the naturally-occurring variability in host–parasite interactions show specific interaction for incompatibility, especially when temperature-sensitivity of the interactions is also considered.

The idea that a host bears a positive function to inhibit development of a parasite is a common idea in plant pathology. The idea that a parasite has a positive function to be unable to develop in a host and that loss of that function leads to the ability to develop in that host, is not such a common idea. Nevertheless, that is the most logical explanation of data from genetic studies of the naturally-occurring variability affecting host-parasite interactions. To explain a positive function on the part of host and parasite genes either to prevent the establishment of a compatible relationship or to destroy a compatible relationship once established, it becomes necessary to postulate a "basic compatibility" upon which the gene–for–gene relationship is built (Ellingboe, 1976). If we assume that a basic compatibility between host and parasite evolved first, then it is easy to rationalize the evolution of a gene–for–gene relationship in which specific recognition is necessary for incompatibility.

Why do we not see genetic variability affecting basic compatibility? Will variability affecting basic compatibility be lethal to the parasite? There are reasons to expect that such mutations would be lethal to the parasite. At least some of the genes affecting basic compatibility in the parasite would be predicted to have a positive function for compatibility. If their interaction occurs with specific genes in the host one would expect a pattern which is the reciprocal of the pattern given in Table 3. Attempts to produce conditional lethals (temperature-sensitive) mutations in two non-mildew pathogens have been successful (Ellingboe and Gabriel, 1977). Mutations in *Colletotrichum lindemuthianum* that are not temperature sensitive on agar medica but are temperature sensitive in the host can produce disease at the permissive temperature but not at the non-permissive temperature. This is consistent with the expected behaviour of genes whose function is necessary for compatible host–parasite relations. The temperature-sensitive interactions of the induced mutations followed a pattern which was the reciprocal of temperature-sensitive gene–for–gene interactions (Loegering, 1966). Attempts are now being made to produce temperature sensitive mutations of *E. graminis* f. sp. *tritici* to test the possibility that this approach will permit the recovery of mutations of genes whose function is necessary for successful parasitism and pathogenesis.

If the concept of at least two types of interactions for any one host and parasite is correct, it will change the approaches to studies of parasitism and pathogenesis. A realization that only one kind of variability is present in natural populations of a host and a parasite in mildew diseases must affect interpretation of published data and the planning of future experiments. For example, the concept of induced resistance in a host has been discussed and argued over many years. The phenomena of "induced accessibility" (Ouchi *et al.*, 1974a, b, 1976) is difficult to interpret. It is easy to speculate how inoculation with a race of a pathogen which is incompatible with a host line will prevent infection by a later inoculation with a race of the same pathogen, which would otherwise have given a compatible relationship with the same host line. It is not so easy to explain the demonstration that inoculation with a race of *E. graminis* f. sp. *hordei* which was compatible with a barley line affected a subsequent inoculation with a race of the same species which was incompatible with the host. The relationship between the strain of the parasite used in the first inoculation and the host line, determined the relationship between the host and the strain used in the second inoculation. The effect was only over very short distances and seemed to be irreversible. The greatest effect was when both first and second infections were in the same cell (Ouchi *et al.*, 1976). The time between the first and second inoculations was suffici- ently short to require that the mutual recognition for compatibility or incom- patibility must have been made very early in the interactions between host and parasite (Ouchi *et al.*, 1976). The fact that "resistance" or "accessibility" can be brought about depending on the sequence of inoculations of two strains of a parasite on a single host line is most consistent with the idea that both are induced phenomena and that the genes controlling these interactions are not constitutive in function. The observation that the effect of the first inoculation affects other species of parasites suggests that it is not the genes which determine the specificity of the interactions (the gene–for–gene relationships) which are involved directly but the genes which are secondarily involved in the changes induced by the initial inoculation. My speculation is that the observations are the result of alterations in function of the genes controlling basic compatibility. By the use of conditional mutations it should be possible to subject this hypothesis to a critical test.

    Research reported herein has been supported in part by Grant AI 06420 from the National Institutes of Health and Grants PCM 73–02043–A01 and OIP 75–02090 from the National Science Foundation. Michigan Agri- cultural Experiment Station, Journal Article No. 7955.

References

Briggs, F. N. and Stanford, E. H. (1938). *J. Genet.* **37**, 107–117.
Edgar, R. S. and Lielausis, I. (1964). *Genetics* **49**, 649–662.

Ellingboe, A. H. (1972). *Phytopathology* **62**, 401–406.
Ellingboe, A. H. (1975). *Aust. Pl. Path. Soc. Newsletter* **4**, 44–46.
Ellingboe, A. H. (1977). *In* "Physiological Plant Pathology". (R. Heitefuss and P. H. Williams, Eds) Encyclopedia of Plant Physiology, New Series, Vol. 4, 761–778. Springer-Verlag, Berlin, Heidelberg and New York.
Ellingboe, A. H. and Gabriel, D. W. (1977). *In* "Induced mutations against plant diseases". IAEA, Vienna.
Favret, E. A. (1971). *In* "Barley Genetics Vol. II". (R. A. Nilan, Ed.) 457–471. Proc. 2nd. Int. Barley Genet. Symp., Pullman, Washington. Washington State University Press.
Flor, H. H. (1946). *J. Agr. Res.* **73**, 335–357.
Flor, H. H. (1947). *J. Agr. Res.* **74**, 241–262.
Flor, H. H. (1955). *Phytopathology* **45**, 680–685.
Haywood, M. J. (1975). *Ph.D. Dissertation*, 69 pp., Michigan State University.
Hsu, S. C. and Ellingboe, A. H. (1972). *Phytopathology* **62**, 876–882.
Jorgensen, J. H. and Moseman, J. G. (1972). *Can. J. Genet. Cytol.* **14**, 43–48.
Loegering, W. Q. (1966). *Proc. 2nd Int. Wheat Genetics Symp.*, Lund 1963, *Hereditas*, Suppl. Vol. 2, 167–177.
Martin, T. J. (1974). *Ph.D. Dissertation*, 116 pp., Michigan State University.
Martin, T. J. and Ellingboe, A. H. *Phytopathology* **66**, 1435–1438.
McCoy, M. S. and Ellingboe, A. H. (1966). *Phytopathology* **56**, 683–686.
Moseman, J. G. (1956). *Phytopathology* **46**, 318–322.
Moseman, J. G. (1959). *Phytopathology* **49**, 469–472.
Moseman, J. G. and Jorgensen, J. H. (1971). *Crop Sci.* **11**, 547–550.
Moseman, J. G. and Schaller, C. W. (1960). *Phytopathology* **50**, 736–741.
Mount, M. S. and Ellingboe, A. H. (1969). *Phytopathology* **59**, 235.
Ouchi, S., Oku, H., Hibino, C., and Akiyama, I. (1974a). *Phytopath. Z.* **79**, 24–34.
Ouchi, S., Oku, H., Hibino, C., and Akiyama, I. (1974b). *Phytopath. Z.* **79**, 142–154.
Ouchi, S., Oku, H. and Hibino, C. (1976). *Phytopathology* **66**, 901–905.
Person, C. (1967). *Can. J. Bot.* **45**, 1193–1204.
Person, C. and Mayo, G. M. E. (1974). *Can. J. Bot.* **52**, 1339–1347.
Person, C. and Sidhu, G. (1971). *"Mutation breeding for disease resistance"*. IAEA, Vienna.
Shepherd, K. W. and Mayo, G. M. E. (1972). *Science* **175**, 375–380.
Slesinski, R. S. and Ellingboe, A. H. (1971). *Can. J. Bot.* **49**, 303–310.
Stuckey, R. E. and Ellingboe, A. H. (1975). *Physiol. Pl. Path.* **5**, 19–26.

Chapter 8

Accumulation of Solutes in Relation to the Structure and Function of Haustoria in Powdery Mildews

W. R. BUSHNELL[1]* and JOHN GAY[2]†

[1]*Cereal Rust Laboratory, Agricultural Research Service, US Department of Agriculture, University of Minnesota USA*
[2]*Department of Botany, Imperial College of Science and Technology, London*

I. Introduction

The powdery mildew fungi are efficiently organized to take solutes from host epidermal cells and to utilize these solutes rapidly for production of conidio-

* Cooperative investigations, ARS, US Department of Agriculture, and the Department of Plant Pathology, University of Minnesota. Paper No. 1641, Misc. Journal Series, Minnesota Agricultural Experiment Station.

spores. Only a thin mycelial network spreads across the host surface. This network produces numerous haustoria in epidermal cells and large numbers of conidiophores, which in turn, can produce even larger numbers of conidiospores. The mycelium serves as a short conduit for the solutes which move into spores from haustoria. These solutes come from epidermal and mesophyll cells beneath the mildew colony and, if sporulation is abundant, from tissues distant from the colonized tissues.

The role of the haustorium in solute movement has not been defined experimentally. The haustorium provides a structural interface where solutes probably move from host to parasite and, indeed, the word "haustorium" is based on the Latin "haurire", meaning "to drink" (Snell and Dick, 1971). However, the absorption of solutes by the haustorium has only recently been demonstrated directly.

The purpose of this chapter is to relate the structure and function of haustoria and associated host cells to what is known about the flow of substances from host to parasite. We will describe the kinds and amounts of substances that accumulate in mildewed tissues, discuss some translocation pathways and accumulation mechanisms that are probably involved and briefly relate these to other physiological changes in mildewed tissue. The organization of the mildew colony will be described as it relates to absorbing activities of the haustorium. Finally, the haustorium itself will be described, both structurally and functionally, again with respect to the movement of substances between host and parasite.

We will not attempt to cover in detail the infection processes and associated events that precede the formation of haustoria, nor the role of haustoria in resistance and susceptibility. These aspects are discussed in Ellingboe's chapter in this book and in several recent reviews (Aist, 1976; Bushnell, 1972, 1976; Bracker and Littlefield, 1973; Ehrlich and Ehrlich, 1971).

II Accumulation of Organic and Inorganic Substances in Mildewed Tissues

A. Quantities and sources of accumulated substances

Because the mycelium of powdery mildew fungi can be brushed or rubbed from leaves, the total amount of a given element or compound can be determined separately in host and parasite, except for the haustoria that remain in epidermal cells. Unfortunately, the total amounts within the fungus have only rarely been determined because the spores produced by the fungus usually have not been retained. Nevertheless, where large quantities of spores are produced*, much of the organic and inorganic substance of the spores clearly must originate from solutes in tissues distant from the infection site.

1. Nitrogen

The magnitude of accumulation that can be measured when spores are included in samples of mildewed tissues is indicated by the data for total nitrogen in powdery mildew of barley (Fig. 1). In this example, infection was limited

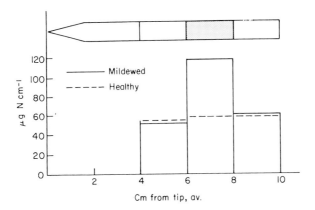

Fig. 1. Accumulation of nitrogen in mildew tissues as shown by the total nitrogen content of mildewed and adjacent non-mildewed segments of attached barley leaves. Twelve days after inoculation with *Erysiphe graminis* f. sp. *hordei*. All spores were included in the mildewed sample. Nitrogen content of the host tissues in mildewed segments is relatively unchanged. (From W. R. Bushnell, unpublished data.)

to 2-cm leaf segments. Total nitrogen increased dramatically during sporulation, nearly doubling. As shown, the nitrogen content of the host tissue in neighbouring segments was relatively unchanged. Likewise, the nitrogen content of the host within the mildewed segment was relatively unchanged (not shown), indicating a large net movement of nitrogen into the mildewed segment from sources outside the sampled parts of the leaf. Host nitrogen sometimes increases in host tissues beneath powdery mildew colonies (Frič, 1964a), especially if leaves are detached (Bushnell, 1967) but such increases are not necessary for accumulation of nitrogen in the mildew fungus.

* Most of the data to be cited on solute accumulation are from studies of *Erysiphe graminis* on wheat or barley in which sporulation is relatively abundant. Although comparative data on accumulation are not available, other powdery mildew fungi may sporulate less and accumulate smaller quantities of solutes. *E. graminis* is known to produce more spores per conidiophore than do other powdery mildew fungi (Hirata, 1967).

Certain soluble nitrogen compounds do consistently increase in mildewed host tissues, principally aspartic and glutamic acids, their corresponding amides, asparagine and glutamine, and ammonia (Fric, 1964a; Sadler and Scott, 1974). Incorporation of ^{14}C from $^{14}CO_2$ into alanine, aspartic acid and glutamic acid is increased (Magyarosy et al., 1976) and radioactively labelled amino acids accumulate (Shaw and Samborski, 1956). By analogy with rust fungi, we can postulate that powdery mildew fungi take up and utilize these amino acids. Each of the amino acids that has been shown to accumulate in mildewed tissues can serve as the principal source of nitrogen for rust fungi grown axenically if a sulphur-containing amino acid such as cysteine is also supplied (Bose and Shaw, 1974; Coffey and Allen, 1973; Harvey and Grasham, 1974; Howes and Scott, 1972; Siebs, 1971). However, the nutritional requirements of the mildew fungi have not been established since they have not yet been grown apart from their hosts.

2. Carbon

The total amounts of carbon in host, parasite, or both have not been determined for powdery mildews, although carbon, like nitrogen, can accumulate in large quantities in fungal spores. Smaller amounts apparently accumulate in the host since the dry weight, which is roughly proportional to carbon content, increased in the host by 17% per unit leaf area according to Shaw and Colotelo (1961) and similar results have been obtained by others (Bushnell, 1967; Frič, 1964a; Johnson et al., 1966; Millerd and Scott, 1963).

Some carbon accumulates in the form of glucose, fructose, sucrose, and starch, primarily in the host (Allen, 1942; Bushnell, 1967; Comhaire, 1963; Frič, 1964a; Hewitt and Ayres, 1976; Shaw et al., 1954). The amounts are small compared to probable quantities of carbon in spores. Allen (1942) found 2·2% of the fresh weight of infected leaves in sucrose, glucose, and starch; Frič (1964a) found 3·5% of the dry weight of host tissues (mycelium removed) in sucrose, fructose and glucose. Both found the largest single fraction to be in sucrose. Amounts of these sugars diminished rapidly as the fungus sporulated.

The carbon compounds entering the mildew fungus can originate from substances translocated from distant tissues or from substances produced locally by photosynthesis. However, the ability of the local mildewed host tissues to produce photosynthate is sharply curtailed as disease progresses, especially when the fungus sporulates. Furthermore, the rate of dark respiration in both fungus and host increases, further reducing the amount of fixed carbon available for accumulation. Thus the ratios of CO_2 uptake (or O_2 evolution) in the light to O_2 uptake (or CO_2 evolution) in the dark usually drop from 15:1 or higher in healthy tissues to about 2:1 in mildewed tissues

(Allen, 1942; Last, 1963; Scott and Smillie, 1966)*. As the rates of photo-synthesis decline toward the compensation point, there is little or no surplus of newly fixed carbon for possible use by the fungus (Allen, 1942) and in some cases the rate of photosynthesis falls below the compensation point (Hewitt and Ayres, 1975). Since dry weights of mildewed host tissues do not decline, any tissues near the compensation point must import carbon from distant tissues for spore production.

Importation of carbon was demonstrated by Edwards (1971) who followed the distribution of radioactivity from $^{14}CO_2$ applied to 3-cm mildewed seg-ments of barley leaves and to uninfected segments above or below the mil-dewed segment (Fig. 2). In comparable segments in healthy leaves, about half of the labelled photosynthate was found in the segment to which the $^{14}CO_2$ had been applied and most of the remainder was exported to roots and shoots. In the mildewed plants, much of the label was again found in the segment which had received the label, but an additional 40% of the total labelled carbon in the plant was found in the mildewed segment if the labelled carbon was applied either to the mildewed segment or to the uninfected segment distal to it. The mildewed segment did not influence export from the segment below it. Thus, the mildewed segment retained its own photosynthate and was also able to capture photosynthate moving basipetally from more apical parts of the leaf. In similar experiments with $^{14}CO_2$ but with powdery mildew infection over the entire leaf blade of barley, Frič (1975) found that infection reduced export from the primary leaf, reduced the amounts of photosynthate entering roots and leaf sheaths but increased the export of photosynthate from healthy secondary leaves.

The ability of powdery mildew fungi to induce a net flow of carbon com-pounds from distant tissue probably depends on the proportion of leaves and plants that are infected, the density of infection, the vigour of the plant and other factors that remain to be evaluated.

3. Inorganic ions

Although measurements of total amounts of individual ions are not usually available, mildewed tissues doubtless accumulate a wide variety of inorganic ions, at least in herbaceous hosts such as wheat, barley, bean, and cucumber. Comhaire (1963) reported that the total ash per unit dry weight increased as

* Photosynthetic capacity decreases to about one-third the normal level (Allen, 1942; Edwards, 1970; Last, 1963; Magyarosy et al., 1976; Paulech, 1966; Sempio, 1950) and respiration generally increases 2–4 fold in the host and 3–10 fold in host and parasite together (Allen and Goddard, 1938; Allen, 1942; Bushnell and Allen, 1962b; Frič, 1964b; Garg and Mandahar, 1976; Millerd and Scott, 1963; Paulech, 1966; Scott, 1965). Hewitt and Ayres (1975) recently reported for oak mildew that the reduction in photosynthesis is accompanied by sharp reduction in photorespiration, so that the actual decline in photo-synthesis is much greater than that estimated by others, whereas the total of dark plus photorespiration does not increase.

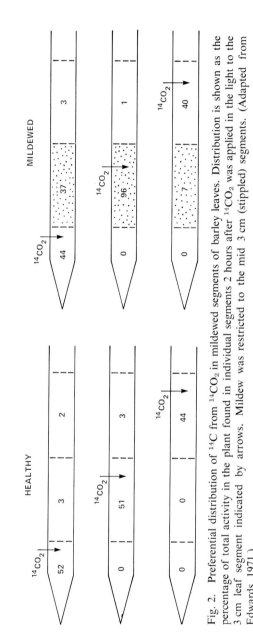

Fig. 2. Preferential distribution of ^{14}C from [14]CO$_2$ in mildewed segments of barley leaves. Distribution is shown as the percentage of total activity in the plant found in individual segments 2 hours after [14]CO$_2$ was applied in the light to the 3 cm leaf segment indicated by arrows. Mildew was restricted to the mid 3 cm (stippled) segments. (Adapted from Edwards, 1971.)

measured in mildewed wheat shoots but as Durbin (1967) cautioned, the data can not be converted to a leaf or area basis. Several investigators have found that radioactive ions accumulate at sites of mildew infection after the ions are supplied to roots, cut ends of leaves, or are injected into tissues away from the diseased tissue as determined in autoradiographs of the mildewed leaves (usually with the fungus in place on the leaf). This occurred with ^{32}P labelled phosphate (Comhaire, 1963; Shaw et al., 1954; Shaw and Samborski, 1956), $^{45}Ca^{++}$ (Shaw and Samborski, 1956; Takanashi and Iwata, 1963) and $^{89}Sr^{++}$ (Takanashi and Iwata, 1963). Applied as a vapour, ^{35}S also accumulated at infection sites (Yarwood and Jacobsen, 1950, 1955).

In contrast to such results with herbaceous plants, ^{35}S vapour did not accumulate in mildewed tissues of several woody perennials (pear, oak and rose; Yarwood and Jacobsen, 1955) and $^{45}Ca^{++}$ did not accumulate and $^{35}SO_4^{--}$ accumulated only in young infections in powdery mildew of apple (Wieneke et al., 1971). This suggests that the ability of powdery mildew fungi to induce accumulation of at least some inorganic ions is less in woody than in herbaceous plants.

B. Translocation and accumulation mechanisms

The preceding section shows that the mildew fungus can be effective in obtaining large quantities of inorganic and organic compounds from the host, some of which can come from host tissues outside the infection site. How, then, does the fungus redirect the movement of substances in the host to produce this accumulation?

1. Circulation patterns

In a review of the movement of solutes in diseased tissues, mostly with rusts, Durbin (1967) emphasized that the influence of the parasite is superimposed on the normal translocation patterns in the host plants. Thus, young expanding leaves must import sugar but begin to export it when they are 25–50% grown (Milthorpe and Moorby, 1969; Crafts and Crisp, 1971). Sugars in basal leaves tend to move to roots; sugars from apical leaves tend to move to growing shoot apices (Crafts and Crisp, 1971). Young leaves also must import minerals but begin to export the mobile elements such as N, P, and K about the time that leaves become fully grown. The leaves continue to import minerals from roots, however. As leaves begin to senesce, amino acids derived from hydrolysed proteins begin to be exported. Most of the organic substances enter and leave leaves by way of the phloem (Milthorpe and Moorby, 1969) and most of the inorganic substances exported from leaves also travel in the phloem (Milthorpe and Moorby, 1969; Wardlaw, 1974).

Because of the continuous recirculation patterns, the parasite can induce net accumulation of a given substance by restricting export, increasing import, or both. From the evidence available in 1967, principally studies on the movement of radioactively labelled substances in rusts, Durbin concluded that inorganic substances accumulate primarily because of the ability of diseased leaves to enhance import; products of photosynthesis accumulate mainly from decreased export.

2. Relations with phloem

How the parasite modulates import and export processes is only partly understood and, indeed, the control of these processes in healthy plants is not well understood. Furthermore, the mechanisms of phloem translocation in healthy plants are subject to continuing experiment and controversy (Milthorpe and Moorby, 1969; Crafts and Crisp, 1971; Wardlaw, 1974). Nevertheless, phloem translocation is usually agreed to be controlled more by events at the source and the sink than by events in the phloem itself (Milthorpe and Moorby, 1969; Crafts and Crisp, 1971; Wardlaw, 1974). In this view, the parasite needs only to enhance the process of phloem unloading near the infection site in order to increase import and to reduce phloem loading in order to decrease export.

The powdery mildew parasite could have a direct influence on loading and unloading processes. These processes are active ones, which can produce movement against a concentration gradient as when sucrose is loaded into the phloem. To our knowledge, no direct studies of phloem loading and unloading have been made with powdery mildews. In an elegant study which could serve as a model for use with powdery mildews, Wolswinkel (1974) showed that the active unloading of phloem is enhanced in stem tissue parasitized by the angiosperm *Cuscuta* spp.

3. Symplast and apoplast

The tissues involved in the movement of substances from the phloem to a powdery mildew fungus are shown in Fig. 3. A typical minor vein is shown, separated by a mesophyll cell from the infected epidermal cell. Interveinal distances in leaves are generally 100–300 μm (Crafts and Crisp, 1971), so that most mesophyll cells are within two to three cell diameters of a vein (Geiger and Cataldo, 1969). The parasite is usually within similar distances from a vein.

Movement of solutes between phloem and other tissues of the leaf is postulated to occur by two systems: the symplast and the apoplast. The symplast consists of the cytoplasm of leaf cells interconnected by plasmodesmata. Movement of amino acids, sugars, inorganic ions and other substances within the symplast is diffusive, following concentration gradients but it is assisted by cytoplasmic streaming and possibly by electrical potentials (Arisz,

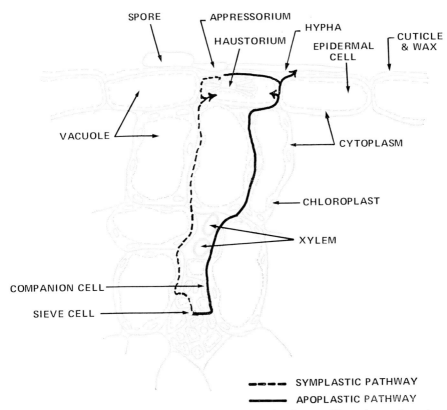

SYMPLASTIC PATHWAY
APOPLASTIC PATHWAY

Fig. 3. Portion of a mildewed leaf in cross section showing possible pathways for substances to move from the phloem (sieve and companion cells) of a minor vein to the mildew fungus. (Vein adapted from Geiger and Cataldo, 1969.) Plasmodesmata interconnect the symplast of adjacent host cells except at the junctions between epidermal and underlying mesophyll cells where plasmodesmata are apparently rare and symplastic continuity is unlikely. The principal pathways from epidermal cells to fungus have not been established but among the alternative pathways shown, the most likely route is from apoplast of epidermis to symplast of epidermis, to cytoplasm of infected cell, to haustorium. Neckbands close off the direct apoplastic pathway to the haustorium from the wall of the infected epidermal cell.

1969; Crafts and Crisp, 1971). Cataldo (1974) has suggested that symplastic movement involves a "transport compartment" consisting of endoplasmic reticulum. In any case, an important property of the symplast is that substances do not have to cross plasma membranes to move from cell to cell. To leave the symplast, on the other hand, substances must cross a plasma membrane (Arisz, 1969), which constitutes a permeability barrier for most substances.

Although mesophyll cells definitely have plasmodesmata between them (Robards, 1976), it seems unlikely that the powdery mildew haustorium has any direct connection with the symplast. As is shown later (Sections IV.B and IV.C), the haustorium lies outside the membrane bounding the cytoplasm of the cell it infects and connections between the membranes of host and pathogen have not been seen. Haustoria may be even further isolated from the symplast of the mesophyll tissues because most powdery mildews are exclusively epidermal parasites. (*Phyllactinia* and *Leveillula* spp., powdery mildew fungi which do not produce epidermal haustoria, are excluded from this discussion.) Plasmodesmata in epidermal cells seem to be limited to connections with adjacent cells of the epidermis, whereas epidermal-mesophyll plasmodesmata were not reported in a recent comprehensive text on intercellular communication in plants (Gunning and Robards, 1976). Systematic electron microscope examination of about 100 independent sections of infected pea leaves showed epidermal-mesophyll plasmodesmata to be virtually absent, at least in the lamina between major veins, although connections were found frequently between adjacent epidermal cells (M. Martin, unpublished). Thus, from the evidence now available, symplastic continuity probably does not exist between epidermal and mesophyll tissues and thence to conducting tissues, except possibly indirectly in the meristematic regions of leaf apices and bases.

The apoplast consists of the intercellular spaces and walls which are outside the symplast. Substances move by diffusion through the apoplast. As shown in Fig. 3, the apoplast provides a possible avenue for solutes to move from phloem to mesophyll cells, to epidermal cells, and to the parasite. Substances in the apoplast could enter the fungus either at the haustorial neck, an unlikely possibility as discussed later (Section IV.C.2), or directly into hyphae if they could pass the cuticle and wax layers on the epidermal surface (Fig. 3). Although cuticle is not very permeable to polar substances (Darlington and Cirulis, 1963), some inorganic ions, water and a wide variety of organic substances can move through cuticle (Crafts and Crisp, 1971; Schönherr, 1976; Yamada *et al.*, 1964). Furthermore, the powdery mildew fungus alters the cuticle layer at sites of penetration as detected histochemically (Kunoh *et al.*, 1975; Kunoh and Akai, 1969) and also alters the wax layer beneath appressoria and hyphae as indicated by the tracks seen by scanning electron microscopy after the fungus has been removed (Day and Scott, 1973; Schwinn and Dahmen, 1973). Thus, some substances may be able to move in the apoplast directly to mildew hyphae.

Most substances, however, probably do not reach the fungus directly via the apoplast, but instead, move from the apoplast into the symplast of the epidermis. Though not shown experimentally, substances which reach the symplast of infected epidermal cells then most likely move within the parietal layer of host cytoplasm (Fig. 3) to the cytoplasm which immediately surrounds the haustorium. A second, but less likely, route is via the vacuole, by which substances would have to cross the host tonoplast to enter the vacuole, diffuse through the vacuolar sap to the vicinity of the haustorium and from there recross the tonoplast to enter the cytoplasm around the haustorium. The final step from cytoplasm to haustorium will be discussed in Section IV.C.

4. Increased permeability

At least a portion of the solutes moving from phloem to mesophyll and across mesophyll cells is in the symplast. Likewise, photosynthate originates in the symplast of mesophyll cells. Does the parasite have a way to increase the permeability of the symplast to one or more solutes to facilitate movement of the solute(s) into the apoplast and thence to haustoria via epidermal cells? That rust and powdery mildew fungi are successful only if they increase the permeability of infected tissue was suggested by the well-known work of Thatcher (1939, 1942) nearly 40 years ago. Studying mesophyll cells, he measured permeability to urea, thiourea, and glucose by determining rates of deplasmolysis caused by influx of the substances into plasmolysed cells over periods of several hours. He also measured permeability to water using the rate of deplasmolysis on hypotonic solutions over periods of a few minutes. He found that the permeability of susceptible host cells to water and solutes was generally increased several-fold in rusts of pea, carnation and wheat. In powdery mildew of swede turnip (rutabaga), he found small increases in permeability to water in the mesophyll tissues beneath mildew colonies but he did not measure permeability to solutes in powdery mildews. Thatcher believed that the increase in permeability was necessary for the fungus to obtain nutrients, partly because permeability did not increase in resistant hosts (Thatcher, 1942, 1943).

Williams (1958) tried to use Thatcher's techniques to study host permeability in powdery mildew of barley. In a careful assessment of methods, he found that variation was too great among samples for valid comparisons between mildewed and healthy tissues. Variation was especially high in mildewed tissue.

Recently, however, Hoppe and Heitefuss (1974a,b,c) partly confirmed Thatcher's conclusions by measuring the amounts of certain solutes that leaked from rusted tissues. Bean rust significantly increased the amounts of amino acids, ions, and sugars that leaked from leaf segments incubated in water on a shaker for eight hours (Hoppe and Heitefuss, 1974a). The increased leakiness related to changes in membrane components, especially phospholipids and glycolipids. Implicated as possible causes of the increased permeability were certain chemical substances, the type depending on whether the leaves were resistant or susceptible (Elnaghy and Heitefuss, 1976a,b).

In a similar way, nitrogenous substances leaked from segments of mildewed barley leaves floated for 3 hours on buffer solutions (Table 1). The leakiness increased at 6 to 8 days after inoculation when the fungus sporulated abundantly. In this experiment, mildew mycelium was brushed from the segments

TABLE 1

Leakiness of powdery mildewed barley tissues as indicated by the percentage of total nitrogen in 1-cm segments that leaked to a washing solution in a 3-hour period after the segments were cut from leaves and mildew hyphae were brushed away (W. R. Bushnell, unpublished data)

Day after inoculation	Percentage N leaked	
	Healthy	Mildewed
4	1·3	2·7
6	1·3	9·5
8	2·8	11·3

before incubation so that some of the leaked material may have been exuded by haustoria which remained in epidermal cells. More and better data on the permeability of infected tissues are needed, particularly with respect to specific substances but, like the rusts, powdery mildews apparently produce a general increase in permeability to solutes in host tissues.

5. Use of specific compounds by the parasite

Another way that powdery mildew fungi apparently promote flow of solutes from their hosts is by converting them to insoluble substances or to soluble substances not utilized by the host. In powdery mildew of barley, Edwards and Allen (1966) determined which substances were radioactive in both host and parasite after $^{14}CO_2$ was fixed photosynthetically in the host for periods of 15–60 min (Table 2). The results indicated that photosynthate entered the fungus as sucrose which was then rapidly converted to mannitol. Small amounts of ^{14}C appeared in trehalose and arabitol, which along with mannitol were the major carbon reserves in mildew spores. None of these compounds was found in healthy leaves. However, small amounts of mannitol were found in host tissues underlying mycelium, possibly from haustoria left in epidermal cells after the fungus was removed. In line with these results, Hewitt and Ayres (1976) found trehalose, arabitol and mannitol to be the principal soluble carbon compounds in mycelium of oak mildew.

Smith et al. (1969) suggested that the conversion of host carbohydrates to polyols and trehalose tends to keep the concentration of soluble host carbohydrates low near the parasite, helping to maintain a concentration gradient which favours movement of host carbohydrates toward the parasite. Similar conversions occur in rust diseases, mycorrhizae and lichens (Smith et al., 1969; Holligan et al., 1973). In all these systems, the host (donor) does not use the fungal substance readily as apparently is the case for mannitol in powdery mildew of barley. Whether flow of the polyol (or trehalose) into the

TABLE 2

Data of Edwards and Allen (1966) which indicate that carbohydrate enters mycelium of *Erysiphe graminis* as sucrose and is then converted principally to mannitol. Shown is the distribution of ^{14}C among ethanol-soluble compounds in mycelium, host leaf tissues underlying mycelium, and healthy leaf tissues 15, 30, and 60 minutes after exposure to $^{14}CO_2$ in the light

Source	Sugar or sugar alcohol	Radioactivity, % of total		
		15 min	30 min	60 min
Mycelium				
	Sucrose	57	41	33
	Mannitol	21	34	45
	Trehalose	1	2	4
	Arabitol	4	6	6
Underlying host				
	Sucrose	67	73	75
	Mannitol	trace	4	4
Healthy host				
	Sucrose	76	83	85
	Mannitol	0	0	0

host is limited by permeability barriers or by lack of specific transport systems is not known.

Long *et al.* (1975) have questioned the interpretation of Edwards and Allen that sucrose is the main form in which carbohydrate enters the powdery mildew fungus. They postulated that sucrose is inverted before uptake by the fungus, based on study of invertase activity in rusts and experimental work with lichens and mycorrhizae. In testing this possibility for oak mildew, Hewitt and Ayres (1976) failed to find evidence for increased invertase activity. They did, however, implicate fructose and glucose as the sugars which are transferred from host to parasite in that both caused a leakage of ^{14}C labelled carbohydrates from mildewed leaf discs to a greater extent than did sucrose. This effect is thought to be specific for the carbohydrates which move from donor to recipient in lichens, mycorrhizae and other biotrophic associations between organisms (Smith *et al.*, 1969). Hewitt and Ayres did not determine directly which carbohydrates were labelled in the fungus after application of $^{14}CO_2$. Since the interface between the host cell and the haustorium probably is isolated from the leaf apoplast (as will be described in Section IV.C.2), the leakage experiment may not reflect events at the host–parasite interface.

6. Accumulation studies by Shaw and co-workers

Shaw and co-workers (Shaw *et al.*, 1954; Shaw and Samborski, 1956) studied the nature of accumulation processes in powdery mildews experimentally.

They showed that a variety of radioactive substances, including ^{14}C labelled sugars, sugar alcohols, amino acids and organic acids, as well as $^{32}PO_4^{3-}$ and $^{45}Ca^{2+}$, accumulated at mildew colony sites as detected in autoradiographs after the substances were injected into leaf tissues or applied to the basal cut ends of detached leaves. They showed that the accumulation did not depend on transpiration by the host plant but did depend on aerobic respiration, indicating that one or more steps in the accumulation process required metabolic energy.

In experiments with radioactive glucose, Shaw and co-workers found that a large part of the accumulation was in the host, in that much label remained after the mycelium had been brushed away. In this case, the increased labelling in the infected host tissue could have resulted from rapid turnover of substances moving toward the fungus instead of an actual increase in total amount (specific activities were not determined). However, accumulation of label also occurred if the mycelium was removed before label was applied, indicating an actual net accumulation in the host.

More significantly, they found that radioactive glucose usually did not accumulate if the epidermis which contained haustoria was stripped away along with the mycelium before the labelled glucose was applied. (Shaw and co-workers did not emphasize this finding; instead they showed an unusual autoradiograph in which accumulation continued after the infected epidermis was removed.) The usual absence of accumulation in this type of experiment suggests that haustoria, or epidermal cells with haustoria, are necessary for accumulation at former sites of mildew colonies, and conversely, that mesophyll tissues beneath the colonies do not have an induced ability to accumulate substances in the absence of the fungus. Haustoria are known to remain alive in epidermal cells after the mycelium is removed (Hirata, 1967). These haustoria possibly influence distribution of solutes in the host after mycelium is removed, perhaps by continuing to take up solutes. In any case, more experiments need to be done to test the possible role of mildewed epidermis in solute accumulation.

C. Physiological changes in host tissues

1. Changes during sporulation

When the fungus sporulates and makes its highest demands for food, host tissues at colony centres become chlorotic. Entire tissues will become chlorotic if the colonies are close together (Allen, 1942; Bushnell, 1967). For both rusts and powdery mildews, such yellowing has been taken to indicate a state of accelerated senescence (Bushnell, 1967; Farkas et al., 1964; Shaw and

Manocha, 1965). In line with this supposition, the amount of protein in mildewed host tissue declines (Johnson et al., 1966; Rubin et al., 1971) and chloroplastic RNA declines (Bennett and Scott, 1971; Callow, 1973), following an early decrease in polysome content of chloroplasts (Dyer and Scott, 1972). Respiratory rates in the host at this time increase 2–4-fold over normal rates (Allen, 1966; Bushnell and Allen, 1962b; Scott, 1965) much as they do in senescing healthy leaf tissues (Allen, 1966; Tetley and Thimann, 1974). The apparent senescence of mildewed tissues may relate to the increased permeability and loss of solutes described earlier, since increased permeability generally accompanies senescence of higher plant tissue.

Instead of senescence, Heath (1974) found that the changes in chloroplasts in chlorotic, rusted cowpea tissues resembled changes that occur in chromoplast formation in ripening fruit. The chloroplasts contained distinctive prolamellar bodies, peripheral vesicles, disorganized thylakoids and carotenoid crystalloids, none of which occurred in healthy senescing leaves. A similar effect on thylakoid stacking was found in powdery mildew of sugar beets (Magyarosy et al., 1976). Like senescent tissue, ripening fruit tissues also have an increased permeability to solutes (Sacher, 1962).

Heath (1974) found that ethylene induced a "ripening" of healthy cowpea leaves much as rust did. Futhermore, rusted leaves evolved 4–5 times more ethylene than did healthy untreated leaves. Ethylene production was also increased substantially in powdery mildew of barley, according to Hislop and Stahmann (1971) but the possible role of ethylene in permeability changes, chlorosis, or other host changes in powdery mildew has not been investigated. Ethylene is not only involved in ripening fruit but can accelerate senescence of leaf tissues (Abeles, 1972; Mayak and Halevy, 1972), it probably can mediate responses to cutting and other forms of injury (Abeles, 1972) and probably can affect secretory phenomena and transport of materials through membranes (Abeles, 1972).

Although ethylene should be studied as it relates to accumulation of solutes in mildewed tissues, other senescence factors may also be involved. Abscisic acid has some ability to accelerate senescence, especially in detached leaves (Colquhoun and Hillman, 1975; Mayak and Halevy, 1972; Milborrow, 1974) and other unidentified senescence factors are present in plants (Lindoo and Nooden, 1976; Milborrow, 1974). Finally, Magyarosy et al. (1976) speculate that the changes in mildewed sugar beets resemble those induced by diuron (3-(3,4-dichlorophenyl)-1,1-dimethylurea) based on reduction of non-cyclic photophosphorylation in the diseased tissues.

An unexplained characteristic of sporulating mildewed tissue, which possibly relates to permeability changes, is that pigments are difficult to extract from such tissues with non-polar solvents. Chlorophylls remain there long after all pigment has been removed from non-infected portions of leaves (Bushnell, 1967; Haspelová-Horvatovičová, 1964).

2. Changes before sporulation—green islands

Preceding sporulation and the induction of chlorosis in host tissues, powdery mildews can induce temporary green islands (Cornu, 1881; Bushnell, 1967; Camp and Whittingham, 1975) in which host tissues have an enhanced ability to accumulate substances from distant tissues. The green islands become visible only if the tissues are incubated under conditions which accelerate senescence, such as low light intensity, darkness, or detachment of leaves. In powdery mildew of barley, the green island can extend 0·4–1·0 mm ahead of hyphal tips at the edges of each mildew colony (Bushnell and Allen, 1962a; Bushnell, 1967). As chlorosis develops at colony centres, each green island becomes a ring, and disappears when colonies coalesce on densely infected leaves. Green islands are not visible in attached leaves growing under good conditions but a zone similar in size and location to the green island stains differentially with IKI, usually more intensely than the leaf background. Cytokinins have been implicated as a possible cause of green islands in powdery mildews (Bushnell and Allen, 1962a; Bushnell, 1967; Engelbrecht, 1968; Mandahar and Garg, 1976; Vizarova, 1973, 1974), and in other diseases (Dekhuijzen and Staples, 1968; Evans and Banerjee, 1973; So and Thrower, 1976) but several other types of substances can also induce green islands in healthy leaves or retard senescence of entire leaves or pieces of leaves (Atkin and Nielands, 1972; Bushnell, 1966; Takegami and Yoshida, 1975; Thimann et al., 1974; Thomas, 1976).

Green islands become sites of accumulation for substances released by senescing parts of leaves. Thus, the nitrogen content of host green island tissue in detached leaves increased by 30% (Bushnell, 1967). Such translocation of amino acids and other substances into an area of delayed senescence within a detached leaf was first described by Mothes (1960) and has been termed the "Mothes effect" (Thimann et al., 1974). This effect can be produced by topical application of cytokinins or other senescence inhibitors to a portion of the leaf. Translocation from yellowing to green parts of such leaves occurs via the phloem (Müller and Leopold, 1966; Nakata and Leopold, 1967) and, like translocation into mildewed tissues, the movement is thought to result from altered source-sink relations instead of a direct effect on the phloem translocation pathway (Brautigäm and Müller, 1975; Schenk, 1974). Thus, protein degradation in the senescing source tissues provides a supply of amino acids, whereas the more juvenile sink tissues maintain a strong ability to accumulate amino acids.

Differential rates of senescence are difficult to induce in the normal, attached, green leaf and any areas of delayed senescence are not sites for accumulation of amino acids or other solutes in large amounts. Repeated topical applications of cytokinin at high concentrations are usually required to delay senescence of attached leaves under good environmental conditions

(Bushnell, 1966; Jacoby and Dagan, 1970; Fletcher *et al.*, 1970). In such cytokinin-treated tissues, the amounts of nitrogen and other substances increase to only a small extent (Adedipe and Fletcher, 1971; Bushnell, 1966; Engelbrecht, 1964). Likewise with powdery mildew, total nitrogen did not increase in host tissues within the zones of influence in attached leaves which correspond to green islands (Bushnell, 1967). The "Mothes effect", therefore, is not a consistent part of accumulation in powdery mildews.

III. Organization of the Mildew Colony

K. Hirata and co-workers have described in meticulous detail the development of hyphae and conidia by *Erysiphe graminis* f. sp. *hordei*, and have shown how these structures relate to haustoria in both location and number (Hirata, 1955, 1960, 1967, 1971, 1973; Nishiyama *et al.*, 1966). These unique studies indicate directly the number and volume of fungal cells that are supported in growth by the absorbing activities of individual haustoria. The growth rates of these cells can therefore be used to estimate flow rates of water and solutes into individual haustoria.

As shown by Hirata (1967) as well as others (Bushnell, 1971; Bushnell and Bergquist, 1975; Ellingboe, 1972; Stanbridge *et al.*, 1971), the mildew fungus germinates and forms an appressorium by 9–10 h after inoculation, penetrates the wall of a host epidermal cell by 10–12 h and then begins to produce a haustorium within that cell. When the haustorium starts to produce its finger-like lobes, about 18 h after inoculation, a hypha begins to grow out from the appressorium (Hirata, 1967; Masri and Ellingboe, 1966). This hypha elongates and branches while additional hyphae sometimes grow from the appressorium (Fig. 4). As the colony continues to grow, it branches repeatedly, so that the rate of total mycelial growth for the entire colony continuously increases. By 45–50 h after inoculation but just before secondary haustoria are produced, Nishiyama *et al.* (1966) found that the total length of hyphae within a colony was 600–1000 μm and the rate of total mycelial growth was 80 μmh^{-1}. Since hyphae are 3 μm in diameter (Hirata, 1967), the increase in hyphal volume was 565 μm^{3}h^{-1}. This, then, was the volume of new cell material being supplied each hour by the single primary haustorium.

Several secondary haustoria are produced 50–70 h after inoculation, usually in the middle of the daily dark period, the time and number depending on environmental conditions and the time of inoculation relative to the daily dark period. Once these secondary haustoria form, the activity of the primary haustorium is difficult to assess but Hirata estimated that its absorbing activity increased for several days (Hirata, 1967, 1971).

The mildew colony continues to grow, producing new batches of haustoria

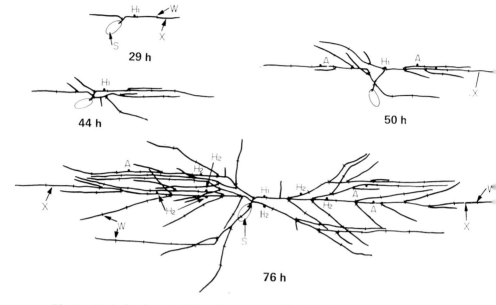

Fig. 4. Typical colonies of *Erysiphe graminis* 29–76 h after inoculation as shown by Hirata (1973). Each colony until 50 h has one primary haustorium (H_1). The colony at 76 h has six secondary haustoria (H_2). A—appressorium; S—conidium; W—septum of hypha; X—axial hypha. Magnification, ×375.

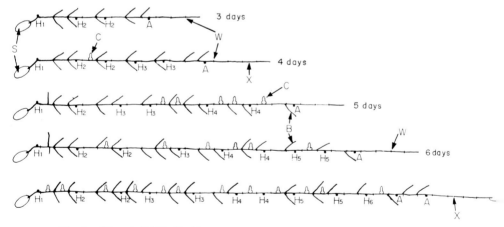

Fig. 5. Hirata's (1973) distribution of branches (B), conidiophores (C), and haustoria (H) on axial hyphae (X) of *Erysiphe graminis* 3–7 days after inoculation. H_1, H_2, to H_6 show the primary, secondary, to sixth set of haustoria, respectively. Conidiophores start to form at four days and eventually nearly equal haustoria in number. S—conidium; W—septum of hypha.

each night, and starts to produce conidiophores and conidia on the fourth day after inoculation, eventually producing 1000 or more haustoria and conidiophores with as many as 80 haustoria in a single epidermal host cell. From analyses of entire colonies, Hirata determined that the proportions of hyphal cells, branches, conidiophores, and haustoria are approximately as shown in Fig. 5 for the hyphae growing on the longitudinal axis of colonies. Haustoria are produced 6–8 hyphal cells behind the growing hyphal tip. Eventually at least one haustorium is produced for every three hyphal cells. The hyphal cells that do not produce haustoria usually produce a hyphal branch, a conidiophore, or both.

Significantly, there was one haustorium for each 1–1·3 conidiophores; i.e., the absorbing activity of each haustorium in the sporulating colony centre provided the substances used by 1–1·3 conidiophores. Hirata observed that each conidiophore produced 8–10 conidia day^{-1} for at least 4–5 days. Each conidium had a volume six times that of hyphal cells, or 3400 μm^3. At the average rate of 0·4 spores h^{-1}, each conidiophore produced 1360 μm^3 of new spore volume h^{-1}. Thus each haustorium supported 1–1·3 times this volume of new growth, about three times the volume of new hyphal growth supported by the primary haustorium at 45–50 h after inoculation.

These values represent our first available estimates of the absorbing capacity of the haustorium. Since water constitutes the major fraction of spores and hyphae, these values indicate primarily the water absorbing capacity of haustoria, without correction for any water lost by transpiration from hyphae and spores, or water entering the fungus by routes other than through haustoria. In a similar way, the absorbing capacity of individual haustoria for carbon or other elements could be estimated from Hirata's growth rate for spores and the content of the element in spores as determined *en masse*.

IV. Structure and Function of the Haustorium and Associated Host Structures

A. Hirata's observations on haustorial structure

Many investigators have described haustoria of powdery mildew fungi as these structures have appeared in fixed condition by light microscopy. Most often cited is the work of Smith (1900) who provided the first detailed study of haustorium development. Smith, however, failed to recognize that haustoria of all powdery mildew fungi are lobed and he also confused development of papillae with development of the extrahaustorial membrane. The first descriptions of powdery mildew haustoria by light microscopy that conform to the view we now have by electron microscopy were provided by Hirata

1937), who later expanded his observations in a series of papers related to his studies of mildew colonies (Hirata, 1958, 1959, 1960, 1971, 1973; Hirata and Kojima, 1962).

1. The haustorium

Hirata described the two types of powdery mildew haustoria as shown in Fig. 6, that of *Erysiphe graminis* and that of other mildews as represented by

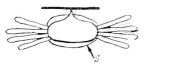

Fig. 6. The two types of powdery mildew haustoria as shown by Hirata and Kojima (1962). Left—haustorium of *Erysiphe graminis*. S—sac. Right—haustorium of *Sphaerotheca fuliginea.*

Sphaerotheca fuliginea (Hirata, 1937, 1967; Hirata and Kojima, 1962). Both have an ellipsoidal central body, but unlike the straight, finger-like branches of *E. graminis*, those of *S. fuliginea* extend convolutely over the surface of the body, giving the entire structure a compact form. Hirata noted that the two types of branching are homologous in that both develop from the two ends of the long axis of the ellipsoidal central body. The form of the compact type of haustorium has been confirmed by examination of swollen haustorial complexes isolated from pea (Gil and Gay, 1977a), beet, apple and vine (Manners and Gay, 1977) and supported by electron microscope observations of *Erysiphe cichoracearum* (Gil, 1976; McKeen *et al.*, 1966), *Erysiphe polygoni* (Stavely *et al.*, 1969), *Sphaerotheca fuliginea* (Dekhuijzen and van der Scheer, 1967, 1969), and *Sphaterotheca pannosa* (Perera and Gay, 1976).

Hirata focused his further studies on haustoria of *E. graminis*. He found that the haustorium of this fungus required 4–5 days to develop its full length of 160 μm (Nishiyama *et al.*, 1966) and to develop its full complement of lobes, about ten at each end. This was a longer time than indicated by others (Masri and Ellingboe, 1966; Bushnell, 1971), who did not continue measurements beyond 2–3 days. The 4–5 day period of haustorial growth appeared to be a time of active absorption of water and solutes by the haustorium in that attached hyphae grew vigorously at that time. Others have also concluded that young haustoria can take up solutes (Bushnell, 1971, 1972), especially Ellingboe and co-workers (Hsu and Ellingboe, 1972; Mount and Ellingboe, 1969; Slesinski and Ellingboe, 1971; Stuckey and Ellingboe, 1975). Using techniques described in Chapter 7, they showed that ^{35}S applied as

SO_4^{2-} to cut ends of detached barley or wheat leaves started to move rapidly into *E. graminis* at 16–20 h after inoculation. This occurred after the central body of the haustorium had been produced and as the haustorial lobes were beginning to form, whereas very little ^{35}S entered the fungus before the haustorium had been produced. Similar results were obtained with ^{32}P, apparently applied as PO_4^{3-} (Mount and Ellingboe, 1969).

Hirata (1967) found that the entire haustorium would swell or shrink, depending on the condition of the hyphae associated with it. The young, primary haustorium attached to vigorously growing hyphae was shrunken. The demands of actively growing hyphae for water apparently slightly exceeded the absorbing capacity of the haustorium until it shrank slightly, thereby decreasing its water potential.* The haustorium attached to hyphae growing poorly because of resistance, fungicide treatments, or other factors, was swollen. If hyphae were removed from the leaf surface, the haustoria that were left in epidermal cells would remain living and also become swollen. The swelling and shrinking of the haustorium as described by Hirata have not been confirmed by other workers. Hirata's figures indicate that the changes in haustorial size reflect changes in the haustorial cell, a phenomenon distinctly different from changes in the haustorial sac which also can swell and contract as will be described below.

2. The haustorial sac

Hirata found that a membrane-like structure would separate from the host side of the haustorial wall after the haustorium became fully grown (Fig. 7). The separation started along small portions of the central body and later grew to include the entire central body and portions of haustorial fingers,

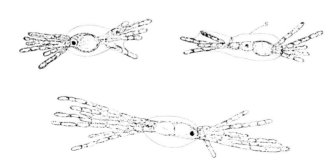

Fig. 7.   The haustorial sac associated with haustoria of *Erysiphe graminis*. S—sac. (From Hirata, 1967.)

* As used in this chapter water potential is the negative of diffusion pressure deficit and osmotic potential is the negative of osmotic pressure.

either singly or in groups. Initially he called this structure a "vesicle" (Hirata, 1937), but later used "sac" or "sack" because the entire structure formed a sealed semipermeable compartment which would expand or contract osmotically when "pulled" from the host cell and placed in water or glucose solutions, respectively (Hirata and Kojima, 1962; Hirata, 1967). The sac would separate from young haustoria when hyphal growth was poor, or when hyphae were removed entirely with a damp cotton swab (Hirata, 1967). Lithium salt solutions applied to the cut ends of detached leaves also caused any sacs to expand (Hirata, 1959).

Hirata (1967) interpreted the sac to be a thickened derivative of the host plasma membrane (ectoplast). He believed that its separation probably signalled the end of nutrient flow from host to haustorium but was uncertain whether the separation was a cause or an effect of weak absorption by the haustorium. In any case, the expanded sac apparently served to separate the haustorium from host cytoplasm, especially when haustoria were dead. The dead haustoria could coexist with living haustoria in living host cells for long periods. The sacs around the dead haustoria were expanded and retained their semipermeability.

Hirata (1967, 1971) suspected that the ability of the host to form the sac in response to the invading fungus was favoured by calcium ions. Solutions of calcium salt applied to the cut ends of detached leaves increased leaf susceptibility and the cells of the epidermis with high amounts of calcium, especially the auxiliary (subsidiary) cells near stomates, were more susceptible to mildew than were cells with smaller amounts of calcium as detected cytochemically. Lithium salts were antagonistic to the effects of calcium. Hirata speculated that calcium favoured sac formation by increasing the viscosity of cytoplasm but he did not show that the increased susceptibility was, in fact, related to enhanced sac formation.

3. Deposits around the haustorium

If haustoria were allowed to remain in cells after hyphae were removed from the leaf surface, the haustorium became swollen and rich in protoplasm and any young haustoria actually continued to grow toward their full size. After 2–3 weeks of incubation, a membranous substance was deposited around the haustorium, starting on the central body, extending to the fingers and eventually enclosing almost the entire haustorium (Fig. 8). Similar deposits formed when colony development was halted by treatment with fungicides or free water. The deposits did not appear to injure living haustoria, although they were sometimes seen around dead ones.

Similar structures have been found in electron microscope studies of *S. pannosa* on rose (Perera and Gay, 1976) and *E. cichoracearum* on Michaelmas Daisy (*Aster nova belgii*) (Crisp and Gay, unpublished). The encasements

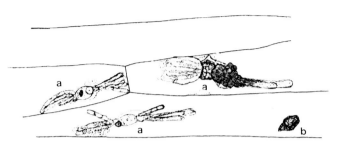

Fig. 8. Membranous deposits around living (a) and dead (b) haustoria of *Erysiphe graminis*. (From Hirata, 1971.)

develop as cup-shaped extensions of the collar, and finally enclose the whole haustorium (Fig. 9). Similar development has been reported in several downy mildew and rust infected plants (Bracker and Littlefield, 1973) and has been shown to be associated with resistance to cowpea rust (Heath, 1971; Heath and Heath, 1971).

B. The fine structure of the haustorium and its interface with the host

The haustorium is an elaborately shaped branch system arising from an appressorium (Fig. 17a). As Hirata recognized, the expanded haustorial form characteristic of gramineous mildews and the compact form of other mildews are homologous. The origins of the lobes are easily seen in haustoria isolated from infected leaves (Figs 19a,c). In both types, one group of lobes is terminal and the other arises near the junction of the neck with the body. The whole haustorium of *E. graminis* alters its orientation at about the time that the lobes arise (Bushnell, 1971) so that its axis lies parallel to the epidermal surface. This does not occur in the compact type.

Electron microscope studies show that the structures of the two types of haustoria are closely similar (Bracker and Littlefield, 1973). As in all vegetative cells in the Erysiphales, the haustorial body has a single nucleus (Figs 14, 15) and the septum at its junction of the neck is perforate, providing a continuous path to the appressorium and other hyphal compartments (Fig. 14). The haustorial cytoplasm lacks a large vacuole but several small ones are common. The cytoplasm is usually rich with ribosomes and sometimes glycogen granules are present (Perera and Gay, 1976). A feature which is conspicuous and probably significant in relation to haustorial physiology, is the high frequency of mitochondria in the lobes and the peripheral cytoplasm of the body (Figs 14, 15, 16).

In all powdery mildew haustoria examined by electron microscopy (Bracker and Littlefield, 1973; Ehrlich and Ehrlich, 1971; Gil and Gay, 1977a; Kunoh,

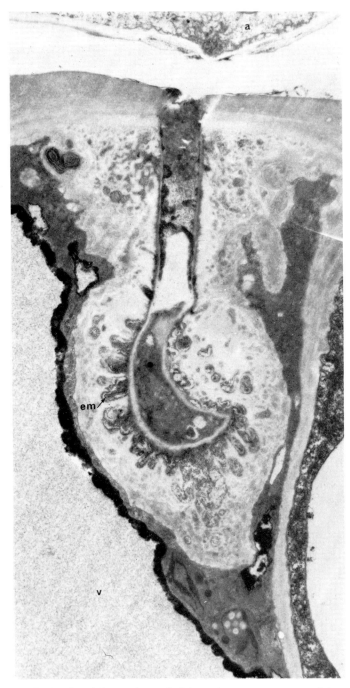

Fig. 9. Poorly developed haustorium of *Sphaerotheca pannosa* encased in collar-like material. The position of the extrahaustorial membrane (em) over the lobes indicates that the encasement develops as an overgrowth from the collar. The vacuole (v) is lined with a dense deposit which probably contains phenolic compounds. The appressorium (a) has become detached from the epidermis. Magnification, ×11 000. (From Perera and Gay, 1976.)

Figs. 10–11. Living haustoria of *Erysiphe graminis* f. sp. *hordei*.

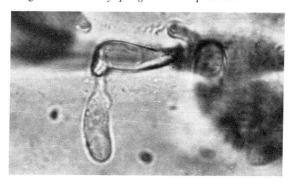

Fig. 10. Young, unlobed haustorium has invaginated the parietal cytoplasm of the host cell and extended into the host vacuole. The invaginated cytoplasm forms a sac which is not visible against the haustorial wall. 18 h after inoculation. Bright field microscopy; magnification, ×1000. (Adapted from Bushnell, 1971.)

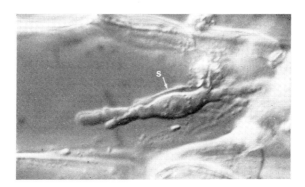

Fig. 11. Growing haustorium has produced finger-like lobes which extend into host vacuole. The sac (s) is visible along one side of the central body of the haustorium but not along young lobes from which it rarely separates. 41 h after inoculation. Differential interference contrast; magnification, ×1000. (W. R. Bushnell, unpublished.)

1972a; McKeen, 1974; McKeen *et al.*, 1966; Perera and Gay, 1976; Stavely *et al.*, 1969), the haustorium and its lobes lie in a pocket of membrane (the extrahaustorial membrane*) which is almost certainly an invagination of the plasma membrane of the host cell (Figs 14, 15). However, there has been difficulty in observing the membrane around the haustorial neck because dense amorphous substances, which are especially abundant in cereals, fill

* Terminology here follows that of Bushnell (1972). Structures around the haustorium have also been designated as the "sheath" or the "encapsulation", which have included what is here termed the extrahaustorial matrix and sometimes also the extrahaustorial membrane (Bracker and Littlefield, 1973; Ehrlich and Ehrlich, 1971).

the cylindrical channel between the neck and collar around it (Fig. 15). Even in the most favourable circumstances continuity has not been established because of a special structure (the B band) in the neck region which is discussed below. In any event the extrahaustorial membrane is the boundary to the host cytoplasm adjacent to the haustorium. It is separated from the vacuolar membrane of the host cell by a layer of host cytoplasm but this is often thin (Fig. 14), especially in cereals (Figs 15, 16), so that by light microscopy the three separate layers cannot easily be resolved and thus have been considered as one entity, the haustorial sac (Bushnell, 1971).

The zone between the extrahaustorial membrane and the haustorial wall is the extrahaustorial matrix. This matrix contains amorphous material (Figs 14, 15), which is usually granular but the granules often differ in frequency from one haustorium to another or may even be absent. The haustorial wall is distinguishable from the matrix (Figs 14, 15, 16) but sometimes its matrical margin is ill-defined. Reports of mitochondria within the matrix (Kunoh 1972b; Kunoh and Ishizaki, 1973) have not been substantiated.

Because the haustorium is separated from the host cytoplasm, in a way it fundamentally resembles a hypha in contact with a host cell. However, the interface is highly specialized and much of the following discussion concerns the distinctness of the extrahaustorial membrane and matrix, which are adjacent to the haustorium, from the plasma membrane and cell wall around other regions of the same cell.

C. Physiology of the host-parasite interface

1. Living haustoria, hyphae and host cells in a microculture system

A microculture system was developed by Bushnell et al. (1967) in which living, functional haustoria could be observed by light microscopy. The system consisted of microcolonies of *Erysiphe graminis* f. sp. *hordei* growing on a single layer of barley epidermal tissue and mounted to facilitate microsurgery and other experimental procedures on individual host cells and haustoria while hyphal growth was being monitored. As will be described briefly here, the system proved valuable for observing the haustorial sac before and after various treatments and for studying the effects on colony development of removing the host protoplast or the haustorium.

(a) The haustorial sac.

The development of young haustoria in the microculture system was obscured by temporary but massive, aggregates of host cytoplasm (Bushnell, 1971; Bushnell and Zeyen, 1976). As the young haustorium emerged from this aggregate, it was covered by an extremely thin layer of host cytoplasm, so thin and so close to the haustorial wall that it was difficult or impossible to see (Fig. 10). It separated from the haustorial wall under a variety of treatments, much as Hirata had described for the haustorial sac. The component layers of the invaginated cytoplasm usually could not be resolved by light microscopy but, as noted earlier, electron microscopy clearly shows that the host tonoplast and plasma membranes are present, enclosing a layer of cytoplasm (Bracker and Littlefield, 1973). All three components moved together when the structure expanded or contracted osmotically. The entire structure was termed the "sac" by Bushnell and co-workers (Bushnell, 1971, 1976; Sullivan et al., 1974) since it behaved the same as the sac which Hirata had regarded as a single membrane.

Hoskin (1969) carefully studied the position of the haustorial sac in the microculture system with respect to growth of hyphae attached to haustoria. As Hirata had concluded from his study of fixed specimens, Hoskin found that the haustorial sac was usually pressed against the haustorial wall when the hyphae were growing. Thus she could not see the sac around the primary haustorium of 45–50% of 2-to 3-day-old colonies because the sac was located against the wall. In the remaining colonies, only a small portion of the sac was visible where it had moved away from the haustorial wall, either at the base of fingers or along one side of the central body of the haustorium (Fig. 11). She clearly showed that the 3-day colonies were, in fact, growing at the time of observation.

Hoskin (1969) noted that the sac was never visible on the apical 66% of haustorial fingers (Fig. 11), regions of new and possibly continuing growth of the young haustoria under study. As we will see, the sac adheres to fingertip regions under a variety of treatments that separate the sac from the central body of the haustorium.

In further confirmation of Hirata's work Hoskin studied the osmotic behaviour of the haustorial sac. It would contract when host cells were plasmolysed on sucrose osmotica, so that any visible portions of the sac disappeared as they touched the haustorial wall. If she incised the host cell with a microneedle so that water or osmotica came in contact with the haustorial apparatus, the sac would contract on solutions hypertonic to the host cell and expand on solutions hypotonic to it, indicating that the solution in the space between the haustorial wall and the sac (the extrahaustorial matrix) was approximately equal to the host in osmotic potential. The contraction or expansion occurred during only 2–3 sec on highly hyper- or hypotonic osmotica, indicating that the sac was highly permeable to water. Once the sac was allowed to expand, the haustorium degenerated within 1·5 h. If the sac were not allowed to expand, the haustorium would remain alive in hypertonic osmotica for several hours. Regardless of the osmotic regime, the sac did not separate from the apical parts of haustorial fingers.

By examining specimens before and after treatment, Hoskin (Hoskin, 1969; Bushnell, 1971) assessed the effects of several chemical fixatives on the haustorium and sac. OsO_4 gave the most life-like results, although it caused some expansion or contraction of the sac of individual specimens. Some treatments of $KMnO_4$ produced an immediate and pronounced expansion of the sac, an effect which apparently had exaggerated the separation between the sac and the haustorial wall in the otherwise excellent EM work by Bracker (1968) and which was avoided in a follow-up study with better fixation (Bracker and Littlefield, 1973). In spite of pronounced expansion around the central body on $KMnO_4$, the sac generally remained close to the wall of the haustorial fingertips.

The observations of living specimens in the microculture system, in combination with Hirata's findings, indicate the following facts about the haustorial sac of E. graminis.

1. The sac is highly permeable to water and less so to sucrose.
2. The solution between the sac and the haustorium is normally in osmotic equilibrium with the host cell. If solutes and water enter and leave the haustorium by way of the sac, they must, as Hoskin (1969) concluded, enter and leave the haustorium in proportions that maintain the osmotic potential of the contents of the sac at a more or less constant value, either equal to or slightly below that of the host cell. Assuming that the sac itself exerts negligible wall pressure, it follows that the water potential within the sac virtually equals the water potential of the host cell.
3. The solution is completely separate from the host cytoplasm or vacuole; no channels or openings allow mass flow through the sac to the solution.
4. Any space between the sac and the haustorial wall contains no more than trace quantities of structural or solid material.
5. The sac is surprisingly tough, resistant to a variety of treatments as others (Bracker, 1968; Dekhuijzen and van der Scheer, 1969) have concluded for the extrahaustorial membrane.
6. The sac has a much tighter association with the young parts of haustorial fingers than elsewhere, a fact of uncertain functional significance.

(b) Removal of haustoria.

The effect of removing the primary haustorium on mildew colony growth was assessed in the microculture system (Bushnell et al., 1967; Bushnell, 1971). The lower wall of an epidermal cell containing the primary haustorium of a young colony was incised with a microneedle and the haustorium was then excised, leaving only part of the haustorial neck in place. The colonies immediately stopped growing but resumed growth a few hours later and grew very slowly for the next 1–1·5 days (Fig. 12). About 66% of the colonies

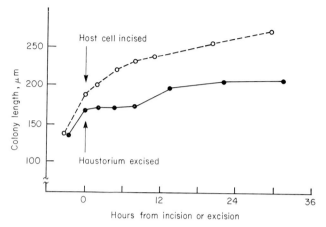

Fig. 12. Growth of *Erysiphe graminis* f. sp. *hordei* after host protoplasts were destroyed by incision (open circles) and after the single haustorium of each colony was excised (closed circles). Averages of 16–18 colonies, all supplied 0·5 M sucrose and 0·05% yeast extract. (Adapted from Bushnell *et al.*, 1967.)

exhibited this delayed, slow growth, elongating 7–26 μm in experiments conducted free from contaminating organisms (Bushnell, 1971). Growth was better with, than without, glucose or sucrose applied to the incised cells but was not improved by addition of yeast extract or a peptone rust culture medium (Bushnell, 1971).

Thus, the mildew hyphae could remain alive and grow in trace amounts without haustoria but the growth rate without haustoria was much less than when haustoria were attached to colonies, either in normal or incised host cells. Apparently, the haustorium provides essential processes for the normal rapid growth of colonies which cannot be replaced by ordinary nutrient media.

(c) Removal of host protoplast.

Hoskin (1969) tried to remove the host protoplast from haustoria of *E. graminis* by plasmolysis, a procedure that had worked for rust and downy mildew fungi and had helped demonstrate that haustoria invaginate host protoplasts (Fraymouth, 1956; Thatcher, 1943). At best, only portions of the *E. graminis* haustorium, usually parts of fingers, appeared to extrude from the shrinking protoplast. The protoplast seemed to be attached to the base of haustorial fingers or the central body (possibly by a neckband–see Section IV.C.2). Rarely, the main protoplast moved away, leaving irregular remnants of the protoplast on the haustorium but the protoplast never separated cleanly from the haustorium.

Bushnell *et al.* (1967) and Sullivan *et al.* (1974) tried to remove the host protoplast with microneedles using a variety of osmotic media made from inorganic salts, sucrose, or both. Host cells were either incised immediately after osmoticum was applied, or after a period of equilibration. The haustoria degenerated if the osmoticum was not low enough in osmotic potential to keep the sac from expanding. The haustoria survived for 2 h or more while hyphae often continued to grow at near-normal rates if the osmotic potential was low enough to keep the sac from expanding (Fig. 12). In these cases, living cytoplasm in a surprising array of configurations tended to collect on or near the haustorium after incision (Fig. 13). Most of this cytoplasm could be swept away with repeated movements of the microneedle but cytoplasm still tended to collect in a layer on the host cell wall around the haustorial neck and frequently in small quantities on the haustorial apparatus. The haustorial sac remained in place. The time of continued streaming or non-random vibratory movement in the remnant cytoplasm on or near the haustorium generally correlated with the time of hyphal growth. As the amount of cytoplasm was reduced, the amount of growth was reduced but growth persisted for 1·5 h (colonies grew 8–17 μm) when the haustorial sac was left with only trace amounts of remnant cytoplasm upon it. So far, no way has been devised to remove all the cytoplasm without injury to the haustorium. Clearly the intact host protoplast is not required for near normal functioning of haustoria and hyphae. However, as Sullivan *et al.* (1974) concluded, their results do indicate ". . . that the host cytoplasm had an active role in maintenance and function of the haustorium in the open cells".

2. Studies of isolated haustorial complexes

Considerable evidence on the specialization of the host-parasite interfacial structures has been obtained by Gil and Gay (1977a) who used an isolation technique modified from one devised by Dekhuijzen (1966). They isolated haustorial structures from pea infected with *Erysiphe pisi*. The leaves were brushed to remove the external mycelium and conidia, washed with water and then were comminuted to release the structures which were purified by filtration and centrifugation. The structures which were isolated comprised haustoria with the intact extrahaustorial membrane and matrix around them and the term "haustorial complex" was introduced for this composite structure (Figs 17b, 18a,b,c). Unlike the sac in the microculture system, host cytoplasm does not remain around the isolated complex. The haustorial body of the isolated complex retains its cytoplasm because the pore in the septum at the junction with the neck seals when the mycelium is detached and, apart from removal of the neck cytoplasm, isolation does not damage the structure. Its respiration rate is comparable with that of a yeast cell.

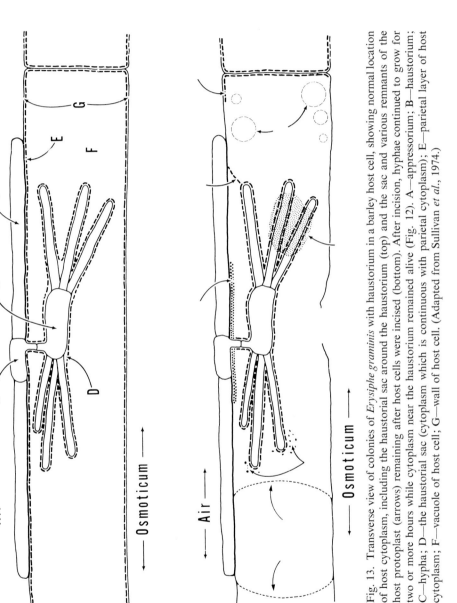

Fig. 13. Transverse view of colonies of *Erysiphe graminis* with haustorium in a barley host cell, showing normal location of host cytoplasm, including the haustorial sac around the haustorium (top) and the sac and various remnants of the host protoplast (arrows) remaining after host cells were incised (bottom). After incision, hyphae continued to grow for two or more hours while cytoplasm near the haustorium remained alive (Fig. 12). A—appressorium; B—haustorium; C—hypha; D—the haustorial sac (cytoplasm which is continuous with parietal cytoplasm); E—parietal layer of host cytoplasm; F—vacuole of host cell; G—wall of host cell. (Adapted from Sullivan *et al.*, 1974.)

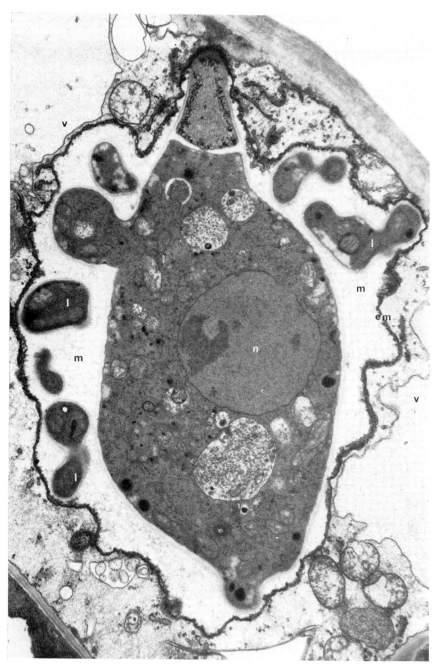

Fig. 14. Haustorium of *Erysiphe pisi* in cell of *Pisum sativum* showing the perforate septum between neck and body, the single nucleus (n), and peripheral mitochondria. Lobes are shown attached at the proximal and distal ends of the haustorium and sections of other lobes (l) are shown in the extrahaustorial matrix (m). The matrix contains small irregularly arranged granules and is separated from the host cytoplasm by the extrahaustorial membrane (em). The membrane is close to the tonoplasts of vacuoles (v) on each side of the haustorial complex. Magnification, ×13 000. (From Gil and Gay, 1977a.)

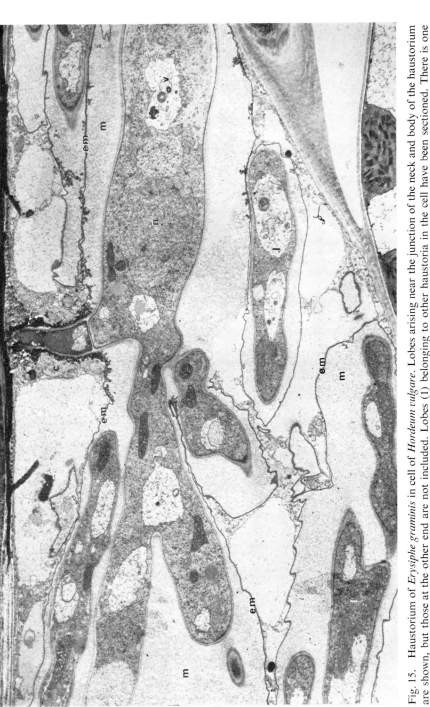

Fig. 15. Haustorium of *Erysiphe graminis* in cell of *Hordeum vulgare*. Lobes arising near the junction of the neck and body of the haustorium are shown, but those at the other end are not included. Lobes (l) belonging to other haustoria in the cell have been sectioned. There is one nucleus (n) and a small vacuole (v) in the haustorial body and mitochondria are prominent in the lobes. The extrahaustorial membrane (em) continues into the channel between the neck and collar (c), and it encloses the granular matrix (m) which is often continuous around several lobes (left). Magnification, ×10 500. (From Sargent and Gay, unpublished.)

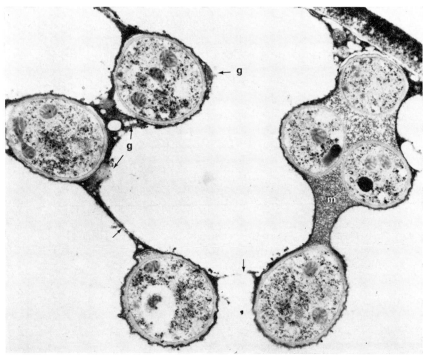

Fig. 16. Transverse sections of haustorial lobes of *Erysiphe graminis*. As in Fig. 15, the extrahaustorial matrix (m) is continuous between several lobes and its granules and ribosomes within the lobes are stained as a result of ruthenium red used in fixation. Thin strands of cytoplasm (arrows) extend between groups of lobes. Golgi bodies (g) are present but are not conspicuous because membranes are not stained after this fixation. Magnification, ×13 600. (From Sargent and Gay, unpublished.)

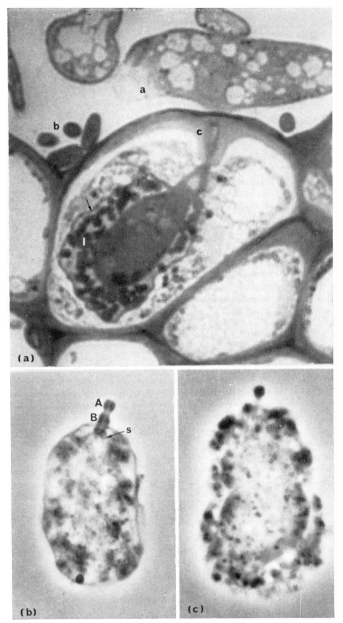

Fig. 17a. Light micrograph of a sectioned, infected epidermal cell of *Pisum sativum* showing a haustorium within the cell, and aerial hyphae, the appressorium (a), and bacteria (b) outside the cell. A broadly based collar (c) surrounds the haustorial neck, and haustorial lobes (l) extend from the body into the matrix. The extrahaustorial membrane is arrowed. During preparation the host cytoplasm has contracted slightly leaving a space against the outer cell wall. Magnification, ×2200. (From Gil, 1976.)

Fig. 17b. Living haustorial complex isolated from infected *Pisum sativum* and suspended in isolation medium (phosphate buffer, pH 6·7, 0·1 M; sucrose, 0·1 M; and $MgSO_4$, 1·5 mM). The extrahaustorial matrix is shrunken so that the extrahaustorial membrane invests the lobes closely and they are unclear. The A and B neckbands (A and B) and the septum (s) are shown. Magnification, ×2580. (From Gil and Gay, 1977a.)

Fig. 17c. As Fig. 17b, but treatment with NaOH has dissolved the extrahaustorial membrane and the B neckband, leaving the haustorium invested only by its lobes. Magnification, ×2580. (From Gil and Gay, 1977a.)

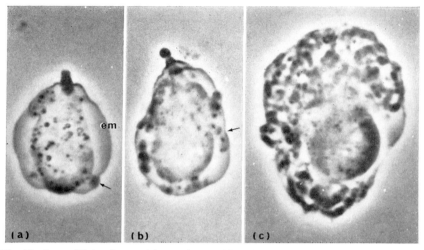

Fig. 18a, b, c. Living haustorial complexes with different degrees of lobe development isolated from infected *Pisum sativum* and suspended in water. The extrahaustorial membrane (em) is distended. Note its attachment (arrowed) to the lobes. Magnification, ×2000. (From Gil and Gay, 1977a.)

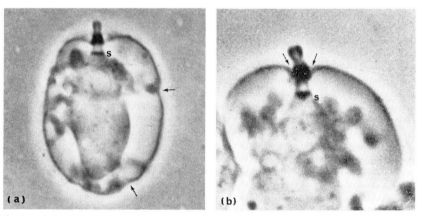

Fig. 19a, b. Living haustorial complexes isolated from infected *Pisum sativum* and swollen in water. The A neckband is at the broken end of the neck, and the distended extrahaustorial membrane is attached (arrows) to one edge of the B neckband (19b) and the haustorial lobes (19a). The septum (s) is at the junction of the neck and body. Magnifications: 19a, ×2080; 19b, ×3350. (From Gil and Gay, 1977a.)

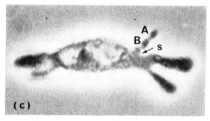

Fig. 19c. Living immature haustorium of *Erysiphe graminis* isolated from *Hordeum vulgare* and mounted in water. The lobes are only partly developed and the extrahaustorial membrane has been lost. The A and B neckbands (A and B) and the septum (s) are shown. Magnification, ×1450. (From Manners and Gay, 1977.)

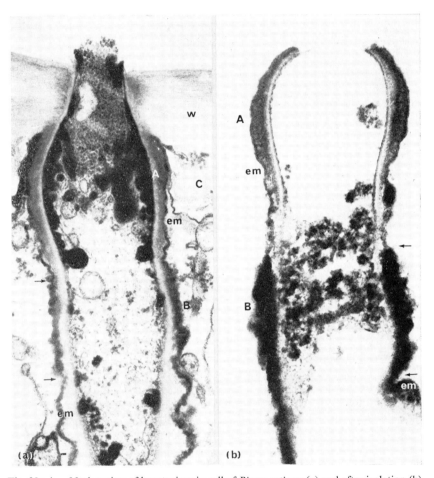

Fig. 20a, b.   Neck region of haustorium in cell of *Pisum sativum* (a) and after isolation (b). The A neckband (A) is barrel-shaped. It adjoins the inner surface of the cell wall (w) and is surrounded by the collar (c) (a). The extrahaustorial membrane (em) and fungal plasma membrane respectively cover and line its surfaces (b). The extrahaustorial membrane joins both edges (arrowed) of the B band (B). Magnifications: 20a, ×23 700; 20b, ×41 700. (From Gil and Gay, 1977a.)

220   W. R. BUSHNELL AND JOHN GAY

(a) Neckbands.

Study of haustorial complexes has shown that the extrahaustorial matrix, which technically is an apoplastic region, is isolated from the apoplast of the rest of the leaf. Light microscopy of unfixed haustorial complexes and electron microscopy show two annular structures in the neck region (Figs 17b, 19a,b, 20a,b). These are distinguished as A and B bands. The A band adjoins the inner face of the host cell wall of infected cells (Fig. 20a) and usually the neck breaks at this level when the complexes are released (Fig. 20b). The B band is separate and nearer the haustorial body. When isolated haustorial complexes are placed in water, they swell distending the extrahaustorial membrane and it is then clear that this membrane is firmly attached to the B band (compare Fig. 17b with 19a and 19b). Since the body does not swell, the lobes and their attachments to the body become conspicuous (Figs. 18b, 19a).

These experiments show that the extrahaustorial matrix contains a solution and is sealed to form an enclosed compartment, similar to that demonstrated in the osmotic experiments described earlier (Section IV.C.1.a) but without the uncertainties introduced by the possible presence of host cytoplasm, host tonoplast, or structural complexities around the haustorial neck. In addition, these experiments show clearly that the seal is accomplished by a junction between the extrahaustorial membrane and the neck at the B neckband. This junction is not permeable to solutes within the extrahaustorial matrix. Thus, matrical solutes are unable to pass from the haustorium along the membrane surfaces or through the neck cell wall to the apoplast (i.e. cell walls and intercellular spaces) of the leaf. Conversely, it is inferred that molecules of similar character are unable to enter the haustorium from the general apoplast of the leaf. Although a physiological sealing function has not been demonstrated for the A band, it shows structural features closely resembling those of the primary Casparian Band of a root endodermis (Figs 20a,b) and probably also provides a barrier around the haustorial neck. These observations indicate that substances entering or leaving the haustorium must pass through the cytoplasm of the infected cell. Similar neckbands have been shown in the necks of powdery mildew haustoria in rose (Perera and Gay, 1976)*, barley (Manners and Gay, 1977 and Fig. 19c; Sargent and Gay, unpublished), sugar beet, apple, and vine (Manners and Gay, 1977) and Lamium album (Bonner and Gay, unpublished). Thus, taking all the evidence together, prevention or limitation of apoplastic movement past the necks of haustoria is likely to be a mechanism common to powdery mildew infection. Heath (1976) has recently provided evidence that a neckband likewise prevents apoplastic movement past the neck of a mature rust haustorium.

* Both bands were illustrated but only the B band was recognized by the authors.

(b) The extrahaustorial membrane.

The extrahaustorial membrane differs structurally from the plasma membrane lining the cell wall of the infected cell. Bracker (1968) showed that the extra-haustorial membrane in infected barley was thicker than the normal plasma membrane and that it resisted damage by fixatives which damaged other cellular membranes. Infected pea and the isolated haustorial complexes have also shown the increased thickness (23 instead of 10 nm) (Fig. 21a) and, further, that the membrane resists treatment with a variety of agents which solubilize cell membranes (Gil and Gay, 1977a). Triton X-100*, Teepol, saponin and deoxycholate did not disrupt it or alter its appearance by electron microscopy, although NaOH did remove the entire membrane (Fig. 17c). When electron microscope preparations were treated with reagents to localize polysaccharides, the extrahaustorial membrane stained heavily (Fig. 22) but the material could be removed if the haustorial complexes were subjected to a mixture of the pectinase and cellulase enzymes normally used to dissolve plant cell walls and release protoplasts. This reduced the thickness of the extrahaustorial membrane to approximately that of the undifferentiated portion of the plasma membrane (Figs 21b, c) and also rendered it susceptible to osmotic or detergent lysis.

The untreated extrahaustorial membrane of isolated haustorial complexes is relatively impermeable to water. Thus, the complexes expand in width only very slowly in water, over a period of several hours (Fig. 23). However, complexes swollen in water remain stable for many days and the swelling is reversed by the addition of a sucrose osmoticum. Detergent and enzyme treatments progressively increase the permeability of the extrahaustorial membrane so that expansion of the complex in water can occur in less than 30 min (Fig. 23). The thickenings of the membrane therefore appear to decrease its permeability to water.

The low permeability of the extrahaustorial membrane in complexes of E. pisi isolated from pea contrasts with the relatively high permeability indicated for the membrane around haustoria of E. graminis in partially dissected coleoptile cells or epidermal cells stripped from leaves of barley (Hoskin, 1969). The extrahaustorial membrane of E. graminis, as part of the sac, expanded in a few seconds on water as described in Section IV.C.1.a. The difference in permeability could relate to differences in fungal species, in age of haustorium, or possibly in experimental procedures. In any case, we need to determine which condition represents the normal, functional extra-haustorial membrane.

Another important distinctive character of the extrahaustorial membrane

* Mention of a trademark or proprietary product does not constitute a guarantee or warranty of the product by the US Department of Agriculture and does not imply its approval to the exclusion of other products that may also be suitable.

222

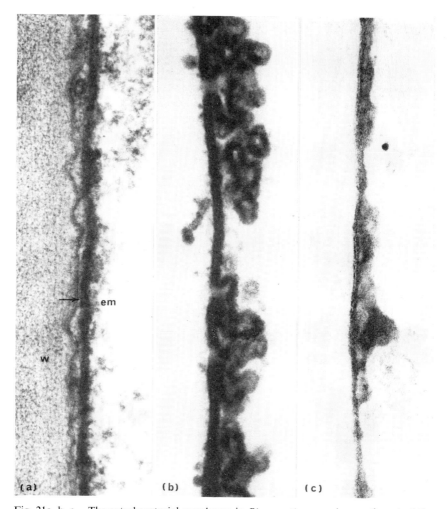

Fig. 21a, b, c. The extrahaustorial membrane in *Pisum sativum*. a. A smooth part of the membrane (em) where it lies adjacent to the host plasma membrane (arrowed) lining the cell wall (w), allowing direct comparison of membrane thicknesses. b. Convoluted part without treatment. c. Convoluted part after treatment with cell wall degrading enzymes. Magnifications: 21a, ×112 000; 21b, ×105 000; 21c, ×118 000. (From Gil and Gay, 1977a.)

has been demonstrated by freeze-fracture studies of isolated haustorial complexes and infected epidermal cells (Gil and Gay, 1977b). The A surface (the aspect of the inner leaflet which faces away from the host cytoplasm) of the extrahaustorial membrane showed crater-like structures representing the edges of broken microvilli projecting into the extrahaustorial matrix (Figs

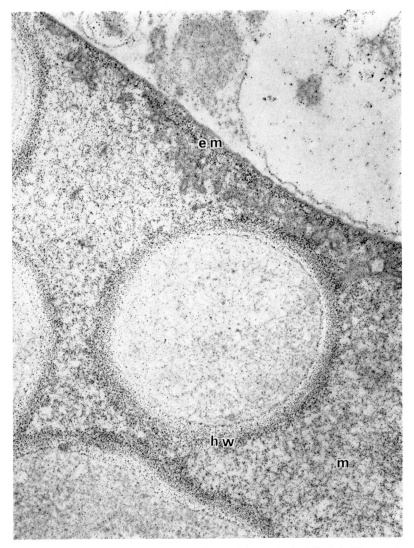

Fig. 22. Section of margins of the extrahaustorial matrix in infected *Pisum sativum* stained for polysaccharides (Sanitar and Boatman, 1972). The extrahaustorial membrane (em) and haustorial lobe walls (hw) are densely stained, and the matrix (m) shows a weaker positive reaction. Magnification, ×38 900. (From Gil, 1976.)

24b, c). However, it was devoid of the small particles found on the A surface of other parts of the plasma membrane of infected cells (Fig. 24a). Particles of this type are characteristic of plasma membranes of walled cells and they are believed to be responsible for the formation of microfibrils in the walls

(Preston, 1974). Their absence from the membrane adjacent to the extra-haustorial matrix indicates that this host function is locally suppressed by the invading fungus. Confirmation is obtained in the absence of fibrils from the

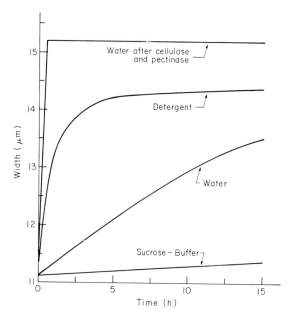

Fig. 23. Changes in width of isolated haustorial complexes of *Erysiphe pisi* suspended in sucrose-buffer isolation medium (0·1 M sucrose, 3 mM MgSO$_4$, 0·1 M PO$_4$ buffer, pH 7), water, detergent (5% Teepol), or water after treatment with cellulase and pectinase. (From Gil and Gay, 1977a.)

matrix. Thus, it is concluded that development of cellulose-producing structures is prevented in the specialized host plasma membrane accommodating the haustorium.

(c). The extrahaustorial matrix.

Although the extrahaustorial matrix lacks evidence of cellulose, it stains with the periodic acid silver-methanamine procedure and fluoresces in UV with Calcofluor (Gil and Gay, 1977a, 1977b) (Figs 22, 25a, b). The enzyme mixture which destabilizes the extrahaustorial membrane removes the matrix and ruthenium red stains it (Caporali, 1960; Sargent and Gay, unpublished) (Fig. 16). From such evidence, Gil and Gay (1977a) concluded that in infected pea, the matrix probably contains pectic and hemicellulosic substances. As in

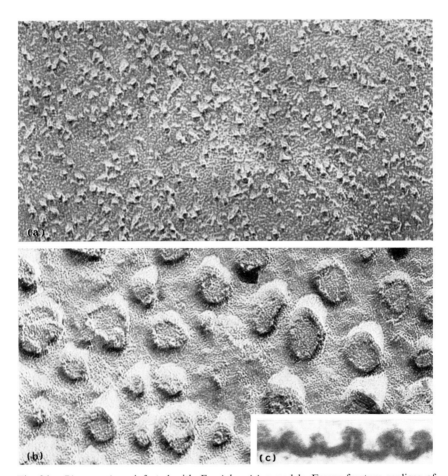

Fig. 24. *Pisum sativum* infected with *Erysiphe pisi*. a and b. Freeze fracture replicas of the A surface (facing away from the cytoplasm) of the plasma membrane lining the wall of an infected cell (a), and of the extrahaustorial membrane (b). c. Section of extrahaustorial membrane for comparison with b. The extrahaustorial membrane (b) lacks the particles present in the plasma membrane (a) thought to be responsible for cellulose microfibril synthesis, but shows many crater-like structures which represent broken villi projecting into the extrahaustorial matrix and which are shown in section (c). Magnifications: 24a and 24b, × 143 000; 24c, × 107 000. (24a, 24b from Gil and Gay, 1977b; 24c from Gil and Gay, 1977a.)

E. graminis infections (Bracker and Littlefield, 1973) (Fig. 16), they found signs of intensive Golgi activity in the adjacent host cytoplasm. Sites on the extrahaustorial membrane suggesting fusion with Golgi vesicles were also found (Gil and Gay, 1977b). Thus, they hypothesized that powdery mildew haustoria cause a major derangement of cell wall metabolism with inhibition

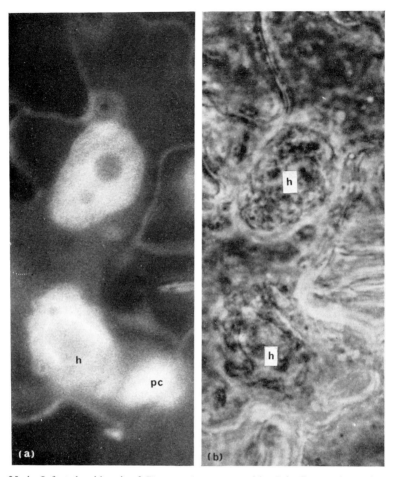

Fig. 25a,b. Infected epidermis of *Pisum sativum* mounted in Calcofluor and examined in dark ground u.v. light (a) and transmitted tungsten light (b). The fluorescence indicates the presence of β-linked polysaccharides (h—haustorium, pc—papilla and/or collar). Magnification, ×1360. (From Gil and Gay, 1977b.)

of cellulose synthesis and increase of Golgi-mediated secretion of pectic and hemicellulosic substances which accumulate as part of the extrahaustorial matrix. The observation of Hirata (1967, 1971) that calcium ions increase the likelihood of haustorium establishment may find its explanation in this mechanism because this element has been shown to enhance the fusion of vesicles with the plasma membrane (Douglas, 1974).

The swelling of the extrahaustorial matrix of isolated haustorial complexes on water occurs without any sign of rupture of the matrix, leading Gil and

Gay (1977a) to conclude that the extrahaustorial matrix is fluid, much as had been concluded from the experiments of Hoskin with the microculture system. Gil and Gay also noted that the fluid matrix may therefore provide an efficient transfer pathway from host to parasite.

Divergent opinions have been expressed regarding the volume of the extrahaustorial matrix in powdery mildews (Bracker and Littlefield, 1973) and its volume may increase with haustorial age as the lobes become more extensive (Figs 18a, b, c). However, results by J. Manners (personal communication) with isolated haustorial complexes, like those of Hoskin (Section IV.C.1.a) with the microculture system, indicate that the matrix is normally in a shrunken condition. Thus, Manners found from swelling experiments that the matrix had an osmotic potential no lower than that of 50 mM sucrose. Since this osmotic potential is greater than the potentials of solutions which plasmolyse host or fungal cells, the matrix in infected cells will tend to be in a shrunken condition and, indeed, this was observed. The lower osmotic potential recorded by Hoskin is possibly due to the ease with which slight changes in matrical volume can be detected along the smooth surface of the haustorial body of *E. graminis* (see Section IV.C.1.a).

(d) Lobe attachments.

The lobes of *E. pisi* haustoria are attached to the surface of the extrahaustorial membrane (Gil and Gay, 1977a) (Figs 18a, 18b, 19a), much as indicated earlier for *E. graminis*. Freeze-fracture studies (Gil and Gay, 1977b) showed that the membrane and lobe walls have special areas which may be the attachment structures and in section some lobe apices have peculiar staining characteristics. These contacts may provide another specialized pathway between host and parasite.

The structural studies of isolated haustorial complexes reviewed here indicate that the striking attribute of a cell infected with a powdery mildew fungus is the new polarity induced in it by the pathogen. The plasma membrane and adjacent structures, both external to it (in the extrahaustorial matrix) and internal to it (in the host cytoplasm), on the haustorial face are distinctly different from the membrane and adjacent structures on normal surfaces of the cell. From a nutritional viewpoint, the cell may be compared with one in an intestinal epithelium, with one face taking materials from the source (probably the apoplast only), while the other secretes particular elaborated compounds (possibly mainly in exocytotic vesicles) into the extrahaustorial matrix, to be taken subsequently into the haustorial cytoplasm. Clearly the polarization is caused by the haustorium but because cell membranes usually possess fluid properties it is surprising that the special features of the host plasma membrane do not become distributed uniformly throughout the membrane. It was argued (Gil and Gay, 1977a) that since the transition in structure occurs at the A band around the haustorial neck, that structure prevents the constituents of the regions of the plasma membrane adjacent to the haustorium and the cell wall from entering one another's domains. In turn, the particular physiological activities of the two faces of the infected cell are maintained in relation both to the haustorium and to other cells of the host.

(e) Direct measurement of solute uptake by haustoria.

The isolated haustorial complex also provides a means of direct study of the nutritional role of haustoria and such preliminary studies have been carried out by Gil and Gay (1977b) and J. M. Manners of Imperial College. The former authors examined the uptake of fluorescein diacetate which is converted to fluorescein by cellular esterases. Infected epidermis, freed from mycelium, was removed from pea leaves and mounted in the fluorogen. The haustoria fluoresced in ultraviolet light but no fluorescence was detectable, even after prolonged incubation, when isolated haustorial complexes were mounted directly in the substrate. It was deduced that the cytoplasm of the infected cell is necessary for uptake. The same conclusion has been inferred by Manners and Gay (unpublished), who studied uptake by haustoria from illuminated leaves exposed to $^{14}CO_2$ and by isolated haustorial complexes in sugar solutions. The mycelium was removed from infested pea leaves and after exposure to $^{14}CO_2$, the haustorial complexes were isolated. Labelled C had entered them rapidly (Table 3), an observation which is

TABLE 3

In vivo uptake of labelled carbon compounds by haustoria in detached pea leaves exposed to $^{14}CO_2$ in the light, as determined in a haustorial complex fraction isolated therefrom. (J. M. Manners and J. L. Gay, unpublished data)

Time of exposure to $^{14}CO_2$	Haustorial complex fraction cpm 100 μg^{-1} protein	Equivalent fraction from uninfected plant cpm 100 μg^{-1} protein
5 min	1691	207
20 min	4410	521

compatible with the uptake of labelled C into attached mycelium reported by Edwards and Allen (1966). Manners took advantage of the stability of the extrahaustorial membrane in media without osmotica and in detergents, and suspended the fraction in water with Triton X–100 to disrupt any contaminating chloroplasts and other organelles of host origin. Furthermore, he found that the CO_2 evolved from the isolated complexes was labelled and thus deduced that photosynthetic products had entered the haustorium and were not confined to the matrix which lacks cellular organelles. These results provide the first direct demonstration of the role of haustoria in nutrition of obligate parasites.

Manners and Gay also examined uptake of glucose and sucrose by haustorial complex fractions *in vitro*. The characteristics of glucose uptake in their studies are listed below.

1. Detergent insensitive.
2. Relatively insensitive to DNP.
3. Relatively insensitive to pH between pH 4–9.
4. Linearly related to concentration of glucose.

The results indicate that the sugars entered by slow diffusion instead of by an active process and, indeed, comparison with yeast showed that on a protein basis, uptake by the haustorial complex fraction was 70 times slower (Manners and Gay, unpublished). Although the number of microbial contaminants was very small, experiments showed that they accounted for half of the uptake of glucose. Thus, the true rate of uptake by the haustorial complex *in vitro* is extremely slow. This is in accord with the observations that the respiration rate of haustorial complexes is unaffected by exogenous substrates (Dekhuijzen, 1966; Manners and Gay, unpublished).

These results with sugar uptake are in accord with the results of the experiments with fluorescein diacetate, with the relationship between mycelial growth and the amount of host cytoplasm remaining in incised host cells (Section IV.C.1.c) and with the hypothesis proposed by Gil and Gay (1977b) on the basis of structural evidence. All indicate strongly that the cytoplasm of the infected cell is necessary for the haustorium to obtain nutrients.

V. Concluding Statement

The key to the movement of solutes from host to powdery mildew fungus lies in processes at the interface between the haustorium and the cytoplasm of the host. Techniques already in use, especially parallel studies of individual haustoria *in situ* and of large numbers of haustoria isolated *in vitro*, show promise to provide an understanding of these processes. However, such techniques have only begun to be utilized. Evidence is far from complete for virtually all aspects of haustorium-solute relations, particularly with respect to possible involvement of specific transport systems across membranes.

Nevertheless, we offer the following tentative scheme for the movement of solutes from a host plant to a powdery mildew fungus during sporulation when the amount of solutes entering the fungus is large:

A general increase in the permeability of host plasma membranes apparently occurs in the mesophyll of infected leaves, increasing the leakage of solutes from the symplast of the mesophyll into the apoplast. Substances move by way of the apoplast from the mesophyll to the epidermis from which small amounts of solute possibly enter hyphae directly through the leaf surface. Flow from the apoplast to the haustorium is blocked by haustorial neckbands, so that substances must enter the cytoplasm (the symplast) of the epidermal cell before entering the haustorium. Solutes probably travel within the epidermal cell to the face of the haustorium by cytoplasmic streaming, although some substances may enter the vacuole of the host cell and diffuse therein toward the haustorium where they re-enter host cytoplasm. In either case, the last step involving the host is a transfer of solutes from the host cytoplasm surrounding the haustorium, across the extrahaustorial membrane, into the extrahaustorial matrix. Available evidence indicates that such a transfer requires living host cytoplasm, at least for some substances. The extremely thin parts of the matrix at the tips of haustorial lobes may provide a specialized pathway between host and parasite.

The extrahaustorial membrane is uniquely differentiated compared to the normal plasma membrane, being thicker and stronger than the normal membrane and lacking the particles associated with synthesis of cellulose microfibrils. The haustorium probably causes a derangement of cell wall metabolism in the part of the host cell facing the haustorium, suppressing synthesis of cellulose microfibrils. Golgi activity is increased and contents of vesicles move by exocytosis into the extrahaustorial matrix, thereby apparently increasing secretion of pectic and hemicellulosic substances into the matrix. How solutes move from the matrix into the haustorium itself has not been studied but fungal enzymes possibly digest the pectic and hemicellulosic substances and the products then enter the haustorium. In any case, the matrix is completely sealed by neckbands so that matrical substances cannot leak into the host apoplast.

Once the solutes enter the fungus, they are incorporated into insoluble fungal substances or into soluble substances such as mannitol which the host does not normally utilize. Solutes are translocated through the septal pore in the haustorial neck into hyphal cells and thence either to conidiophores where they are utilized in synthesis of new conidiospores, or to a lesser extent, to growing hyphal tips. The continual removal of solutes by the fungus creates concentration gradients which in part sustain solute flow from host to fungus. These gradients tend to reduce phloem loading or increase phloem unloading, in effect keeping the infected tissues a strong sink which competes successfully with shoots and roots for host solutes.

Dedication and Acknowledgements

This chapter is respectfully dedicated to Dr K. Hirata (Amano), for his pioneering descriptive studies of the haustorium and its relationship to the powdery mildew colony.

W. R. B. wishes to thank Mrs M. H. Bailey, Miss D. Drennan, Mrs S. Carr, Mr D. Moore, and Mrs P. H. Long for technical assistance. He also thanks the National Science Foundation for partial support of research described here through grants GB–2002 and GB–5344X.

J. L. G. wishes to thank colleagues at Imperial College, Dr F. Gil, Mr J. M. Manners, Miss Marion Martin, Dr Caroline Sargent, Mr P. Bonner, and Miss Janet Crisp, for permission to refer to unpublished results. He also wishes to acknowledge financial support by Imperial Chemical Industries, the Agricultural Research Council, and the Science Research Council.

References

Abeles, F. B. (1972). *A. Rev. Pl. Physiol.* **23**, 259–292.
Adedipe, N. O. and Fletcher, R. A. (1971). *Can. J. Bot.* **49**, 59–61.
Aist, J. R. (1976). *A. Rev. Phytopath.* **14**, 145–163.
Aist, J. R. and Israel, H. W. (1976). *Phytopathology,* in press.
Allen, P. J. (1942). *Am. J. Bot.* **29**, 425–435.
Allen, P. J. (1966). *Phytopathology* **56**, 255–260.
Allen, P. J. and Goddard, D. R. (1938). *Am. J. Bot.* **25**, 613–621.
Arisz, W. H. (1969). *Acta Bot. Neerl.* **18**, 14–38.
Atkin, C. L. and Neilands, J. B. (1972). *Science* **176**, 300–301.
Bennett, J. and Scott, K. J. (1971). *FEBS Lett.* **17**, 93–95.
Bose, A. and Shaw, M. (1974). *Can. J. Bot.* **52**, 1183–1195.
Bracker, C. E. (1968). *Phytopathology* **58**, 12–30.
Bracker, C. E. and Littlefield, L. J. (1973). *In* "Fungal Pathogenicity and the Plant's Response", (R. J. W. Byrde and C. V. Cutting, Eds), 159–318. Academic Press, London and New York.
Bräutigam, E. and Müller, E. (1975). *Biochem. Physiol. Pflanzen.* **167**, 29–39.
Bushnell, W. R. (1966). *Can. J. Bot.* **44**, 1485–1493.
Bushnell, W. R. (1967). *In* "The Dynamic Role of Molecular Constituents in Plant-Parasite Interaction", (C. J. Mirocha and I. Uritani, Eds), 21–39. Am. Phytopathol. Soc., St. Paul, Minn.
Bushnell, W. R. (1971). *In* "Morphological and Biochemical Events in Plant-Parasite Interaction", (S. Akai and S. Ouchi, Eds), 229–254. Phytopathol. Soc. Jap., Tokyo.
Bushnell, W. R. (1972). *A. Rev. Phytopath.* **10**, 151–176.
Bushnell, W. R. (1976). *In* "Specificity in Plant Diseases", (R. K. S. Wood and A. Graniti, Eds), 131–145. Plenum, New York.
Bushnell, W. R. and Allen, P. J. (1962a). *Pl. Physiol.* **37**, 50–59.
Bushnell, W. R. and Allen, P. J. (1962b). *Pl. Physiol.* **37**, 751–758.
Bushnell, W. R. and Bergquist, S. E. (1975). *Phytopathology* **65**, 310–318.
Bushnell, W. R. and Zeyen, R. J. (1976). *Can. J. Bot.* **54**, 1647–1655.
Bushnell, W. R., Dueck, J. and Rowell, J. B. (1967). *Can. J. Bot.* **45**, 1719–1732.

Callow, J. A. (1973). *Physiol. Pl. Path.* **3**, 249–257.
Camp, R. R. and Whittingham, W. F. (1975). *Am. J. Bot.* **62**, 403–409.
Caporali, M. L. (1960). *C. R. hebd. Séanc., Acad. Sci.*, Paris **250**, 2415–2417.
Cataldo, D. A. (1974). *Pl. Physiol.* **53**, 912–917.
Coffey, M. D. and Allen, P. J. (1973). *Trans. Br. mycol. Soc.* **60**, 245–260.
Colquhoun, A. J. and Hillman, J. R. (1975). *Z. Pflanzenphysiol.* **76**, 326–332.
Comhaire, F. (1963). *Lejeunia, Rev. Bot.* **15**, 1–87.
Cornu, M. (1881). *C. R. hebd. Séanc. Acad. Sci.*, Paris **93**, 1162–1164.
Crafts, A. S. and Crisp, C. E. (1971). "Phloem Transport in Plants". W. H. Freeman, San Francisco.
Darlington, W. A. and Cirulis, N. (1963). *Pl. Physiol.* **38**, 462–467.
Day, P. R. and Scott, K. J. (1973). *Physiol. Pl. Path.* **3**, 433–435.
Dekhuijzen, H. M. (1966). *Neth. J. Pl. Path.* **72**, 1–11.
Dekhuijzen, H. M. and Scheer, C. van der. (1967). *Neth. J. Pl. Path.* **73**, 121–125.
Dekhuijzen, H. M. and Scheer, C. van der. (1969). *Neth. J. Pl. Path.* **75**, 169–177.
Dekhuijzen, H. M. and Staples, R. C. (1968). *Contrib. Boyce Thompson Inst.* **24**, 39–52.
Douglas, W. W. (1974). *In* "Calcium and Cell Regulation", (R. M. S. Smellie, Ed.), 1–28. Biochemical Society, London.
Durbin, R. D. (1967). *In* "The Dynamic Role of Molecular Constituents in Plant-Parasite Interaction", (C. J. Mirocha and I. Uritani, Eds), 80–99. Am. Phytopathol. Soc., St. Paul, Minn.
Dyer, T. A. and Scott, K. J. (1972). *Nature, Lond.* **236**, 237–238.
Edwards, H. H. (1970). *Pl. Physiol.* **45**, 594–597.
Edwards, H. H. (1971). *Pl. Physiol.* **47**, 324–328.
Edwards, H. H. and Allen, P. J. (1966). *Pl. Physiol.* **41**, 683–688.
Ehrlich, M. A. and Ehrlich, H. G. (1971). *A. Rev. Phytopath.* **9**, 155–184.
Ellingboe, A. H. (1972). *Phytopathology* **62**, 401–406.
Elnaghy, M. A. and Heitefuss, R. (1976a). *Physiol. Pl. Path.* **8**, 253–267.
Elnaghy, M. A. and Heitefuss, R. (1976b). *Physiol. Pl. Path.* **8**, 269–277.
Engelbrecht, L. (1964). *Flora (Jena)* **154**, *Abt. A.* 57–69.
Engelbrecht, L. (1968). *Flora (Jena)* **159**, *Abt. A.* 208–214.
Evans, R. L. and Banerjee, K. (1973). *Absts. Second Int. Congr. Pl. Path.*, Abst. 102.
Farkas, G. L., Dézsi, L., Horváth, M., Kisbán, K. and Udvardy, J. (1964). *Phytopath. Z.* **49**, 343–354.
Fletcher, R. A., Hofstra, G. and Adedipe, N. O. (1970). *Physiol. Plant.* **23**, 1144–1148.
Fraymouth, J. (1956). *Trans. Br. mycol. Soc.* **39**, 79–107.
Frič, F. (1964a). *Biologia (Bratislava)* **19**, 597–610.
Frič, F. (1964b). *Phytopath. Z.* **51**, 101–117.
Frič, F. (1975). *Phytopath. Z.* **84**, 88–95.
Garg, I. D. and Mandahar, C. L. (1976). *Phytopath. Z.* **85**, 298–307.
Geiger, D. R. and Cataldo, D. A. (1969). *Pl. Physiol.* **44**, 45–54.
Gil, F. (1976). Ultrastructural and physiological properties of haustoria of powdery mildews and their host interfaces. PhD. thesis: University of London.
Gil, F. and Gay, J. L. (1977a). *Physiol. Pl. Path.* **10**, 1–12.
Gil, F. and Gay, J. L. (1977b). Ultrastructural specialization of the interface between *Erysiphe pisi* and *Pisum sativum*. In preparation.
Gunning, B. E. S. and Robards, A. W. (1976). "Intercellular Communication in Plants: Studies on Plasmodesmata". Springer-Verlag, Berlin, Heidelberg, New York.
Harvey, A. E. and Grasham, J. L. (1974). *Phytopathology* **64**, 1028–1035.
Haspelová-Horvatovičová, A. (1964). *In* "Host-Parasite Relations in Plant Pathology", (S. Király and G. Ubrizsy, Eds), 243–247. Hung. Acad. Sci., Budapest.
Heath, M. C. (1971). *Phytopathology* **61**, 383–388.

Heath, M. C. (1974). *Can. J. Bot.* **52**, 2591–2597.
Heath, M. C. (1976). *Can. J. Bot.* **54**, 2484–2489.
Heath, M. C. and Heath, I. B. (1971). *Physiol. Pl. Path.* **1**, 277–287.
Hewitt, H. G. and Ayres, P. G. (1975). *Physiol. Pl. Path.* **7**, 127–137.
Hewitt, H. G. and Ayres, P. G. (1976). *New Phytol.* **77**, 379–390.
Hirata, K. (1937). (In Japanese, English summary). *Ann. Phytopathol. Soc. Jap.* **6**, 319–334.
Hirata, K. (1955). (In Japanese, English summary). *Ann. Phytopathol. Soc. Jap.* **20**, 73–76.
Hirata, K. (1958). (In Japanese, English summary). *Ann. Phytopathol. Soc. Jap.* **23**, 139–144.
Hirata, K. (1959). (In Japanese, English summary). *Bull. Fac. Agr. Niigata Univ.* **11**, 34–42.
Hirata, K. (1960). (In Japanese). *Trans. mycol. Soc. Jap.* **2**, 2–6.
Hirata, K. (1967). *Mem. Fac. Agr. Niigata Univ.* **6**, 207–259.
Hirata, K. (1971). *In* "Morphological and Biochemical Events in Plant-Parasite Interaction", (S. Akai and S. Ouchi, Eds), 207–228. Phytopathol. Soc. Jap., Tokyo.
Hirata, K. (1973). *Shokubutsu Byogai Kenkyu* **8**, 85–101.
Hirata, K. and Kojima, M. (1962). *Trans. mycol. Soc. Jap.* **3**, 43–46.
Hislop, E. C. and Stahmann, M. A. (1971). *Physiol. Pl. Path.* **1**, 297–312.
Holligan, P. M., Chen, C. and Lewis, D. H. (1973). *New Phytol.* **72**, 947–955.
Hoppe, H. H. and Heitefuss, R. (1974a). *Physiol. Pl. Path.* **4**, 5–10.
Hoppe, H. H. and Heitefuss, R. (1974b). *Physiol. Pl. Path.* **4**, 11–23.
Hoppe, H. H. and Heitefuss, R. (1974c). *Physiol. Pl. Path.* **4**, 25–36.
Hoskin, D. A. (1969). The haustorial sac of *Erysiphe graminis*. M.S. Thesis: University of Minnesota.
Howes, N. K. and Scott, K. J. (1972). *Can. J. Bot.* **50**, 1165–1170.
Hsu, S. C. and Ellingboe, A. H. (1972). *Phytopathology* **62**, 876–882.
Jacoby, B. and Dagan, J. (1970). *Physiol. Plant. Pl.* **23**, 397–403.
Johnson, L. B., Brannaman, B. L. and Zscheile, Jr., F. P. (1966). *Phytopathology* **56**, 1405–1410.
Kunoh, H. (1972a). *Bull. Fac. Agr. Mie Univ.* (*Japan*) **44**, 141–224.
Kunoh, H. (1972b). *Ann. phytopathol. Soc. Japan* **38**, 357–358.
Kunoh, H. and Akai, S. (1969). *Mycopath. Mycol. appl.* **37**, 113–118.
Kunoh, H. and Ishizaki, H. (1973). *Ann. phytopathol. Soc. Japan* **39**, 42–48.
Kunoh, H., Ishizaki, H. and Kondo, F. (1975). *Ann. phytopathol. Soc. Japan* **41**, 33–39.
Last, F. T. (1963). *Ann. Bot. Lond.* **27**, 685–690.
Lindoo, S. J. and Nooden, L. D. (1976). *Pl. Physiol.* **57**, Supp. 27, Abstr. 143.
Long, D. E., Fung, A. K., McGee, E. E. M., Cooke, R. C. and Lewis, D. H. (1975). *New Phytol.* **74**, 173–182.
Magyarosy, A. C., Schürmann, P. and Buchanan, B. B. (1976). *Pl. Physiol.* **57**, 486–489.
Mandahar, C. L. and Garg, I. D. (1976). *Phytopath. Z.* **85**, 86–89.
Manners, J. M. and Gay, J. L. (1977). *Physiol. Pl. Path.* **11**, 261–266.
Masri, S. S., and Ellingboe, A. H. (1966). *Phytopathology* **56**, 389–395.
Mayak, S. and Halevy, A. H. (1972). *Pl. Physiol.* **50**, 341–346.
McKeen, W. E. (1974). *Can. J. Microbiol.* **20**, 1475–1478.
McKeen, W. E., Smith, R. and Mitchell, N. (1966). *Can. J. Bot.* **44**, 1299–1306.
Milborrow, B. V. (1974). *A. Rev. Pl. Physiol.* **25**, 259–307.
Millerd, A. and Scott, K. J. (1963). *Aust. J. biol. Sci.* **16**, 775–783.
Milthorpe, F. L. and Moorby, J. (1969). *A. Rev. Pl. Physiol.* **20**, 117–138.
Mo hes, K. (1960). *Naturwissenschaften* **47**, 337–351.
Mount, M. S. and Ellingboe, A. H. (1969). *Phytopathology* **59**, 235.
Müller, K. and Leopold, A. C. (1966). *Planta* **68**, 186–205.
Nakata, S. and Leopold, A. C. (1967). *Am. J. Bot.* **54**, 769–772.
Nishiyama, K., Naito, H. and Hirata, K. (1966). (In Japanese, English summary). *Niigata Agr. Sci.* **18**, 26–34.

Paulech, C. (1966). *Biologia (Bratislava)* **21,** 321–327.
Perera, R. and Gay, J. L. (1976). *Physiol. Pl. Path.* **9,** 57–65.
Preston, R. D. (1974). "The Physical Biology of Plant Cell Walls". Chapman and Hall, London.
Robards, A. W. (1976). *In* "Intercellular Communication in Plants: Studies on Plasmodesmata", (B. E. S. Gunning and A. W. Robards, Eds), 15–57. Springer-Verlag, Berlin, Heidelberg, New York.
Rubin, B. A., Axenova, V. A. and Hyen, N. D. (1971). *Acta Phytopath. Hung.* **6,** 61–64.
Sacher, J. A. (1962). *Nature, Lond.* **195,** 577–578.
Sadler, R. and Scott, K. J. (1974). *Physiol. Pl. Path.* **4,** 235–247.
Sanitar, P. D. and Boatman, E. S. (1972). *J. Bacteriol.* **111** 267–271.
Schenk, E. (1974). *Acta. Bot. Neerl.* **23,** 739–746.
Schönherr, J. (1976). *Planta* **128,** 113–126.
Schwinn, F. J. and Dahmen, H. (1973). *Phytopath. Z.* **77,** 89–92.
Scott, K. J. (1965). *Phytopathology* **55,** 438–441.
Scott, K. J. and Smillie, R. M. (1966). *Pl. Physiol.* **41,** 289–297.
Sempio, C. (1950). *Phytopathology* **40,** 799–819.
Shaw, M. and Colotelo, N. (1961). *Can. J. Bot.* **39,** 1351–1372.
Shaw, M. and Manocha, M. S. (1965). *Can. J. Bot.* **43,** 1285–1292.
Shaw, M. and Samborski, D. J. (1956). *Can. J. Bot.* **34,** 389–405.
Shaw, M., Brown, S. A. and Rudd-Jones, D. (1954). *Nature, Lond.* **173,** 768–769.
Siebs, E. (1971). *Phytopath. Z.* **72,** 97–114.
Slesinski, R. S. and Ellingboe, A. H. (1971). *Can. J. Bot.* **49,** 303–310.
Smith, D., Muscatine, L. and Lewis, D. (1969). *Biol. Rev.* **44,** 17–90.
Smith, G. (1900). *Bot. Gaz. (Chicago)* **29,** 153–184.
Snell, W. H. and Dick, E. A. (1971). "A Glossary of Mycology", Revised Edition, p. 74. Harvard University Press, Cambridge, Mass.
So, M. L. and Thrower, L. B. (1976). *Phytopath. Z.* **86,** 252–265.
Stanbridge, B., Gay, J. L. and Wood, R. K. S. (1971). *In* "Ecology of Leaf Surface Microorganisms", (T. F. Preece and C. H. Dickinson, Eds), 367–379. Academic Press, London and New York.
Stavely, J. R., Pillai, A. and Hanson, E. W. (1969). *Phytopathology* **59,** 1688–1693.
Stuckey, R. E. and Ellingboe, A. H. (1975). *Physiol. Pl. Path.* **5,** 19–26.
Sullivan, T. P., Bushnell, W. R. and Rowell, J. B. (1974). *Can. J. Bot.* **52,** 987–998.
Takanashi, K. and Iwata, Y. (1963). (In Japanese, English summary). *Nogyo Gijutsu Kenkyusho Hokoku* C, **15,** 83–98.
Takegami, T. and Yoshida, K. (1975). *Pl. and Cell Physiol.* **16,** 1163–1166.
Tetley, R. M. and Thimann, K. V. (1974). *Pl. Physiol.* **54,** 294–303.
Thatcher, F. S. (1939). *Am. J. Bot.* **26,** 449–458.
Thatcher, F. S. (1942). *Can. J. Res.* C, **20,** 283–311.
Thatcher, F. S. (1943). *Can. J. Res.* C, **21,** 151–172.
Thimann, K. V., Tetley, R. R. and Thanh, T. V. (1974). *Pl. Physiol.* **54,** 859–862.
Thomas, H. (1976). *Pl. Sci. Lett.* **6,** 369–377.
Vizárová, G. (1973). *Proc. Res. Inst. Pomoc., Poland,* E, **3,** 559–564.
Vizárová, G. (1974). *Phytopath. Z.* **79,** 310–314.
Wardlaw, I. F. (1974). *A. Rev. Pl. Physiol.* **25,** 515–539.
Wieneke, J., Covey, Jr., R. P. and Benson, N. (1971). *Phytopathology* **61,** 1099–1103.
Williams, P. G. (1958). The application of plasmolytic methods to studies in barley mesophyll. M. S. Thesis: University of Sydney.

Wolswinkel, P. (1974). *Acta. Bot. Neerl.* **23**, 177–188.
Yamada, Y., Wittwer, S. H. and Bakovac, M. J. (1964). *Pl. Physiol.* **39**, 28–32.
Yarwood, C. E. and Jacobson, L. (1950). *Nature, Lond.* **165**, 973–974.
Yarwood, C. E. and Jacobson, L. (1955). *Phytopathology* **45**, 43–48.

Pertinent literature published since completion of the manuscript

Ayres, P. G. (1976). Patterns of stomatal behaviour, transpiration, and CO_2 exchange in pea following infection by powdery mildew (*Erysiphe pisi*). *J. exp. Bot.* **27**, 1196–1205.
Ayres, P. G. (1977). Effects of powdery mildew *Erysiphe pisi* and water stress upon the water relations of pea. *Physiol. Pl. Path.* **10**, 139–145.
Manners, J. M. and Gay, J. L. (1978). Uptake of ^{14}C photosynthates from *Pisum sativum* by haustoria of *Erysiphe pisi*. *Physiol. Pl. Path.* **12**, 215–223.
Mendgen, K. (1977). Reduced lysine uptake by bean rust haustoria in a resistant reaction. *Naturwissenschaften* **64**, S.438.
Sargent, C. and Gay, J. L. (1977). Barley epidermal apoplast structure and modification by powdery mildew contact. *Physiol. Pl. Path.* **11**, 195–205.

Chapter 9

Breeding for Resistance to Powdery Mildew in the Temperate Cereals

R. A. McINTOSH

Plant Breeding Institute, University of Sydney, N.S.W. Australia. 2006

I. Introduction

The basic objective in cereal breeding is to produce cultivass which will give consistently high yields of grain of good quality when grown on a national scale (Hayes, 1972).

One major variable affecting consistent performance of cultivars is the occurrence of plant diseases. Of the cereal mildews resulting from infection by various "form species" of *Erysiphe graminis* viz. the powdery mildews of

barley, wheat, oats and rye, that of barley is of the greatest relative importance, whereas mildew of wheat is a problem to a lesseer xtent in certain
cooler or irrigated areas. Economically important occurrences of oat mildew
appear to be confined to relatively small areas, in western Europe and South
America and occasional isolated instances of losses have been reported in
North America (Simons and Murphy 1961). Mildew of rye is relatively
unimportant (Dickson, 1956).

     Disease incidence and intensity in any one area varies with climatic con-
ditions and the likelihood, or degree, of mildew is difficult to predict. Hence
choice of cultivars is limited to the assessment of information available from
prior crop seasons. This problem is accentuated in areas, such as Europe,
where crops grown in one country are affected by spore showers originating
in another. For example, Hermansen and Stix (1974) and Hermansen et al.
(1975) present evidence in support of an United Kingdom origin of E.
graminis hordei and other pathogens occurring in Denmark but apparently
not overwintering there. Attempts to prevent overwintering of mildew in
Denmark by the prohibition of winter-, or autumn-sown barleys were
frustrated by wind-blown spores arriving from outside the country.

     Following the introduction of modern short-statured "green revolution"
cereal types and the agronomic methods associated with their use, there have
been some reports of increased mildew incidence. Much of this increase is
undoubtedly related to greater use of nitrogen fertilizers and irrigation.
Normally, the powdery mildew diseases are considered low temperature
diseases, hence the greatest losses tend to be confined to cooler, moister areas
of production such as western, northern and central Europe, eastern United
States and Canada, and New Zealand. In Australia where spring cereals are
late autumn- or winter-sown, widespread mildew infections on barley and
wheat may occur in early growth stages but as temperatures increase and
tillers elongate, levels of disease rapidly decline such that at flowering, mildew
is usually confined to a few pustules on the lower stems and leaves. It has been
assumed that these early mildew infections have no effects on yielding ability
under Australian conditions. In the United Kingdom Griffiths et al. (1975)
and Jenkyn (1974) reported significant effects on yield parameters of barley
resulting from mildew infection at early growth stages.

     Jenkins (1973) estimated for the years 1967–73 that 25–57% of spring
barley fields in England and Wales had sufficiently high levels of mildew
incidence to warrant control by chemical means. Furthermore, he stated that
in 1973, chemical control was applied to 30% of the barley crop. Because of
the economic benefits of chemical control it is evident that breeding programmes aimed at producing mildew resistant barleys need no further
justification.

     The major disadvantage with mildew resistance to date has been the brief
time periods during which resistance genes have remained effective before

the pathogens gained virulence. On the other hand, the possibility of pathogens becoming resistant, or tolerant, to the effects of fungicides is also significant. Wolfe and Dinoor (Jenkins, 1973) reported tolerance to ethirimol in *E. graminis* f. sp. *hordei*. Probably the greatest potential for both genetic and chemical control is the use of less susceptible cultivars which require lower amounts, or less frequent applications, of chemicals.

Losses in wheat attributable to mildew are much lower than in barley. King (1973) estimated annual losses in England at 2-5% for the years 1970–73. However he also pointed out that while various chemicals and spray procedures were satisfactory for control of barley mildew they were not as effective for control of wheat mildew. Furthermore, due to the erratic nature of both diseases, spray applications for prophylactic purposes could not be recommended. Hence farmers were required to recognize the critical disease levels and crop growth stages at which sprays should be applied. Additionally some damage to crops occurs during spray application from surface-drawn machines.

In oats, losses in grain yield of up to 39% have been reported (Hayes, 1970).

Once a decision is made to breed for resistance to a particular disease the breeder has to assess all known data regarding the disease before deciding upon the best means of producing resistance. Breeding for resistance demands the simultaneous study of both hosts and pathogens. If pathogen populations change, or disease intensities increase in response to new host cultivars or new cultural methods, it may be necessary to change further the host genotypes. With some diseases, resistance breeding is a routine procedure. For example, certain Australian wheat cultivars have remained resistant to the flagsmut disease caused by *Urocystis agropyri* Koern. since they were first produced in the 1920s. There are no confirmed instances of virulence for such resistances in Australian populations of the flagsmut organism. Although flagsmut resistance is not simply inherited (McIntosh, 1968) selection for resistance in sequential generations during the breeding programme presents little difficulty to obtaining resistant segregates. Many sources of resistance are available but the degree of genetic variation among sources is unknown. In contrast, resistances in cereals to the mildews and rusts have been relatively short-lived and breeders have often experienced loss of resistance in cultivars shortly after introduction to commercial production.

The expression of disease in a particular environment is dependent upon the genotypes of both hosts and pathogens. The fact that a particular disease is an agricultural problem implies that the prevailing environment is conducive to disease in a sufficient number of crop seasons over a significantly wide area. Moreover, for economically important levels of disease to develop, initial inoculum must be present in adequate quantities. The survival of mildews between crop seasons depends upon conidial infections of volunteer plants, or on the cleistothecial stages but autumn-sown cereals may contribute to the winter survival of larger amounts of inoculum. Jenkins and Storey (1975) indicated that losses were greater in spring barley when sown alongside winter barley. On the other hand, from three years' observations Johnson *et al.* (1971) had suggested that mildew epidemics were not necessarily dependent on high levels of overwintering inoculum. Severe local epidemics developed even in years when winter survival was low but when subsequent conditions favoured disease development. These observations were supported by Jenkyn (1976).

II. The Gene-for-Gene Model in Breeding for Disease Resistance

Host reaction and pathogen virulence or avirulence are determined from the varying levels of disease. Disease intensity is always relative and varies from no disease, which in the presence of inoculum implies host resistance and pathogen avirulence, to maximum disease as expressed by a standard susceptible cultivar. If host–pathogen interactions resulting in maximum disease intensity are described as compatible then interactions which deviate from this can be described as incompatible. Consequently, incompatibility covers a wide range of disease expressions varying from only slight reductions in pustule number or size to no apparent disease symptoms, the actual expression depending on the genotypes of host and pathogen, and on the environment. In studying variability in the host species, the genotypes of the pathogen are held constant, or are assumed to be relatively constant and in studying variability in pathogens the genotypes of hosts are constant.

Genetic principles established by Flor (1956) in the flax rust disease (*Melampsora lini:Linum usitatissimum*) have been demonstrated for the mildew diseases of barley and wheat (reviewed by Moseman, 1966) and can be assumed to apply to the majority, if not all, of plant disease systems. On the "gene-for-gene" model (Table 1) the expression of incompatibility is dependent on the concurrent presence of alleles for resistance in the host and corresponding alleles for avirulence in the pathogen. Critical assessment of the model for biotrophic pathogens suggests that incompatibility is dependent upon interactions of active gene products irrespective of dominance in host or pathogen. For example the deficiency of alleles for resistance to rusts and mildew, whether dominant or recessive, in wheat results in compatible interactions.

Certain generalities can be drawn from the one-gene and two-gene models in Table 1.

1. Each interacting gene pair in host and pathogen produces a level of incompatibility (infection type) which may be characteristic and may assist in the identification of the gene(s) involved. Some interacting gene

TABLE 1

Genetic interactions of homozygous hosts and haploid pathogens assuming one (a) and two (b) gene models

a) 1-gene model. R and r = alleles for resistance and susceptibility.
A and a = corresponding alleles for avirulence and virulence with respect to host allele R.
I = incompatibility, C = compatibility.

| | | Pathogen genotypes | |
		A	a
Host genotypes	RR	$\dfrac{A}{RR} = I$	$\dfrac{a}{RR} = C$
	rr	$\dfrac{A}{rr} = C$	$\dfrac{a}{rr} = C$

b) 2-gene model. $R1$ and $R2$ = resistance alleles at different host loci.
$A1$ and $A2$ = corresponding pathogen alleles for avirulence.
$I1$ and $I2$ = incompatibilities expressed.

| | | Pathogen genotypes | | | |
		A1A2	A1a2	a1A2	a1a2
Host genotypes (Haploid)	R1R2	$\dfrac{A1A2}{R1R2} = I1 + I2^a$	$\dfrac{A1a2}{R1R2} = I1$	$\dfrac{a1A2}{R1R2} = I2$	$\dfrac{a1a2}{R1R2} = C$
	R1r2	$\dfrac{A1A2}{R1r2} = I1$	$\dfrac{A1a2}{R1r2} = I1$	$\dfrac{a1A2}{R1r2} = C$	$\dfrac{R1r2}{a1a2} = C$
	r1R2	$\dfrac{A1A2}{r1R2} = I2$	$\dfrac{A1a2}{r1R2} = C$	$\dfrac{a1A2}{r1R2} = I2$	$\dfrac{r1R2}{a1a2} = C$
	r1r2	$\dfrac{A1A2}{r1r2} = C$	$\dfrac{A1a2}{r1r2} = C$	$\dfrac{a1a2}{r1r2} = C$	$\dfrac{a1a2}{r1r2} = C$

[a] I1 and I2 may be characterized by distinctive interactions (infection types). I1 + I2 is at least as incompatible as the more incompatible interaction.

pairs produce very intense levels of incompatibility characterized by "hypersensitivity", whereas others may produce relatively small differences from compatibility. Moreover, such incompatibilities may occur at either seedling or post-seedling growth stages. Therefore some genes for resistance or avirulence may be very difficult to classify but the genetic systems involved are similar to those in which incompatible interactions are readily recognized.

2. When two or more interacting gene pairs are present the level of in-compatibility is similar to, or greater than, that of the least compatible of the individual infection types. Hence the presence of some resistance genes may be masked by the presence of others producing lower infec-tion types. Frequently, pathologists realize the existence of additional genes only when virulence develops in the pathogen for the gene interacting to produce the lowest level of incompatibility.

3. The interactions shown in Table 1b provide the genetic bases for the identification of pathogen strains and for host genotype surveys. If the genotypes of host testers are known and if they carry alleles for resis-tance relative to an arbitrary genetic base considered to be the "universal suscept" then the genotypes of various random, or non-random, isolates of the pathogen can be determined. Traditionally strain or race des-ignations of pathogens have described various combinations of genes for virulence and avirulence, often without relevance to breeding pro-grammes but Wolfe and Schwarzbach (1975) and Wolfe et al. (1976) have emphasized greater need for study of the individual frequencies of genes for virulence, or avirulence, relating to those host genes either present, or under consideration for use in commercial cultivars. These workers showed that combined virulence frequencies in European populations of Erysiphe graminis f. sp. tritici often can be predicted from the individual virulence frequencies indicating either mutational or recombinational equilibria.

III. Types of Resistance

The objective of the plant breeder is to produce host genotypes which are resistant to all variants of the pathogen. That is, he must produce and identify host lines whose resistance gene, or gene combinations, are not matched with virulence genes in the pathogen. This requirement is true irrespective of the type of resistance or the growth stage at which it is ex-pressed. Resistance is usually considered in two categories for which a number of contrasting terms have been used, for example, specific and non-specific (Watson, 1970a,b), general (Caldwell, 1968), or generalized (Smith and Smith, 1970), vertical and horizontal (Van der Plank, 1968; Robinson, 1973) and differential and uniform (Zadoks, 1972) resistance. However, it is usually difficult to determine whether resistance not matched by virulence in current isolates of the pathogen will be permanently effective. Hence a decision to place a certain instance of resistance into the second of these contrasting categories may be difficult. For this reason pathologists at the Plant Breeding Institute, Cambridge (Johnson and Taylor, 1973) describe instances where resistance has remained effective over a number of years as "durable" resistance. If a particular resistance has remained effective under prevailing conditions for relatively long periods of time, there may be an increased likelihood that it will remain effective in the future.

The expression of resistance differs from the extremes of apparent immunity, or hypersensitivity, to levels only very slightly different from susceptible standards and detectable only by expensive epidemiological methods (Zadoks, 1972). In the case of the rusts and mildews high levels of resistance expressed as hypersensitivity and obviously reduced pustule sizes are frequently selected on the basis of seedling tests. It is virulence for these resistances that has received most attention by those suggesting alternative means of producing resistance.

Zadoks (1972) described two general test methods for detecting resistance of the non-hypersensitive or intermediate type. In the monocyclic test, inoculation occurs only once and results of a single infection cycle are assessed by qualitative (infection type) methods, or by quantitative methods such as number of pustules per unit area, latent period, or sporulation rate. The polycyclic test attempts to measure the rate of increase of disease within a particular genotype, or mixture of genotypes, from an initial infected focus. Most greenhouse tests involve monocyclic testing. Zadoks considered that plant breeders' nurseries, where disease spreads mainly from susceptible spreaders and susceptible segregates, rather than from within partially resistant segregates, as continuous monocyclic tests. Van der Plank (1968) argued that such conditions provide levels of inoculum that overwhelm useful levels of intermediate but uniform resistance recognizable under conditions of polycyclic testing. While this may be true, the plant breeder requires levels of disease whereby the more resistant segregates can be rapidly distinguished from susceptible counterparts in large populations. Furthermore, for reasons of physical space and economy each breeding line, at the early stages of selection at least, must be confined to a relatively small area. As the number of lines remaining after each phase of selection becomes less, the degree and scope of testing may increase.

With intermediate resistances some yield losses will be inevitable under conditions most favourable for disease development. Losses in spring cereals grown near infected winter cereals may be greater in comparison with those grown in isolation. The breeder therefore must consider the level at which such losses can be tolerated. Simmonds (1973) predicted that under conditions where resistance to a number of diseases is required, breeders will tend to discard lines which are excessively susceptible to individual diseases rather than select only those which are highly resistant. However, at all times genotypes being assessed must be considered in comparison with cultivars they are likely to replace and if mildew resistance is a major objective then only those lines with resistance superior to a minimum standard should be retained.

Tolerance to disease has been demonstrated in some systems (Schafer, 1971) but is usually difficult to distinguish from slow and late disease development. Hayes (1972) suggests that yield tests should be part of the disease assessment process such that selection will then favour any mechanism which permits maximum yield.

IV. Genetic and Breeding Structures

A. Hosts

Of the four temperate winter cereals, wheat, barley and oats are inbreeders whereas most cultivars of rye are outbreeding. Bread wheat, *Triticum aestivum*, is an allohexaploid species (genomes AABBDD) with 21 chromosome pairs whereas the durum and emmer wheats, *T. turgidum*, are allotetraploids with 2n = 28. There are no wild forms of bread wheat. The most widespread forms of cultivated oats, *Avena sativa* and *Avena byzantina*, are allohexaploid (genomes AACCDD) with 2n = 42, but tetraploid, *Avena abyssinica* (AABB) and diploid, *Avena strigosa* (A_sA_s), forms are also grown (Rajhathy and Thomas, 1974). Wild oat forms occur at all three levels of ploidy.

Cultivated barley, *Hordeum vulgare* and rye, *Secale cereale*, are both diploids with 2n = 14. Wild species closely related to both crops occur in the respective centres of origin.

B. Pathogens

Erysiphe graminis is an heterothallic Ascomycete. The vegetatively produced haploid conidia are dispersed by air movement. However, sexual hybridization is common, and the cleistothecial stage is important in survival of the pathogen. The sexual stage of *E. graminis* f. sp. *avenae* is unknown in North America and possibly does not occur in South America (Moseman, personal communication).

V. Genetics of Host Resistance

The bulk of the available genetic information on mildew resistance in cereals involves resistance which is effective in the seedling, or early growth, stages and usually controlled by one, or a few, pairs of genes.

A. Barley

The genetics of resistance to mildew in barley were reviewed by Moseman (1966) and Wolfe (1972). Wiberg (1974) conducted comprehensive tests on 164 host genotypes with up to 94 Scandinavian cultures of the pathogen and discussed the known genetics of resistance in many of the genotypes. Resistance to at least some variants of *E. graminis hordei* is determined by alleles located in at least seven loci, some resistance alleles of which are designated *Ml-a* (Algerian), *Ml-at* (Atlas), *Ml-g* (Goldfoil), *Ml-h* (Hanna), *Ml-k* (Kwan), *Ml-p* (Psaknon) and *ml-o*. Moseman and Jørgensen (1971) list eleven alleles for resistance occurring at, or near, the *Ml-a* locus and Clifford (1975) refers to *Ml-a12* (*Ml-as*) in cultivar Sultan. Jørgensen and Moseman (1972) were unable to obtain recombinants from repulsion phase crosses involving alleles *Ml-a*, *Ml-a3*, *Ml-a8*, and *Ml-a9*. However, from recombination studies and correlated behaviour with many pathogen cultures, the data of Wiberg (1974) suggest that the gene present in HOR 1063 may be identical with *Ml-a4* present in cultivar No. 22. The same gene appeared to be present in several cultivars possessing *Ml-a7* or *Ml-a9*. At least five loci involved in mildew reaction appear to be located in the long arm of chromosome 5, whereas two loci, *Ml-g* and *ml-o*, are located in chromosome 4 (Wolfe, 1972; Jørgensen, 1971).

Wolfe (1972) discussed the sequential changes in European *E. graminis* f. sp. *hordei* populations following the cultivation of cultivars with resistance genes derived from Weihenstephan CP127422 (*Ml-g*), Ragusa b (*Ml-h* + ?) and *Hordeum spontaneum nigrum* (*Ml-a6*). Variants of the fungus virulent on plants having a combination of all three genes for resistance rapidly evolved and their fitness was unaffected. On the other hand combined virulence for *Ml-g* and *Ml-a12* appeared to occur with lower than expected frequencies (Wolfe and Schwarzbach 1975).

The main form of incomplete or adult plant resistance to mildew so far used in barley was derived from *Hordeum laevigatum*. This resistance, originally thought by some workers to be non-specific and present in cultivars Vada, Minerva and Bomi, appears to be conditioned by a single allele. Virulence is common throughout Europe and Israel (Wolfe, 1972; Fischbeck *et al.*, 1976). Hence the type of resistance does not necessarily indicate the complexity of inheritance or the likelihood of non-specificity.

Jørgensen (1971) and Wolfe (1972) reported that the recessive *ml-o* allele occurred in a number of European and Ethiopian barleys. Jørgensen (1974a) indicated that ten independently induced mutant lines also carried *ml-o* but recombination studies have shown that susceptible recombinants from at least some heteroallelic crosses may be obtained (Jørgensen 1976). Apparently *ml-o* can be separated from the necrotic leaf flecking with which it is associated in both natural and induced mutant lines (Jørgensen, 1976).

B. Wheat

The early literature was reviewed by Moseman (1966). Resistance alleles in bread wheat have been designated at seven loci *viz. Pm1* (chromosome 7AL), *Pm2* (5DS), *Pm3* (1AS), *Pm4* (2AL), *Pm5* (7BL), *Pm6* (2BL) and *Pm7* (4Aβ) (McIntosh, 1973, 1975). Three resistance alleles, *Pm3a, Pm3b* and *Pm3c*, have been described at the *Pm3* locus and *Pm4* is closely linked, or allelic, with *Mle* derived from the tetraploid carthlicum wheat (McIntosh, unpublished). Baier *et al.* (1973) located *Mle* in chromosome 2A.

Virulence in *E. graminis* f. sp. *tritici* has been reported for plants carrying all resistance alleles except *Pm6* derived from the tetraploid, *Triticum timopheevi*, and *Pm7* transferred to wheat from Rosen rye. *Pm6* usually occurs in coupling with *SrTt1* for resistance to *Puccinia graminis* f. sp. *tritici* but occasional recombinations have been identified. According to Swedish workers (Svensson, personal communication) the presence of *SrTt1* has detrimental effects upon certain agronomic characters. These defects were absent in recombinants carrying only *Pm6*. Maris Huntsman which probably carries *Pm6* does not possess *SrTt1* (McIntosh, unpublished).

A further resistance allele, *Mlr*, derived from Petkus rye is present in several wheats grown in Europe and the USSR (Zeller, 1973) but virulent variants in the pathogen have been reported (Johnson *et al.*, 1971; MacKey, 1974).

Smith and Smith (1970) considered that post-seedling resistances in the New Zealand wheats, Dreadnought and Arawa, were of the generalized type. They suggested that resistance was determined by many genes possibly with additive effects. Smith and Wright (1974) listed a series of cultivars grown in New Zealand in decreasing order of resistance namely, Dreadnought, Arawa, Gamenya, Kopara 73, Kopara, Karamu, Aotea, Cross 7–61, Hilgendorf-61, Tainui, Raven.

Shaner and Finney (1975) considered that slow mildewing in Knox wheat, thought to be non-specific because of its long term effectiveness, had low-to-moderate heritability. Ellingboe (1975) cited an instance of slow mildewing in Genesee wheat controlled by a single allele but some isolates of the pathogen were virulent. From greenhouse infection studies the inheritance of slow mildewing in Genesee was thought to be genetically complex and the simple nature of inheritance was demonstrated only under more precisely controlled conditions.

Ellingboe (1975) described situations in which cultures with apparently independently derived virulence for *Pm4* showed longer incubation periods or fewer pustules per unit area when infecting a genotype possessing *Pm4* compared with a near-isogenic line possessing the contrasting allele *pm4*. Under some conditions, therefore, resistance genes may produce a degree of "ghost resistance" (Hayes, 1972) to cultures considered to be virulent. Furthermore, the expression of ghost resistance may vary among independently derived mutants.

C. Oats

Few instances of seedling resistance to *E. graminis* f. sp. *avenae* in cultivated hexaploid oats have been reported. Apart from three genotypes showing differential reactions to the pathogen in Wales, Hayes and Jones (1966) were unable to find high levels of resistance. Certain genotypes reported to be resistant in Holland or United States were susceptible. Two forms of the diploid, *Avena strigosa*, were resistant.

Hayes and Jones (1966) reported varying levels of adult plant resistance among hexaploid oat cultivars. Maldwyn was the most resistant and genetic studies indicated that resistance was determined by a number of genes. Hayes (1970) considered adult plant resistance in Maldwyn and Maelor to be non-specific.

VI. Breeding Methods

Breeding involves two phases (Frey, 1976):
1. the creation of genetic variability;
2. selection among the genotypes produced.

A. Creation of genetic variability

The usual method of creating genetic variability in inbreeding plants is hybridization of the various genotypes chosen as parents. Usually at least one parent is a current cultivar of proven performance and other parents are chosen to complement the various deficiencies. Hybrids may range from single crosses involving two parents to complex intercrosses involving many genotypes. In instances where the main character of interest is disease resistance the backcross method has been commonly employed. Because backcrossing is very conservative some breeders prefer to backcross only once or twice in the hope that sufficient variability can be retained in order to gain further improvement in a number of characters. However, if only limited backcrossing occurs then one major advantage of the backcross method, the reduced need for detailed performance testing, is lost.

Genetic variability may be produced with mutagens. In mutation breeding adapted agronomic types may be used as base populations. In many instances where induced mutations for disease resistance have been claimed, subsequent studies suggested that contamination and outcrossing within the base populations may have been responsible. Jørgensen (1974b) discussed some of the methods by which this problem can be reduced. In the case of barley the same resistance allele, *ml–o*, was produced in several independent studies.

B. Selection

After hybridization, or mutagen treatment, breeding populations must pass through a number of self-pollinating generations in order to achieve homozygosity, the usual genetic state of inbreeding species apart from commercial hybrids. In pedigree breeding programmes, selection for characters of high heritability is practised during the early generations of inbreeding while selection for characters of low heritability is delayed until later generations. Disease resistance is usually one of the characters included in the high heritability group. Selection may be conducted on a single plant basis, or in F3 and subsequent generations, on a family or progeny basis. Subsampling and replicated testing is possible in the F3 and subsequent generations.

In contrast to the pedigree method, the bulk method may be utilized. This involves the growing and harvesting of materials in bulk for a number of generations during the inbreeding process. Artificial or natural selection may be applied to bulk populations in order to modify gene frequencies. Derera and Bhatt (1973) described mechanical mass selection as a means of shifting gene frequencies in the desired direction. One effect of mildew is reduced grain size (Griffiths *et al.*, 1975), hence mechanical selection of larger seeds in bulks harvested from diseased plots should assist in the selection of genotypes with mildew resistance or tolerance.

The method of single seed descent (SSD) (Brim, 1966) has received some recent attention and should be of considerable use for the evaluation of disease resistances by polycyclic testing methods. In this method, one or two seeds are harvested and grown from each F2 plant and thereafter one seed is grown from each plant in each generation of inbreeding. No selection is practised during the inbreeding process. This method lends itself to systems permitting single plants to be grown in minimal areas since few seeds are required from each plant, and to rapid generation turnover. Since entire populations are available at the end of the inbreeding process the additive genetic variation within and between cross populations can be estimated. Near homozygous SSD populations in F6 or F7 can be increased to desired quantities for replicated testing.

The stages at which selection for disease resistance will occur can be specified by the breeder and will depend upon the type of resistance required. Disease tests can be carried out in the actual breeding nursery where yield and quality tests are being conducted and if disease is endemic and occurs each year this will be the case. Alternatively, disease testing can be carried out on subsamples from the same populations at many field sites, or in the greenhouse and laboratory. If selection is to be conducted at one site using only natural populations of pathogens then genetic variation with respect to the pathogen may be very limited. In some instances increase of less frequent variants of the pathogen can be favoured by using selective spreader genotypes which are susceptible only to those variants. The purposeful release of selected cultures in disease nurseries is apparently not a common practice in breeding for mildew resistance. Although the breeder may believe that particular sources of resistance are non-specific the only measure of non-specificity is durability over time or space. On the other hand if resistance is non-specific then by definition, multistrain testing is not required. However, under such circumstances multilocational testing would provide information regarding the effects of genotype environment interactions on disease reaction as well as durability. Obviously, multilocational testing on a global basis is expensive and depends upon the goodwill of those involved. Therefore it is usually limited to the testing of advanced selections.

C. Recording disease reactions

Disease reactions can be recorded on the basis of pustule size and type (infection type) or on the number and distribution of pustules at various growth stages. Smith and Smith (1970) used a modified Cobb scale in order to identify four to five grades of mildew severity. In cultivaral comparisons they sampled six tillers per genotype at ear emergence and determined percentages of infected areas on the four leaves immediately below the flag leaf. Greenhouse and field ratings were correlated. In a breeding programme it may be sufficient to classify breeding lines in relation to known genotypic standards and to select on this basis.

VII. Detection of Sources of Resistance

In order to find new sources, or types of resistance, breeders must continue to survey genotypes of the host species. Krull and Borlaug (1970) stress the value of large diverse collections of wheat, oats and barley in this respect and indicate the need for "increased stability and breadth" of disease resistance without specifying how this might be achieved apart from a warning regarding the dangers of "seedling hypersensitive genes". According to Van der Plank (1975) and Robinson (1973), considerable levels of resistance to most diseases can be assembled by intercrossing and selection among current cultivars displaying slightly reduced levels of disease. The latter author suggested "random polycross" methods to produce maximum levels of genetic variability followed by very strong selection pressures for resistance but with an avoidance of specific resistance.

In addition to collections of cultivated genotypes, wild species producing fertile hybrids with cultivated genotypes of barley and hexaploid oats are available. Fischbeck *et al.* (1976) described resistance to selected European isolates of *E. graminis* f. sp. *hordei* in genotypes of *H. spontaneum* collected in Israel. Hayes (1970) indicated that *Avena sterilis* was being used as a source of resistance to *E. graminis* f. sp. *avenae*. Wild forms of hexaploid wheat do not occur but genes for specific resistance to mildew have been transferred from cultivated tetraploid wheats. Additionally, chromosome engineering techniques have been developed for the transfer of genes from chromosomes that normally do not pair with those of wheat (Sears, 1972, 1973). However, as sources of resistance become taxonomically more distant from the cultivated species the probability that derived resistances will be based on single genes with major effects will increase and it is now quite apparent that genes for disease resistance derived from related species can be overcome by pathogens as readily as those in the host species (Watson, 1970a). Such introgressed resistance genes simply add to the overall host variability rather than offering more permanent types of protection.

VIII. Deployment of Resistance Genes

The recent southern corn leaf blight epidemic in North America illustrated the hazards of genetic uniformity in major crop plants and prompted much discussion (e.g. Horsfall *et al.*, 1972). Although it may be difficult to decide the attributes of a genetic hazard in advance, there is no doubt that the widespread use of cultivars with identical genes for disease resistance constitutes vulnerability. Horsfall *et al.* (1972) and Day (1973, 1974) drew attention to the fact that with most crops the area cultivated at any time was represented by very few cultivars. Macer (1975) chose the example of Cappelle Desprez in the UK to show how the use of a single genotype in crosses with other parents can lead to derived cultivars having a narrow genetic base.

It is common for breeders and pathologists to extrapolate from experiences based on one disease, or one particular set of circumstances, to another disease or another set of circumstances. Most successes, or failures, in producing disease resistances have been based empirically on trial and error rather than being preconceived in the laboratory. For example, the successful introduction of flagsmut resistance in Australia was based on the use of the first resistance sources encountered but when the same approach was made with the rust and mildew diseases, resistance was ephemeral. At that time the type, mechanisms and genetics of resistance were not considered.

It is now generally accepted that disease resistances based upon single alleles in hosts, especially if grown in large areas, are likely to be ephemeral although there are notable exceptions (Watson, 1970a). Furthermore, as stressed by Van der Plank (1968) the longer a particular resistance gene remains effective the more it will be used by breeders. This results in greater exposure of the gene to rare virulent variants and hence to increased likelihood that potential losses from disease will occur as a consequence. In Australia, wheat breeders are rapidly increasing the use of gene *Sr26* for resistance to *Puccinia graminis* f.sp. *tritici* despite warnings that its continued effectiveness cannot be guaranteed. An International Survey of Gene Virulence in this pathogen has indicated that *Sr26*, originally transferred to wheat from *Agropyron elongatum*, confers stem rust resistance on a global basis (Watson, personal communication). Similarly, the use of genes *Pm6* in wheat for resistance to *E. graminis* f. sp. *tritici* and *ml–o* in barley for resistance to *E. graminis* f. sp. *hordei* will undoubtedly occur provided they remain effective and provided genetically associated agronomic defects, when they occur, can be deleted or suppressed. On the other hand, if such genes do confer long-lasting resistance, breeders not using them will be at a disadvantage since most breeding organizations justify their activities through financial profits or a successful record of new cultivars grown over large areas. A decision to use such resistance genes permits increased flexibility with respect to the many other genetic variables considered by the breeder.

Provided the hazards to the use of identifiable resistance genes, in contrast to polygenes, are accepted there appears to be a number of ways in which such genes might be managed in order to extend their effective usefulness, or to decrease the magnitude of losses if pathogens do change. All strategies are based on genetic heterogeneity of the host species, but also, more attention is now being drawn to the use of integrated pest or disease control (Van Emden, 1974) where several approaches are combined. However, the need for integrated control has been recognized for many years. The barberry eradication schemes in North America and earlier in western Europe, undoubtedly delayed the beginnings of wheat stem rust epidemics as well as contributing to the reduction of dangers of sexual recombination involving recessive alleles for virulence in the dikaryotic pathogen. In North America, barberry eradication was combined with breeding for resistance. In areas where winter survival of powdery mildew in the vegetative phase is important, efforts to reduce the amount of inoculum during this period could be worthwhile. This could be achieved by increased resistance at early growth stages in cultivars for autumn sowing, or by chemical control.

In addition to the systems of disease resistance based on identifiable genes, either at seedling or post seedling growth stages, it is commonly accepted that polygenic systems affect the degree of disease and disease losses occurring in susceptible genotypes. Furthermore, it is usually accepted that a worthwhile breeding strategy is to combine identifiable gene systems with such polygenic systems. Other workers (e.g. Robinson, 1973) suggest that these polygenic systems can be additively accumulated to produce non-damaging levels of disease. Van der Plank (1968) presents examples where, following the occurrence of virulence in pathogens, previously resistant commercial cultivars were not only extremely susceptible but were more susceptible than the genotypes from which the resistances were derived. This indicated that the donor genotypes carried additional genes for resistance which were not transferred to the newer commercial genotypes. Van der Plank designated this failure to recover the entire resistance of donor genotypes as the "Vertifolia" effect. Despite an awareness of its occurrence, it is likely that breeders of commercial cultivars will continue to use identifiable genes in various ways and will wait for further guidelines from research, or for proven parental lines to be produced by programmes such as those proposed by Robinson and Chiarappa (1975).

The first level at which genetic heterogeneity can be obtained is on the farm where farmers might arrange to grow at least three or four cultivars especially concentrating on genetic differences with respect to the known hazards. In the case of wheat at least, this objective may be only partially achievable since similar related quality types tend to be widely grown on a regional basis.

Despite Clifford's (1975) arguments to the contrary, a programme aimed at the use of multigenic resistances to rust in the northern wheat belt of Australia has been largely successful. This strategy is based upon the findings that most virulence changes in *P. graminis* f. sp. *tritici* are caused by single gene mutations. In the case of mildews a multigenic strategy is expected to be less successful because sexual recombination as well as mutation in the pathogen will be involved in the production of new virulence combinations. Hence the components of a multigenic mildew resistance must consist of genes conferring resistance to all, or most, variants of the pathogen. Furthermore, the successful use of a multigenic resistance is dependent upon the absence of cultivars with resistance based upon the individual gene components of that resistance. For example it is suggested that in wheat a very useful resistance might be *Pm4* + *Pm6* since virulence for the former is still comparatively rare and is unknown for the latter, but the simultaneous widespread use of cultivars possessing only *Pm4* or *Pm6* would provide the "stepping-stones" necessary for the virulence genes to accumulate. Because of distinctive seedling infection types such a combination of genes is easily achieved by producing *Pm6 Pm6* homozygotes in *Pm4 pm4* heterozygotes. Alternatively, rare cultures virulent on seedlings with *Pm4* could be used in the laboratory in order to classify for *Pm6*.

Van der Plank's (1968, 1975) arguments in support of resistance gene management based on a theory that variants of the pathogen with virulence for certain host genes have reduced survival abilities in the absence of non-selective hosts are being increasingly questioned (Nelson, 1973; Wolfe, 1973; Clifford, 1975). The evidence for stabilizing selection with respect to *Sr6* in North America appears to be unfounded since Roelfs (1974) presented data indicating that populations of *Puccinia graminis* f. sp. *tritici* and *P. recondita* in Mexico and southern Texas were distinctive from those in Northern Texas and Oklahoma. There is no evidence for changes in race constitution or virulence gene frequencies, from these latter populations and those in Canada. If virulence and aggressiveness, or competitive ability are independent attributes or are not necessarily correlated, then some variants of the pathogen may be relatively unfit and hence will decline in the absence of selective hosts but independent mutations or genetic recombinations could be equally fit, or occasionally better fit, than the progenitor types. From population studies Wolfe and Schwarzbach (1975) describe an instance in *E. graminis* f. sp. *hordei* where the combined virulence for two barley resistance genes, *Ml–g* and *Ml–a12*, has failed to reach frequencies predicted from random recombination and suggest that the strategic use of such genes may have a place in disease control. On the other hand, this reduced frequency of double virulence may be temporary and new variants with the expected combined virulence frequencies may occur in future. As these authors suggest, a more detailed understanding of variation in pathogen populations is needed.

Regional gene deployment has been suggested as a means of obtaining genetic diversity over wide areas. The successful use of this method depends upon knowledge of the genetic components of resistance but frequently this is not available. Additionally, cultivars produced in one area are often successfully grown in other areas. Cultivation of these might be prohibited in rigid systems of gene deployment. In any case, in areas such as Europe where many countries are involved such systems would be difficult to manage and impossible to control. The agreed restriction of known resistance sources to designated breeding institutions may be a practical compromise.

A further means of obtaining genetic diversity is the use of multilines, or mixtures of agronomically similar genotypes differing in genes for disease resistance. Since multiline breeding depends on backcrossing the method is conservative and appears to be restricted to host genotypes which will be grown very widely and for relatively long periods of time. There are two major contrasting philosophies regarding the development and use of multiline cultivars. The first depends on the use of mixtures of genotypes with resistance to all prevailing variants of the pathogen (Borlaug, 1953; Breth, 1976) whereas the second depends upon the use of mixtures of genotypes resistant to some components of the pathogen population but susceptible to others (Browning and Frey, 1969; Frey, 1976).

The development of multiline cultivars resistant to all components of a particular pathogen population has been advocated and pursued in relation to rust control in Mexico (Breth, 1976). Multilines of this type are unlikely to be severely damaged because virulence is unlikely to develop for all components in any one season. When virulence for one host component does occur, that component is replaced in the mixture in the following season. This approach avoids extreme losses and is particularly applicable where cultivars, such as the "8156" derivatives, are grown over very wide areas. The CIMMYT method of producing multiline components is based on the crossing scheme (A × B) × (A × C) where A is the current agronomic type and B and C are disease resistance sources. In the following generations disease resistant A-like segregates are selected for yield testing. Approximately 250 different lines were produced and distributed by CIMMYT for local testing. It is apparent that the individual lines carry multigenic resistances but since no comprehensive list of genotypes is available, the degree of heterogeneity within and between lines with respect to all three wheat rusts is not known. The release of multilines, Miramar 63 and Miramar 65, in Columbia was claimed to result in successful control of losses to wheat stripe rust (Breth, 1976).

Successful reductions in losses attributable to oat crown rust have been claimed by Iowa workers who have released eleven multiline oat cultivars for commercial cultivation in midwestern USA (Frey *et al.*, 1976). In this programme single resistance alleles are added to acceptable adapted genotypes by five generations of backcrossing. This method of multiline use depends upon stabilizing selection in the pathogen favouring variants with relatively few genes for virulence such that each variant will attack mainly one component of the host multiline. Hence the increase of the particular pathogen variant is buffered by the presence of resistant host genotypes. The survival and epidemiology of other pathogen components will be similar such that the overall effect should be a decrease in disease as measured by spore frequencies in the air above plots or by yield. The value of this method rests firmly on the assumptions of stabilizing selection. Frey *et al.* (1976) noted the repeated disappearance of "race 264A"—a variant with multiple virulence— as evidence for stabilization in the pathogen. The question to be asked is whether all independently evolving instances of "race 264A" and other complex pathogen variants, behave in this way. If multilines have reduced losses to oat crown rust in midwestern USA, what are the comparative loss estimates for non-multiline oats grown in the same area?

Johnson and Allan (1975) suggest that induced resistance to virulent variants caused by prior inoculation with avirulent types may have a role in retarding disease spread in multilines. This aspect needs further investigation under field conditions.

IX. Conclusions

Insufficient population data, not only for the cereal mildew pathogens but for most plant pathogens, are available to suggest the best means of deploying host variability in disease control. Until better guidelines are available, plant breeders in the field, where a particular disease is only one problem, will continue to produce cultivars as they have done in the past. It will be the task of geneticists and pathologists working in conjunction with breeders to demonstrate which particular strategies should be used in commercial breeding programmes. Some suggested approaches are very difficult even if only one disease is considered a problem but in many situations the breeder has to cope with more than one disease or pest. Hence decisions made with respect to one often result in compromises relating to others.

The following resolution, was presented at the Third International Barley Genetics Symposium in 1975.

To increase the durability of the resistance of barley varieties to pathogens such as *Erysiphe graminis* or *Puccinia hordei*, this symposium considers that it is essential to have as wide a range as possible of different resistances in the commercial crop, both at the national, and the farm, level. In this sense it is not necessary to consider different categories of resistance. High levels of resistance should be released, however, only in varietal backgrounds which have a low level of susceptibility in the absence of the newly introduced resistance.

Pathogen populations should be characterised in terms of virulences and virulence frequencies with respect to potentially useful sources of host resistance, rather than by physiologic races.

References

Baier, A. C., Zeller, F. J., Oppitz, K. and Fischbeck, G. (1973). *Z. PflZucht.* **70**, 177–194.
Borlaug, N. E. (1953). *Phytopathology* **43**, 467.
Brim, C. A. (1966). *Crop Sci.* **6**, 20.
Breth, S. A. (1976). *In* CIMMYT Today No. 4. 11 pp. published by Centro Internacional de Mejoramiento de Maiz y Trigo, Mexico.
Browning, J. A. and Frey, K. J. (1969). *A. Rev. Phytopath.* **7**, 355–382.
Caldwell, R. (1968). *In Proc. 3rd Int. Wheat Genetics Symp.* (K. W. Finlay and K. W. Shepherd Eds.). 263–272. Aust. Acad. Science.
Clifford, B. C. (1975). *Rep. Welsh Pl. Breed. Stn.* **1974**, 107–113.
Day, P. R. (1973). *A. Rev. Phytopath.* **11**, 293–312.
Day, P. R. (1974). Genetics of host-parasite interaction. W. H. Freeman and Co. Reading.
Derera, N. F. and Bhatt, G. M. (1973). *In Proc. 4th Int. Wheat Genet. Symp.* (Sears, E. R. and Sears, L. M. S. Eds). 495–498. Univ. Missouri.
Dickson, G. D. (1956). Diseases of Field Crops. McGraw Hill Book Co. New York.
Ellingboe, A. H. (1975). *Aust. Pl. Path. Soc. Newsletter* **4**, 44–46.
Fishbech, G., Schwarzbach, E., Sobel, Z. and Wahl, I. (1976). *Z. PflZücht.* **76**, 163–166.

Flor, H. H. (1956). *Adv. Genet.* **8**, 29–54.
Frey, K. J. (1976). *Egypt. J. Genet. Cytol.* **5**, 184–206.
Frey, K. J., Browning, J. A. and Simons, M. D. (1976). *In* Induced mutations for disease resistance in crop plants (1975). International Atomic Energy Agency Tech. Document 181, 101–109.
Griffiths, E., Jones, D. G. and Valentine, M. (1975). *Ann. appl. Biol.* **80**, 343–349.
Hayes, J. D. (1970). *Rep. Welsh Pl. Breed. Stn.* **1919–1969**, 101–167.
Hayes, J. D. (1972). *Ann. appl. Biol.* **72**, 140–144.
Hayes, J. D. and Jones, I. T. (1966). *Euphytica* **15**, 80–86.
Hermansen, J. E. and Stix, E. (1974). Yb. R. Vet. Agric. Univ. Copenhagen, 87–100.
Hermansen, J. E., Ulrip, T. and Prahm, L. (1975). Yb. R. Vet. Agric. Univ. Copenhagen, 17–30.
Horsfall, J. G. (1972). Chairman's Report. "Genetic vulnerability of major crops". Rep. Nat. Acad. Sci. USA No. 2030, 300 pp.
Jenkins, J. E. E. (1973). *Proc. 7th Br. Insectic. Fungic. Conf.* 781–790.
Jenkins, J. E. E. and Storey, I. F. (1975). *Pl. Path.* **24**, 125–134.
Jenkyn, J. F. (1974). *Ann. appl. Biol.* **78**, 281–288.
Jenkyn, J. F. (1976). *Pl. Path.* **25**, 34–43.
Johnson, R. and Allen, D. J. (1975). *Ann. appl. Biol.* **80**, 359–363.
Johnson, R. and Taylor, A. J. (1973). *Rep. Pl. Breed. Inst.* **1972**, 136–139.
Johnson, R., Scott, P. R. and Wolfe, M. S. (1971). *Rep. Pl. Breed. Inst.* **1970**, 100–110.
Jørgensen, J. H. (1971). *Hereditas* **69**, 298.
Jørgensen, J. H. (1974a). *In* Induced mutations for disease resistance in crop plants, International Atomic Energy Agency Vienna, 67.
Jørgensen, J. H. (1974b). *In* Induced mutations for disease resistance in crop plants, International Atomic Energy Agency Vienna, 57–66.
Jørgensen, J. H. (1976). *In* Induced mutations for disease resistance in crop plants (1975). International Atomic Energy Agency Technical Document 181, 129–140.
Jørgensen, J. H. and Moseman, J. G. (1972). *Can. J. Genet. Cytol.* **14**, 43–48.
King, J. E. (1973). *Proc. 7th Br. Insectic Fungic. Conf.* 771–780.
Krull, C. F. and Borlaug, N. E. (1970). *In* "Genetic resources in plants". O. H. Frankel and E. Bennett (Eds). 427–439. IBP Handbook No. 11. Blackwell, Oxford.
Macer, R. C. F. (1975). *Trans Br. mycol. Soc.* **65**, 351–374.
McIntosh, R. A. (1968). Ph.D. Thesis. Univ. of Sydney, 184 pp.
McIntosh, R. A. (1973). *In Proc. 4th Int. Wheat Genetics Symp.* (Sears, E. R .and Sears L. M. S. Eds) 892–937. Univ. of Missouri.
McIntosh, R. A. (1975). *Cereal Res. Comm.* **3**, 69–71.
Mackey, J. (1974). *In* Induced mutations for disease resistance in crop plants, International Atomic Energy Agency Vienna, 9–22.
Moseman, J. G. (1966). *A. Rev. Phytopath.* **4**, 269–290.
Moseman, J. G. and Jørgensen, J. H. (1971). *Crop Sci.* **11**, 547–550.
Nelson, R. (1973). "Breeding plants for disease resistance. Concepts and applications". Penn. State Univ. Press.
Rajhathy, T. and Thomas, H. (1974). Genetics Soc. Canada Misc. Publications No. 2.
Robinson, R. A. (1973). *Rev. Pl. Path.* **52**, 483–501.
Robinson, R. A. and Chiarappa, L. (1975). *FAO Plant Protection Bull.* **23**, 125–129.
Roelfs, A. P. (1974). *Pl. Dis. Reptr* **58**, 806–809.
Schafer, J. F. (1971). *A. Rev. Phytopath.* **9**, 235–252.
Sears, E. R. (1972). *In* Stadler Symp. (Kimber, G. and Redei, G. D. Eds), **4**, 23–38. (Univ. of Missouri).
Sears, E. R. (1973). *In Proc. 4th Int. Wheat Genetics Symp.* (Sears, E. R. and Sears, L. M. S. Eds). 191–207. Univ. of Missouri.

Shaner, G. and Finney, R. (1975). *Proc. Am. Phytopath. Soc.* **2**, 49.
Simmonds, N. W. (1973). *Phil. Trans. R. Soc.* B, **267**, 145–156.
Simonds, M. D. and Murphy, H. C. (1961). *In* "Oat and oat improvement". (Coffman, F. A. Ed.). 330–390. Am. Soc. Agron.
Smith, H. C. and Smith, M. (1970). *N. Z. Wheat Rev.* **11**, 54–61.
Smith, H. C. and Wright, G. M. (1974). *N. Z. Wheat Rev.* **12**, 30–34.
Van der Plank, J. E. (1968). "Disease resistance in plants". Academic Press, New York.
Van der Plank, J. E. (1975). *Aust. Pl. Path. Soc. Newsl.* **4**, 27–30.
Van Emden, H. F. (1974). Pest control and its ecology. Edward Arnold Ltd. London.
Watson, I. A. (1970a). *In* "Genetic resources in plants". (Frankel, O. H. and Bennett, E. Eds). 441–457. IBP Handbook No. 11 Blackwell, Oxford.
Watson, I. A. (1970b). *Ann. Rev. Phytopath.* **8**, 209–230.
Wiberg, A. (1974) *Hereditas* **78**, 1–40.
Wolfe, M. S. (1972). *Rev. Pl. Path.* **51**, 507–522.
Wolfe, M. S. (1973). *Ann. appl. Biol.* **75**, 132–136.
Wolfe, M. S. and Schwarzbach, E. (1975). *Phytopath. Z.* **82**, 297–307.
Wolfe, M. S., Barrett, J. A., Shattock, R. C., Shew, D. S. and Whitbread, R. (1976). *Ann. appl. Biol.* **82**, 369–374.
Zadoks, J. C. (1972). *In* "The way ahead in plant breeding". (Lupton, F., Jenkins, G. and Johnson, R. Eds). 89–98 *Proc. 6th Cong. of Eucarpia*, Cambridge.
Zeller, F. J. (1973). *In Proc. 4th Int. Wheat Genet. Symp.* (Sears, E. R. and Sears, L. M. S. Eds) 209–221, Univ. of Missouri.

Chapter 10

Chemical Control of Powdery Mildews

K. J. BENT

Imperial Chemical Industries Ltd., Jealott's Hill Research Station, Bracknell, England

I. Introduction

The serious economic damage to crops caused by the powdery mildews is referred to at many points in this book. This damage may be manifested as a direct fall in yield, as a suppression or distortion of plant growth which in the longer term affects yield, as a spoilage of fruit or as a disfigurement of ornamental plants. These effects are often mixed. The first serious attempts to stop damage caused by powdery mildews were probably made in the early nineteenth century, when sulphur dusting came into use for the control of mildew on fruit trees and, a little later, on grape vines. Since that time the usage of fungicides against powdery mildews has increased enormously, in overall amount, in the number and variety of fungicides used and in the range of crops which are treated. At present, sales of fungicides for use primarily against powdery mildews amount to some £120 m out of a total world fungicide market of about £650 m (B. G. Lever, pers. comm.). Application is predominantly on barley, apples and grape-vines. About 20 different fungicides are in use, and sulphur still accounts for 25–30% of the total expenditure on powdery mildew fungicides.

The value of other methods of control, such as planting resistant varieties, pruning and burning mildewed shoots, and removing volunteer host plants, has also been increasingly recognized. However, it is the application of chemicals, either with or without the concomitant use of other measures, that forms the principal defence against powdery mildews throughout the world.

In this chapter we consider in turn the various types of powdery mildew fungicide. Their chemistry, mode of action and practical value and limitations are described briefly. We start with sulphur, the oldest material, move on to other surface fungicides which first came into use in the 1950s, and then discuss the systemic fungicides which have appeared within the last decade. There follow sections which deal with general aspects of mode of action, methods of application, problems of adaptation to fungicides and current and future trends in the chemical control of powdery mildews.

II. Powdery Mildew Fungicides

A. Sulphur

The use of sulphur has an interesting history, which is reviewed by McCallan (1967) and Sharvelle (1969). Sulphur was used in ancient times as a medicine and as a fumigant. However the first mention of sulphur as a fungicide appears to be a recommendation in 1802 by William Forsyth, gardener to His Majesty, that mildew on fruit trees could be controlled by the application of a concoction of sulphur, tobacco, lime and elder buds (Forsyth, 1802). Other horticulturalists recommended the use of sulphur and lime water (Weighton, 1814) and sulphur and soap (Robertson, 1821) for treating fruit trees. The appearance of grape powdery mildew in Europe around 1845 was followed by a great increase in the use of sulphur. To this day elemental sulphur remains the predominant fungicide for the control of grape powdery mildew and it is also still used on apples, peaches and other crops. Sulphur treatment is cheap.

In many circumstances it gives adequate disease control, provided it is applied frequently.

Sulphur is applied mainly as a fine dusting powder at rates which vary between 5 and 40 kg sulphur ha^{-1}. A small proportion of kaolin or bentonite is commonly added to prevent the tendency of the particles to clump together. Wettable powder and colloidal liquid formulations are also available; these usually contain 60–80% sulphur. They are diluted with water and applied in high-volume sprays at about 1000–10 000 ppm sulphur, or in sprays at lower volumes at correspondingly increased concentrations. Lime-sulphur ("Eau Grison"), made by boiling an aqueous suspension of calcium hydroxide with sulphur, is a concentrated solution of calcium polysulphides together with a little calcium thiosulphate. On mixing with water and exposure to air the product decomposes to a fine suspension of sulphur particles. It has been used for about 100 years, especially in orchard sprays. Its use has declined greatly because compared with the simple sulphur formulations it is more odorous, less compatible with other pesticides and often more phytotoxic.

Sulphur is also applied by volatilization. The practice of painting the hot-water pipes of grape-vine houses has been followed for over one hundred years. This method is reasonably effective although the vapour tends to control mildew better on the upper leaf surfaces than on the under-sides. Various types of lamps and hot-plates have been devised for vaporizing sulphur (see Martin, 1964).

Sulphur exerts its fungicidal action at the surface of the leaves, stems, flowers or fruits to which it is applied. It is re-distributed over such surfaces to a limited extent by vaporization and also by the action of rain and dew. It does not penetrate into plants or move through them to an extent sufficient to protect the new parts of the plant which appear after treatment and which are often highly susceptible to powdery mildew. It must be applied repeatedly to protect these new tissues and to offset losses caused by weathering. On grape-vines 8–12 applications per season are commonly made. Coatings of sulphur on treated plants will for some time protect them from infection by hindering or preventing the germination of powdery mildew spores which alight on them. The effective period of protection can vary greatly according to the particular crop, to the weather and to the intensity of the inoculum but it is often 10–15 days. Because powdery mildews grow mainly on the plant surface, applied sulphur can come into direct contact with existing mycelium and suppress its growth and sporulation. However, this curative action is transient and rather feeble compared with that of most other powdery mildew fungicides. Sulphur tends to work better in warmer countries. This phenomenon may result from an increase in vapour action from localized deposits. Other fungicides such as bupirimate and quinomethionate have a more marked vapour action but do not show such a clear enhancement under warmer conditions.

Sulphur does penetrate into plants to a limited degree, and can cause damage. Rapid "scorching" (direct local injury) of leaves tends to occur in hotter climates. Physiological effects, especially premature drop of leaves or fruit or increased russeting in apples can also be induced, even under temperate conditions. Certain varieties of apples (e.g. Stirling Castle) and gooseberries (e.g. Leveller) are particularly vulnerable to sulphur damage and are said to be sulphur-shy. Grape-vines are rarely affected adversely by sulphur.

Applications of sulphur will control a number of other fungal diseases, for example apple scab (*Venturia inaequalis*) and peach leaf-curl (*Taphrina deformans*). But in general powdery mildews are controlled best. Sulphur also has a useful action against spider-mites and scale insects.

The mode of action of sulphur has been the subject of many studies made in the 1920s and 1930s (see Martin, 1964, and Tweedy, 1969, for detailed reviews). However, it is still not fully understood. Various workers have suggested that sulphur itself is the primary toxicant, or that it must first be oxidized to sulphur dioxide or trioxide or to sulphuric acid, or conversely that it must be reduced to hydrogen sulphide. There may be several mechanisms of action, acting together. However, the bulk of the evidence does suggest that elemental sulphur *per se* plays a direct and predominant role. Very little of this mode of action work has been carried out with powdery mildews, because they are difficult to handle experimentally. The ultimate biochemical sites of action of sulphur and the basis of the special sensitivity of powdery mildew fungi have not been explained.

B. Dinitrophenols

Dinocap is second to sulphur in general importance as a powdery mildew fungicide on horticultural crops (little use of either has been made on cereals). Dinocap was first synthesized as an acaricide, in the course of an investigation into non-phytotoxic analogues of dinitro-orthocresol (DNOC). It does have useful acaricidal properties but following the discovery of its fungicidal action in 1949 it has been used primarily for control of powdery mildews. It does not have appreciable effects on other classes of fungi. Dinocap is in fact a mixture containing 65–70% of 2,6-dinitro-4-octylphenylcrotonate (dinocap-4) and 30–35% of 2,4-dinitro-6-octylphenylcrotonate (dinocap-6). The fungicidal activity resides mainly in dinocap-4, whereas dinocap-6 is mainly acaricidal (Kirby and Hunter, 1965; Byrde *et al.*, 1966; Kirby *et al.*, 1966).

Dinocap is formulated mainly as a 25% w/w wettable powder or as a 50% w/v emulsifiable liquid. High-volume sprays, i.e. sprays to run-off, are usually applied at concentrations of 200–250 ppm dinocap at 10–14 day intervals, or at 100–125 ppm at shorter intervals. These rates are much lower than those used for sulphur but dinocap is much more expensive and hence the treatments are more costly. The product gradually displaced sulphur in the 1950s and 1960s, in situations where sulphur gave unsatisfactory control or caused damage. For example it became the predominant treatment for powdery mildew in dessert apples in the UK, its acaricidal action being a useful bonus. It is also used widely on cucurbits, gooseberries, strawberries, roses and chrysanthemums. Usage on grape-vines is small and on cereals probably nil.

Dinocap-4   Binapacryl   Dinobuton

Binapacryl and dinobuton are very closely related to dinocap. Introduced in the early 1960s (Emmel and Czech, 1960; Pianka and Smith, 1965), they have a similar spectrum of antimicrobial action, being active specifically against powdery mildews and against spider-mites. In some situations they are rather more effective than dinocap against mildews, although they are also more expensive. Binapacryl is used mainly on apples and dinobuton has a smaller usage in glasshouse crops.

The dinitrophenols are all surface fungicides. However, they have a suppressant effect on existing infections as well as a protectant action. Simpler dinitrophenols are well-known as inhibitors of oxidative phosphorylation in plant and animal tissues and in micro-organisms; they "uncouple" the linkage between respiration and the storage of energy in phosphate bonds. Esterification to the crotonate and the addition of the octyl side-chain in some way confer specificity of action towards powdery mildews and spider-mites. Possibly these fungicides are not taken up in toxic amounts by other organisms. However, their mechanisms of toxicity or selectivity seem not to have been studied to any extent in fungi.

C. Quinomethionate

Formerly known as oxythioquinox, this compound was introduced in 1962. It gives good control of powdery mildew and spider-mites when applied in programmes of repeated sprays, on crops such as currants, gooseberries, strawberries and cucurbits and it is used both in the glasshouse and the field. It has a protectant action against some other diseases, such as leaf spot of blackcurrants (*Pseudopeziza ribis*). High-volume sprays are generally applied at 100–150 ppm quinomethionate. It is often more effective than sulphur or dinocap but it is also more expensive and it can cause damage to certain crops (e.g. russeting of apples). Quinomethionate is a surface fungicide, having protectant, curative and antisporulant actions (Sasse, 1960). Its mechanism of fungitoxicity appears not to have been investigated.

Quinomethionate

D. Drazoxolon

This is an isoxazolone which is active against a very wide range of fungi (Geoghegan, 1967). With regard to most diseases it acts as a surface protectant, but it does exert, in addition, a good curative action against powdery mildews.

Drazoxolon

Introduced in the late 1960s, drazoxolon came into commercial use in some countries as a "col" formulation (an aqueous suspension of fine particles) for the dual control of powdery mildew and leaf-spot of blackcurrants, and for control of apple and rose powdery mildews. Although drazoxolon is superior to dinocap for mildew control and also has some effect on apple scab, its usage on apples has been limited by a tendency to cause russeting, particularly under cool moist conditions at the time of fruit formation. Drazoxolon is also highly effective as a seed treatment against *Pythium* and other diseases, and it is used commercially as a grass seed treatment in the UK. Its biochemical mode of action is unknown.

E. Ditalimfos

This organophosphorus compound was first developed as a fungicide in the mid 1960s, and came into commercial use some ten years later. It bears a structural resemblance to folpet. It is a surface fungicide and possesses both protectant and curative activities against powdery mildews (Tolksmith, 1966). It also has a protectant action against some other diseases including apple scab and rose black-spot (*Diplocarpon rosae*). It is coming into increasing use as a spray treatment on a range of crops, including apples, cereals and ornamentals. Its mode of action is unknown.

Ditalimfos

F. Other surface fungicides

The list of powdery mildew fungicides has now grown so long that it must suffice to make only brief mention of products that are relatively new and about which information is limited, or that have been withdrawn from use, or that are used primarily to control other types of disease. Fluotrimazole, halacrinate and nitrotal-isopropyl are powdery mildew fungicides which have been introduced in Europe recently, mainly as sprays to control the cereal or apple mildews. Chlorquinox was used for several years as a spray treatment specifically to control barley mildew but has now been withdrawn.

Fluotrimazole   Halacrinate   Nitrotal-isopropyl

Chlorquinox   Piperalin   Chlorothalonil

Piperalin is used in the USA for the control of powdery mildew on roses and other ornamental plants. Products which are most highly active against a range of other diseases but which can give a useful suppression of powdery mildews include thiram, folpet and copper fungicides, and also chlorothalonil, a more recent introduction. In contrast it may be noted that maneb, zineb, captan and captafol are completely ineffective against powdery mildews; sometimes they can encourage mildew growth, presumably by removing competitive fungi. Spraying petroleum oil to control spidermites in orchards and glasshouses gives some control of powdery mildews, although this is generally insufficient to be of practical value. Solutions of various surfactants, or soaps, or even washing soda have some curative action on mildewed plants. Didecyldimethylammonium bromide (DDAB) is used to a small extent commercially to control apple scab and mildew. The special use of high concentrations of surfactants in dormant season sprays is considered later.

G. 2-Amino-pyrimidines

Discovered in the mid-1960s, this family of systemic fungicides has produced three commercial fungicides, dimethirimol, ethirimol and bupirimate.

Dimethirimol   Ethirimol   Bupirimate

In common with most other systemic fungicides, these compounds are readily translocated upwards in the xylem, but are not moved out of treated leaves and are not transported downwards in the phloem. They can be applied to roots, either by soil incorporation or by seed treatments; alternatively they can be used as sprays.

Dimethirimol was introduced first, mainly for the control of cucurbit powdery mildews (Elias et al., 1968). One application of 0·25 g a.i. as an aqueous soil drench around the base of a large cucumber plant in a commercial glasshouse conferred complete protection on the whole plant for at least six weeks. This led to exceptionally good yields and allowed the grower to stop repeated spraying (a tedious, inefficient and unpleasant procedure in cucumber houses). Unlike most powdery mildew fungicides, dimethirimol has no acaricidal action and the product can be used in houses where biological control of mites with the predator *Phytoseiulus persimilis* is being practised. Application to the soil of open-air melon beds, by incorporating granules in the irrigation furrows, also gives long-lasting protection against mildew. Spray applications of dimethirimol are more effective than those of surface fungicides such as sulphur or dinocap, with regard both to protectant and curative action but repeated sprays are necessary in order to control mildew on the new leaves as they appear.

Ethirimol is particularly effective against powdery mildew of cereals (Bebbington *et al.*, 1969) and it has come into extensive use for the control of barley powdery mildew in Europe. It is applied mainly as a seed treatment formulated as a 58% w/v aqueous suspension. Ethirimol passes through the soil into the roots and thence into the foliage, which is protected from mildew for many weeks after sowing. This control of mildew from the earliest stages enhances root growth and tillering (Brooks, 1971); hence yield increases are commonly 10–12% and sometimes higher. Ethirimol is also used as a spray on wheat and barley. On barley the critical timing of sprays to hit early mildew infections is required in order to match the effects of the seed treatment and this is sometimes hard to attain in practice.

Bupirimate, a sulphamate derivative of ethirimol, has been introduced recently as a spray treatment to control mildews of apples, roses, cucurbits and other horticultural crops (Bent, 1974; Finney *et al.*, 1975). It is much less effective than dimethirimol and ethirimol as a root treatment. However, it is more active than these as a spray against certain powdery mildews such as those of apple and rose. The marked vapour action and translaminar effects of bupirimate spray deposits no doubt contribute to its good performance.

The pyrimidines have a direct action on powdery mildew fungi. They can inhibit spore germination *in vitro*. When applied to roots, dimethirimol and ethirimol exert effects at the surface of the leaves, inhibiting mildew development prior to penetration of the host by the fungus (Bent, 1970). Other effects include a "repellant" action on vegetative powdery mildew hyphae, which, instead of being adpressed, grow away from the leaf surface, and an anti-sporulant action (Fig. 1). The biochemical action of the amino-pyrimidines is not understood. It has been suggested, mainly on the basis of reversal experiments, that they inhibit folic-acid-mediated C-1 transfer reactions (Sampson, 1969).

H. Pyridine and pyrimidine carbinols

There has as yet been little commercial use of members of this group of fungicides. However, the very promising performance of the pyrimidine

A

Fig. 1. Effects of an ethirimol spray (50 ppm) on barley powdery mildew under glass-house conditions. A: untreated leaf. B: leaf sprayed 24 h before inoculation; note patches of spray deposit, repression of overall growth of mildew, inhibition of germtube and

B

appressorium development, apparent repellant action on hyphae giving aerial growth, and absence of sporulation. (SEM photographs prepared by C. A. Hart and A. M. Skidmore, ICI Plant Protection Division.)

carbinols in field trials has attracted much attention. The published data have referred mainly to three closely related fungicides, parinol, triarimol and fenarimol.

Parinol   Triarimol   Fenarimol

Parinol was the first of these to be reported (Brown *et al.*, 1967; Thayer *et al.*, 1967). It is highly active against powdery mildews, but has relatively little effect on other pathogens. Its translaminar action appears particularly good. Parinol has been introduced commercially in the USA, for use on roses, zinnias and non-bearing fruit crops.

Triarimol was the next to appear, showing a more potent action and a broader spectrum of activity than parinol (Gramlich *et al.*, 1969). Excellent field control of several powdery mildews was obtained from sprays containing unusually low concentrations of fungicide (20–40 ppm), in repeated high-volume applications. Good control of some other diseases, including apple scab and leaf spot diseases of several crops caused by *Cercospora* spp., was also obtained. Triarimol showed a limited degree of systemic movement via the xylem and marked curative and translaminar effects. Commercial development ceased in 1972, in the light of some adverse effects obtained in long-term toxicological tests. Fenarimol is now at an advanced stage of testing; it appears to offer similar possibilities for disease control to those shown by triarimol (Brown *et al.*, 1975).

Triarimol has been found not to affect germination of powdery mildew spores but to interfere with the development of haustoria (Brown, 1970). Recent data suggest that triarimol inhibits certain stages of sterol biosynthesis in sensitive fungi (Ragsdale and Sisler, 1973; Ragsdale, 1975). It affects de-methylation reactions which are catalysed by mixed function oxidases. Concurrently fatty acids accumulate and these may be involved in the fungi-toxic action. It is notable that resistance to triarimol and fenarimol of fungal mutants obtained in the laboratory also extends to triadimefon and to triforine (Sherald and Sisler 1975; Fuchs and Viets-Verweij, 1975; de Waard and Sisler, 1976). A common mode of action is indicated by this and other evidence.

I. Benzimidazoles

This important group of systemic fungicides contains carbendazim (formerly known as MBC or BCM) and compounds such as benomyl and thiophanate-methyl which are readily converted to carbendazim. Two related compounds, thiabendazole and fuberidazole may also be included in the group.

Carbendazim   Benomyl   Thiophanate-methyl

Benomyl came into commercial use in the late 1960s. It gives protectant and curative control of a wide range of fungal pathogens, including powdery mildews in general, apple scab, *Cercospora* leaf spots, *Botrytis* diseases and many others (Delp and Klopping, 1968). Like almost all other systemic fungicides it is not effective against Phycomycetes. It is translocated in the transpiration stream of the plant, that is to say it will move from the root to the shoot and from the base to the tips of leaves but not in the reverse directions. The conversion of benomyl to carbendazim occurs readily in plants and much of the applied fungicide appears in fact to be translocated in the form of carbendazim (Peterson and Edgington, 1970, 1971). Thiophanate-methyl, introduced shortly after benomyl, is also converted to carbendazim, but at a slower rate (Selling *et al.*, 1970). More recently carbendazim itself has been made available as a commercial fungicide. All three fungicides are formulated as wettable powders, and are sprayed at high-volume concentrations of 150–300 ppm (benomyl and carbendazim) or 250–400 ppm (thiophanate-methyl). They are used to control powdery mildews of apples, cucurbits, roses and other ornamental plants. More recently they have been recommended for use on wheat, sometimes as mixtures with dithiocarbamates, to give control of powdery mildew, *Septoria* spp. and eye-spot (*Cercosporella herpotrichoides*) (Obst, 1975; Cock, 1975).

For control of powdery mildews the benzimidazoles are applied almost entirely as sprays. These are repeated at the normal intervals used for surface fungicides. Applied in this way they usually give better disease control than the surface fungicides. Lengthening spray intervals, for example, from 10–14 days to 21–28 days, has proved feasible for certain diseases (e.g. apple scab and sugar-beet leaf-spot) but not for powdery mildews. With the latter, a considerable loss of disease control usually results if the spray intervals are extended. The use of thiophanate-methyl as a seed treatment for barley, to control a range of pathogens including powdery mildew, has recently been recommended in the UK. Several other carbendazim-generating fungicides have been reported in the past few years but they have not yet attained the prominence of the three discussed here.

Some major problems have been raised by the adaptation of fungi to the benzimidazole fungicides and these are considered in Section V. We may note here, however, that resistant strains obtained from the laboratory or from the field are usually resistant to all the carbendazim-generators and also to thiabendazole and fuberidazole (e.g. Bollen and Scholten, 1971). These two latter fungicides whilst not producing carbendazim, bear an obvious relationship to it in chemical structure and in spectrum of biological activity as well as in the behaviour of the resistant mutants. Thiabendazole and fuberidazole are not used much commercially to control powdery mildews; they are active against these fungi although not so highly as the carbendazim group. Thiabendazole is used mainly as a preservative for fruits and vegetables and fuberidazole as a component of cereal seed treatments to control soil-borne diseases.

Thiabendazole   Fuberidazole

The benzimidazole fungicides have little effect on spore germination but strongly inhibit hyphal and haustorial growth. Morphological aberrations occur in the hyphae. Their biochemical mode of action has not been studied in powdery mildews but studies on other sensitive fungi indicate that carbendazim inhibits mitosis, probably by interfering with the assembly and functioning of spindle microtubules (Clemons and Sisler, 1971; Davidse, 1973). There is evidence that carbendazim binds to tubulin, the component protein of the microtubules, in extracts from sensitive fungi. Interestingly, tubulin preparations from insensitive fungal species or strains or from porcine brain show a much lower affinity for carbendazim (Davidse, 1975; Davidse and Flach, 1976) Thus specific effects on microtubules, either nuclear or cytoplasmic or both, may well account for the potent and selective fungitoxicity of the benzimidazole fungicides.

J. Morpholines

Dodemorph and tridemorph are two closely related systemic fungicides which are highly active against powdery mildews (Pommer *et al.*, 1969). They are also active against certain other fungi; for example, dodemorph will give control of rose rust (*Phragmidium mucronatum*) and tridemorph will control leaf-spot of bananas (*Cercospora musae*).

Dodemorph, R = Cyclododecyl   Tridemorph, R = C₁₃H₂₇

Tridemorph is used widely in Europe as a spray treatment against barley powdery mildew; it has both protectant and eradicant effects. It is also used to some extent on wheat, although there are problems of phytotoxicity on certain varieties. Tridemorph is not used against grape, apple, cucurbit or most other powdery mildews, largely because of phytotoxicity problems. Dodemorph is mainly used against powdery mildew and other diseases of roses and other ornamental plants.

The biochemical action of these fungicides is not understood. The presence of the long alkyl chain suggests that, like other surface-active materials, they affect the permeability of cell membranes (Sijpestein, 1972). However there is some evidence to indicate that inhibition of protein synthesis is a more likely mode of action for tridemorph (Fisher, 1974).

K. Triforine

Triforine is a piperazine derivative, bearing a certain structural relationship to chloraniformethan.

Triforine

It is highly active against powdery mildews in general and also against apple scab, rose black spot (*Diplocarpon rosae*), rose rust and a number of other diseases (Schicke and Veen, 1969; Fuchs *et al.*, 1971). It is systemic, moving in the xylem system, and has protectant, curative and translaminar effects. It has come into use in the past few years as a spray for the control of powdery mildews of apples, wheat, barley, cucumbers and roses, giving concomitant control of other diseases. Recently it has been recommended as a seed treatment for the control of barley powdery mildew; it is applied to seeds in a solvent, which is said to induce direct uptake through the seed coat and thus to obviate possible difficulties of movement through the soil to the roots.

Triforine is known to affect ergosterol synthesis in fungi (Sherald and Sisler, 1975). It is notable that triforine-resistant fungal mutants are resistant also to triarimol and to triadimefon, which appear to have the same mode of action (Sherald and Sisler, 1975; Buchenauer, 1977; Buchenauer and Grossman, 1978).

L. Pyrazophos

This pyrazolopyrimidine is one of the small number of powdery mildew fungicides which contain phosphorus (others are ditalimfos and triamiphos).

Pyrazophos

It is absorbed by foliage and stems and translocated upwards in the transpiration stream. Root uptake is relatively poor. Applied as sprays at 7–14 day intervals it gives good control of a wide range of powdery mildews and it is being used increasingly on apples, hops, cucurbits and other horticultural crops and ornamentals. It appears to have little or no effect on diseases other than powdery mildews. The mode of action is not clear but there are data to suggest that pyrazophos is metabolically converted in the fungus to another product which is the active fungitoxicant (de Waard, 1974).

M. Other systemic fungicides

The main groups of systemic fungicides which have come into practical use against powdery mildews have been described in the previous sections. A large number of other compounds have been reported in the scientific or the patent literature to be systemic fungicides active against powdery mildews. Only a few of these can be mentioned here.

The introduction of benomyl, carboxin, dimethirimol and tridemorph into agriculture in 1968–69 can be said to mark the start of the "systemic era" of fungicides. However the discovery of these compounds was preceded by studies on other experimental fungicides which revealed the feasibility of a systemic action and stimulated further research. Amongst these earlier systemic fungicides were griseofulvin, triamiphos, procaine and 6-azauracil (see reviews by Woodcock, 1972, and Wain and Carter, 1972). For various reasons, none of these compounds has achieved any real practical importance, although triamiphos and griseofulvin were used commercially for a few years, for control of powdery mildew on ornamental plants. Chloraniformethan was applied for several years in Europe in sprays to control powdery mildew of barley, but is no longer used.

Griseofulvin   Triamiphos   Procaine

6-aza-uracil   Chloraniformethan   Triadimefon

Imazalil

Several fungicides are at an advanced stage of testing against powdery mildews. Fenarimol has already been mentioned. Another example is triadimefon (Kaspers *et al.*, 1975). This has a fairly broad spectrum of activity, which includes powdery mildews and rusts in general, apple scab, and cereal smuts (*Ustilago* spp.) and bunts (*Tilletia* spp.). It has given promising results in spray trials against mildew and rusts of wheat and barley, and against apple mildew and scab. Applied as a seed treatment it can control powdery mildew of barley for long periods but marked retardation of seedling emergence and growth has been encountered. Like the pyrimidine carbinols and triforine, it has been found to inhibit ergosterol synthesis in fungi (Buchenauer, 1977; Buchenauer and Grossmann, 1978). Cross-resistance between these fungicides has been mentioned above. Another common feature between triadimefon and triarimol is that both can inhibit the synthesis of gibberellic acid in plants and this probably explains the stunting of growth observed with both fungicides under certain situations. Imazalil is another new systemic fungicide which is active against powdery mildews and a wide range of other diseases, and has given good control of several powdery mildews in spray trials.

III. Mode of Action

We have discussed briefly the mode of action of some individual fungicides or fungicide groups. In this section some general features are considered. All the fungicides appear to work by exerting a fungitoxic action, either directly or after conversion to another active product. The idea of using compounds which act by changing the metabolism of the host or by interfering with the host-parasite relationship is an attractive one (Dimond, 1972) but no compound of this type has yet emerged in practice.

The first step in the action of a fungicide is its movement from the initial deposit to the fungus. Systemic fungicides applied to the soil or seed may need to reach the roots. Passage through soil can present difficulties. For example, less ethirimol is taken up when treated seed is sown in soils of very high organic content (Graham-Bryce and Coutts, 1971). Subsequent translocation upward to the shoot occurs in the apoplast (in practical terms the transpiration stream, or the xylem); the efficiency of this process is affected by the rate of transpiration and also by the presence of barriers to movement, as yet not well defined, which appear to exist in the roots, stems and leaves of plants and especially in woody plants (Crowdy, 1972, 1973; Shephard, 1973). The ways in which chemical and physical factors determine the efficiency of uptake and translocation of particular molecules are still not well understood. It will be seen from the previous sections that the systemic group and the non-systemic group of mildew fungicides each contain a bewildering variety of chemical structures; there are also wide variations in physical properties within each group.

When a fungicide reaches the leaf epidermis from within, it will probably be taken up by the haustoria of existing powdery mildews or by the developing haustoria from new infections. Inhibition or distortion of haustorial growth is often observed (see Sijpesteijn, 1972). When spores alight on a leaf treated systemically with benomyl, tridemorph, triforine, triarimol or triamiphos, they germinate and produce a normal growth tube and appressorium. Fungicidal action starts only after the fungicide penetrates the epidermis (Schluter and Weltzien, 1971). After systemic treatment with ethirimol, dimethirimol or pyrazophos, however, spores either fail to germinate or to produce appressoria (Bent, 1970; Schluter and Weltzien, 1971). In such cases the fungicide must pass outwards through the surface of the leaf and travel either directly from the leaf surface to the spore, or through a water film if there is one, or by vapour action through minute air-spaces to the spore.

Spray deposits on leaves are generally discontinuous. Since powdery mildews germinate without free water, a chemical must move to the spores through either the plant tissue or the air. There is evidence to suggest that many if not all powdery mildew fungicides do exert some degree of localized vapour action (Bent, 1967). Re-distribution by local systemic action is often also important, and is easier in the leaves of herbaceous plants than in woody plants (Shephard, 1973). Translaminar activity is an obvious advantage; it has been demonstrated in a few cases, for example for bupirimate (Finney et al., 1975) and for benomyl and related fungicides (Solel, 1970). It may in fact be shown by most if not all systemic fungicides and possibly by some that are not systemic. None of the present systemic fungicides re-distributes from sprayed leaves to the terminal buds in the way certain herbicides do and hence either repeated sprays or a continual feed-in from a soil reservoir is required to protect new growth.

Levels of accumulation of systemic fungicides in plant tissues are generally low, of the order of one or two parts per million in leaves and lower in fruits and grain. Thus residue problems are no greater and in some situations are less than with the surface fungicides.

The biochemical mode of action of many powdery mildew fungicides is unknown and in no case is this process completely understood. We have seen that members of the carbendazim group appear to attack the microtubules of sensitive fungi; there is reason to believe that griseofulvin also acts in this way (Bent and Moore, 1966; Gull and Trinci, 1973). Triarimol, triforine and triadimefon all affect ergosterol synthesis and possibly other aspects of lipid metabolism in fungi. The fact that many fungi do not respond to powdery mildew fungicides is not understood. In particular the basis of the resistance of phycomycete fungi to almost all systemic fungicides is unknown. Selectivity between fungus and host has not been studied with the powdery mildews. However, the limited data obtained with other types of disease suggest that selective accumulation in the pathogen and relative exclusion from the living parts of the host often operates.

IV. Methods of Application

This is a large subject and only a few aspects can be mentioned here. Methods of application essentially concern timing and placement. The optimal timing of applications depends greatly on the particular crop diseases, on the properties of the chemical and on environmental conditions. The timing of powdery mildew treatments may be less critical than timing against other organisms, for example apple scab or potato blight (*Phytophthora infestans*). Mildews are mostly debilitating rather than being lethal or having devastating effects on the quality of the product and a slightly mistimed spray is not likely to cause acute losses. Also the curative effect of many mildew fungicides allows any infection which arises as the result of bad timing to be suppressed at a later stage. Indeed Evans *et al.* (1970) have reported that a compound completely lacking in protectant activity but having a marked curative action gave good control of apple mildew after repeated field application on a conventional spray schedule. Nevertheless, good timing is important to secure optimal benefits. It is extremely difficult to draw out general guidelines on spray timings (for example it differs between wheat and barley mildew) and empirical results of many field trials probably provide the best indications. It is possible to forecast periods of high disease risk from meteorological data, for example for barley mildew (Polley and Smith, 1973). However, such forecasting systems are not well-proven or used much in practice. Farmers prefer either to use a seed treatment that will give protection throughout the period of highest risk to the crop, or to wait until a small amount of mildew is visible and then spray, possibly repeating once if the disease builds up again well before the crop senesces. On apples and grapes, however, regular routine sprays are usually applied, irrespective of conditions and inoculum pressures; this approach is more expensive but has the merits of reliability and simplicity. Another approach is to attack the pathogen during the dormant season in perennial crops or whilst it is maintained on another host in the case of annuals (such as weeds or volunteer plants). Powdery mildew fungicides are usually ineffective when applied in the dormant season to perennials but the use of single winter sprays containing high concentrations of certain surfactants is a promising technique for the control of mildew overwintering in apple buds (Burchill and Frick, 1972; Hislop and Clifford, 1976).

Equipment for the application of fungicides is a large topic of its own; Evans (1968) gives a useful account. In general, powdery mildew fungicides are applied with conventional machinery. Special developments are the introduction of a lancet applicator to apply dimethirimol solution at the base of cucumber plants, and the modification of seed treatment machinery to apply the relatively large amounts of ethirimol (c. $8g \, kg^{-1}$ seed of formulated product) required on barley seed. Similarly, conventional formulation techniques (see Evans, 1968) have been used for the mildew fungicides. Wettable powders and emulsifiable concentrates are the commonest formulations. Added surfactants sometimes assist the performance of powdery mildew fungicides, especially on cereal crops. The basis of these enhancements is not fully understood and effects on both initial distribution and on uptake of the fungicide are probably involved. Individual surfactants vary greatly in their effects, sometimes independently of their wetting properties and the best surfactant for one fungicide is often found not to be the best for another.

V. Adaptation of Powdery Mildews

Until 1969 no cases of acquired resistance to fungicides had been observed amongst the powdery mildews. Also the phenomenon was rare in other types of fungus. In that year strains of cucumber mildew resistant to benomyl were isolated from field plots in New York State, in which disease control by this fungicide had been poor (Schroeder and Provvidenti, 1969). Similar findings were reported from Israel, on musk-melons (Netzer and Dishon, 1970). Subsequently cucurbit powdery mildews resistant to benomyl have been found in several countries, both in glasshouse and field crops, and disease control has become inadequate (see review of Fehrmann, 1976). Bent et al. (1971) reported the appearance of strains of cucumber mildew relatively resistant to dimethirimol from glasshouses in Holland, Germany and England. These strains were associated with instances of poor disease control; the effect was widespread in Holland, where the product was withdrawn from use. Performance in outdoor cucurbit crops in Israel and Spain has remained good. Triforine-resistant strains of mildew have been isolated from outdoor plots of gherkins in the USA (Gilpatrick, and Provvidenti, 1973).

This rapid onset of resistance and loss of disease control, which with benomyl extends to several other types of disease, has caused much concern. Presumably it is related to the specific biochemical modes of action of the systemic fungicides, in contrast to the surface fungicides which tend to act as general enzyme inhibitors. Repeated fungicide applications, the closed environment of the glasshouse and abundant sporulation probably comprise optimal conditions for the onset of resistance. It is notable that fairly large variations in sensitivity to ethirimol have been detected in mildew isolates from barley fields in the UK and that the sensitivity of mildew in treated fields was somewhat lower than in untreated fields (Wolfe, 1973, 1975; Shephard et al., 1975). Nevertheless effective mildew control has been maintained and there is no obvious correlation between levels of sensitivity and levels of mildew control or yields. The situation in extensive crops such as barley may well differ from that in horticultural crops; also the use of ethirimol in winter barley crops has been strongly discouraged, in order to reduce the period of selection pressure, and this has probably helped to prevent the build-up of less sensitive strains. Making such a break in the annual cycle of use is only one way of reducing the risk of resistance problems. Alternate treatment with unrelated fungicides is another useful approach, as demonstrated by studies with a joint programme of dimethirimol and benomyl application in a cucumber glasshouse (Ebben and Spencer, 1973). The policy of "ringing the changes" between fungicides is being adopted increasingly by growers. Application of mixed sprays is also a possibility, although this would be more expensive unless half-strength concentrations are used, and the response to these is hard to judge.

VI. Trends in Chemical Control

The very large increase in our armoury of powdery mildew fungicides which has taken place over the past ten years will be abundantly obvious from this chapter. Standards of mildew control on all crops have risen steadily and continue to increase. There is much room for improvement, however. Levels of control are still far from adequate; apple crops in the UK, wheat in Eastern Europe and peppers and tomatoes infected with *Leveillula taurica* in Mediterranean countries are but a few of the difficult cases. The requirement for 10–12 repeated treatments in grape and apple crops also demands attention. Highly active chemicals with a more persistent action are required. Clearly, redistribution in the phloem would be a great advantage. Dormant season applications could become widely used in apple and possibly in other perennial crops, although expense and phytotoxicity may be limitations. The use of injection techniques to give long-lasting control in tree crops is another interesting possibility. Reducing periods of selection pressure to a minimum, particularly in glasshouse crops, and avoidance of the continuous use of single products are likely to increase. Integrated control measures, that include biological techniques are likely to acquire an increased importance, especially in orchards and glasshouses and to demand a more careful planning of fungicide selection and usage.

References

Bebbington, R. M., Brooks, D. H., Geoghegan, M. J. and Snell, B. K. (1969). *Chemy Ind.* 1512.
Bent, K. J. (1967). *Ann. appl. Biol.* **60**, 251–263.
Bent, K. J. (1970). *Ann. appl. Biol.* **66**, 103–113.
Bent, K. J. (1974). *Proc. Am. Phytopath. Soc. 1*, 21.
Bent, K. J. and Moore, R. M. (1966). *In* "Biochemical Studies of Antimicrobial Drugs".

(B. A. Newton and P. E. Reynolds, Eds) 82–110. Sixteenth Symposium of the Society for General Microbiology. Cambridge University Press: Cambridge.

Bent, K. J., Cole, A. M., Turner, J. A. W. and Woolner, M. (1971). *Proc. 6th Br. Insectic. Fungic. Conf.* **1**, 274–282.

Bollen, G. J. and Scholten, G. (1971). *Neth. J. Pl. Path.* **72**, 187–193

Brooks, D. H. (1971). *Ann. appl. Biol.* **70**, 149–156.

Brown, I. F. (1970). *Proc. 7th Int. Cong. Plant Protection* 206.

Brown, I. F., Whaley, J. W., Taylor, H. M. and van Heynigen, E. M. (1967). *Phytopathology* **57**, 805.

Brown, I. F., Taylor, H. M. and Hall, H. R. (1975). *Proc. Am. Phytopath. Soc.* **2**, 31.

Buchenauer, H. (1977). *Pestic. Biochem. Physiol.* **7**, 309–320.

Buchenauer, H. and Grossman, F. (1976). *Neth. J. Pl. Path.* in press.

Burchill, R. T. and Frick, E. L. (1972). *Pl. Dis. Reptr* **56**, 770–772.

Byrde, R. J. W., Clifford, D. R. and Woodcock, D. (1966). *Ann. appl. Biol.* **57**, 223–230.

Clemons, G. P. and Sisler, H. D. (1971). *Pest. Biochem. Physiol.* **1**, 32–43.

Cock, L. J. (1975). *Proc. 8th Br. Insectic. Fungic. Conf.* **3**, 859–869.

Crowdy, S. H. (1972). *In* "Systemic Fungicides". (R. W. Marsh, Ed.). 92–115. Longman, London.

Crowdy, S. H. (1973). *Proc. 7th Br. Insectic. Fungic. Conf.* **3**, 831–840.

Davidse, L. C. (1973). *Pestic. Biochem. Physiol.* **3**, 317–325.

Davidse, L. C. (1975). *In* "Microtubules and Microtubule Inhibitors". (M. Borgers and M. de Brabander, Eds) 483–495. North-Holland, Amsterdam.

Davidse, L. C. and Flach (1976). *J. Cell. Biol.* in press.

Delp, C. J. and Klopping, H. L. (1968). *Pl. Dis. Reptr* **52**, 95–99.

Dimond, A. E. (1972). *In* "Systemic Fungicides". (R. W. Marsh, Ed.) 116–131. Longman, London.

Ebben, M. H. and Spencer, D. M. (1973). *Proc. 7th Br. Insectic. Fungic. Conf.* **1**, 211–216.

Elias, R. S., Shephard, M. C., Snell, B. K. and Stubbs, J. (1968). *Nature, Lond.* **219**, 1160.

Emmel, L. and Czech, M. (1960). *Anz. Schädlingsk.* **33**, 145.

Evans, E. (1968). "Plant Diseases and their Chemical Control". Blackwells, Oxford.

Evans, E., Marshall, J., Couzens, B. J. and Runham, J. L. (1970). *Ann. appl. Biol.* **65**, 473.

Fehrmann, H. (1976). *Phytopath. Z.* **86**, 144–185.

Finney, J. R., Farrell, G. M. and Bent, K. J. (1975). *Proc. 8th Br. Insectic. Fungic Conf.* **2**, 667–673.

Fisher, D. J. (1974). *Pestic. Sci.* **5**, 219–224.

Forsyth, W. (1802). *In* "A Treatise on the Culture and Management of Fruit Trees". Nichols, London.

Fuchs, A. and Viets-Verweij, M. (1975). *Meded. Fac. Landbouwwet. Gent.* **40**, 699–706.

Fuchs, A., Doma, S. and Voros, J. (1971). *Neth. J. Pl. Path.* **77**, 42–54.

Geoghegan, M. J. (1967). *Proc. 4th. Br. Insectic. Fungic. Conf.* **2**, 451–460.

Gilpatrick, J. D. and Provvidenti, R. (1973). *Proc. 2nd Intern. Cong. Pl. Path., Minneapolis,* Abstract 0780.

Graham-Bryce, I. J. and Coutts, J. (1971). *Proc. 6th Br. Insectic. Fungic. Conf.* **2**, 419–426.

Gramlich, J. V., Schwer, J. F. and Brown, I. F. (1969). *Proc. 5th. Br. Insectic. Fungic. Conf.* **2**, 576–584.

Gull, K. and Trinci, A. P. J. (1973). *Nature, Lond.* **244**, 292–294.

Hislop, E. C. and Clifford, D. R. (1976). *Ann. appl. Biol.* **82**, 557–568.

Kaspers, H., Grewe, F., Brandes, W., Scheinpflug, H. and Büchel, K. H. (1975). *Proc. 8th Int. Pl. Prot. Cong. Sect 3(2)* 398–401.

Kirby, A. H. M. and Hunter, L. D. (1965). *Nature, Lond.* **208**, 189–190.

Kirby, A. H. M., Frick, E. L. and Gratwick, M. (1966). *Ann. appl. Biol.* **57**, 211–221.

McCallan, S. E. A. (1967). *In* "Fungicides". (D. C. Torgeson Ed.) **1**, 1–37.

Martin, H. (1964). "Scientific Principles of Crop Protection". 5th ed. Edward Arnold, London.
Netzer, D. and Dishon, I. (1970). *Pl. Dis. Reptr* **54**, 909–912.
Obst, A. (1975). *Proc. 8th Br. Insectic. Fungic. Conf.* **3**, 953–966.
Peterson, C. A. and Edgington, L. V. (1970). *Phytopathology* **60**, 475–478.
Peterson, C. A. and Edgington, L. V. (1971). *Phytopathology*, **61**, 91–94.
Pianka, M. and Smith, C. B. F. (1965). *Chemy Ind.*, 1216.
Polley, R. W. and Smith, L. P. (1973). *Proc. 7th. Br. Insectic. Fungic. Conf.* **2**, 373–378.
Pommer, E. H., Otto, S. and Kradel, J. (1969). *Proc. 5th Br. Insectic. Fungic. Conf.* **2**, 347–353.
Ragsdale, N. N. (1975). *Biochim. Biophys. Acta* **380**, 81–96.
Ragsdale, N. N. and Sisler, H. D. (1973). *Pestic. Biochem. Physiol.* **3**, 20–29.
Robertson, J. (1821). *Trans. Hort. Soc. London* **5**, 175–185.
Sampson, M. J. (1969). *Proc. 5th Br. Insectic. Fungic. Conf.* **2**, 746–751.
Sasse, K. (1960). *Hofchenbr. Bayer Pflschutz-Nachr.* **13**, 197.
Schicke, P. and Veen, K. H. (1969). *Proc. 5th Br. Insectic. Fungic. Conf.* **2**, 569–575.
Schluter, K. and Weltzien, H. C. (1971). *Meded. Fac. Landbouwwet. Gent.* **36**, 1159–1165.
Schroeder, W. T. and Provvidenti, R. (1969). *Pl. Dis. Reptr* **52**, 630–632.
Selling, H. A., Vonk, J. W. and Sijpesteijn, A. K. (1970). *Chemy Ind.*, 1625–1626.
Sharvelle, E. (1969). "Chemical Control of Plant Diseases". 1–16. University Publishing, Texas.
Shephard, M. C. (1973). *Proc. 7th Br. Insectic. Fungic. Conf.* **3**, 841–850.
Shephard, M. C., Bent, K. J., Woolner, M. and Cole, A. M. (1975). *Proc. 8th Br. Insectic. Fungic. Conf.* **1**, 59–66.
Sherald, J. L. and Sisler, H. D. (1975). *Pest. Biochem. Physiol.* **5**, 477–488.
Sijpesteijn, A. K. (1972). In "Systemic Fungicides". (R. W. Marsh, Ed.) 132–155, Longmans, London.
Solel, Z. (1970). *Phytopathology* **66**, 1186–1189.
Thayer, P. L., Ford, D. H. and Hall, H. R. (1967). *Phytopathology* **57**, 833–836.
Tolksmith, H. (1966). *Nature, Lond.* **211**, 522.
Tweedy, B. G. (1969). *In* "Fungicides". Vol. 2. (D. C. Torgeson Ed.) 119–145. Academic Press: New York and London.
Waard, M. A. de (1974). *Meded. Landbhoogesch. Wageningen*, 14–74.
Waard, M. A. de and Sisler, H. D. (1976). *Meded. Fac. Landbouwwet. Gent* **41**, 571–578.
Wain, R. L. and Carter, G. A. (1972). *In* "Systemic Fungicides". (R. W. Marsh, Ed.). 6–23. Longman, London.
Weighton, D. (1814). *Mem. Caledonian Hort. Soc.* **1**, 131–142.
Wolfe, M. S. (1973). *Proc. 7th Br. Insectic. Fungic. Conf.* **1**, 11–19.
Wolfe, M. S. (1975). *Proc. 8th Br. Insectic. Fungic. Conf.* **3**, 813–822.
Woodcock, D. (1972). *In* "Systemic Fungicides". (R. W. Marsh, Ed.). 34–85. Longman, London.

Chapter 11

Biology and Pathology of Cereal Powdery Mildews

J. F. JENKYN and A. BAINBRIDGE

Plant Pathology Department, Rothamsted Experimental Station, Harpenden, England

I. Introduction

Powdery mildew of cereals and grasses is caused by *Erysiphe graminis* DC. It is restricted to the temperate cereal species including wheat, barley, oats, rye and their hybrids. We have found no reports that it or any other member of the Erysiphaceae are pathogenic on warm-climate cereals such as millet, sorghum, rice and maize.

Yarwood, writing in 1957, commented on the vast literature dealing with the powdery mildews. This is now equally true of the literature dealing with *E. graminis* alone, so that it is difficult to review adequately all the available information. We have not attempted to cover all aspects of the biology of the disease in equal detail. Several are considered elsewhere in this volume or in other reviews, to which we have referred where appropriate. Whenever possible we have consulted the original papers but when these were unobtainable we have relied upon abstracts in the *Review of Plant Pathology*.

II. Distribution and Importance

The pathogen occurs in most, if not all, parts of the world where the host cereals are grown (Anon, 1967) but is not everywhere considered to be a major problem. Identifying those regions of the world where *E. graminis* is actually or potentially an important pathogen is not easy because the scant information from many areas is inconclusive. Prevalence may also be affected by local conditions, leading to atypical and misleading reports. Nevertheless the disease generally seems to cause most damage in temperate latitudes. It is seldom serious in hot, dry regions although it can be important where altitude or maritime influences have a moderating effect on climate. Thus, it is a major pathogen in much of Europe (Hermansen, 1968; Ríman, 1973; King, 1973; Yakubtsiner *et al.*, 1974; Obst, 1975), in the south-western parts of the USSR, in the region of the Black Sea and Caspian Sea (Mehtieva, 1956; Simonyan, 1959; Nikulina and Chumakov, 1968) and in western Turkey (Bremer *et al.*, 1947). In North America it is damaging in British Columbia (Cherewick, 1944) and the maritime provinces of Canada (Newton and Cherewick, 1947; Johnston, 1974), in the south-eastern United States (Taylor *et al.*, 1949; Lowther, 1950) and on the East and West coasts (Mackie, 1928; Moseman, 1954; Briggle and Vogel, 1968; Anon, 1968a) but not in the drier Great Plains (Salmon and Throckmorton, 1929; Moseman, 1954) or prairies (Buchannon and Wallace, 1962). From South America the reports implicating *E. graminis* as a major pathogen seem to be from temperate areas in the south and the Andes (Barducci, 1942; Sarasola *et al.*, 1946; Vallega and Stillger, 1947). It is not normally a problem on dryland wheat in South Africa (Gorter, 1955) or in the main wheat belt of Australia (Carne and Campbell, 1924; Adam and Colquhoun, 1937; Anon, 1954) but can cause losses in some coastal areas (Simmonds, 1961; Anon, 1962). Reports from India suggest that *E. graminis* is unimportant in the plains but more serious near the Himalayas (Mehta, 1930).

As the cultivation of host cereals, especially wheat, spreads to new areas, the distribution of *E. graminis* as an important pathogen may increase as, for example, in parts of India (Patil *et al.*, 1969). Furthermore many of the newly introduced semi-dwarf wheat varieties, which are replacing traditional varieties in many areas, are very susceptible to foliar diseases (Eshed and Wahl, 1975) which may also be encouraged by the denser crop canopies which these varieties produce (Briggle and Vogel, 1968).

In those parts of Europe where mildew is a problem, yield losses in experiments can exceed 20% in barley (Last, 1955a; 1957; Johansen, 1963; Slootmaker and van Essen, 1969; Evans and Hawkins, 1971; Kaspers and Kolbe, 1971; Brooks, 1972), wheat (Vik, 1937; Ts'ao, 1960; Mráz, 1969; Picco *et al.*, 1971) and oats (Griffiths, 1958; Lawes and Hayes, 1965; Jung and Bedford, 1971; Anon, 1975). Average losses are certainly much smaller but there have been few attempts to estimate them on a regional or national basis. Annual surveys of foliar diseases on spring barley and winter wheat in England and Wales have been made since 1967 and 1970 respectively. Based on these data, estimated losses up to 1973 ranged from 6–14% on barley and from 2–5% on wheat (King, 1973).

In the USA losses can also exceed 20% (Newton and Cherewick, 1947; Hebert *et al.*, 1948; Schaller, 1951; Leukel and Tapke, 1956; Anon, 1968a; Shaner, 1972) but, because many cereals are grown where the disease is not a problem, the estimates of national loss, made by the United States Department of Agriculture, are less than 1% for both wheat and barley (Anon, 1965). In the province of Ontario, Canada and in New Zealand losses due to mildew on wheat, based on disease surveys, were also estimated to be less than 2% (James, 1971; Smith and Smith, 1974) but in localities where the disease can be severe, losses greater than 25% have been recorded (Johnston, 1972; Smith and Wright, 1974).

III. Biology and Epidemiology

A. Description

Hyphal growth is superficial and haustoria are normally restricted to the epidermal cells. Individual colonies are typically elliptical but if numerous they may fuse to form a dense mycelial mat over the leaf surface. Colonies when young are usually white but become grey or even reddish-brown when sporulating. Conidial chains develop from a characteristic basal cell (Salmon, 1903; Plumb and Turner, 1972; Aust and Hindorf, 1975). The spore chain, described by Hammett and Manners (1973), usually contains about eight immature but readily discernible spores, firmly attached to one another. Mature spores at the apex of the chain are readily detached by wind. They are ellipsoidal and hyaline, with little surface ornamentation and are 25–30 × 12–15 μm.

Globose cleistothecia usually are formed in the ageing mycelial mat. They are black, with simple appendages, interwoven with the mycelium, and are about 200 μm in diameter. Each contains up to 25 ellipsoidal to cylindrical asci. Ascospores, usually 8 but occasionally 4 per ascus, are, like conidia, elliptical and hyaline and are 20–24 × 10–14 μm.

Conidia germinate on most surfaces but germ-tube growth is commonly disorientated on water and other non-living materials (Corner, 1935; Dickinson, 1949) so that germ tubes grow away from the surface. Normal appressoria can develop not only on susceptible and resistant cultivars of appropriate species (Paulech, 1968) but also on closely related species which are hosts for other formae speciales of *E. graminis* (White and Baker, 1954; Yang and Ellingboe, 1972) and even on non-graminaceous species (Corner, 1935; Staub et al., 1974). They are rarely formed on non-living materials and probably develop as a response to the physical properties of leaf surfaces (Yang and Ellingboe, 1972).

B. Penetration and host reaction

The processes of penetration are not fully understood although penetration of the cuticle is probably mechanical while that of the underlying cell wall requires enzymes (Stanbridge et al., 1971; McKeen et al., 1969). On barley leaves, up to 90% of the initial infections are of auxiliary cells (Hirata and Togashi, 1957).

Detectable changes in host cells are induced around penetration sites even before successful infections have established and are probably associated with resistance mechanisms. Prior to the formation of the infection peg, aggregation of host cytoplasm occurs beneath the appressorium and electron dense material is deposited on the inner surface of the cell wall to form the papilla (Edwards and Allen, 1970; Stanbridge et al., 1971; McKeen and Rimmer, 1973; Bushnell and Bergquist, 1975). Differential staining reveals a "halo" around the infection peg (Lupton, 1956) and this clearly results from changes in the cell wall. Enzymes produced by the pathogen are probably responsible for cellulose breakdown in this area (McKeen et al., 1969) but accumulation of calcium, manganese and silicon around the infection peg probably represents a host response (Kunoh et al., 1975; Kunoh and Ishizaki, 1975; 1976).

Once the infection peg has penetrated the cell wall and papilla, to form the digitate haustorium, the relationship between host and pathogen probably becomes even more complex. The structure and function of haustoria have been reviewed by Hirata (1971), Bushnell (1971; 1972) and Scott (1972). Studies of haustorial morphology, using the electron microscope (Ehrlich and Ehrlich, 1963; Bracker, 1968; McKeen, 1974), show that the host plasmalemma invaginates in advance of the developing haustorium to form an extra haustorial plasmalemma, which, with time, becomes thicker, more amorphous and more electron dense than the original membrane. A thin layer of cytoplasm around the young haustorium increases in volume as the haustorium ages and becomes almost electron transparent. Organelles and endoplasmic reticulum become widely separated.

Haustoria have long been assumed to be absorbing organs and this is supported by the work of Mount and Ellingboe (1969) and Slesinski and Ellingboe (1971) who found that the rate of transfer of ^{32}P and ^{35}S from host to pathogen increased sharply once haustoria formed. Furthermore concentrations of acid phosphatase, which may be important in the transfer of metabolites from host to pathogen, have been detected in and on haustoria (Atkinson and Shaw, 1955).

After the formation of the first haustorium, hyphae begin to grow from the appressorium over the leaf surface (Masri and Ellingboe, 1966a). Haustoria are essential for continued hyphal growth (Bushnell et al., 1967) and further epidermal cells are penetrated as the colony spreads over the leaf surface.

C. Resistance to infection

Resistance to E. graminis may be expressed at various stages in the infection process. Where appressoria form, resistance may be conferred by exclusion of the pathogen and by distortion and destruction of haustoria or through collapse of infected and adjacent cells after haustoria have formed (Cherewick, 1944; White and Baker, 1954; Lupton, 1956; Masri and Ellingboe, 1966b). Seedling leaves are usually more susceptible than leaves of older plants (Graf-Marin, 1934) and on barley the susceptibility of different parts of the same leaf may differ (Russell et al., 1975; 1976).

The biochemical changes associated with resistance mechanisms operating after cell penetration has occurred are little understood. It has been suggested that the redox potential of host tissue is important for the pathogen (Benada, 1966) but a direct relationship between redox potential and resistance seems unlikely (Benada, 1972). Interactions between phenols and peroxidase systems may be involved in these resistance mechanisms (Frič, 1969; Hislop and Stahmann, 1971) and the biophysical state of host tissue could affect resistance indirectly by influencing the activity of these systems (Benada, 1972).

D. Effects of environmental conditions

Effects of environment on powdery mildews in general were discussed in reviews by Yarwood (1957) and Schnathorst (1965). The effects of some environmental factors on powdery mildew of barley were considered by Aust (1973a).

The optimum temperature for infection by and growth of *E. graminis* is relatively low at 15–20°C but the range over which it can develop is wide (Cherewick, 1944). Infection of barley can occur at temperatures at least as low as 5°C (Ward and Manners, 1974) and spores can germinate near freezing point (Cherewick, 1944). The upper limit for infection is about 30°C (Akai, 1952). At this temperature, germination may be prevented (Leinonen and Kurkela, 1969) or much reduced (Manners and Hossain, 1963) but if it occurs and appressoria form, haustoria may fail to develop (Aust, 1974b).

Powdery mildews generally can tolerate very dry conditions (Boughey, 1949) and although the optimum relative humidity for germination of *E. graminis* is 100% (Manners and Hossain, 1963; Zaracovitis, 1964), free water is detrimental. Grainger (1947) observed no germination below 85% relative humidity but others have reported germination at relative humidities near zero (Cherewick, 1944; Manners and Hossain, 1963).

The water content of mildew spores can be almost 75% of their fresh weight (Yarwood, 1950; Somers and Horsfall, 1966). It is probable that the water required for germination is provided internally (Yarwood, 1957) and it has been suggested that lipids in the cell wall form a barrier which limits water loss (Johnson *et al.*, 1976). However, water content depends on the humidity conditions under which spores are produced (Somers and Horsfall, 1966) and this probably explains why those produced at high humidities germinate better in dry conditions than those produced at low humidities (Prabhu *et al.*, 1962; Aust, 1973b; Ward and Manners, 1974). It seems likely that differences in the conditions under which inoculum is produced may explain some of the conflicting results which have been reported. Although spores produced at high humidities may germinate better than those produced at low humidities, Ward and Manners (1974) found that their ability to survive drying conditions was not improved.

Although there are few data on the relationships between humidity and growth after germination, Manners and Hossain (1963) found that there was little or no elongation of germ tubes at humidities below 98%. The rate at which spores are produced is also much less at very low humidities than in saturated air (Ward and Manners, 1974).

Short periods of darkness before collection of spores decrease their ability to germinate and infect (Ward and Manners, 1974) and the formation of appressoria is delayed by both darkness and high light intensity (Masri and Ellingboe, 1963; Aust, 1974a). Long periods of continuous darkness or of very low light intensity, may prevent or very much reduce disease development but only if the treatment begins before inoculation, suggesting that the effect is indirect and operating through changes in host metabolism (Trelease and Trelease, 1929; Volk, 1931; Cherewick, 1944; Pratt, 1944). In contrast, exposure to darkness immediately after inoculation was found to stimulate infection (Sempio, 1939). Disease development may also be decreased by high light intensity and by continuous light (Volk, 1931; Sempio, 1943) and is less with long than with short day lengths (Grainger, 1947; Domsch, 1954).

Light of wavelengths between 550 nm and 750 nm was found by Sempio and Castori (1952) to depress germination but shorter wavelengths (450–550 nm) stimulated it. The fungus is most likely to be damaged by ultraviolet radiation during long-distance dispersal. However, it seems to be most sensitive to ultraviolet during primary infection, i.e. between germination and the establishment of the first haustorium (Mount and Ellingboe, 1968), so that less ultraviolet radiation is needed to prevent spores infecting than to kill established infections (Hey and Carter, 1931) or to kill spores (Moseman and Greeley, 1966). Radiation of wavelength 254 nm is more damaging than that of 350 nm (Janitor, 1975).

Ozone also decreases infection more than germination (Heagle and Strickland, 1972).

E. Ascospore production

Cleistothecia are usually formed in the mycelial mat after conidial production has ceased (Foëx, 1923; Smedegård-Petersen, 1967) but the conditions favouring their development are imperfectly understood. Graf-Marin (1934) found that cleistothecia were produced only on old leaves and that numbers increased with increasing age of the plant at the time of inoculation but he also found that starvation of the fungus, through starvation of the host plant, did not play an important role in their formation. However, formation of cleistothecia on wheat leaves, which were not fully developed, was observed by Cherewick (1944). He showed that their production was stimulated by fluctuating temperatures, long bright days and by the application of nitrogen fertilizer. There was no effect of moisture on production of cleistothecia. In contrast, maturation of cleistothecia and the production of ascospores do seem to be induced by high moisture and especially by free water (Graf-Marin, 1934; Cherewick, 1944; Turner, 1956; Moseman and Powers, 1957; Arya, 1964; Smedegård-Petersen, 1967). There are few data showing the effects of temperature on maturation but the optimum is probably less than 20°C (Moseman and Powers, 1957; Smedegård-Petersen, 1967).

Differentiation does not seem to occur until cleistothecia are in a "ripe-to-spore" condition (Turner, 1956). Cleistothecia taken from green leaves do not produce ascospores when wetted unless they have first been dried (Cherewick, 1944; Turner, 1956; Smedegård-Petersen, 1967) suggesting that drying may induce the "ripe-to-spore" state.

Ascospore discharge is also favoured if cleistothecia are kept wet and by temperatures near 20°C (Moseman and Powers, 1957). Ascospores are actively discharged (Smedegård-Petersen, 1967; Koltin and Kenneth, 1970) and presumably wind dispersed.

Cleistothecia occur on all the cereal hosts. In some situations a large proportion may fail to mature (Turner, 1956) but, given the right conditions, the majority seem to be potentially fertile (Smedegård-Petersen, 1967; Koltin and Kenneth, 1970).

F. Pathogenic specialization

The number of graminaceous species attacked by *E. graminis* is large but since the beginning of this century, when Marchal (1902) distinguished seven formae speciales, it has been recognized that there is marked specificity of parasitism in this species (Salmon, 1904a; Reed, 1909). Mühle and Frauenstein (1970), for example, found that mildews from barley, wheat and rye could not attack any of 27 grasses tested. However, the host range is often not as narrow as the forma specialis epithet implies. Hardison (1944) showed that isolates of *E. graminis*, including those from cereals, frequently attack species in more than one genus. Thus, different isolates from *Agropyron repens* which could infect wheat, barley or *Elymus* sp. could be labelled, *E. graminis* f. sp. *agropyri*, f. sp. *tritici*, f. sp. *hordei* or f. sp. *elymi*. In Israel many grass species, from four tribes, were found to be hosts of mildew from wheat, barley and oats (Eshed and Wahl, 1970). Nevertheless, strains of mildew from wheat, barley, oats and rye are not normally able to cross-infect (Cherewick, 1944) and in New Zealand powdery mildew does not occur on oats although both barley and wheat are attacked (Smith *et al.*, 1973). Salmon (1905) found that mildew conidia from wheat were able to germinate on barley leaves and penetrate epidermal cells but haustoria usually failed to develop and little or no hyphal growth was observed. However, Moseman *et al.* (1965) showed that conidia of *E. graminis* f. sp. *hordei* were produced on wheat and those of

E. graminis f. sp. *tritici* on barley provided the plants were already infected with the respective compatible formae speciales. In similar experiments Tsuchiya and Hirata (1973) were able to infect mildewed barley leaves with numerous other species from within the Erysiphaceae. Moseman *et al.* (1965) concluded, on the basis of pathogenicity tests, that nuclear exchange did not occur between isolates and that the normally incompatible isolate had infected following physiological changes in the host induced by the compatible isolate. However, Naito and Hirata (1969) observed fusion of germ tubes with hyphae and the transfer of nuclei when barley leaves already infected with *E. graminis* f. sp. *hordei* were inoculated with conidia of *E. graminis* f. sp. *agropyri*. Conidia pathogenic on *Agropyron* were produced on these leaves and were considered to contain nuclei of the forma specialis *agropyri*. Either dual infection or nuclear exchange would explain the occurrence of mixed cultures on some hosts, found by Hardison (1944).

Specialization of parasitism within *E. graminis* f. sp. *hordei* and *E. graminis* f. sp. *tritici* was first reported in 1930 (Mains and Dietz, 1930; Waterhouse, 1930) and since then numerous physiologic races have been identified in all the major cereal growing areas (Moseman, 1966; Wolfe, 1972).

The sexual process leading to the formation of cleistothecia provides the opportunity for recombination of genetic material and the production of new races with different combinations of virulence genes (Moseman, 1956; Powers and Moseman, 1957; Al-Ani, 1969). Although Cherewick (1944) reported homothallism in *E. graminis* f. sp. *hordei* and *E. graminis* f. sp. *tritici*, more recent evidence suggests that the species is heterothallic (Powers and Moseman, 1956; Moseman, 1966; Wolfe, 1972).

Leijerstam (1972) was unable to obtain crosses between f. sp. *tritici*, f. sp. *secalis* and f. sp. *agropyri* and although Hiura (1962) did succeed in obtaining cleistothecia from a cross between forms of mildew from wheat and barley, the ascospores were not viable. However, the same worker (Hiura, 1964) did obtain viable ascospores from crosses between f. sp. *tritici* and f. sp. *agropyri* and some of the progeny were pathogenic on both hosts.

Conidia of the different formae speciales are morphologically indistinguishable but those of *E. graminis* f. sp. *hordei* contain proline whereas those of *E. graminis* f. sp. *tritici* do not (Sommereyns and Parmentier, 1967; Sommereyns, 1968) and this may provide a method for distinguishing between these two forms.

G. Production and dispersal of conidia

The morphology and cytology of conidial development was described by Foëx (1911–13; 1924). Their production is continuous (Hammett and Manners, 1973; Pady *et al.*, 1969) but is maximal at about 20°C and 100%

relative humidity. Light intensity and photoperiod seem to have no effect except when very little light is supplied (Ward and Manners, 1974) but under these conditions fungal growth could be affected indirectly through effects on host metabolism. In contrast there is often marked diurnal periodicity in spore concentration above crops during dry weather, with the maximum occurring soon after midday (Hirst, 1953; Sreeramulu, 1964; Hammett and Manners, 1971; Eversmeyer and Kramer, 1975). Hammett and Manners (1971) found that high concentrations at this time often result from the coincidence of those factors positively correlated with release of conidia; under the conditions prevailing in southern England, these were high wind speed, dry leaf surfaces, high temperatures and low relative humidity. However, these factors are often strongly correlated with one another and in wind tunnel tests they found that temperature and relative humidity had no direct effect on spore release (Hammett and Manners, 1974). Onset of wind was more important than continued wind. Rainfall, which acts independently of the other factors, initially increases spore concentration (Hirst, 1953; Hammett and Manners, 1971) but this is quickly followed by a substantial decrease in numbers as rain removes spores from the atmosphere (Hirst, 1953; May, 1958). After heavy rain, numbers recover only slowly because mycelium and conidiophores are damaged (Yarwood, 1936) and may take 3–5 days to reach pre-rain levels (Hirst, 1953; Sreeramulu, 1964). During wet weather, the pattern of spore release is therefore variable and shows no regular periodicity (Hammett and Manners, 1971).

As is to be expected, the concentration of E. graminis spores in the air varies seasonally, broadly reflecting the incidence of disease (Last, 1955b; Gregory and Hirst, 1957; Sreeramulu, 1964; Jenkyn, 1974a; Eversmeyer and Kramer, 1975).

An active discharge mechanism for conidia of some members of the Erysiphaceae was described by Hammarlund (1925) but E. graminis appeared to be an exception. Under still conditions a mat of mature spores develops, suggesting that spore release is probably passive. Hammett and Manners (1973; 1974) found that wind speeds of about 1 m sec^{-1} created sufficient drag to remove some conidia from clamped leaves in a wind tunnel but main spore release required higher wind speeds. Wind speeds within cereal crop canopies are usually less than 1 m sec^{-1} but Bainbridge and Legg (1976) found that even in light winds the acceleration as leaves flapped, generated sufficient force to dislodge conidia and concluded that most spores are probably shaken free. Impact of raindrops also shakes leaves but rapid air movement in advance of the impacting drop may also dislodge spores (Hirst and Stedman, 1963).

Concentrations of conidia within infected crops can be very large but those spores which escape from the crop canopy are rapidly dispersed by turbulent air movement so that concentrations above the crop may be relatively small

(Last, 1955b). In crops with a dense canopy it is probable that the majority of the conidia produced never escape from the crop.

Close to a mildew source, the gradient of spore deposition per unit area is typically very steep (Bainbridge and Cross, 1974) and in the early stages of infection can result in distinct disease gradients (Yarham and Pye, 1969; Jenkyn and Bainbridge, 1974). Gradients probably indicate that the majority of spores are deposited within very short distances (Gregory, 1973) but the rapid upward dispersal of some spores would also contribute to a steep gradient.

Spores of many fungi can be present in air up to considerable altitudes in the troposphere (Hirst and Hurst, 1967; Gregory, 1973) and viable conidia of barley mildew have been detected at least as high as 1500 m (Johansen et al., 1973). At these heights, air movement may be sufficient to ensure transport over some hundreds of kilometres within 24 hours (Hirst et al., 1967; Hermansen et al., 1975). Hermansen and Wiberg (1972) concluded that barley in the Faeröes was infected each year by inoculum from other parts of Europe, a distance of at least 300 km. Viable conidia, thought to have come from Britain, have been detected in Denmark, some 500 km distant (Hermansen et al., 1975).

Fungal spores may be deposited by impaction, sedimentation or in falling rain. Conidia of E. graminis are of the size classed by Gregory (1973) as impactors and among seedlings or at the tops of canopies, where wind speeds often exceed 2 m sec^{-1}, impaction is probably important. However, at those wind speeds common within canopies, impaction will be inefficient and sedimentation is probably the major deposition mechanism. In laboratory experiments, spore deposition was less on leaves of erect varieties than on leaves of more prostrate varieties (Russell, 1975a). Conidia of E. graminis are non-wettable and adhere to the surfaces of water droplets (Hammett and Manners, 1973). Davies (1961) found that non-wettable spores on water drops are left behind as the drops roll over hydrophobic leaf surfaces, so spores captured by rain may be deposited in this way.

H. Disease development in crops

Mildew can develop quickly when environmental conditions are favourable for fungal growth (Aust, 1973c). In the field, pustules may be clearly visible on seedlings within five days of inoculation (Jenkyn, 1973). Pustules at this stage have usually begun to sporulate and as spores may be produced in large numbers (Ward and Manners, 1974) and are easily dispersed, the disease can rapidly become severe. Benada (1962) considered that two generations of mildew could be sufficient to cause an epidemic.

E. graminis is able to grow at relatively low temperatures, so that where winters are mild, the disease may be active for much of the year (Padalino *et al.*, 1970; Yarham *et al.*, 1971). It can be severe on adult plants and on seedlings of both autumn- and spring-sown crops. However, because the fungus cannot tolerate temperatures near 30°C, development of the disease in areas with hot dry summers is restricted to winter and early spring and it is usually unimportant on adult plants (Carne and Campbell, 1924; Bremer *et al.*, 1947).

In most cereal growing areas of the world, the fungus has to survive a period of adverse weather conditions. The critical period may be either a hot, dry summer or a cold winter.

In areas with hot, dry summers, where host plants do not survive, cleistothecia are probably the only means of oversummering (Koltin and Kenneth, 1970). In Israel, where the fungus apparently has a wider host range than elsewhere, wild grasses are important alternate hosts of cereal mildew. Susceptible grasses are infected by ascospores in autumn and the disease then spreads to emerging cereal crops (Eshed and Wahl, 1970; 1975). Indeed the life-cycle of *E. graminis* appears to be well adapted to conditions prevailing in the centres of origin of the cereal hosts (Koltin and Kenneth, 1970).

It was once commonly believed that in cooler areas overwintering was also by means of cleistothecia but it is now known that even in these areas asco-spores are usually shed during late summer or autumn and that the pathogen overwinters by other means (Cherewick, 1944; Smith and Blair, 1950; Moseman and Powers, 1957; Turner, 1956; Ts'ao, 1960; Smedegård-Petersen, 1967; Aleksandrov, 1968). Ascospores are mainly responsible for transmitting infection to autumn-sown cereals or stubble volunteers, again indicating an oversummering role (Gorlenko, 1940; Turner, 1956; Smedegård-Petersen, 1967). Unusually, cleistothecia may overwinter to produce asco-spores in spring (Foster and Henry, 1937) but this probably occurs only where dry weather in autumn is rapidly followed by temperatures low enough to prevent maturation of cleistothecia (Smedegård-Petersen, 1967). It is possible that some cleistothecia may also survive on stored straw (Smith and Blair, 1950; Tursumbaev, 1973) but the practical importance of these in most areas is probably small (Hermansen, 1964).

It seems clear that the fungus usually overwinters vegetatively on living host tissue. In those areas where winters are cold, the fungus probably survives mainly as dormant mycelium (Smiljaković, 1967; Aleksandrov, 1968). In such areas survival of the host, and hence the pathogen, may be aided if crops are covered by snow (Johnston, 1974). Where winters are less severe and temperatures are not continuously below freezing the fungus can be active and produce conidia during mild periods. In cold weather many infected leaves may die, causing a large decrease in surviving inoculum (Hermansen, 1964; Yarham *et al.*, 1971; Jenkyn, 1976a) but survival on lightly infected leaves is probably sufficient to ensure carry-over (Yarham *et al.*, 1971; Jenkyn, 1976a).

At temperatures above freezing conidia do not survive for long; at 18–20°C, survival is only for about two days. Survival is much longer where temperatures are kept continuously below freezing and at very low temperatures may be sufficient to allow overwintering (Hermansen, 1966). In the maritime provinces of Canada, Johnston (1974) found that some conidia produced on infected winter wheat seedlings remained viable for some months where temperatures stayed below freezing. However, the proportion which remained viable declined steadily through the winter. Increased germination in spring was probably due to the production of new conidia but it is not clear whether new infections had developed to produce these conidia or whether they formed on overwintering mycelium.

I. Relative importance of overwintering sources

Where winter cereals are commonly grown they are almost certainly the most significant overwintering hosts. Infection of spring crops close to winter barley is commonly much earlier and becomes severe much more rapidly than infection of crops distant from winter barley (Pape and Rademacher, 1934; Stephan, 1957; Johansen, 1963; Hermansen, 1964; Stapel, 1966; Yarham and Pye, 1969; Yarham et al., 1971). In Denmark, where the area sown with winter barley in the mid-1960s was small but was increasing, the potential importance of this crop in the carry-over of leaf diseases was considered sufficient to justify a ban on its cultivation (Stapel and Hermansen 1968; Hermansen et al., 1975).

The importance of survival on other overwintering hosts is difficult to estimate and may vary from region to region. Volunteer seedlings might be expected to be important. However, Hermansen (1964) found that amounts of mildew on overwintering barley volunteers were small and considered them to be unimportant as sources of inoculum for spring crops in Denmark. He concluded that, in the absence of winter barley, initial inoculum was derived from sources outside the country. In Britain, where conditions are often more suitable for the survival of volunteers, Turner (1956) concluded that they transmitted infection from stubble to winter cereals and that they could also survive as overwintering hosts in clover leys. However, Yarham and Pye (1969) found little evidence of gradients of infection in spring crops adjacent to fields with volunteers.

Not all of the many susceptible grass species which grow in Israel occur in the other major cereal-growing areas, where the fungus needs to survive through a winter. In such areas some cross infection to grass species can occur (Hardison, 1944; Grasso, 1964) but the extent is uncertain. It seems likely, however, that, at least in Europe, grasses are much less important than winter cereals as a source of primary inoculum (Frauenstein, 1970).

It has been shown that wheat and barley can be infected by formae speciales *hordei* and *tritici* respectively provided the plants are already infected with the compatible form (Moseman and Greeley, 1964; Moseman *et al.*, 1965). This phenomenon has not been demonstrated in the field but it is perhaps possible that one winter cereal may be an overwintering host for mildew from other cereals.

IV. Effects on the Physiology of the Host

Changes in the metabolism of diseased leaves are complex. They have been discussed in detail by Scott (1972) and will be considered only briefly here. There is a substantial increase in respiration and in the activity of the pentose phosphate pathway (Shaw and Samborski, 1957; Scott and Smillie, 1966). Some of the increase in respiration is due to fungal activity but most is attributable to a stimulation of host respiration and although not restricted to cells invaded by the pathogen, the respiratory increase is confined to tissues immediately underlying the infected areas (Allen and Goddard, 1938; Bushnell and Allen, 1962b; Scott and Smillie, 1966). There is a large decrease in photosynthetic activity (Last, 1963) following the degradation of chloroplasts (De Diego Calonge and Rubio-Huertos, 1965; Camp and Whittingham, 1975) and the increase in respiration may be linked to this (Allen, 1942; Scott and Smillie, 1966). Dyer and Scott (1972) found a decrease in chloroplast polysomes 24 h after inoculation suggesting that chloroplast breakdown begins soon after infection. Similarly, Bennett and Scott (1971) showed a complete loss of chloroplast ribosomes and ribosomal RNA within nine days of inoculation. In contrast, total RNA content of infected leaves is increased (Malca *et al.*, 1964; Zscheile *et al.*, 1969) apparently as a result of an increase in the number of cytoplasmic ribosomes (Bennett and Scott, 1971). Subsequently, RNA decreases more rapidly in diseased than in healthy leaves (Malca *et al.*, 1964; Plumb *et al.*, 1968). Infection seems to have little, if any, effect on DNA content (Plumb *et al.*, 1968).

Although RNA content is increased, at least initially, there is no evidence of an increase in protein content (Millerd and Scott, 1963) and indeed Zscheile (1974) detected a decrease. However, protein composition is changed (Johnson *et al.*, 1966) and amino acids not present in healthy tissues also appear (Zscheile, 1974). General changes in nitrogen metabolism result in the accumulation of amides and ammonium ions, and the evolution of appreciable amounts of ammonia gas (Sadler and Scott, 1974).

Infection with powdery mildew also alters the hormone balance of cereal plants. Indole-acetic acid (IAA) activity in diseased leaves may fall initially but it subsequently increases to exceed that of healthy leaves (Shaw and

Hawkins, 1958; Vizárová, 1973). The increase in IAA is apparently due to the inhibition of IAA-oxidase by phenolic compounds, at least some of which may be produced by the fungus (Vizárová and Janitor, 1968; Vizárová, 1973; Frič, 1974). Auxins are known to affect nutrient mobility (Osborne, 1967) so IAA may be at least partly responsible for the movement of compounds to infection sites. Shaw and Samborski (1956), for example, observed accumulation of various carbon compounds, including sugars, amino and organic acids and phenols, as well as phosphate and calcium, around mildew pustules. Initially, there is enhanced starch synthesis around mildew colonies (Thrower, 1964) leading also to starch accumulation but six to eight days after inoculation it has largely disappeared (Allen, 1942; Bushnell and Allen, 1962a).

Although mildew is restricted to the aerial parts of the plant it does affect root growth (Last, 1962a; Paulech, 1969). Elongation and branching are both decreased (Vizárová and Minarčic, 1974) as is mitotic cell division in apical root meristems (Minarčic and Paulech, 1975). The latter may result from an increase in cytokinin level (Vizárová and Minarčic, 1974) or from a reduction in the supply of assimilates to the roots (Minarčic and Paulech, 1975).

A commonly observed effect on senescing leaves is the formation of "green islands" (Bushnell and Allen, 1962a). These may be areas of chlorophyll retention (Bushnell, 1967), although Allen (1942) observed that they developed in previously chlorotic tissue suggesting that chlorophyll was reformed. Green islands also form when extracts of diseased tissue or of mildew conidia are applied as drops to detached leaves (Bushnell and Allen, 1962a; Vizárová, 1974). Spot treatment of detached leaves with cytokinins gives similar areas of chlorophyll retention (Gunning and Barkley, 1963; Vizárová, 1975) suggesting that they may be involved in the formation of green islands. Cytokinins have a similar effect to IAA in causing accumulation of compounds at the site of application (Gunning and Barkley, 1963).

The tendency for compounds to accumulate at infection sites is presumably responsible, at least in part, for observed changes in the patterns of translocation in plants. Edwards (1971) showed that the assimilated carbon which accumulated at infection sites on primary leaves of barley came predominantly from that part of the leaf acropetal to pustules and that there was little interference with translocation of carbon compounds from basipetal parts of the infected leaf. He also showed that in primary leaves, most of the assimilate from the leaf tip is normally translocated to the roots so that mildew infection inhibits carbon flow to the roots more than to the shoot. Similarly, the proportion of radiocarbon recovered from the roots of spring wheat, following application of $^{14}CO_2$ to the leaves, was found to be decreased by infection with mildew (Lupton and Sutherland, 1973). Frič (1975) again found that infection of primary leaves of barley caused retention of photo-assimilate in these leaves but, unlike Edwards (1971), found evidence that export from uninfected second leaves was increased.

Although few experiments have investigated the effects of mildew on uptake of inorganic nutrients by roots, it has been established that transport of P and K into barley leaves and uptake of ^{32}P from liquid culture by whole wheat plants are increased by infection (Priehradný and Ivanička, 1970; Comhaire, 1963). However, Majerník (1965) found that although uptake of ^{35}S was doubled after four days, it was subsequently decreased. Nitrate concentrations in stem tissue may also be increased by mildew (Jenkyn, 1977) but it is not known whether this is due to increased uptake or de- creased assimilation. Bushnell (1967) however, found no increase in total nitrogen content of mildewed barley leaves. Uptake of some nutrients can be much affected by transpiration rate (Russell and Barber, 1960) but other mechanisms are almost certainly also involved. Indeed in the early stages of mildew infection, stomata close and transpiration is decreased. The sub- sequent increase in transpiration is mainly due to cuticular water loss (Majerník, 1966; Šprochová, 1967; Priehradný, 1975; Martin et al., 1975).

Summarizing the available evidence, Durbin (1967) concluded that leaves infected with obligate parasites both import more and export less inorganic and organic solutes than do comparable healthy leaves. The roots compensate for the loss of materials from infected leaves by obtaining an increased proportion of photosynthate from the apex of the plant. The large increase in water loss from diseased tissue at the time of sporulation creates a general water stress in the whole plant because water moves towards infected tissue.

V. Effects on Crop Growth and Yield and on Grain Quality

The disruption in host physiology caused by mildew can result in considerable changes in host development. Most of the experiments to investigate these effects on crop growth and yield have been done on barley and have used fungicides to control the disease. Detailed growth analyses of healthy and diseased barley plants were done in glasshouse experiments by Last (1962a). Although the effects of the disease are often particularly severe under glass- house conditions, many of his observations have since been supported by results from the field.

Surprisingly, root growth is often affected more than shoot growth (Paulech, 1969) and may be decreased by up to 50% (Last, 1962a; Brooks, 1972), resulting in smaller root:shoot ratios. In our experience, seedlings are seldom killed by mildew in the field although the disease may decrease winter hardiness (Månsson, 1955). In experiments done by Finney and Hall (1972), diseased seedlings produced fewer tillers than unsprayed seedlings. Tillers on diseased plants are also less likely to survive and produce ears (Paulech, 1969). The relative importance of effects on tiller production and tiller survival is uncertain but it is certainly a commonly observed effect of the disease in the field that ear number per plant is decreased (Benada, 1961; Kaspers and Kolbe, 1971; Brooks, 1972; Finney and Hall, 1972; Rea and Scott, 1973; Jenkyn, 1974b; Klose *et al.*, 1975). (A personal communication from Professor G. Fischbeck, confirmed that "Bestandesdichte", in their paper (Klose *et al.*, 1975), should be translated as number of shoots per unit area.) Shoot growth is also affected and diseased crops are often shorter than sprayed (Benada, 1961; Last, 1962a; Jenkyn, 1974b).

There is no clear evidence that mildew affects leaf size but it can retard leaf development (Paulech, 1969). The disease greatly accelerates leaf senescence and thus affects the total green leaf area of shoots (Last, 1962a; Large and Doling, 1962; Brooks, 1972; Simkin and Wheeler, 1974). This combined with the frequent decrease in tiller number can result in the leaf area index being severely affected (Last, 1962a; Rea and Scott, 1973; Jenkyn, 1976b). Brooks (1972) found that individual leaves on mildew affected plants weighed less than corresponding leaves from sprayed plants, supporting the observations of Last (1962a) that infected leaves have smaller dry weights per unit area than healthy leaves. Leaf efficiency is also decreased so that infected plants have smaller net assimilation rates (Last, 1962a; 1963; Finney *et al.*, 1976).

The components of grain yield (Thorne, 1974), namely: number of ears per unit area, number of grains per ear and grain size can all be affected by mildew. Effects on ear number, already referred to, may be considerable and increases of 20% or more where mildew is controlled have been reported (Last, 1957; Kaspers and Kolbe, 1971; Brooks, 1972; Finney and Hall, 1972; Rea and Scott, 1973). It is more difficult to measure the effect of mildew on numbers of grains per ear because ear size is so variable but it is usually much smaller than the effect on ear number. Where differences have been detected in the field the increase obtained by controlling mildew has usually been about 5% but occasionally the effects have been larger (Schaller, 1951; Kaspers and Kolbe, 1971; Brooks, 1972). The disease frequently affects grain size and increases of 5–10% where it is controlled are common (Schaller, 1951; Last, 1957; Kaspers and Kolbe, 1971; Brooks, 1972; Mundy and Page, 1973; Jenkyn, 1974b).

A complicating factor when examining the effects on yield components may be the tendency for them to be inversely correlated (Thorne, 1974). For example, in an experiment with winter barley, Finney and Hall (1972) found that the average number of grains per ear and to a lesser extent, grain size, were decreased by fungicide, probably because mildew control greatly increased ear number. Similarly, Kaspers and Kolbe (1971) found that grain size in winter barley was decreased where ear number was greatly increased by fungicide treatment.

In practice there may be important differences in the development of mildew and in its control on winter and spring barley crops but in the previous sections we have not attempted to distinguish between them. However, in both, the stage at which mildew develops is important in determining which components of yield are most affected by the disease. Generally, early mildew has most effect on ear number and late mildew on grain size although when early mildew is very severe it can so affect crop growth that grain filling is also impaired (Schaller, 1951; Rea and Scott, 1973; Jenkyn, 1974b; Griffiths *et al.*, 1975).

Mildew clearly affects quality of barley grain by decreasing grain size and therefore starch:fibre ratio. No effect on nitrogen content was found by Hanf *et al.* (1970) nor by Mundy and Page (1973). However, the results of Laurence (1970) suggest that there may be changes in amino-acid composition.

There have been fewer experiments to examine the effects of mildew on wheat than on barley and many have used broad spectrum fungicides so it is often not possible to be sure that all effects are due to mildew control. However, there is good evidence from experiments comparing varieties which differ in susceptibility to mildew, or using more specific fungicides, that losses can exceed 20% (Hebert *et al.*, 1948; Ts'ao, 1960; Johnston, 1972; Smith and Wright, 1974). There are few data showing the effects of mildew control on components of yield in wheat although increases in grain size have sometimes been reported (Kaspers and Kolbe, 1971; Johnston, 1972; Nissinen, 1973; Morel *et al.*, 1975). The number of grains per ear may also be affected (Kaspers and Kolbe, 1971) but we have found no data showing a clear effect on ear number. Ts'ao (1960) recorded that mildew decreased both starch and protein content by about 7·5%.

Losses can be equally severe in oats (Griffiths, 1955; Lawes and Hayes, 1965; Jung and Bedford, 1971). In an experiment using isogenic spring oat lines, resistant or susceptible to mildew, Lawes and Hayes (1965) found that severe mildew decreased panicle number, number of grains per panicle and thousand grain weight by 23, 18 and 6% respectively causing a total yield loss of 39%. Straw yield was affected less than grain yield, being decreased by 26%.

Schicke *et al.* (1971) obtained increases in yield of up to 11% by controlling mildew on winter rye and the data of Kaspers and Kolbe (1971) suggest that all components of yield may be affected.

VI. Relationship between Mildew Severity and Yield Loss

An essential requisite for establishing a relationship between mildew severity and yield loss, and for its subsequent use, is a reliable method for assessing disease. Large and Doling (1962; 1963), after a detailed study of cereal growth and mildew development devised a descriptive key for assessing this disease in the field. Data obtained from barley, oat and wheat trials over several years, using lime-sulphur to control mildew, were then used to investigate the relationship between mildew at completion of earing (Growth Stage 58–59; Zadoks *et al.*, 1974) and yield loss. They concluded that for barley and oats, percentage yield loss was about 2·5 times the square root of the percentage leaf area infected. For wheat it was about twice the square root of percentage mildew.* Klose *et al.* (1975) obtained a result similar to that of Large and Doling with their data for spring barley.

Using linear scales, the barley data of Large and Doling (1962) show a non-linear relationship, with proportionately greater yield loss per unit of mildew when disease is slight than when it is severe. Interestingly, Last (1963) showed that decreased photosynthetic activity of mildewed barley leaves could also be related to the square root of the percentage leaf area affected.

Much variability in their data was unaccounted for by the equations derived by Large and Doling, suggesting that the accuracy with which yield losses can be estimated is not great (Church, 1971). In one series of trials throughout England, between 1969 and 1971, there was little agreement between predicted and measured yield losses for the barley varieties tested (Doodson and Saunders, 1969; Little and Doodson, 1971). However, in another series of trials, Doling *et al.* (1971) obtained very good agreement between estimated and measured yield losses for both wheat and barley.

In the trials reported by Large and Doling, mildew seldom exceeded 20%. Results from our experiments suggest that where mildew is more severe, yield losses are less than the square root relationship implies. Furthermore, the Large and Doling equation relates yield loss to mildew assessed at a single, late growth stage although the pattern of mildew development can differ from year to year (Doling *et al.*, 1971) and small amounts of mildew during early growth stages can have relatively large effects on crop growth and yield (Brooks, 1972; Jenkyn, 1974b). The importance of assessing mildew through the season is supported by the results of Jenkins and Storey (1975), who related yields to a cumulative mildew index. A quadratic curve fitted to these data accounted for 76% of the total variation.

A different approach to investigating disease-loss relationships has been described by Richardson *et al.* (1975). They measured disease on individual shoots and showed that amounts of mildew on barley were significantly related to number of grains per ear and grain weight. However, the relationships accounted for only about 10% of the total variation and the method, as described, cannot take account of effects on yield, resulting from decreases in ear number, which can be a major cause of loss. Ear size, even in the absence of mildew, can be very variable. Parmentier and Rixhon (1973a) also found a

* An error easily made when using this equation is to estimate average loss from the average percentage mildew. Correctly, individual disease assessments should be transformed and then averaged.

relationship between ear size and mildew, on winter wheat, but concluded that there were physiological differences between tillers and those with large ears were more resistant than small-eared tillers.

It is clear that the relationship between mildew and yield loss in cereals remains imperfectly understood. However, more accurate models will probably require more detailed disease measurement not only for their derivation but also for their use and this may limit the extent to which such equations can be usefully applied.

VII. Interactions with other Pathogens

There are several reports describing interactions between cereal mildew and other pathogens. It appears that when the host is normally susceptible to the interacting pathogens, the levels of one or both diseases are usually decreased, suggesting competition, but incidence of pathogens to which the host is normally resistant, including other powdery mildews, may be increased. Thus, in barley crops where mildew is controlled, the incidence of *Rhyncho-sporium secalis* (Kampe, 1974) and *Puccinia hordei* (Honecker, 1943; Little and Doodson, 1972) can increase. Simkin and Wheeler (1974) found that on barley leaves simultaneously inoculated with *E. graminis* and *P. hordei*, incidence of each was approximately halved. On wheat, severe mildew may also decrease development of brown rust (*Puccinia triticina*) (Manners and Gandy, 1954). However, on wheat varieties normally resistant to particular races of rust, the presence of mildew can make susceptible those parts of the leaf which are close to mildew pustules (Johnston, 1934; Mains, 1934; Roberts, 1936; Manners and Gandy, 1954). Wheat leaves normally resistant to *Septoria tritici* may also become susceptible if pre-inoculated with mildew (Brokenshire, 1974). In contrast, susceptibility of wheat to mildew was decreased if plants were infected with *Tilletia caries* (Sempio, 1938).

Infection of oat plants with barley yellow dwarf virus (BYDV) may make them more susceptible to infection with mildew (Potter, 1974). There is evidence that *E. graminis* can transmit strains of Tobacco Mosaic virus (TMV) to *Chenopodium* spp. (Yarwood, 1971; Yarwood and Hecht-Poinar, 1973) and that the virus can be extracted from conidia (Nienhaus, 1971). Although TMV can infect both barley and wheat (Hamilton and Dodds, 1970; Paulsen *et al.*, 1975) it is not a recognized pathogen of cereals and as yet there is no evidence that *E. graminis* is a vector of pathogenic cereal viruses.

VIII. Effects of Cultural Conditions on Incidence of Disease

A. Host nutrition

Although many plant nutrients can influence the susceptibility of cereals to mildew, reports on their effects are often contradictory. This may sometimes be due to interactions between elements; supplying one, for example, may lead to a deficiency in others (Schaffnit and Volk, 1927).

Of the major plant nutrients, nitrogen seems to have the most consistent effect and usually increases the incidence of mildew (see, for example, Spinks, 1913; Schaffnit, 1922; Trelease and Trelease, 1928; Grainger, 1947; Last, 1953; Mygind, 1970). Although nitrogen fertilizers may alter the microclimate within crops, a direct effect of nitrogen on susceptibility has been demonstrated (Bainbridge, 1974). Nitrogen increases susceptibility at whatever growth stage it is applied. However, Last (1954) found that in pot experiments with winter wheat, nitrogen applied at different times before flag leaf emergence increased disease more, the later it was applied. Parmentier (1959) also found that nitrogen applied at tillering increased disease more than nitrogen applied at seedling emergence. Incidence of the disease in cereal crops grown in rotations may be increased if there are nitrogen residues from a preceding crop (Parmentier and Rixhon, 1973b) or fallow (Glynne, 1959).

Dentler (1958) reported that pathogenicity of *E. graminis* was weakened after passing a number of generations on barley liberally supplied with nitrogen but strengthened if the host was given liberal potassium.

Where mildew is controlled, yield response per unit of applied nitrogen is often increased (Dilz and Schepers, 1972; Jenkyn and Moffatt, 1975) and the amount of nitrogen giving best yield may be altered (Jenkyn and Finney, 1977). Response to nitrogen applied at different times can also be affected by mildew. Hebert *et al.* (1948) found that nitrogen top dressings for winter wheat were best applied early if mildew was not controlled but late applications gave best yields where a sulphur dust was applied.

Most reports on the effect of potassium suggest that plants which are deficient in this element are very susceptible to mildew and applications of potassium fertilizers often decrease the disease (Spinks, 1913; Stuch, 1926; Schaffnit and Volk, 1927; van Poeteren, 1935; Olson, 1956; Glynne, 1959). Last (1962b), however, found no effect of potassium on susceptibility and Grainger (1947), working with oats, found that this element increased the disease.

The effect of phosphate is also inconsistent; although application of this nutrient especially when correcting a deficiency, can decrease disease (Spinks, 1913; Schaffnit and Volk, 1927; Grainger, 1947; Last, 1962b) this is not always so and susceptibility may even be increased (Stuch, 1926; Glynne, 1959).

Among other elements, calcium added to soil had little effect on susceptibility (Sadowski, 1966) although susceptibility of detached leaves is increased if they are supplied with calcium (Hirata, 1971). Vlamis and Yarwood (1962) found that application of lime did increase susceptibility of barley to mildew but they considered that this was due to a reduced uptake of manganese. Mildew resistance of wheat is increased by manganese applied to either soil or seed (Colquhoun, 1940; Zubko, 1961).

Boron may decrease mildew, especially on barley (Eaton, 1930), although Yarwood (1938) found that this element had no effect on the susceptibility of oats. It may also stimulate formation of perithecia (Cherewick, 1944).

Salmon (1904b) reported that copper did not control E. graminis but barley plants deficient in this element do seem to be more susceptible than plants adequately supplied (Olsen, 1939).

Resistance may be increased by application of silicon (Lowig, 1933; Germar, 1934; Wagner, 1940), cadmium (Sempio, 1936; Meyer, 1950) and lithium (Spinks, 1913; Kent, 1941; Smith and Blair, 1950; Carter and Wain, 1964). Spinks (1913) found that lead and zinc had little effect on susceptibility of wheat although Hewitt and Jones (1951) reported that zinc deficiency increased the resistance of oats. Surprisingly, Cherewick (1944), who tested many of these elements, was unable to demonstrate that any of them had any effect on the susceptibility of either wheat or barley.

Crops grown in soil to which dung has been frequently applied, may have relatively little mildew (Spinks, 1913; Glynne, 1959). This may be because dung supplies many of the elements known to increase resistance but the slow release of nitrogen to plants may also be a factor in retarding mildew development. Dung will also increase the available water content of most soils and this in turn might affect resistance (Yarwood, 1949).

It is clear that nitrogen frequently increases susceptibility under normal agricultural conditions. In contrast many of the other elements, most of which seem to increase resistance, appear to have significant effects only when supplied to plants otherwise deficient in that element. Lithium is probably an exception in that it is not known to be required for plant growth but the amounts necessary to obtain good mildew control are often phytotoxic. It seems on present evidence, therefore, that while correction of nutrient deficiencies may improve both crop growth and yield, further adjustment of fertilizer use, apart from avoiding excess nitrogen, does not offer an effective means of disease control.

B. Sowing date and crop density

Incidence of mildew can be much affected by sowing date. In temperate areas, late-sown spring cereals usually develop more disease than those sown early (Last, 1957) and there is a corresponding decrease in yield (Mórász, 1970). Autumn-sown crops commonly have more disease on seedlings the earlier they are sown (Gram and Rostrup, 1923; Månsson, 1955) but after winter the earlier sown have less than late-sown crops (Bremer et al., 1947; Mygind, 1970; Jenkyn, 1976a). Crops sown late in autumn or early in spring will be exposed to little inoculum when they first emerge and the seedlings may therefore escape disease. As later formed leaves are usually more resistant than seedling leaves (Graf-Marin, 1934; Jones and Hayes, 1971), early-sown crops may be resistant by the time conditions have become suitable for rapid disease development. However, it is not clear whether such interactions between plant age and inoculum density provide an adequate explanation for the large differences in disease which may be evident later in crop growth. Last (1957) suggested that increased susceptibility of late-sown spring crops might be related to their faster growth rate, in spring and early summer.

In some circumstances, adjustments to sowing date might decrease the damage done by mildew. However, in many areas, sowing date is largely determined by factors outside the grower's control. Furthermore, because time of sowing can have a large effect on yield, an alteration in sowing date, even if it decreases disease, may itself cause a loss in yield.

Altering plant spacing will influence the environment within the crop and this might be expected to affect the development of the disease. Jenkyn (1970) reported that wheat sown at a wide row-spacing had more mildew than a narrow spaced crop, perhaps because leaves remained green for longer, although crop density may also have influenced inoculum dispersal within the crop. However, Månsson (1955) found that decreasing seed rate resulted in less disease. Similarly, Smith and Blair (1950) using extreme differences in seed rate (c. 3 and 79 plants m^{-1} of row) reported that there was six times more disease at the large rate than at the small rate. In an experiment described by Last (1957), in which seed rate had no effect on mildew, plants at the smaller seed rate produced more tillers, so that numbers of shoots per unit area and hence crop density were similar at both seed rates.

C. Irrigation

Irrigation can be expected to alter both the physical environment and host physiology and has been reported to increase the incidence of mildew in South Africa (Gorter, 1955). This is in accord with the results of Tapke (1951) who showed that liberal watering of glasshouse plants during the pre-inoculation period can increase their subsequent susceptibility to mildew. Other experiments have shown the opposite effect of soil moisture and indicate that susceptibility is increased if plants are grown in soils at low rather than high moisture contents (Volk, 1931; Yarwood, 1949). Availability of water will also affect nutrient uptake and Tapke (1951) further demonstrated that watering had little affect on susceptibility when plants were grown in an infertile soil. Roder (1967) found that results from experiments to test the effects of soil moisture content could be very variable and also differed according to host species.

It is not surprising that observations on the effects of agronomic practices on incidence of mildew so often produce conflicting information. Cultural treatments usually modify many environmental factors, each of which may have distinct and perhaps opposite effects on disease development. Thus the net effect of altering agronomic practices may differ according to the conditions prevailing when the change is imposed.

IX. Control

A. Cultural methods

While the cultural treatments discussed in the previous section can influence the incidence of mildew in crops, there is generally little scope for achieving significant decreases in the damage done by cereal mildew solely by altering these treatments.

Any practice which decreases sources of primary inoculum is desirable but the effectiveness of any such measure will depend on the strength of the source and the area over which the measure is applied. For pathogens like *E. graminis*, with a fast "infection rate", large differences in the amounts of initial inoculum are necessary to affect the development of an epidemic significantly (van der Plank, 1963).

Because the pathogen survives on crop debris as cleistothecia, destruction of crop remains will only be effective in preventing spread to autumn-sown crops if done before ascospore discharge. Crops sown after ascospore discharge will not be at risk from this source and surveys of spring barley in Britain have shown that mildew is not increased if barley crops are grown in succession (King, 1972).

Crops sown very close to any strong source, for example spring barley near to winter barley, are likely to develop severe disease in the seedling stage and this may be important in determining yield. However, because the disease can develop so rapidly, proximity to winter barley may have little effect on incidence of mildew at later growth stages (Yarham *et al.*, 1971). To decrease the amount of disease significantly, it is generally necessary to eliminate major sources from a large area (Lester, 1971) so that amounts of primary inoculum are very small. In the Faeröes, for example, the disease on barley becomes prevalent only on adult plants because overwintering sources are rare or absent and primary inoculum is derived from distant sources elsewhere in Europe (Hermansen and Wiberg, 1972).

Where the major primary sources are internal, sanitation, to be effective, may require legislation, as in Denmark, where the cultivation of winter barley is prohibited.

B. Biological control

Many members of the Erysiphaceae are parasitized by *Ampelomyces* spp. (*Cicinnobolus* spp.) (Rogers, 1959; Clare, 1964; Anon, 1968b), although we have not found reports that it can be a parasite of *E. graminis*. However, Yarwood (1957) considered that control of mildews using this hyper-parasite was unlikely to be successful because the host fungus and the parasite are favoured by different weather.

Grebenchuk (1965) reported that infection of barley by mildew was decreased if the plants were sprayed with dung infusion, containing mycolytic bacteria, or with pure cultures of *Trichoderma viride*. Culture filtrates of *Trichothecium roseum* (Darpoux and Faivre-Amiot, 1952) are also active against cereal mildews as are many other antibiotics (Davis *et al.*, 1960; Rhodes *et al.*, 1961; Ubrizsy and Voros, 1967; Savel'yev and Polyakova, 1972) but none have yet been developed for commercial use.

C. Resistant varieties

The genetics of host resistance and pathogen virulence have been extensively studied and have been reviewed by Moseman (1966) and Wolfe (1972). For many years, the use of resistant varieties has been the principal method of controlling *E. graminis* on cereals. The resistance conferred by the so called major genes, most commonly used in resistance breeding to date, has seldom remained effective for more than a few years after widespread adoption of a variety (Doling, 1969) because new races of the pathogen arise and spread through the population very rapidly. Up to the end of 1966, for example, 72 races of *E. graminis* f. sp. *hordei* had been identified in Europe (Johnson and Wolfe, 1968). Leijerstam (1972) estimated that the mutation rate towards virulence could be as high as 2000 locus^{-1} ha^{-1} day^{-1}. Nevertheless, numerous resistance genes have been identified so that breeders have been able to develop a succession of new varieties, thus achieving useful control.

Multiline varieties (i.e. varieties consisting of a number of lines homogeneous for most characteristics but differing in the resistance genes they contain) have been little used to control mildew. They may provide a useful means for further exploiting major genes, especially those which are allelic and, therefore, more difficult to combine in other ways. A convenient alternative to true multilines may be to grow mixtures of existing varieties (Lang *et al.*, 1975; Wolfe *et al.*, 1976) although it may not always be easy to devise mixtures which are agronomically acceptable in other ways. A possible risk associated with the widespread culture of mixtures or multilines is that selection of complex races, combining many virulence genes, may be accelerated but present evidence suggests that this is not an inevitable consequence (Wolfe *et al.*, 1976).

Varieties which are susceptible to infection by *E. graminis* may nevertheless differ very much in the extent to which the disease develops. These less complete types of resistance, variously known as field, generalized, partial, race non-specific, or durable resistance are the subject of increasing interest by plant breeders and geneticists (Jones and Hayes, 1971). Such resistance, which is often expressed in the adult plant rather than in seedlings (Macer, 1964; Hayes and Jones, 1966; Shaner, 1973a), may affect the pathogen at any stage in its life cycle. Infection, colony growth and sporulation can all be delayed or decreased, as may haustorial size and efficiency (Carver and Carr, 1974; 1975; Shaner, 1973b; Russell, 1974; 1976), leading to a smaller rate of epidemic development (Shaner, 1973b). Partial resistance has been transmitted to new varieties in breeding programmes (Roberts and Caldwell, 1970; Jones, 1974) and it can indeed be very much more durable than typical race specific resistance (Roberts and Caldwell, 1970; Shaner, 1973a). However, there is some evidence that adult plant resistance may be partly race specific and that strains of the pathogen able to overcome the resistance may sometimes be selected (Jones, 1974; Russell, 1975b).

Some varieties are apparently more tolerant to mildew infection than others and suffer less yield loss for similar amounts of disease (Little and Doodson, 1972; Jenkins and Storey, 1975), a potentially important character if it can be incorporated in new varieties (Jones, 1974).

D. Fungicides

Chemical control of cereal mildews is discussed here and is considered also in Chapter 10 by K. J. Bent.

Various materials, especially elemental sulphur or compounds containing sulphur, have long been known to have activity against mildews (Sempio, 1932; Yarwood, 1951; Hashioka and Morihashi, 1952; Crosier and Szkolnik, 1956; Mässing, 1959). However, many of these fungicides are predominantly protectant and their persistence and activity are often relatively small, so that useful control may require that they be applied frequently (Sempio and Castori, 1949; Anon, 1956; Robayo *et al.*, 1967; Nikulina and Chumakov, 1968). Although routine application of fungicides to cereals has been practised in some areas (Brooks, 1970) and has proved valuable in research, it has, in the past, generally been considered uneconomic for commercial cereal crops.

The situation has been radically altered in recent years by the development of very active fungicides, some of which are partially systemic, including the benzimidazoles (Delp and Klopping, 1968), ethirimol (Bebbington *et al.*, 1969), tridemorph (Kradel *et al.*, 1969b), triforine (Schicke and Veen, 1969) and, more recently, triadimefon (Frohberger, 1973). Furthermore, large changes in the value of cereal grain have made disease control even more profitable.

The use of fungicides to control foliar diseases on commercial cereal crops is most common in Western Europe where the practice has become widespread. In many of these countries over 20% of the wheat and barley acreage was treated in 1975 and in some the figure exceeded 50% (Obst, 1975; King, 1977a,b). Most of the information concerning the use of fungicides against cereal mildew therefore relates to countries in Western Europe.

Among the systemic fungicides, some are relatively specific against mildew (Dickinson, 1973) and have often proved more effective against mildew on barley than on other cereals. Where mildew is the predominant disease and especially on barley, these specific fungicides may provide the only protection needed. However, where diseases other than mildew are also prevalent, which seems to be particularly common on wheat, there may be a greater need for fungicides with a broad spectrum of activity. This need may be met by the benzimidazole group of compounds and by the more recently developed triadimefon but, increasingly, mixtures of fungicides, which often include non-systemics, are being used (Douchet *et al.*, 1973; Lescar *et al.*, 1973; Meeus and Haquenne, 1973; Morel *et al.*, 1975; Obst, 1975; Phillips and Frost, 1975; Smith *et al.*, 1975).

Some of the systemic fungicides used against leaf diseases, including mildew, can be applied as seed dressings, although ethirimol is the only one which has so far been used extensively (Graham-Bryce, 1973). This fungicide is reversibly adsorbed by most soils so that it persists in an available form for much of the life of the host plant (Brooks, 1970; Graham-Bryce and Coutts, 1971). However, it is strongly bound in soils with large amounts of organic matter and may then be less effective. Efficient uptake by roots depends on adequate soil moisture and as a method of control it is much more suited to some areas than to others (Brooks, 1970). There are obvious advantages in applying fungicides as seed dressings since the problem of correctly timing the application is avoided and no extra farming operation, which may cause crop damage, is necessary. However, the compound can be present in the plant for some time before protection is needed and concentration may decline later in the season when disease is still increasing. If the material is sufficiently persistent, in soil or plant, this need not be a serious disadvantage.

The mildew fungicides which have been used as seed dressings can also be applied as sprays and many chemicals are only available as sprays. If economic or environmental considerations militate against more than one spray application, timing is important because differences in date of application may very much alter the efficiency of disease control and the increases in yield obtained (Evans and Hawkins, 1971; Cock, 1975; Jenkins and Storey, 1975; Bainbridge and Jenkyn, 1976). Where the economics of control are such that more than one spray is justified, timing is obviously less important. Farmers may then combine a seed dressing with a later spray or use a number of sprays, ensuring that the first is applied at the very early stages of disease development or even before the disease is seen.

Most of the information relating to the timing of sprays for mildew control has been obtained from experiments on spring barley. These show that a single spray is best applied during the very early stages of disease development (Kradel et al., 1969a; Evans and Hawkins, 1971; Jung and Bedford, 1971; Marshall, 1972; Mundy and Owers, 1974; Neuhaus and Reich, 1975). Not only does this appear to have most effect on the epidemic but also ensures protection of the crop during the seedling stage when damage by mildew can be disproportionately large.

Various attempts have been made to develop methods which forecast the best time to apply a single spray to achieve maximum yield increase. The first was suggested by Rosser (1969) who considered data from experiments in which barley was sprayed with drazoxolon. He suggested that epidemic development followed a period when the mean maximum temperature exceeded 20°C for seven days, but in practice this criterion seems to have proved to be too stringent. Observations at Rothamsted suggested that the concentrations of spores in the air above crops might reflect the amount and the activity of the fungus, to which spray times could be related (Bainbridge and Hirst, 1972; Jenkyn, 1972; Bainbridge and Alsop, 1973). This approach was taken further by Polley and King (1973) who related phenology of barley mildew, and spore catch, to meteorological data. They formulated weather criteria which identify periods when the probability of infection is high.

Even when meteorological data suggest that sprays are required, an assessment of the amount of disease present in a crop may be necessary before deciding whether or not to act on the warning (Polley and Smith, 1973). Furthermore, it is increasingly important that farmers should examine crops regularly if modern husbandry methods are to be used to maximum advantage and it is possible that timing sprays solely on the basis of amounts of disease in the crop may be sufficiently accurate for practical purposes (Jenkins and Storey, 1975).

Spring barley crops near to a strong source of inoculum, such as winter barley, usually need to be sprayed earlier than those distant from such sources (Jenkins and Storey, 1975). The criteria used for identifying best spray dates for spring barley can probably be applied to winter barley in the post-winter period (Neuhaus and Reich, 1975). However, early-sown winter barley (and other cereals) may become severely infected in autumn. Mildew control at this time can result in yield increases (Hall, 1971; Brooks, 1972; Finney and Hall, 1972; Bainbridge et al., 1977), although other experiments suggest that autumn control is not always beneficial (Jung and Bedford, 1971).

In those areas where fungicides are commonly used on cereal crops, oats are generally less important than wheat or barley and as a consequence much less work has been done on the control of oat mildew. However, from the limited evidence available it seems likely that timing of sprays for oats can be based on similar criteria to those used for barley (Anon, 1975).

The most widely used fungicides at the present time tend to be less effective against wheat mildew than against barley mildew. Furthermore, mildew on wheat often occurs with other diseases so that broad spectrum fungicides are often used. There is, therefore, little information on the timing of sprays specifically for the control of mildew although Serra et al. (1973) found that a single spray of tridemorph at about Growth Stage 31 could be as effective as a two-spray treatment (at Growth Stage 31 and Growth Stage 54). However, Jeffrey et al. (1975) found that fluotrimazole was equally effective whether applied at Growth Stage 31–32 or at Growth Stage 50. Experience with broad spectrum fungicides, often applied to control a complex of leaf diseases, suggests that application at about the time of ear emergence is often best (Lescar et al., 1973; Anon, 1975; Obst, 1975; Anon, 1976). Earlier sprays are often applied to control foot-rot diseases and these may, incidentally, have a useful effect against mildew (Rapilly, 1970; Obst, 1975).

One factor affecting the use of some fungicides may be the development of pathogen strains tolerant to the compounds used in their control (Wolfe, 1971). Strains of E. graminis tolerant to the benzimidazole fungicides have been reported (Vargas, 1972) but those showing tolerance to ethirimol have been most extensively studied. Within E. graminis tolerance can differ very much between isolates and to decrease selection of the most tolerant forms, use of ethirimol on winter barley is no longer recommended (Wolfe and Dinoor, 1973; Shephard et al., 1975). Ability of ethirimol-tolerant strains to compete, in the absence of ethirimol, seems to be variable (Wolfe and Dinoor, 1973; Hollomon, 1975) but tolerance may be linked to virulence genes in the pathogen (Wolfe and Dinoor, 1973). To delay selection of tolerant strains it has not only been suggested that different fungicides be used on winter and spring crops but also that in any disease control programme, a range of unrelated chemicals be used (Evans, 1971; Wolfe and Dinoor, 1973; Wolfe, 1975).

Tolerance of the host to phytotoxic effects of the fungicide may also vary. Siddiqui and Haahr (1971) reported that tridemorph induced chlorosis on some wheat mutants but not on others and suggested that breeders should select for resistance to phytotoxicity of fungicides.

Other materials, not normally used as fungicides, may also show fungicidal activity. Examples include calcium cyanamide (Meyer-Hermann, 1935), chlordane (Scharen, 1962) and metasystox (Ryan and Clare, 1972) as well as some wetting agents (Parmentier, 1956; Somers et al., 1967). Although such incidental activity of agricultural chemicals has not in the past been considered as part of disease control programmes, the recently introduced wild oat herbicide, difenzoquat, will control powdery mildew during early crop growth (Cyanamid of Great Britain Ltd., pers. comm.) and may influence timing of later applied fungicides.

Although some of the modern fungicides can provide such effective control of mildew that very susceptible varieties, having other desirable characteristics, may be grown on a large scale (Gilmour, 1975), sole reliance on fungicides is not desirable. It can be costly to the grower and there is a risk that the compound will become less effective if mutant strains of the pathogen are selected, as commonly happens with genetically resistant varieties. There is, therefore, increasing emphasis on integrated control. The value of combining partial resistance and fungicides has been stressed by a number of workers (Macer, 1964; Clifford et al., 1971; Jones, 1974; Wolfe, 1975) but it is likely that the control of mildew will, increasingly, come to be regarded as only one part of a comprehensive crop protection programme.

References

Anon. (1954). Agric. Gaz. N.S.W. 65, 526–527.
Anon. (1956). A. agric. Suisse (70), N.S. 5, 399–513.
Anon. (1962). Rep. Waite agric. Res. Inst. 1960–61.
Anon. (1965). Agric. Handb. USDA No. 291.
Anon. (1967). CMI Desc. Path. Fungi and Bacteria, Set 16, Sheet 153.
Anon. (1968a). Agric. Handb. USDA No. 338.
Anon. (1968b). Index of fungi 3, 457.
Anon. (1975). Rep. Sci. Arm, ADAS, Min. Ag. Fish. Fd 1973.
Anon. (1976). Rep. Sci. Serv., ADAS, Min. Ag. Fish. Fd 1974.
Adam, D. B. and Colquhoun, T. T. (1937). J. Dep. Agric. S. Aust. 40, 787–792.
Akai, S. (1952). Agric. Hort. Tokyo 27, 1135.
Al-Ani, H. Y. (1969). Z. PflKrankh. PflSchutz. 76, 551–556.
Aleksandrov, I. N. (1968). Mikol. i Fitopatol. 2, 475–480.
Allen, P. J. (1942). Am. J. Bot. 29, 425–435.

Allen, P. J. and Goddard, D. R. (1938). *Science* **88**, 192–193.
Arya, H. C. (1964). *Indian Phytopath.* **17**, 27–34.
Atkinson, T. G. and Shaw, M. (1955). *Nature, Lond.* **175**, 993–994.
Aust, H. J. (1973a). *Inaugural Diss., Justus Liebig Univ., Giessen* 110 pp.
Aust, H. J. (1973b). *Phytopath. Z.* **76**, 179–181.
Aust, H. J. (1973c). *25th Int. Symp. Phytopharm. Phytiatry* **1973**, 1477–1483.
Aust, H. J. (1974a). *Z. PflKrankh. PflSchutz.* **81**, 114–118.
Aust, H. J. (1974b). *Z. PflKrankh. PflSchutz.* **81**, 597–601.
Aust, H. J. and Hindorf, H. (1975). *Z. PflKrankh. PflSchutz.* **82**, 549–552.
Bainbridge, A. (1974). *Pl. Path.* **23**, 160–161.
Bainbridge, A. and Alsop, J. (1973). *Rep. Rothamsted exp. Stn* **1972**, part 1, 132.
Bainbridge, A. and Cross, P. (1974). *Rep. Rothamsted exp. Stn* **1973**, part 1, 128.
Bainbridge, A. and Hirst, J. M. (1972). *Rep. Rothamsted exp. Stn* **1971**, part 1, 137.
Bainbridge, A. and Jenkyn, J. F. (1976). *Ann. appl. Biol.* **82**, 477–484.
Bainbridge, A. and Legg, B. J. (1976). *Trans. Br. mycol. Soc.* **66**, 495–498.
Bainbridge, A., Finney, M. E. and Jenkyn, J. F. (1977). *Rep. Rothamsted exp. Stn* **1976**, part 1, 257–258.
Barducci, T. B. (1942). *Mems a. Dep. Genética Vegetal, Estac. exp. agríc. Molina*, Lima, Peru 1941.
Bebbington, R. M., Brooks, D. H., Geoghegan, M. J. and Snell, B. K. (1969). *Chemy Ind.* **1969**, 1512.
Benada, J. (1961). *Ann. Acad. tchécosl. Agric.* **34**, 1329–1342.
Benada, J. (1962). *Ann. Acad. tchécosl. Agric.* **35**, 1287–1304.
Benada, J. (1966). *Phytopath. Z.* **55**, 265–290.
Benada, J. (1972). *Proc. Int. Symp. Breeding and Productivity of Barley*, Kroměříž **1972**, *Istanbul*, B, II, 553–561.
Bennett, J. and Scott, K. J. (1971). *FEBS Letters* **17**, 93–95.
Boughey, A. S. (1949). *Trans. Br. mycol. Soc.* **32**, 179–189.
Bracker, C. E. (1968). *Phytopathology* **58**, 12–30.
Bremer, H., İşmen, H., Karel, G., Özkan, H. and Özkan, M. (1947). *Rev. Fac. ci. Univ. Istabul*, B, 122–172.
Briggle, L. W. and Vogel, O. A. (1968). *Euphytica* **17**, Suppl. 1, 107–130.
Brokenshire, T. (1974). *Trans. Br. mycol. Soc.* **63**, 393–397.
Brooks, D. H. (1970). *Outl. Agric.* **6**, 122–127.
Brooks, D. H. (1972). *Ann. appl. Biol.* **70**, 149–156.
Buchannon, K. W. and Wallace, H. A. H. (1962). *Can. J. Pl. Sci.* **42**, 534–536.
Bushnell, W. R. (1967). *In* "The Dynamic Role of Molecular Constituents in Plant-Parasite Interaction". (C. J. Mirocha and I. Uritani, Eds). The American Phytopath. Soc., Minnesota.
Bushnell, W. R. (1971). *In* "Morphological and Biochemical Events in Plant-Parasite Interaction". (S. Akai and S. Ouchi, Eds). Phytopath. Soc. Japan, Tokyo.
Bushnell, W. R. (1972). *A. Rev. Phytopath.* **10**, 151–176.
Bushnell, W. R. and Allen, P. J. (1962a). *Pl. Physiol.* **37**, 50–59.
Bushnell, W. R. and Allen, P. J. (1962b). *Pl. Physiol.* **37**, 751–758.
Bushnell, W. R. and Bergquist, S. E. (1975). *Phytopathology* **65**, 310–318.
Bushnell, W. R., Dueck, J. and Rowell, J. B. (1967). *Can. J. Bot.* **45**, 1719–1732.
Camp, R. R. and Whittingham, W. F. (1975). *Am. J. Bot.* **62**, 403–409.
Carne, W. M. and Campbell, J. G. C. (1924). *W. Aust. Dept. Agric.* Bull. No. 121.
Carter, G. A. and Wain, R. L. (1964). *Ann. appl. Biol.* **53**, 291–309.
Carver, T. L. W. and Carr, A. J. H. (1974). *Rep. Welsh Pl. Breed. Stn* **1973**, 70.
Carver, T. L. W. and Carr, A. J. H. (1975). *Rep. Welsh Pl. Breed. Stn* **1974**, 90–92.
Cherewick, W. J. (1944). *Can. J. Res.*, C, **22**, 52–86.

Church, B. M. (1971). *In* "Crop Loss Assessment Methods". (L. Chiarappa, Ed.) FAO, Comm. Agric. Bureaux, Slough, England.
Clare, B. G. (1964). *Univ. Queensland Papers, Dept. Bot.* **4**, 145–149.
Clifford, B. C., Jones, I. T. and Hayes, J. D. (1971). *Proc. 6th Br. Insectic. Fungic. Conf.* **1**, 287–294.
Cock, L. J. (1975). *Experiments and Development in the Eastern Region, ADAS, Min. Ag. Fish. Fd* **1974**, 125–128.
Colquhoun, T. T. (1940). *J. Aust. Inst. agric. Sci.* **6**, 54.
Comhaire, F. (1963). *Lejeunia*, N.S. **15**, 89 pp.
Corner, E. J. H. (1935). *New Phytol.* **24**, 180–200.
Crosier, W. and Szkolnik, M. (1956). *Pl. Dis. Reptr* **40**, 337–339.
Darpoux, H. and Faivre-Amiot, A. (1952). *Phytiat.-Phytopharm.* **1**, 21–24.
Davies, R. R. (1961). *Nature, Lond.* **191**, 616–617.
Davis, D., Chaiet, L., Rothrock, J. W., Deak, J., Halmos, S. and Garber, J. D. (1960). *Phytopathology* **50**, 841–843.
De Diego Calonge, F. and Rubio-Huertos, M. (1965). *An. Inst. bot. A. J. Cavanilles* **23**, 212–224.
Delp, C. J. and Klopping, H. L. (1968). *Pl. Dis. Reptr* **52**, 95–99.
Dentler, J. (1958). *Z. Acker-u. PflBau* **105**, 89–107.
Dickinson, C. II. (1973). *Trans. Br. mycol. Soc.* **60**, 423–431.
Dickinson, S. (1949). *Ann. Bot.* **13**, 89–104.
Dilz, K. and Schepers, J. H. (1972). *Stikstof* No. 71, 452–458.
Doling, D. A. (1969). *Pl. Path.* **18**, 56–61.
Doling, D. A., Saunders, P. J. W. and Doodson, J. K. (1971). *Ann. appl. Biol.* **67**, 1–11.
Domsch, K. H. (1954). *Arch. Mikrobiol.* **19**, 287–318.
Doodson, J. K. and Saunders, P. J. W. (1969). *Proc. 5th Br. Insectic. Fungic. Conf.* **1**, 1–7.
Douchet, J. P., Lhoste, J. and Quere, G. (1973). *Compt. rendu. de journée d'étude sur la lutte contre les maladies des céréales Versailles*, 119–133.
Durbin, R. D. (1967). *In* "The Dynamic Role of Molecular Constituents in Plant-Parasite Interaction". (C. J. Mirocha and I. Uritani, Eds). The American Phytopath. Soc., Minnesota.
Dyer, T. A. and Scott, K. J. (1972). *Nature, Lond.* **236**, 237–238.
Eaton, F. M. (1930). *Phytopathology* **20**, 967–972.
Edwards, H. H. (1971). *Pl. Physiol.* **47**, 324–328.
Edwards, H. H. and Allen, P. J. (1970). *Phytopathology* **60**, 1504–1509.
Ehrlich, H. G. and Ehrlich, M. A. (1963). *Phytopathology* **53**, 1378–1380.
Eshed, N. and Wahl, I. (1970). *Phytopathology* **60**, 628–634.
Eshed, N. and Wahl, I. (1975). *Phytopathology* **65**, 57–63.
Evans, E. (1971). *Pestic. Sci.* **2**, 192–196.
Evans, E. J. and Hawkins, J. H. (1971). *Proc. 6th Br. Insectic. Fungic. Conf.* **1**, 33–41.
Eversmeyer, M. G. and Kramer, C. L. (1975). *Phytopathology* **65**, 490–492.
Finney, J. R. and Hall, D. W. (1972). *Pl. Path.* **21**, 73–76.
Finney, M. E., Bainbridge, A. and Thorne, G. N. (1976). *Rep. Rothamsted exp. Stn* **1975**, part 1, 248.
Foëx, E. (1911–1913). *Annls Éc. natn. Agric.* Montpellier **11–12**, 246–265.
Foëx, E. (1923). *Rept. Intern. Conf. Phytopath. and Econ. Entom.*, Holland **1923**, 184–190.
Foëx, E. (1924). *Bull. Soc. mycol. Fr.* **15**, 166–176.
Foster, W. R. and Henry, A. W. (1937). *Can. J. Res., C*, **15**, 547–559.
Frauenstein, K. (1970). *NachrBl. dt. PflSchutzdienst, Berl.* N. F. **24**, 47–51.
Frič, F. (1969). Biológia, Bratislava **24**, 54–69.
Frič, F. (1974). *Phytopath. Z.* **80**, 67–75.
Frič, F. (1975). *Phytopath. Z.* **84**, 88–95.

Frohberger, P. E. (1973). *Mitt. biol. BundAnst.Ld- u.Forstw., Berl.* **151**, 61–74.

Germar, B. (1934). *Z. PflErnähr. Düng. Bodenk.* **35**, 102–115.

Gilmour, J. (1975). *Proc. 8th Br. Insectic. Fungic. Conf.* **1**, 189–195.

Glynne, M. D. (1959). *Pl. Path.* **8**, 15–16.

Gorter, G. J. M. A. (1955). *Fmg in S. Afr.* **30**, 281–282.

Gorlenko, M. V. (1940). *C. R. Acad. Sci. URSS*, N.S. **27**, 866–870.

Graf-Marin, A. (1934). *Cornell Agric. Exp. Stn Mem.* **157**, 48 pp.

Graham-Bryce, I. J. (1973). *Proc. 7th Br. Insectic. Fungic. Conf.* **3**, 921–932.

Graham-Bryce, I. J. and Coutts, J. (1971). *Proc. 6th Br. Insectic. Fungic. Conf.* **2**, 419–430.

Grainger, J. (1947). *Trans. Br. mycol. Soc.* **31**, 54–65.

Gram, E. and Rostrup, S. (1923). *Tidsskr. for Planteavl.* **24**, 236–307.

Grasso, V. (1964). *Ricerca scient.* **5**, 243–248.

Grebenchuk, E. A. (1965). *Vest. khar'kov. Inst., Biol. Ser.* 1. **1965**, 53–56.

Gregory, P. H. (1973). "The Microbiology of the Atmosphere". (2nd ed.). Leonard Hill Books, Aylesbury, England.

Gregory, P. H. and Hirst, J. M. (1957). *J. gen. Microbiol.* **17**, 135–152.

Griffiths, D. J. (1955). *J. agric. Soc. Univ. Coll. Wales* **36**, 25–29, 31–32.

Griffiths, D. J. (1958). *Rep. Welsh Pl. Breed. Stn 1950–1956*, 75–78.

Griffiths, E., Jones, D. G. and Valentine, M. (1975). *Ann. appl. Biol.* **80**, 343–349.

Gunning, B. E. S. and Barkley, W. K. (1963). *Nature, Lond.* **199**, 262–265.

Hall, D. W. (1971). *Proc. 6th Br. Insectic. Fungic. Conf.* **1**, 26–32.

Hamilton, R. I. and Dodds, J. A. (1970). *Virology* **42**, 266–268.

Hammarlund, C. (1925). *Hereditas* **6**, 1–126.

Hammett, K. R. W. and Manners, J. G. (1971). *Trans. Br. mycol. Soc.* **56**, 387–401.

Hammett, K. R. W. and Manners, J. G. (1973). *Trans. Br. mycol. Soc.* **61**, 121–133.

Hammett, K. R. W. and Manners, J. G. (1974). *Trans. Br. mycol. Soc.* **62**, 267–282.

Hanf, M., Kradel, J. and Menck, B. (1970). *Z. Acker-u. PflBau* **131**, 19–27.

Hardison, J. R. (1944). *Phytopathology* **34**, 1–20.

Hashioka, Y. and Morihashi, T. (1952). *Contr. Hokuriku agric. Exp. Stn* 7 pp.

Hayes, J. D. and Jones, I. T. (1966). *Euphytica* **15**, 80–86.

Heagle, A. S. and Strickland, A. (1972). *Phytopathology* **62**, 1144–1148.

Hebert, T. T., Rankin, W. H. and Middleton, G. K. (1948). *Phytopathology* **38**, 569–570.

Hermansen, J. E. (1964). *Acta Agric. scand.* **14**, 33–51.

Hermansen, J. E. (1966). *Årsskr. K.Vet.-Landbohøjsk.* **1966**, 61–67.

Hermansen, J. E. (1968). *Friesia* **8**, 1–206.

Hermansen, J. E. and Wiberg, A. (1972). *Friesia* **10**, 30–34.

Hermansen, J. E., Torp, U. and Prahm, L. (1975). *Årsskr. K. Vet.-Landbohøjsk.* **1975**, 17–30.

Hewitt, E. J. and Jones, E. W. (1951). *Rep. agric. hort. Res. Stn Univ. Bristol* **1950**, 56–63.

Hey, G. L. and Carter, J. E. (1931). *Phytopathology* **21**, 695–699.

Hirata, K. (1971). *In* "Morphological and Biochemical Events in Plant-Parasite Interaction". (S. Akai and S. Ouchi, Eds). Phytopath. Soc. Japan, Tokyo.

Hirata, K. and Togashi, K. (1957). *Ann. phytopath. Soc. Japan* **22**, 230–236.

Hirst, J. M. (1953). *Trans. Br. mycol. Soc.* **36**, 375–393.

Hirst, J. M. and Hurst, G. W. (1967). *Symp. Soc. gen. Microbiol.* No. 17, 307–344.

Hirst, J. M. and Stedman, O. J. (1963). *J. gen. Microbiol.* **33**, 335–344.

Hirst, J. M., Stedman, O. J. and Hurst, G. W. (1967). *J. gen. Microbiol.* **48**, 357–377.

Hislop, E. C. and Stahmann, M. A. (1971). *Physiol. Pl. Path.* **1**, 297–312.

Hiura, U. (1962). *Phytopathology* **52**, 664–666.

Hiura, U. (1964). *Ber. Ohara Inst. landw. Biol.* **12**, 131–132.

Hollomon, D. W. (1975). *Proc. 8th Br. Insectic. Fungic. Conf.* **1**, 51–58.

Honecker, L. (1943). *Z. PflZücht.* **25**, 209–234.

James, W. C. (1971). *Can. Pl. Dis. Surv.* **51**, 39–65.

Janitor, A. (1975). *Česká Mykologie* **29**, 35–45.

Jeffrey, R. A., Rowley, N. K. and Smailes, A. (1975). *Proc. 8th Br. Insectic. Fungic. Conf.* **2**, 429–436.

Jenkins, J. E. E. and Storey, I. F. (1975). *Pl. Path.* **24**, 125–134.

Jenkyn, J. F. (1970). *Rep. Rothamsted exp. Stn* **1969**, part 1, 151.

Jenkyn, J. F. (1972). *Rep. Rothamsted exp. Stn* **1971**, part 1, 137.

Jenkyn, J. F. (1973). *Ann. appl. Biol.* **73**, 15–18.

Jenkyn, J. F. (1974a). *Ann. appl. Biol.* **76**, 257–267.

Jenkyn, J. F. (1974b). *Ann. appl. Biol.* **78**, 281–288.

Jenkyn, J. F. (1976a). *Pl. Path.* **25**, 34–43.

Jenkyn, J. F. (1976b). *Ann. appl. Biol.* **82**, 485–488.

Jenkyn, J. F. (1977). In "Fertilizer Use and Plant Health". Proc. 12th Colloquium Int. Potash Inst., Izmir **1976**, 119–128.

Jenkyn, J. F. and Bainbridge, A. (1974). *Ann. appl. Biol.* **76**, 269–279.

Jenkyn, J. F. and Finney, M. E. (1977). *Rep. Rothamsted exp. Stn* **1976**, part 1, 258.

Jenkyn, J. F. and Moffatt, J. R. (1975). *Pl. Path.* **24**, 16–21.

Johansen, H. B. (1963). *Åsskr. K. Vet.-Landbohøjsk.* **1963**, 54–63.

Johansen, H. B., Hermansen, J. E. and Hansen, H. W. (1973). *Åsskr. K. Vet.-Landbohøjsk.* **1973**, 86–94.

Johnson, D., Weber, D. J. and Hess, W. M. (1976). *Trans. Br. mycol. Soc.* **66**, 35–43.

Johnson, L. B., Brannaman, B. L. and Zscheile, F. P. (1966). *Phytopathology* **56**, 1405–1410.

Johnson, R. and Wolfe, M. S. (1968). *Rep. Pl. Breed. Inst.* **1966–67**, 113–114.

Johnston, C. O. (1934). *Phytopathology* **24**, 1045–1046.

Johnston, H. W. (1972). *Can. Pl. Dis. Surv.* **52**, 82–84.

Johnston, H. W. (1974). *Can. Pl. Dis. Surv.* **54**, 71–73.

Jones, I. T. (1974). *Rep. Welsh Pl. Breed. Stn* **1973**, 63–64.

Jones, I. T. and Hayes, J. D. (1971). *Ann. appl. Biol.* **68**, 31–39.

Jung, K. U. and Bedford, J. L. (1971). *Proc. 6th Br. Insectic. Fungic. Conf.* **1**, 75–81.

Kampe, W. (1974). *NachrBl. dt. PflSchutzdienst, Berl.* **26**, 148–150.

Kaspers, H. and Kolbe, W. (1971) *PflSchutz-Nachr. Bayer* **24**, 327–362.

Kent, N. L. (1941). *Ann. appl. Biol.* **28**, 189–209.

King, J. E. (1972). *Pl. Path.* **21**, 23–35.

King, J. E. (1973). *Proc. 7th Br. Insectic. Fungic. Conf.* **3**, 771–780.

King, J. E. (1977a). *Pl. Path.* **26**, 8–20.

King, J. E. (1977b). *Pl. Path.* **26**, 21–29.

Klose, A., Fischbeck, G. and Diercks, R. (1975). *Z. PflKrankh. PflSchutz.* **82**, 467–475.

Koltin, Y. and Kenneth, R. (1970). *Ann. appl. Biol.* **65**, 263–268.

Kradel, J., Efflland, H. and Pommer, E. H. (1969a). *Gesunde Pfl.* **21**, 121–124.

Kradel, J., Pommer, E. H. and Effand, H. (1969b). *Proc. 5th Br. Insectic. Fungic. Conf.* **1**, 16–19.

Kunoh, H. and Ishizaki, H. (1975). *Physiol Pl. Path.* **5**, 283–287.

Kunoh, H. and Ishizaki, H. (1976). *Physiol. Pl. Path.* **8**, 91–96.

Kunoh, H., Ishizaki, H. and Kondo, F. (1975). *Ann. phytopath. Soc. Japan* **41**, 33–39.

Lang, R. W., Holmes, J. C., Taylor, B. R. and Waterson, H. A. (1975). *Expl Husb.* No. 28, 53–59.

Large, E. C. and Doling, D. A. (1962). *Pl. Path.* **11**, 47–57.

Large, E. C. and Doling, D. A. (1963). *Pl. Path.* **12**, 128–130.

Last, F. T. (1953). *Ann. appl. Biol.* **40**, 312–322.

Last, F. T. (1954). *Ann. appl. Biol.* **41**, 381–392.

Last, F. T. (1955a). *Pl. Path.* **4**, 22–24.

Last, F. T. (1955b). *Trans. Br. mycol. Soc.* **38**, 453–464.

Last, F. T. (1957). *Ann. appl. Biol.* **45**, 1–10.

Last, F. T. (1962a). *Ann. Bot.* **26**, 279–289.
Last, F. T. (1962b). *Pl. Path.* **11**, 133–135.
Last, F. T. (1963). *Ann. Bot.* **27**, 685–690.
Laurence, R. C. N. (1970). *Prog. Rep. Res. Dev.- HGCA* **1969–1970**, 39–41.
Lawes, D. A. and Hayes, J. D. (1965). *Pl. Path.* **14**, 125–128.
Leijerstam, B. (1972). *Meddn St. VäxtskAnst.* **15**, 231–248.
Leinonen, S. and Kurkela, T. (1969). *Ann. Bot. Fenn.* **6**, 340–343.
Lescar, L., Bouchet, F. and Faivre-Dupaigre, R. (1973). *Proc. 7th Br. Insectic. Fungic. Conf.* **1**, 97–104.
Lester, E. (1971). *Proc. 6th Br. Insectic. Fungic. Conf.* **3**, 643–647.
Leukel, R. W. and Tapke, V. F. (1956). *Fmrs' Bull. US Dep. Agric.* 2089, 28 pp.
Little, R. and Doodson, J. K. (1971). *Proc. 6th Br. Insectic. Fungic. Conf.* **1**, 91–97.
Little, R. and Doodson, J. K. (1972). *J. natn. Inst. agric. Bot.* **12**, 447–455.
Lowig, E. (1933). *Ernähr. Pfl.* **29**, 161–167.
Lowther, C. V. (1950). *Phytopathology* **40**, 872.
Lupton, F. G. H. (1956). *Trans. Br. mycol. Soc.* **39**, 51–59.
Lupton, F. G. H. and Sutherland, J. (1973). *Ann. appl. Biol,.* **74**, 35–39.
Macer, R. C. F. (1964). *Rep. Pl. Breed. Inst.* **1962–63**, 31–32.
Mackie, J. R. (1928). *Phytopathology* **18**, 901–910.
Mains, E. B. (1934). *Phytopathology* **24**, 1257–1261.
Mains, E. B. and Dietz, S. M. (1930). *Phytopathology* **20**, 229–239.
Majerník, O. (1965). *Phytopath. Z.* **54**, 202–206.
Majerník, O. (1966). *Phytopath. Z.* **53**, 145–153.
Malca, I., Zscheile, F. P. and Gulli, R. (1964). *Phytopathology* **54**, 1112–1116.
Manners, J. G. and Gandy, D. G. (1954). *Ann. appl. Biol.* **41**, 393–404.
Manners, J. G. and Hossain, S. M. M. (1963). *Trans. Br. mycol. Soc.* **46**, 225–234.
Månsson, T. (1955). *Sver. Utsädesför. Tidskr.* **65**, 220–241.
Marchal, E. (1902). *C. r. hebd. Séanc. Acad. Sci., Paris* **135**, 210–212.
Marshall, J. (1972). *Ag Tec*, Spring **1972**, 36–40.
Martin, T. J., Stuckey, R. E., Safir, G. R. and Ellingboe, A. H. (1975). *Physiol. Pl. Path.* **7**, 71–77.
Masri, S. S. and Ellingboe, A. H. (1963). *Phytopathology* **53**, 882.
Masri, S. S. and Ellingboe, A. H. (1966a). *Phytopathology* **56**, 304–308.
Masri, S. S. and Ellingboe, A. H. (1966b). *Phytopathology* **56**, 389–395.
Mässing, W. (1959). *Anz. Schädlingsk.* **32**, 155–156.
May, F. G. (1958). *Q.J.R. met. Soc.* **84**, 451–458.
McKeen, W. E. (1974). *Can. J. Microbiol.* **20**, 1475–1478.
McKeen, W. E. and Rimmer, S. R. (1973). *Phytopathology* **63**, 1049–1053.
McKeen, W. E., Smith, R. and Bhattacharya, P. K. (1969). *Can. J. Bot.* **47**, 701–706.
Meeus, P. and Haquenne, W. (1973). *Parasitica* **29**, 71–83.
Mehta, K. C. (1930). *Agric. J. India* **25**, 283–285.
Mehtieva, N. A. (1956). *Proc. Acad. Sci. Azerb. S.S.R.* **12**, 217–224.
Meyer, H. (1950). *Phytopath. Z.* **17**, 63–80.
Meyer-Hermann, K. (1935). *Dt. landw. Presse* **62**, 27.
Millerd, A. and Scott, K. J. (1963). *Aust. J. biol. Sci.* **16**, 775–783.
Minarčic, P. and Paulech, C. (1975). *Phytopath. Z.* **83**, 341–347.
Mórász, S. (1970). *Növenytermelés* **19**, 215–222.
Morel, J. L., Gilchrist, A. J. and Sparrow, P. R. (1975). *Proc. 8th Br. Insectic. Fungic. Conf.* **2**, 413–420.
Moseman, J. G. (1954). *Pl. Dis. Reptr* **38**, 163–166.
Moseman, J. G. (1956). *Phytopathology* **46**, 318–322.
Moseman, J. G. (1966). *A. Rev. Phytopath.* **4**, 269–290.

Moseman, J. G. and Greeley, L. W. (1964). *Phytopathology* **54**, 618.
Moseman, J. G. and Greeley, L. W. (1966). *Phytopathology* **56**, 1357–1360.
Moseman, J. G. and Powers, H. R. (1957). *Phytopathology* **47**, 53–56.
Moseman, J. G., Scharen, A. L. and Greeley, L. W. (1965). *Phytopathology* **55**, 92–96.
Mount, M. S. and Ellingboe, A. H. (1968). *Phytopathology* **58**, 1171–1175.
Mount, M. S. and Ellingboe, A. H. (1969). *Phytopathology* **59**, 235.
Mráz, F. (1969). *Ochr. Rost.* **5**, 1–7.
Mühle, E. and Frauenstein, K. (1970). *Theor. appl. Genet.* **40**, 56–58.
Mundy, E. J. and Owers, A. C. (1974). *Expl Husb.* No. 26, 14–21.
Mundy, E. J. and Page, R. A. (1973). *Expl Husb.* No. 24, 94–104.
Mygind, H. (1970). *Tidsskr. PlAvl* **74**, 177–195.
Naito, H. and Hirata, K. (1969). *Niigata agric. Sci.* **21**, 29–36.
Neuhaus, W. and Reich, R. (1975). *NachrBl. dt. PflSchutzdienst*, DDR **29**, 161–164.
Newton, M. and Cherewick, W. J. (1947). *Can. J. Res.*, C, **25**, 73–93.
Nienhaus, F. (1971). *Virology* **46**, 504–505.
Nikulina, N. K. and Chumakov, A. E. (1968). *Zashch. Rast.*, *Mosk.* **13**, 12.
Nissinen, O. (1973). *Maatalous. Aikakausk.* **45**, 461–467.
Obst, A. (1975). *Proc. 8th Br. Insectic. Fungic. Conf.* **3**, 953–966.
Olsen, C. (1939). *C.r. Lab. Carlsberg*, *Sér. chim.* **23**, 37–44.
Olson, L. (1956) *Växtskyddsnotiser* **5–6**, 80–85.
Osborne, D. J. (1967). *Symp. Soc. exp. Biol.* No. 21, 305–321.
Padalino, O., Antonicelli, M. and Grasso, V. (1970). *Phytopath. Mediter.* **9**, 122–135.
Pady, S. M., Kramer, C. L. and Clary, R. (1969). *Phytopathology* **59**, 844–848.
Pape, H. and Rademacher, B. (1934). *Angew. Bot.* **16**, 225–250.
Parmentier, G. (1956). *Parasitica* **12**, 74–86.
Parmentier, G. (1959). *Parasitica* **15**, 71–72.
Parmentier, G. and Rixhon, L. (1973a). *Parasitica* **29**, 107–113.
Parmentier, G. and Rixhon, L. (1973b). *Parasitica* **29**, 129–133.
Patil, D. K., Hegde, R. K. and Govindu, H. C. (1969). *Mysore J. agric. Sci.* **3**, 238–239.
Paulech, C. (1968). *Biológia*, Bratislava **23**, 281–288.
Paulech, C. (1969). *Biológia*, Bratislava **24**, 709–719.
Paulsen, A., Niblett, C. L. and Willis, W. G. (1975). *Pl. Dis. Reptr* **59**, 747–750.
Phillips, W. H. and Frost, A. J. P. (1975). *Proc. 8th Br. Insectic. Fungic. Conf.* **2**, 437–443.
Picco, D., Ottolini, P. and D'Aragona, G. M. (1971). *Notiz. Mal. Piante* **84**, 5–29.
Plumb, R. T. and Turner, R. H. (1972). *Trans. Br. mycol. Soc.* **59**, 149–150.
Plumb, R. T., Manners, J. G. and Myers, A. (1968). *Trans. Br. mycol. Soc.* **51**, 563–573.
Polley, R. W. and King, J. E. (1973). *Pl. Path.* **22**, 11–16.
Polley, R. W. and Smith, L. P. (1973). *Proc. 7th Br. Insectic. Fungic. Conf.* **2**, 373–378.
Potter, L. R. (1974). *Rep. Welsh Pl. Breed Stn* 1973 79–80.
Powers, H. R. and Moseman, J. G. (1956). *Phytopathology* **46**, 23.
Powers, H. R. and Moseman, J. G. (1957). *Phytopathology* **47**, 136–138.
Prabhu, A. S., Rajendran, V. and Prasada, R. (1962). *Indian Phytopath.* **15**, 280–286.
Pratt, R. (1944). *Bull. Torrey bot. Club* **71**, 134–143.
Priehradný, S. (1975). *Phytopath. Z.* **83**, 109–118.
Priehradný, S. and Ivanička, J. (1970). *Biológia*, Bratislava **25**, 471–476.
Rapilly, F. (1970). *Phytiat.-Phytopharm.* **19**, 185–203.
Rea, B. L. and Scott, R. K. (1973). *Proc. 7th Br. Insectic. Fungic. Conf.* **1**, 29–37.
Reed, G. M. (1909). *Bull. Torrey bot. Club* **36**, 353–388.
Rhodes, A., Fantes, K. H., Boothroyd, B., McGonagle, M. P. and Crosse, R. (1961). *Nature, Lond.* **192**, 952–954.
Richardson, M. J., Jacks, M. and Smith, S. (1975). *Pl. Path.* **24**, 21–26.
Ríman, L. (1973). *Zb. Ref. z I. Konf. o Ochrobách Obo-nín.*, Bratislava, 173–176.

Robayo, G., Orjuela, J. and Thurston, H. D. (1967). *Agricultura trop.* **23**, 455–464.
Roberts, F. M. (1936). *Ann. appl. Biol.* **23**, 271–301.
Roberts, J. J. and Caldwell, R. M. (1970). *Phytopathology* **60**, 1310.
Roder, W. (1967). *NachrBl. dt. PflSchutzdienst, Berl.* N. F. **21**, 201–206.
Rogers, D. P. (1959). *Mycologia* **51**, 96–98.
Rosser, W. R. (1969). *Proc. 5th Br. Insectic. Fungic. Conf.* **1**, 20–24.
Russell, G. E. (1974). *Rep. Pl. Breed. Inst.* **1973**, 151.
Russell, G. E. (1975a). *Phytopath. Z.* **84**, 316–321.
Russell, G. E. (1975b). *Rep. Pl. Breed. Inst.* **1974**, 138–139.
Russell, G. E. (1976). *Rep. Pl. Breed. Inst.* **1975**, 132.
Russell, G. E., Andrews, C. R. and Bishop, C. D. (1975). *Ann. appl. Biol.* **81**, 161–169.
Russell, G. E., Andrews, C. R. and Bishop, C. D. (1976). *Ann. appl. Biol.* **82**, 467–476.
Russell, R. S. and Barber, D. A. (1960). *A. Rev. Pl. Physiol.* **11**, 127–140.
Ryan, C. C. and Clare, B. G. (1972). *Aust. Pl. Path. Soc. Newsl.* **1**, 20.
Sadler, R. and Scott, K. J. (1974). *Physiol. Pl. Path.* **4**, 235–247.
Sadowski, S. (1966). *Zesz. nauk. wyższ. Szk. roln. Olsztyn.* **22**, 151–156.
Salmon, E. S. (1903). *Beih. bot. Zbl.* **14**, 261–317.
Salmon, E. S. (1904a). *New Phytol.* **3**, 55–60.
Salmon, E. S. (1904b). *Annls mycol.* **2**, 70–99.
Salmon, E. S. (1905). *New Phytol.* **4**, 217–222.
Salmon, S. C. and Throckmorton, R. I. (1929). *Bull. Kans. agric. Exp. Stn* No. 248, 80–82.
Sarasola, J. A., Favret, E. A. and Vallega, J. (1946). *Rev. Argent. Agron.* **13**, 256–276.
Savel'yev, V. F. and Polyakova, E. Y. (1972). *Zroshuvane Zemlerobstvo* No. 13, 41–43.
Schaffnit, E. (1922). *Landw. Jbr* **57**, 259–283.
Schaffnit, E. and Volk, A. (1927). *Forschn Geb. PflKrankh., Berl.* **3**, 1–45.
Schaller, C. W. (1951). *Agron. J.* **43**, 183–188.
Scharen, A. L. (1962). *Phytopathology* **52**, 173–174.
Schicke, P. and Veen, K. H. (1969). *Proc. 5th Br. Insectic. Fungic. Conf.* **2**, 569–575.
Schicke, P., Adlung, K. G. and Drandarevski, C. A. (1971). *Proc. 6th Br. Insectic. Fungic. Conf.* **1**, 82–90.
Schnathorst, W. C. (1965). *A. Rev. Phytopath.* **3**, 343–366.
Scott, K. J. (1972). *Biol. Rev.* **47**, 537–572.
Scott, K. J. and Smillie, R. M. (1966). *Pl. Physiol.* **41**, 289–297.
Sempio, C. (1932). *Annali Tec. agr.* **5**, 4–60.
Sempio, C. (1936). *Riv. Patol. veg., Padova* **26**, 201–278.
Sempio, C. (1938). *Riv. Patol. veg., Padova* **28**, 377–384.
Sempio, C. (1939). *Riv. Patol. veg., Padova* **29**, 1–69.
Sempio, C. (1943). *Riv. Biol.* **35**, 11 pp.
Sempio, C. and Castori, M. (1949). *Riv. Biol.* **41**, 163–172.
Sempio, C. and Castori, M. (1952). *Riv. Biol.* **42**, 287–293.
Serra, G., Lartaud, G. and Marchand, D. (1973). *Compt. rendu de journée d'étude sur la lutte contre les maladies des céréales Versailles*, 193–200.
Shaner, G. (1972). *Pl. Dis. Reptr* **56**, 358–362.
Shaner, G. (1973a). *Phytopathology* **63**, 867–872.
Shaner, G. (1973b). *Phytopathology* **63**, 1307–1311.
Shaw, M. and Hawkins, A. R. (1958). *Can. J. Bot.* **36**, 1–16.
Shaw, M. and Samborski, D. J. (1956). *Can. J. Bot.* **34**, 389–405.
Shaw, M. and Samborski, D. J. (1957). *Can. J. Bot.* **35**, 389–407.
Shephard, M. C., Bent, K. J., Woolner, M. and Cole, A. M. (1975). *Proc. 8th Br. Insectic. Fungic. Conf.* **1**, 59–66.
Siddiqui, K. A. and Haahr, V. (1971). *Naturwissenschaften* **58**, 415–416.
Simkin, M. B. and Wheeler, B. E. J. (1974). *Ann. appl. Biol.* **78**, 237–250.

Simmonds, J. H. (1961). *Rep. Dep. Agric. Stk Qd* 1960–61, 23–25.
Simonyan, S. A. (1959). *Trudý bot. Inst., Erevan* **12**, 93–148.
Slesinski, R. S. and Ellingboe, A. H. (1971). *Can. J. Bot.* **49**, 303–310.
Slootmaker, L. A. J. and van Essen, A. (1969). *Neth. J. agric. Sci.* **17**, 279–282.
Smedegård-Petersen, V. (1967). *K.Vet.-og Landbohøisk. Aarsskr.* **1967**, 1–28.
Smiljaković, H. (1967). *Zb. Rad. Zavoda Strana Zita. Kragujevac* **2**, 149–157.
Smith, H. C. and Blair, I. D. (1950). *Ann. appl. Biol.* **37**, 570–583.
Smith, H. C. and Smith, M. (1974). *N.Z. J. exp. Agric.* **2**, 441–445.
Smith, H. C. and Wright, G. M. (1974). *N.Z. Wheat Rev.* No. 12, 30–34.
Smith, H. C., Smith, M. and Marshall, M. I. (1973). *N.Z. J. exp. Agric.* **1**, 387–389.
Smith, J. M., Manning, T. H. and Urech, P. A. (1975). *Proc. 8th Br. Insectic. Fungic. Conf.* **2**, 421–428.
Sommereyns, G. (1968). *Parasitica* **24**, 43–44.
Sommereyns, G. and Parmentier, G. (1967). *Parasitica* **23**, 79–89.
Somers, E. and Horsfall, J. G. (1966). *Phytopathology* **56**, 1031–1035.
Somers, E., Pring, R. J. and Byrde, R. J. W. (1967). *J. Sci. Fd Agric.* **18**, 153–155.
Spinks, G. T. (1913). *J. agric. Sci.*, Camb. **5**, 231–247.
Šprochová, H. (1967). *Phytopath. Z.* **60**, 177–180.
Sreeramulu, T. (1964). *Trans. Br. mycol. Soc.* **47**, 31–38.
Stanbridge, B., Gay, J. L. and Wood, R. K. S. (1971). *In* "Ecology of Leaf Surface Micro-organisms". (T. F. Preece and C. H. Dickinson, Eds) 367–379. Academic Press, London.
Stapel, C. (1966). *Tidsskr. Landøkon.* **1966**, 67–84.
Stapel, C. and Hermansen, J. E. (1968). *Tidsskr. Landøkon* **1968**, 218–230.
Staub, T., Dahmen, H. and Schwinn, F. J. (1974). *Phytopathology* **64**, 364–372.
Stephan, S. (1957). *NachrBl. dt. PflSchutzdienst, Berlin* **11**, 169–177.
Stuch (1926). PflBau **3**, 93–95.
Tapke, V. F. (1951). *Phytopathology* **41**, 622–632.
Taylor, J. W., Rodenhiser, H. A. and Bayles, B. B. (1949). *Agron. J.* **41**, 134–135.
Thorne, G. N. (1974). *Rep. Rothamsted exp. Stn* **1973**, part 2, 5–25.
Thrower, L. B. (1964). *Phytopath. Z.* **51**, 425–436.
Trelease, S. F. and Trelease, H. M. (1928). *Bull. Torrey bot. Club* **55**, 41–68.
Trelease, S. F. and Trelease, H. M. (1929). *Bull. Torrey bot. Club* **56**, 65–92.
Ts'ao, S. H. (1960). *Mém. Inst. agron. Léningr.* **80**, 133–139.
Tsuchiya, K. and Hirata, K. (1973). *Ann. phytopath. Soc. Japan* **39**, 396–403.
Turner, D. M. (1956). *Trans. Br. mycol. Soc.* **39**, 495–506.
Tursumbaev, A. (1973). *Mikol. i Fitopat.* **7**, 46–48.
Ubrizsy, G. and Vörös, J. (1967). *Acta phytopath. hung.* **2**, 379–387.
Vallega, J. and Stillger, G. (1947). *Rev. Invest. agric., B. Aires* **1**, 269–278.
van der Plank, J. E. (1963). "Plant Diseases: Epidemics and Control". Academic Press, London, 349 pp.
van Poeteren, N. (1935). *Versl. PlZiekt. Dienst Wageningen* **80**, 108 pp.
Vargas, J. M. (1972). *Phytopathology* **62**, 795.
Vik, K. (1937). *Meld. Norg. LandbrHøisk.* **17**, 435–495.
Vizárová, G. (1973). *Biologia Pl.* **15**, 270–273.
Vizárová, G. (1974). *Phytopath. Z.* **79**, 310–314.
Vizárová, G. (1975). *Biologia Pl.* **17**, 380–382.
Vizárová, G. and Janitor, A. (1968). *Phytopath. Z.* **62**, 311–318.
Vizárová, G. and Minarčic, P. (1974). *Phytopath. Z.* **81**, 49–55.
Vlamis, J. and Yarwood, C. E. (1962). *Pl. Dis. Reptr* **46**, 886–887.
Volk, A. (1931). *Phytopath. Z.* **3**, 1–88.
Wagner, F. (1940) *Phytopath. Z.* **12**, 427–479.

Ward, S. V. and Manners, J. G. (1974). *Trans. Br. mycol. Soc.* **62**, 119–128.
Waterhouse, W. L. (1930). *Proc. Linn. Soc. N.S.W.* **55**, 596–636.
White, N. H. and Baker, E. P. (1954). *Phytopathology* **44**, 657–662.
Wolfe, M. S. (1971). *Proc. 6th Br. Insectic. Fungic. Conf.* **3**, 724–734.
Wolfe, M. S. (1972). *Rev. Pl. Path.* **51**, 507–522.
Wolfe, M. S. (1975). *Proc. 8th Br. Insectic. Fungic. Conf.* **3**, 813–822.
Wolfe, M. S. and Dinoor, A. (1973). *Proc. 7th Br. Insectic. Fungic. Conf.* **1**, 11–19.
Wolfe, M. S., Minchin, P. N. and Wright, S. E. (1976). *Rep. Pl. Breed. Inst.* **1975**, 130–131.
Yakubtsiner, M. M., Markhaseva, V. A., Tron', E. A. and Belyakova, E. M. (1974).
    Byull. vses. Ordena Lenina Inst. Rasteniev. Imeni N.I. Vavilova **38**, 3–11.
Yang, S. L. and Ellingboe, A. H. (1972). *Phytopathology* **62**, 708–714.
Yarham, D. J. and Pye, D. (1969). *Proc. 5th Br. Insectic. Fungic. Conf.* **1**, 25–33.
Yarham, D. J., Bacon, E. T. G. and Hayward, C. F. (1971). *Proc. 6th Br. Insectic. Fungic.
    Conf.* **1**, 15–25.
Yarwood, C. E. (1936). *Phytopathology* **26**, 845–859.
Yarwood, C. E. (1938). *Phytopathology* **28**, 22.
Yarwood, C. E. (1949). *Phytopathology* **39**, 780–788.
Yarwood, C. E. (1950). *Am. J. Bot.* **37**, 636–639.
Yarwood, C. E. (1951). *Proc. 2nd Int. Congr. Crop Prot., July 1949.*
Yarwood, C. E. (1957). *Bot. Rev.* **23**, 235–301.
Yarwood, C. E. (1971). *Pl. Dis. Reptr* **55**, 342–344.
Yarwood, C. E. and Hecht-Poinar, E. (1973). *Phytopathology* **63**, 1111–1115.
Zadoks, J. C., Chang, T. T. and Konzak, C. F. (1974). *Weed Res.* **14**, 415–421.
Zaracovitis, C. (1964). *Annls Inst. phytopath. Benaki* **6**, 73–106.
Zscheile, F. P. (1974). *Phytopath. Z.* **80**, 120–126.
Zscheile, F. P., Moseman, J. G. and Brannaman, B. L. (1969). *Phytopathology* **59**, 492–495.
Zubko, I. Y. (1961). *In* "Rol' mikroelementov v sel'skom khozya stve" (The Role of Micro-
    elements in Agriculture). 221–228. Moscow University, Moscow.

Chapter 12

Powdery Mildews of Beet Crops

C. A. DRANDAREVSKI

Celamerk, Biolog. Research, 6507 Ingelheim, Germany

I. Distribution

Powdery mildew of beet was first reported in Europe in 1903 (Vanha, 1903). In the years that followed, the fungus was observed in other parts of Europe— in 1920 in Switzerland, 1921 in France and 1927–28 in the USSR (Blumer, 1933). Thereafter the increasingly numerous reports of powdery mildew in Europe are closely linked with the steady spread of the host plant and confirm the presence of the fungus in this area (Canova, 1954; Decoux *et al.*, 1939; Golovin, 1956; Handiquet, 1958; Mourashkinski, 1942; Wenzl, 1957). The fungus was first reported on the American continent by Yarwood (1937) in California. Later the parasite was also observed in Washington and Oregon (Carsner, 1947). In the countries of the Middle East, powdery mildew of sugar beet was not reported until 1958, in Iran, by Viennot-Bourgin and 1963 in Lebanon by Weltzien. At present the fungus has spread with varying severity into all sugar-beet growing areas of Europe. In these regions, the appearance and course of the outbreak depends to a great extent on environmental factors during the growing period. Generally the spread has been more marked towards the South, which is understandable since temperature is a major factor in the development of the fungus.

The latest inventory of the distribution of powdery mildew in the North American beet-growing areas shows that the parasite is to be found everywhere and in recent years outbreaks have been increasingly severe (Ruppel et al., 1975).

The appearance of the fungus in India was reported by Pantanagar (Mukhopadhyay, 1974).

In the beet-growing areas of western and southwestern Asia the parasite finds conditions very favourable to its development and so it is of great economic significance (Ahmadi-Nejad, 1973; Bachthaler, 1958; Christias, 1964; Minnassian, 1967; Weltzien 1963). By means of extensive literature searches, Drandarevski (1969c) established that the disease was on the increase. This prediction can be substantiated today, when the striking spread of the fungus has brought its economic significance, as well as the problem of its control, to the foreground.

I have not found reports of the fungus in South America, Australia or eastern Asia.

II. Morphology and Taxonomy

The oidial form of the organism is the most widespread. The ectoparasitic, hyaline mycelium spreads over the leaves and forms simple conidiophores. The individual conidia are usually almost cylindrical, but elongated, ellipsoid forms are known. In sheltered situations the linked conidia form long chains which can contain as many as eight conidia. Blumer's (1952) view that the mass of conidia is subject to important modifications as a result of environmental factors was confirmed experimentally by Drandarevski (1969a) in the case of *Erysiphe betae*. The size of the conidia depends on the temperature and relative humidity of the air. Smaller conidia are produced at low temperatures (40·1 × 14·2 μm) in a dry atmosphere (45·1 × 15·3 μm) and larger conidia at high temperatures (47·9 × 17·4 μm) and in moist air (49·4 × 17·7 μm). A high percentage of *E. betae* conidia germinate to produce a germ tube and an appressorium.

The perithecia embedded in the ectoparasitic mycelium arise singly, either evenly distributed or in small groups. In seed beet they are found on leaves, stems and seed heads. Most perithecial colonies have a diameter of 6–9 mm

(Mamluk and Weltzien, 1973b). After appearing as small white balls, the perithecia gradually take on a darker colouring as they ripen: from light yellow shortly after formation through light brown to a deep black when ripe. The increase in size takes place at the brown-black stage (Mamluk and Weltzien, 1973c). They are round or slightly depressed. The numerous appendages which resemble mycelium are inserted basally, brown coloured, at least at the base, and display more or less irregular coral-like branching. The size of the perithecia depends, amongst other things, on their density on the leaf surface so that in greater densities smaller perithecia are formed (Junell, 1967a). After comparing the results of various works on taxonomy with his own experimental results Drandarevski (1969a) considered the following scale to be characteristic of *E. betae*.

Diameter of the perithecia	95–120 μm
Length of appendages	100–250 μm
Asci per perithecium	4–6
Length of asci	50–70 μm
Width of asci	30–40 μm
Ascospores per ascus	2–4
Length of ascospores	20–30 μm
Width of ascospores	14–16 μm
Length of conidia	30–50 μm
Width of conidia	13–20 μm

From the many reports in the literature on taxonomy of powdery mildews of beet it is obvious that for a long time there was no agreement as to the position of the fungus within the family of the Erysiphaceae.

On the basis of the oidium form, the parasite was reported from Portugal, Switzerland, France, Belgium, Italy, USA (California and Washington), Iran and Germany under the names *Erysiphe communis* Lev., *Oidium erysiphoides* Fr., *Microsphaera betae* Vanha, Oidium of sugar beet or *Erysiphe polygoni* D.C. Even the investigations in which perithecia were produced failed to give unanimous results. So the fungus was described as *Microsphaera betae* (Vanha, 1903; Neuwirth, 1930), *Erysiphe polygoni* (Newodowski, 1918; Savulescu and Sandu-Ville, 1929; Mourashkinski, 1942; Blumer, 1967), *Erysiphe communis* (Jaczewski, 1927; Muravjev, 1927; Wenzl, 1957; Ruggeri, 1966) and *Erysiphe betae* (Weltzien, 1963).

For about ten years now the fungus has been referred to in specialist literature under the name of *Erysiphe betae* (Vanha) Weltzien, after thorough investigations in Lebanon (Weltzien, 1963), Denmark (Jensen, 1966, 1967), Rumania (Sandu-Ville, 1967), Sweden (Junell, 1967a,b) and Germany (Drandarevski, 1969a) had led to comparable results. A further criterion supporting this nomenclature is the specialization of the fungus on the species *Beta* (Drandarevski, 1969a).

III. Biology

The biology of *E. betae* has been investigated in detail, both in the asexual form and also in the sexual stage of development of the pathogen. The present chapter deals with the climatic requirements, during the separate important development stages of the fungus. The conidial form of *Erysiphe betae* is given special attention as being the most important in practice. The known facts about the separate development stages of the pathogen will also be discussed.

A. Conidium production

Research in protected environments (Drandarevski, 1969b) has shown that conidium production is strongly influenced by atmospheric humidity. Atmospheres that are water saturated (*c.* 100% relative air humidity) and those with a relatively high water content (60–70% relative air humidity) have an inhibiting effect both on the number of conidia formed and on their viability. The highest conidium production and the most viable conidia were seen in a relative air humidity of 30–40%. Moreover increased conidium formation and high conidium viability can be confirmed in the presence of increasing daily variations of temperature up to a variation of 15°C. These results define the fungus as an organism which seems well adapted to the climatic conditions of semi-arid, arid and continental-type climates, in other words one which finds the optimum environmental conditions for its development and spread in warm, dry climates.

Observations of very many conidiophores have shown that conidium formation in *Erysiphe betae* follows a rhythm. Using illustrations Drandarevski (1969b) showed how the upper cell, arising from the division of the mother spore cell of any conidiophore, gradually increases in volume in the course of the day so that on the following morning a conidium is formed from it. Every conidiophore forms a conidium in the space of 24 hours which is fully differentiated and ripe by midday.

B. Biology of germination of the conidia

The maximum germination of *E. betae* conidia takes place in relatively high temperatures and reaches the highest percentage germination at temperatures of around 30°C (Christias, 1964; Minassian, 1967; Drandarevski, 1969b). From the germination data on percentage germination, formation of appressoria and germination with two germ tubes and also germ-tube length it is clear that the optimum temperature for development *in vitro* and *in vivo* is in a temperature range from 20–30°C (See Table 1). The cardinal points for germination lie more or less around 5°, 30° and 35°C (Drandarevski, 1969b).

TABLE 1

Percentage germination (*in vitro* and *in vivo*) of *E. betae* conidia in relation to temperature (from Drandarevski, 1969b)

°C	% Germination	% Appressorium formation	% Appressoria and germ tube	% 2 germ tubes	length of germ tubes in μm
			in vitro		
0	0·0	0·0	0·0	0·0	0·0
5	0·5	0·0	0·0	0·0	3·0
10	7·5	76·0	0·2	0·2	8·6
15	19·9	69·0	0·5	0·3	62·1
20	37·7	72·0	0·4	0·6	86·0
25	47·8	57·0	0·3	0·5	85·4
30	84·6	36·0	0·3	0·1	69·4
35	0·7	0·0	0·0	0·0	3·0
40	0·0	0·0	0·0	0·0	0·0
			in vivo		
0	0·0	0·0	0·0	0·0	0·0
5	2·0	0·9	0·0	0·0	4·6
10	21·1	92·0	0·2	0·1	10·0
15	43·6	98·7	18·8	10·5	56·0
20	64·2	97·0	30·6	22·6	121·8
25	78·6	90·9	16·2	3·6	105·4
30	86·7	85·1	10·4	0·8	72·5
35	2·6	0·0	0·0	0·0	3·0
40	0·0	0·0	0·0	0·0	0·0

The conidia have no special requirements as regards air humidity as distinct from temperature (Zaracovitis, 1964a; Minassian, 1967; Saad and Minassian, 1971). The highest percentage germination is indeed reached in 100% relative air humidity but even at 40% air humidity, more than 70% germination of the conidia is achieved (Drandarevski, 1969b). This characteristic can be explained by the high water content of the conidia, about 63% (Drandarevski, 1969b) and confirms Yarwood's assumption (1950) that the high water content of the conidia allows the powdery mildew fungus to germinate even in low air humidity. Zaracovitis (1964b), as well as Jhooty and

McKeen (1965), classify *E. betae* in the group of true powdery mildew fungi which achieve high percentage germination even when the moisture content of the air is extremely low.

It is known that the conidia of Erysiphaceae in general begin to germinate very quickly (Hammarlund, 1975; Weltzien-Stenzel, 1959; Morrison, 1964; Zaracovitis, 1964b; Masri and Ellingboe, 1966; McCoy and Ellingboe 1966), similarly a high germination speed was ascertained for *E. betae*. Dranda-revski's (1969b) investigations into this speed of germination of the conidia showed that it is heavily dependent on the environmental factors of light, temperature and relative air humidity. As can be seen in Table 2 the interaction of the factors, light, optimum temperature (30°C) and saturated atmosphere (100% relative air humidity) bring about an increase in germination

TABLE 2

Speed of conidial germination of *E. betae* in relation to light, temperature and relative atmospheric humidity (from Drandarevski, 1969b) (% germination)

Hours	Temperature					
	10°C % relative humidity		20°C % relative humidity		30°C % relative humidity	
	0	100	0	100	0	100
with light						
2	—	—	15	11	10	16
4	4	1	38	50	29	50
6	13	9	52	61	50	76
8	20	12	56	69	59	82
10	26	16	59	75	62	90
20	37	31	61	76	67	92
without light						
2	—	—	3	13	2	6
4	—	—	21	26	5	18
6	—	2	28	37	14	38
8	1	3	32	43	24	51
10	1	4	34	49	27	66
20	2	6	35	50	30	92

rate and the attainment of high percentage germination (92%). On the other hand in darkness with a low temperature (10°C) and dry atmosphere (0% relative air humidity) germination is not more than 2%. Similar results were obtained from the study of appressorium formation by the conidia during germination. These results confirm and complete Zaracovitis research (1964a,b; Weltzien, 1965; Minassian, 1967).

A characteristic feature of the germination of *E. betae* conidia is the daily rhythm. As Drandarevski (1969b) was able to show, the germination capacity of the conidia increases very sharply in the period from 08.00–12.00 hours: from 3·9–48·8%. After 12 noon the percentage germination declines and by 20.00 hours only 14·9% of conidia germinate. Moreover the indications of appressorium formation follow the same daily rhythm. Investigations into the causes of the daily germination rhythm show that if the temperature remains constant, the length of the day, in other words the light factor, is of decisive importance. Conidial germination follows the day and night rhythm even when its timing is changed so that the day begins at 20.00 hours and ends at 08.00 (Drandarevski, 1969b). In the open air, where increasing light intensity is accompanied by increasing temperature it may be assumed that the light and temperature factors jointly affect the daily rhythm of conidia germination.

C. Perithecium formation

E. betae perithecia are formed regularly and very liberally in the open (Mamluk and Weltzien, 1973a) but their role in the life cycle of the fungus is not clearly understood. Some authors have suggested that the fruit bodies of *E. betae* enable the fungus to overwinter (Chebolda and Getts, 1967; Weltzien, 1968; Drandarevski, 1969c). However, fruit body formation is significant for another reason: new physiological strains may be produced from the ascospores (Blumer, 1967). These can lead to vigorous spread of the fungus and in some circumstances to less effective control of the disease.

Formation of the fruit bodies of *Erysiphe betae* was originally and exhaustively studied by Mamluk and Weltzien (1973a,b). They paid special attention to the effects of the environment and the significance of the host plant. It was established that no special significance is attributable to the temperature factor in the formation of perithecia. Both in the field and under glass, perithecium formation takes place in a very broad range of temperatures. Daily variations in temperature or even exposure of the fungus culture to "shock cold" do not affect the formation of fruit bodies. Only the timing of the appearance of perithecia is determined by the temperature factor. The time lag between the end of incubation and the appearance of perithecia is shortened from 51–9 days if the temperature rises from 10–22°C.

No biometrically certain effect could be ascertained for the air humidity factor either, although fruit body formation is greater at 30% and 60% relative humidity than at 90%. The two authors considered that these findings confirm the xerophytic properties of *E. betae*.

The effect of light on perithecium formation is treated differentially by Mamluk and Weltzien (1973a). Light intensity (from 500–20 000 lux) and ultraviolet light (maximum emission 253·7, 350 and 350–750 nm) have no

effect. However the period of illumination, that is, the day length, is significant. Thus a short day (8 or 12 h) favours perithecium formation as compared
with a long daylight period (16 or 20 h). The role of environmental factors in
perithecium formation was conceived by Mamluk and Weltzien (1973a) not
as an immediate effect but only as a quantitative influence.

The influence of the host plant on perithecium formation was examined in
the work of Mamluk and Weltzien (1973a,b). It was established that the age
of the leaves attacked by E. betae was significant for the formation of the
higher fruit form: on young leaves (2–6 weeks old) fewer perithecia are formed
than on old leaves (10–14 weeks old). These findings agree with the outdoor
observations of Minassian (1967) and with the generally more severe attacks
on older leaves by E. betae (Wiesner, 1960; Weltzien and Christias, 1963;
Russell 1965, 1966, 1970; Minassian, 1967; Skoyen et al., 1975).

Different species and varieties of beet as host plants do not affect fruit-
body formation in E. betae, since perithecia were formed on all the plants
tested, albeit in varying quantities. The apparently more vigorous formation
of the higher fruit form on fodder beet, Beta vulgaris L. rapa, could not be
confirmed experimentally. On the other hand during the generative phase of
the host plant E. betae forms fruit bodies more vigorously and so the seed
bearing parts of sugar beet appear to be especially favourable to the formation of perithecia (Mamluk and Weltzien, 1973a).

In accordance with the behaviour of other Erysiphaceae (Laubert, 1908;
Woodward, 1927; Gollmick, 1950; Seeliger, 1939; Weltzien and Weltzien,
1962; Laibach, 1930, mentioned in Mamluk and Weltzien, 1973a) perithecium
formation in E. betae is linked with the severity of the attack on the host:
it increases with severe attacks.

Mamluk and Weltzien (1973a) were unable to establish a clearly preferred
site for the appearance of perithecia on sugar beet because they were found on
the petiole, on both sides of the leaf and on the seed heads. However it is
striking that more fruit bodies are formed on the upper than on the lower
side of the beet leaf. This can be explained by the technique of infection adopted
for the experiment by which spores were dusted over the plants and most of
them fell on the upper leaf surfaces.

Whether endogenous factors can possibly trigger the formation of perithecia
was investigated by Mamluk and Weltzien (1973b). They showed that no significance can be attributed to the age of the mycelium in the formation of fruit
bodies since this may be initiated very early, in some cases simultaneously with
conidium formation or it may be significantly later. Nor can the geographical
origin of the race of the fungus explain the inclination of some races to form
fruit bodies, because the fungi used in the experiments originated in various
parts of the Federal Republic of Germany and the Lebanon, yet all behaved
alike.

The results of further experiments by Mamluk and Weltzien (1973b) indicated that the rhythm of the seasons has no effect on perithecium forma-tion either. On the other hand in the course of culture maintenance a decrease in the ability to fructify was observed as the number of infections of the host plants increased. The fungus loses the power to form perithecia after three to eight transfers within 3–10 months. It was established that this phenomenon is irreversible since it was not possible to trigger renewed fructification by varying the biotic and abiotic factors which promote fruit body formation. It is interesting that once perithecium formation has ceased, even combining two or three fungus races does not bring about renewed fructification.

At the present time the state of our knowledge of the laws or rather the trigger mechanism for formation of perithecia of *E. betae* is incomplete. There are unresolved questions which have already been touched on in the work of Mamluk and Weltzien (1973b) and which concern the whole complex problem of the sexuality not only of *E. betae* but of the Erysiphaceae as a whole.

D. Formation and germination of ascospores

As is generally the case for the Erysiphaceae the significance of the ascospores in the life cycle of *E. betae* is not clear. In the field there has been relatively little research because it is time consuming and beset with methodological difficulties. Moreover the experiments carried out so far have not been very encouraging. Tests with ascospores of *E. betae* have been run by Didina (1968) and Saad and Minassian (1971) (both quoted in Mamluk and Weltzien, 1973c) but they were unable to encourage the spores to germinate. Mamluk (1970) in extensive tests was the first to succeed in triggering germination in a few ascospores and thus proved that ascospore germination is indeed possible.

A detailed and systematic study of *E. betae* ascospores can be found in the work of Mamluk and Weltzien (1973c). They established, by studies at the microscope, that the differentiation of ascospores takes place simultaneously with the ripening of the perithecia. Thus the ascospores are fully differentiated when the perithecium becomes dark coloured. Research into the vitality and lifespan of the ascospores shows this to be dependent on temperature condi-tions during storage. Perithecia stored at room temperature and in a refrig-erator at 10°C contain viable ascospores (identified by their ability to take up vital stains) even after 33–37 weeks. Under open-air conditions living ascos-pores cannot be found after 13 weeks. In this research unfortunately the team did not manage to produce germination of the ascospores or infection from them so it is not possible to draw far-reaching conclusions from it. It is surprising, too, that the viability of ascospores stored in open-air conditions can be demonstrated for a relatively long period and such spores germinate the most readily but the active emission of ascospores only takes place from freshly gathered perithecium material. Supplementary investigations seem to

be desirable under those conditions in which perithecia are produced freely
and regularly, that is, in a place where the fungus enjoys optimum conditions
for its complete development. The work of Mamluk and Weltzien (1973a,b,c)
strengthens the hypothesis that the higher fruit form might enable the
fungus—given the infecting power of the ascospores and the presence of host
plants—to survive unfavourable conditions, for example winter or periods in
which no host plants were available but experimental proof of this hypothesis
is still lacking.

E. The course of disease and its symptoms

The development of the disease after infection by *E. betae* has been studied
by Drandarevski (1969b). Under defined conditions the duration of the
periods of incubation and fructification were determined in a temperature
range of 5–35°C. It emerged that from 5–15°C the shortening of the incubation period to 13 days followed an almost linear course. This very pronounced
dependence on temperature was less strongly in evidence from 15–25°C and
reduced the incubation period to only five days. The optimum temperature is
clearly 25°C corresponding to an incubation period of about five days. Further
temperature increase, especially over 30°C causes a sharp lengthening of the
duration of incubation. The temperature dependence of the period of fructification followed a similar pattern, except that no fructification was observed
at 5°C and 35°C.

The optimum temperature ascertained for the development of the disease
was lower by 5°C than the optimum for conidium germination but is still
relatively high and in the author's opinion confirms the high temperature
requirements of *E. betae*.

Infection tests in the open air in the Bonn area of Germany during the
months from April to July 1967 confirm the high temperature requirements of
the fungus. The relationship between the temperature curve (average monthly
values) and the percentage of successful infections is very close and in graph
form the two curves run parallel. The rise in temperature from 3–18°C corresponds to an increase in infection from 40–100% respectively (Drandarevski,
1969b).

According to present knowledge the spread and development of the disease
in the open takes the following course: The conidia which are fully ripe at
midday, are carried by air turbulence around and away from the plant stand.
At this time day temperature and light intensity are at their highest. These
optimum conditions and the diurnal germination rhythm ensure rapid
conidial germination and appressorium formation so that by late afternoon
the infection process is completed. The speed of development of the disease
then depends on temperature conditions.

The disease like all powdery mildew diseases in plants, is observable as the fungus itself. Because of the epiphytic life style of *E. betae*, the fungus covers the upper leaf surface with a dense layer of mycelium, richly endowed with conidiophores. The disease begins with single, generally roundish, white mealy specks of mildew, most often on the upper surface of the leaves. The spots, which can be easily washed off, enlarge, often amalgamate and according to the severity of the attack, cover varying areas of the leaf surface. Attacks on the lower leaf surface are not so often observed. As a rule it is the outer, that is older, and middle leaves of beet plants that are attacked by the fungus. Only rarely and in very severe cases are the younger leaves covered with a layer of fungus. As a rule the perithecia are formed, independently of the time of appearance of the first symptoms, when clearly defined pustules are present: they can be seen with the naked eye as little balls, distinguishable by their darker colour, arranged irregularly on the mycelium.

F. Range of host plants and varietal susceptibility

As an obligate parasite *E. betae* has a restricted range of host plants. Results of trials by Weltzien (1965), Minassian (1967), Drandarevski (1969a) and Saad and Minassian (1971) have agreed that from samples of the species, *Astragalus, Aethionema, Atriplex, Beta, Brassica, Chenopodium, Colutea, Corispermum, Cucumis, Cucurbita, Emex, Euphorbia, Hablitzia, Kochia, Medicago, Phaseolus, Pisum, Polycnemum, Polygonum, Rheum, Ricinus, Rumex, Salicornia, Salsola, Spinacea, Suaeda*, only varieties of *Beta* (*B. cicla, B. diffusa, B. maritima, B. patellaris, B. patula, B. rapa, B. trigyna, B. vulgaris*) are attacked by *E. betae*. These results confirm the investigations of Canova (1954), Coons (1954) and Christias (1964).

The clearly defined monophagous nature of *E. betae* is not surprising, since it is generally characteristic not only of obligate parasites but of Erysiphaceae. At the same time it provides additional confirmation for the identification of the fungus *E. betae*.

In phytopathology and plant protection literature there is little hard information on the resistance of varieties of beet to *E. betae*. This may be because the powdery mildew problem has only become evident with the great increase of sugar beet acreage in the last two decades and because reports have been received of heavy yield losses due to the disease—even heavier in most recent times (Saad and Minassian, 1971; Dzhanuzakov and Lipgardt, 1974; Skoyen *et al.*, 1975).

Weltzien (1965) reports that in the open (Beirut, Lebanon) and under glass all 20 sugar beet varieties tested were highly susceptible to the fungus. Skoyen and his collaborators (1975) also report the sensitivity of cv US 10 which is widely grown in the USA, and severe attacks of mildew on varieties and breeding strains of very varied origins. The same is true of sugar beet cv US 9 (Kontaxis *et al.*, 1974).

Christmann (1976) reports that some of the progeny from cv collections in France and elsewhere show increased susceptibility to mildew. These include the diploid monogerm strains discovered by Savitski in the USA. Families descended from these have, since 1950, been the basis for selection of the French commercial monogerm varieties which were severely attacked by mildew in 1974 and 1975.

On the other hand it has been noticed that, in a stand of plants, heavily infected beet may be found next to others with only slight or no infection (Kontaxis *et al.*, 1974; Skoyen *et al.*, 1975). Russell reports some resistant varieties and crosses (1965, 1966). Skoyen *et al.* (1975) intimate that in the conditions of Salinas Valley, California, in 1974 most European varieties were less severely attacked by *E. betae* than the native varieties and crosses. Moreover preliminary successes in breeding resistant strains are reported by Zhukova (1962, according to Weltzien 1965). These indications make it clear that in the stock of sugar beet there exists genetic material which could be drawn on to breed resistant varieties.

At present it can be stated that all the most economically important varieties of sugar beet are sensitive to *E. betae*.

G. Interaction with other pathogens

In addition to powdery mildew other pathogens attack beet leaves giving rise in practice to mixed infections. In early work (Wiesner, 1960) it was observed that the development of powdery mildew was encountered if the beet was infected with virus yellows disease. This took the form of vigorous development of the fungus and a higher proportion of plants affected in the virus-infected beet.

Russell (1965, 1966) gave details of the effect on the development and severity of the disease produced by previous virus yellows and disease caused by *Peronospora farinosa*. More recently Fradkina and Helman (1975) also reported that in normal growing conditions *E. betae* attacks beet already affected by virus disease earlier and more vigorously. The injury to beet from the simultaneous attacks is greater than the sum of injury from each disease singly.

All these observations suggest that *E. betae* finds more favourable conditions for development on plants which are already infected with other diseases. Further experimentation is needed to gain more precise insight into how and in what this interaction lies.

IV. First Appearance and Epidemology

Particularly vigorous first incidence of the disease is reported from districts in which dry periods dominate during the growing period (Bachthaler, 1958; Viennot-Bourgin, 1958; Poschar and Polevoj, 1959; Weltzien, 1963, 1965) but in the literature there are many reports of heavy incidence of *E. betae* in various sugar beet growing districts with a damp climate.

In Germany *E. betae* was first observed in 1950 in the south of the country and it reappeared there in 1964 (according to Warmbrunn reported by Weltzien 1965) and in both cases it was on dry sites where in those years the growing period had been described as dry. Heavy outbreaks of the disease were also observed in 1959 (Wiesner, 1960, 1967; Drandarevski, 1969c). Both authors described the conditions during the growth period as warm and dry. From Denmark Jensen (1967) reported a severe outbreak and spread of *E. betae* in the extremely hot dry summer of 1959. Russell (1966) described how powdery mildew on sugar beet only appears in warm, dry summers at the end of August/September. Chorin and Palti (1962) gave their experience that in Israel the powdery mildew fungus, including the one on sugar and fodder beet, appears earlier and spreads more widely when temperatures early in the year are in the high range. From California Kontaxis *et al.* (1974) suggested that the severe outbreak of *E. betae* in 1974 could be attributed to the previous mild winter.

The common denominator in all these reports is that the fungus *E. betae* prefers warm, dry districts and similarly can only spread strongly in cooler damp climates when there are spells of warmer dryer weather. These demands on the environmental factors of temperature and relative air humidity for the spread of the disease are characteristic and combine the requirements for the individual stages in the development of the fungus with those for pathogenesis (Drandarevski, 1969b).

At present a generally valid statement cannot be made about the over-wintering of the pathogen in most districts, since the role of the ascospores in nature has not yet been explained (Mamluk and Weltzien, 1973c).

Hypotheses about the overwintering and spread of *E. betae* in the high plain between Lebanon and the Anti Lebanon Mountains are put forward by Weltzien (1965); these may indeed be valid for other comparable climates, characterized by winter rains without either a definite cold season or pronounced rainy periods in summer. In these regions the disease makes an early start since it easily survives the winter period, in Lebanon on seed beet (sown September–October) and in Syria on winter beet (sown December–January). From the Mediterranean region Canova (1954) quoted by Weltzien (1965) reported that *B. maritima* plants develop powdery mildew in the open air and it is conceivable that the fungus overwinters on the wild beet growing there and transfers to the cultivated beet. An epidemic builds up so quickly that a powdery mildew outbreak observed on seed beet at the beginning of April had developed into a 100% attack by the beginning of July. Similarly the fungus appeared on sugar beet (sown in March–April) at the end of June and by the end of July over 80% of the plants were affected (Weltzien 1965). Epidemics are observed in such areas every year and as can be seen from Fig. 1 (after Saad and Minassian, 1971) these could be described as disease explosions.

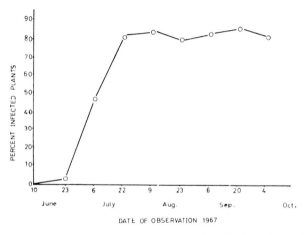

Fig. 1. Duration and intensity of the outbreak of *E. betae* in the Lebanon in 1967 (from Saad and Minassian, 1971).

The fungus does not spread as quickly in continental areas and in those with a damp climate. As a result of hard winters and the absence of host plants over a period of several months the fungus is forced to start again from square one, as it were, each growing season, as far as its development and spread are concerned. Whether the outbreak is sporadic or a widespread epidemic, is decided by the temperature and rainfall in the growing period. As is well known, warmth and drought foster the development of the disease and lead to a severe attack on beet. This is the case in the beet-growing districts of the Southern European and Central Asian parts of the USSR. Every year there is a vigorous spread of powdery mildew here but as a rule only in the second half of the growing period (Fradkina and Helman, 1975). In humid climates environmental conditions during the growing period do not usually favour a vigorous spread of the disease. So in the beet growing districts of Western Europe severe epidemics occur only sporadically in the North (Wiesner, 1960; Jensen, 1966).

A similar situation prevails in the North-west and Central parts of the USA (Ruppel *et al.*, 1975). Here the disease usually appears at the end of the growing period and so does not often cause damage. A study of the reports of the appearance and spread of *E. betae* brings to light a certain pattern. In the southern and generally warmer growing areas the appearance of the disease is reported earlier than in more northerly regions. Jensen (1967) suggested that the fungus is spread from Germany on the wind and this explains its appearance a month later in Denmark. Drandarevski (1969c) confirmed this hypothesis for Europe in travelling south-east from Germany (Austria) and south (Italy). Ruppel *et al.* (1975) described the beginning and course of the epidemic in USA in 1974 and verified the spread from south to north and west to east. The authors also showed this cartographically (see Fig. 2) and illustrated the time factor in the course of the spread of the disease.

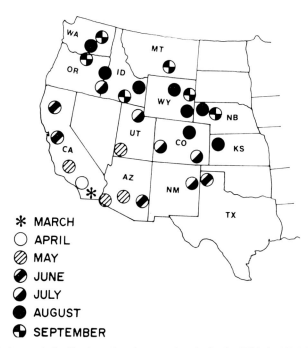

* MARCH
○ APRIL
⊘ MAY
◑ JUNE
◐ JULY
● AUGUST
✪ SEPTEMBER

Fig. 2. Timing and distribution of *E. betae* outbreaks in the USA in 1974 (from Ruppel *et al.*, 1975).

The influence of the altitude of the beet-growing area on the spread of powdery mildew was demonstrated by Drandarevski (1969c) who showed that in higher districts powdery mildew outbreaks are milder. What this vertical spread of powdery mildew is based on is still not clear. It is assumed that the short growing period, low temperatures and intense UV radiation (Blumer, 1967) as well as rainfall may play a decisive part.

Weltzien (1966, 1967) was the first to tackle the epidemiology of *E. betae*. From his own observations and enquiries into the appearance of the fungus in beet-growing areas of central, south-west and south-east Europe and the Near East he was able to get an overall impression of the spread and economic significance of the disease. As a result he was able to indicate the limits of the main areas of distribution and damage from *E. betae*, *Cercospora beticola* and *Peronospora schachtii* (Weltzien 1966, 1967). This line of work was carried further by Drandarevski (1969c) for *E. betae*. By comparing the environmental requirements of the parasites and the climatic characteristics of the corresponding beet-growing districts he was able to draw inferences about the possible spread of the disease. It was possible to define three types of climate which each determined a different course of pathogenesis so giving rise to different distributions of the disease. These types of climate are defined by Drandarevski (1969c):

1. Regular vigorous development of the mildew is furthered by the following type of climate: average monthly temperature rises to over 20°C during the growing period; drought period of 4–5 months. This corresponds roughly to climate types III (Climate Diagrams—World Atlas, Walter and Lieth, 1967) (arid sub-tropical desert, occasional "radiation frosts"), IV (winter rain area, not quite frost-free but no pronounced cold season) and VII (temperate arid zone with hot summers and cold winters)

2. Occasional vigorous development of mildew occurs in the following type of climate: average monthly temperature values 15–20°C during growing period; short drought or dry period; frost and cold spells play a minor part

3. Slight development of the mildew; not economically significant, is found in the following type of climate: average monthly temperature during the growing period 10–20°C; no dry period. This corresponds roughly to the climate types V (warm, temperate always moist zone with pronounced seasonal temperature movements but only occasional frosts), VI (temperate, humid zone with a pronounced but not very long cold season) and VIII (boreal zone with very long cold season, monthly average of the warmest month still over 10°C). Using climate charts (Walter and Lieth, 1967) all beet-growing districts were assigned to the various climate types to provide a survey of the spread of the fungus and to indicate the degree of destructiveness of the disease (Fig. 3).

From this it is possible to designate as districts with the most severe outbreaks of the disease: southern Europe, North Africa, Turkey, the Near

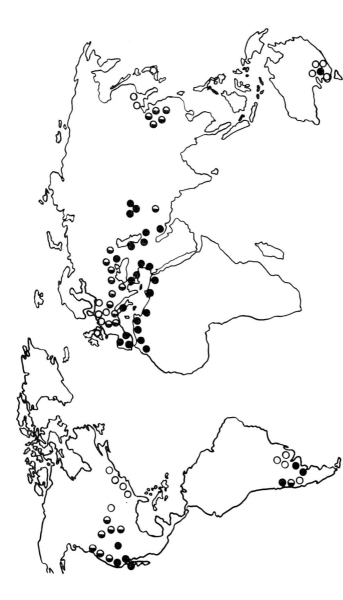

Fig. 3. Distribution and economic significance of *E. betae*. ○ no economic significance (sporadic outbreaks); ◑ occasionally great economic significance (marginal loss districts); ● consistently great economic significance (high loss districts) (from Drandarevski, 1969c).

East, South-west Asia, the southern districts of the Central Asian Soviet republics and South-west USA. It is also possible to indicate those districts which are only slightly at risk from mildew. The advantage of this approach lies in the fact that any district that may be used in the future for growing beet can be assessed for its mildew risk. The accuracy of this procedure has already been confirmed by an outbreak of *E. betae* of the predicted severity in the Western USA (Skojen *et al.*, 1975; Ruppel *et al.*, 1975).

The knowledge of *E. betae* biology and of the spread of the disease allow a detailed assessment of the economic significance of the disease as Weltzien (1972) understands it by distinguishing and delimiting zones of varying disease density.

V. Economic Significance and Control

The question of the economic significance of powdery mildew of beet must be considered in relation to the time of the appearance of the disease and the severity of the attack. In assessing the disease within the larger growing areas, useful conclusions can be drawn if a division into zones of different severity of the disease is undertaken, as proposed by Weltzien (1966, 1967, 1970, 1972). This leads to the identification of districts of highest loss, marginal loss and sporadic outbreaks as elaborated by Drandarevski (1969c).

As can be seen in Fig. 3, the districts in which *E. betae* occurs regularly and in a severe form can also be designated high loss districts (Fig. 3 symbol ●) and it is there that the disease is most significant economically. The beet-growing regions of the eastern Mediterranean, Turkey and the southern steppe regions of the USSR may be taken as typical examples of high loss districts.

The marginal loss districts consist of those regions lying generally around or adjoining high loss districts where the fungus occasionally assumes great economic significance (Fig. 3 symbol ◐). However such districts can include regions which are not adjacent to high loss districts. As an example of such a district the beet-growing districts of central and southern France can be cited.

Finally there are the districts subject to sporadic outbreaks, within which the disease (Fig. 3 symbol ○) has no economic significance and occurs only in mild and isolated outbreaks or more often becomes widely distributed only towards the end of the growing period. This is the situation in the beet-growing areas of England.

Perhaps not unexpectedly most reports of control methods for *E. betae* come from the high and marginal loss districts because only there are control measures urgently needed. Methods for controlling the disease have to be considered separately for the different loss zones because the varying development of the disease requires control measures of varying intensity.

As examples of control of powdery mildew in a high loss district, work from the eastern Mediterranean and southern USSR steppes can be cited. Weltzien (1965) reported practically complete control of the disease in the Lebanon after six applications of fungicide at 14-day intervals between 19 July and 27 September. The effectiveness of this control method is made clear by an increase in beet yield of 23%, in leaf mass of 50% and in sugar of 24%. The active substances sulphur, dinocap and quinomethionate were effective and did not harm the crop but rates are not quoted. Copper oxychloride at 0·3% worked well but crop tolerance was poor and Folpet proved ineffective against *E. betae*. Saad and Minassian (1971), also in the Lebanon, tested 0·5% sulphur in sequential treatments with various time intervals. Four treatments at monthly intervals beginning in late June/early July proved effective and economically worthwhile. In parallel trials dinocap (Karathane 0·05%), quinomethionate (Morestan 0·3%) and sulphur (0·5%) proved reliable and increased root yield by about 32% and leaf mass by 64%. Copper oxychloride and dodine did not increase yields.

Dzhanuzakov and Lipgardt (1974) reported two-year trials in the Dzhambul region (west of Alma Ata and to the south of Lake Balchasch).

The best results were obtained with two treatments with benomyl at 0·2%. Compared with 93% infection of plants in the central plot, only 9% of beet treated with benomyl showed infection and the yield increases amounted to 19·6% roots, 20·7% leaf mass and 27·2% sugar. The benomyl treatment kept the beet free from attack for 30 days and thiophanate methyl kept it free for 25 days. The first application was prophylactic, about a week before the outbreak of the disease and the second as the outbreak started. In these normal field conditions benomyl was 82% effective compared with 72·8% for thiophanate methyl and 36·2% for sulphur.

To sum up, it can be stated that control of powdery mildew of beet in the high loss districts is difficult and can only be achieved by several applications of fungicide. In any case the first treatment must be made shortly before or simultaneously with the appearance of the disease. The number of applications and the timings are determined by the potency of the compound. Four sprayings will be needed with both inorganic and organic but not with systemic fungicides; whereas with systemic fungicides only 2–3 treatments should be sufficient. This statement is not valid for a beet-growing district that has only been designated a high loss district for a relatively short period, as for instance the southern and western regions of the USA (Hills *et al.*, 1975; Ruppel *et al.*, 1975; Skoyen *et al.*, 1975; Kontaxis, 1976), in which control measures against *E. betae* for the years 1974–75 did not lead to results which correspond to those from other comparable regions. Paulus *et al.* (1976) were not able to achieve a yield increase with two sulphur treatments at 44·7 Rg ha^{-1} in their 1974–75 trials in Imperial Valley, California; although the mildew was well controlled. In contrast Kontaxis (1975) suggested

as a result of trials in the same growing areas, that only 16·8 instead of 44·7 kg
ha⁻¹ should be used. In Salinas Valley, California four applications of sulphur
at 10 lb in 200 gal. of water per acre led to an increased sugar yield of 38%
(Skoyen *et al.*, 1975). Positive results were obtained at Davis, California when
five and three sulphur treatments (9·6 kg ha⁻¹) as well as four benomyl
applications (1·1 kg ha⁻¹) increased root yield by about 27% (Hills *et al.*,
1975). Experiences with the systemic fungicide benomyl are also very contradictory and range from very slight to good control (Hills *et al.*, 1975;
Ruppel *et al.*, 1975; Kontaxis, 1976). At present in the western USA there is
no clear cut experience in *E. betae* control from which far reaching conclusions can be drawn.

    E. betae control in a marginal loss area can be illustrated by experiences in
Austria. Three sprayings with copper oxychloride (1·7 kg pure copper ha⁻¹)
succeeded in completely controlling, and two applications in almost completely
controlling, a severe attack according to Graf and Wenzl (1957). This
brought yield increases of 22% for roots, 20% leaf mass and 24% sugar.
Krexner (1975) reported good results with fentin acetate (Brestan 2·2 kg ha⁻¹,
Brestan Conc. 0·8 kg ha⁻¹) and sulphur (Kumulus Super 2·5 kg ha⁻¹) as
well as a combination of benomyl (0·18 kg) and sulphur (2·5 kg ha⁻¹) after two
applications. Lime sulphur spraying mixture and Dinocap-Mittel were not
satisfactory. Krexner (1975) is of the opinion that, where conditions favour
the development of the disease, a second treatment is useful, but it is still
uncertain whether a single application towards the end of August would be
sufficient where conditions are normal. In official recommendations (Anon.,
1975) for beet growing in Austria from one to three sprayings between late
July and early September are recommended for the districts at risk (eastern
Lower Austria and northern Burgenland). Fentinacetate compounds are
recommended for *E. betae* control but it is pointed out that the beet stands are
partially protected against *E. betae* attack by the *Cercospora* treatments (in
early August or late July and mid-August) so that further treatments against
mildew are often unnecessary.

    It is clear that the disease is easier to control in the marginal than in the
high loss areas and that here satisfactory control is possible with just one or
two applications of fungicide. Here too the need to check the spread of the
disease in the early stages is obvious and in this respect the control of *E.
betae* by spraying with fungicides is held to be adequate. In this way marked
yield increases are achieved, varying with the district, and mildew control is
not in doubt. In practice questions are asked more about the timing of
treatment intervals between applications and the effectiveness of newly introduced herbicides in disease control and such questions demand further
investigation. Details about tests of effectiveness and possible eventual use of
further systemic fungicides which show good activity against Erysiphaceae in
general are lacking, though tridemorph, triforine, triadimefon, fenarimol and

pyrazophos are now being evaluated. Control practices for *E. betae* could undergo significant changes as a result of these substances but the known danger of building up resistance to the systemic fungicides must be borne in mind in devising and recommending a control programme.

In connection with powdery mildew of beet there has been little attention given to research and practice in other specialist fields as, for instance, breeding for resistance and in agrotechnique in the wider sense. Indeed research in breeding for resistance is only beginning (Zhukova, 1962; Wenzl, 1976) and at present only isolated pieces of information on the susceptibility of beet strains to disease caused by *E. betae* are available. Only recently have estimates been made of the possibility of influencing the disease by means of mineral and organic fertilization. Movchan (1974), for instance, was able to establish that NPK fertilizer equivalent to 180, 120, 180 kg ha^{-1} respectively and organic fertilizer of 40 tons ha^{-1} reduced the incidence of the disease. Arkusha and collaborators also established a decrease of 11–12% in the disease if 60 kg ha^{-1} NPK and 25 tons ha^{-1} organic manure were applied to the crop. Further increase in fertilizer rates showed a reduction of 12–19% in the disease.

The latest directions taken by research can certainly contribute to the prevention of losses from the disease in the future. These contributions however must not be looked at in isolation but in common with possibilities for chemical control. For the surest way of protecting beet against *E. betae* lies in the integration of various control measures.

VI. Outlook

At the present moment the spread of beet crops is accompanied by severe outbreaks of powdery mildew. The fungus is achieving distressing significance not only in the older areas of cultivation, for example in the USA (Hills *et al.*, 1975; Ruppel *et al.*, 1975; Skoyen *et al.*, 1975; Kontaxis, 1976) but new sites are being reported where yield losses are considerable, for example, in India (Mukhopadhyay, 1974). At the same time it is obvious that most commonly grown varieties are susceptible to the disease since no extensive and consistent breeding work is being done. This situation gives rise to a host of questions demanding the attention of many specialist departments for their elucidation. Only a few of them can be raised here. Without doubt, the contribution of resistance breeding is of great importance in disease control and it is desirable that future work to this end should be intensified. There has already been great activity in breeding for resistance to *E. betae* in France (Wagner, 1976, personal communication) where genetic material from the USA and USSR is being examined.

Further work is needed too in the field of epidemiology inside closed beet growing areas, especially in the marginal districts. Such studies can be used to draw up among other things short and long term forecasts which, according to first reports (Krutova, 1974), should prove useful. Work to elucidate the possibilities for control by applying fungicides is urgently needed. When there is a wide choice of compounds for control of the disease, detailed questions will remain to be answered on matters such as timing of applications and intervals between sprayings under the very varied conditions. The inclusion of a range of new systemic fungicides in the general practice of plant protection is to be expected and will lead to further intensification of powdery mildew control in sugar beet in the future.

References

Ahmadi-Nejad, A., (1973). *Iranian J. Plant Path.* **9**, 28–31.
Anon., (1975). *Pflanzenarzt* **28**, (2), 8–10.
Arkusha, V. E., Rogozhinskii, B. J., Budzherak, A. J. (1975). *Zashch. Rast.* **4**, 52–53.
Bachthaler, G., (1958). Zucker **2**, 2–5.
Blumer, S., (1933). Die Erysiphaceen Mitteleuropas. Zürich.
Blumer, S., (1952). *Ber. schweiz. bot. Ges.* **62**, 384–401.
Blumer, S., (1967). Echte Mehltaupilze (Erysiphaceae). Fischer Verlag Jena.
Canova, A., (1954). *Ann. Sper. Agraria* **8**, 1181–1186.
Carsner, E., 1947. *Phytopathology* **37**, 843.
Chebolda, V. F. and Getts, V. (1967). *Zashch. Rast. Mosk.* **12**, 25–26 (RAM 47, 1973, 1968).
Chorin, M. and Palti, J. (1962). *Israel J. of Agric. Res.* **12**, 153–166.
Christias, C., (1964). Master Thesis Am. Univ. Beirut, Fac. Agric., 70 pp.
Christmann, J. (1976). L'oidium de la Betterave en 1975. *Phytoma-Défense des cultures–Juin*, 5–7.
Coons, G. H., (1954). *Proc. Ann. Soc. Sug. Beet Technol.* **8**, 142–147.
Decoux, L., Vanderwaeren, J. and Roland, G. (1939). Publ. Inst. belge Amel. Betterave **7**, 293–317.
Didina, Caia, (1968). Schriftl. Mitteilung, Ing. Caia Didina. St. Exp. Agric. Podu-Iloae Jasi, Rumänien.
Drandarevski, C. A. (1969a). *Phytopath. Z.* **65**, 54–68.
Drandarevski, C. A., (1969b). *Phytopath. Z.* **65**, 124–154.
Drandarevski, C. A. (1969c). *Phytopath. Z.* **65**, 201–218.
Dzhanuzakov, A. D. and Lipgardt, Y. Y. (1974). *Zashch. Rast.* **10**, 28.
Fradkina, D. L. and Helman, L. V. (1975). *Zashch. Rast.* **10**, 54.
Gollmick, F. (1950). Beobachtungen über den Apfelmehltau *NachrBl. dt. PflSchutzdienst* N. F. **4**, 205–214.
Golovin, P. N. (1956). *Trans. V. L. Komarov Bot. Inst. USSR Acad. Sci.* II, **10**, 179–194.
Graf, A. and Wenzl, H. (1957). *Pflanzenschutzberichte (Wien)* **18**, 81–89.
Hammarlund, C. (1925). *Hereditas* **6**, 1–126.
Haudiquet, R. (1958). *Industria aliment. Paris* **75**, 375–379.
Hills, F. J., Hall, D. H. and Kontaxis, D. G. (1975). *Pl. Dis. Reptr.* **59**, 513–515.
Jaczewski, A. A. (1927). Karmannii opredelitel gribov. 626 S. Leningrad.
Jensen, A. (1966). *Ugeskr. Landm.* **111**, 663–666.
Jensen, A. (1967). *Friesia* **8** (1966), 28–31. Kobenhavn.
Jhooty, J. S. and McKeen, W. E. (1965). *Can. J. Microbiol.* **11**, 531–538.

Junell, L. (1967a). *Svensk bot. Tidskr.* **61**, 210–230.
Junell, L. (1967b). *Symb. bot. upsal.* **19**, 1.
Kontaxis, D. G. (1975). *Calif. Agric.*, August, 15.
Kontaxis, D. G. (1976). *Calif. Agric.* January, 13–14.
Kontaxis, D. G., Meister, H. and Sharma, R. K. (1974). *Pl. Dis. Reptr* **58**, 904–905.
Krexner, R. (1975). *Pflanzenarzt* **28**, 134–137.
Krutova, N. P. (1974). *Byull. vses. nauchno. issled. Inst. Zash. Rast. Nr.* **27**, 57–59. Abstr.
    In Rev. Pl. Path. **55**, (1976). No. 4901.
Laibach, F. (1930). *Jb. wiss. Bot.* **72**, 106–136.
Laubert, R. (1908). *Dt. landw. Presse* **35**, 628–629.
Mamluk, O. F. (1970). *Phytopath. Z.* **67**, 87–88.
Mamluk, O. F. and Weltzien, H. C. (1973a). *Phytopath. Z.* **76**, 221–252.
Mamluk, O. F. (1973b). *Phytopath. Z.* **76**, 285–302.
Mamluk, O. F. (1973c). *Phytopath. Z.* **78**, 42–56.
Marsi, S. S. and Ellingboe, A. H. (1966). *Phytopathology* **56**, 389–395.
McCoy, M. S. and Ellingboe, A. H. (1966). *Phytopathology* **56**, 683–686.
Minassian, V. (1967). Fungicidal control of Erysiphe betae (Van.) Welt. Master Thesis.
    Am. Univ. Beirut, Fac. Agric.
Morrison, R. M. (1964). *Mycologia* **56**, 232–236.
Mourashkinsky, K. E. (1942). Control of diseases of sugar beet in the eastern districts.
    Publs Off. People's Comm. Agric. USSR.
Movchan, J. A. (1974). *Zashch. Rast.* **9**, 59.
Mukhopadhyay, A. N. (1974). *Pesticides* **8**, 40–42.
Muravjev, W. P. (1927). *Morbi Plantarum, Leningrad*, **16**, 175–178.
Neuwirth, F. (1930). *Z. ZuckInd.* (*Prag.*) **55**, 75–79.
Newodowski, G. S. (1918). NachrBl. (Vestn.) Bot. Garten Tiflis **26**, Nach Muravjev (1927).
Paulus, A. O., Kontaxis, D. G. and Nelson, J. A. (1976). *Calif. Agric.* February, 21.
Poschar, Z. A. and Polevoj, B. B. (1959). *Zuckerrübenbau*, Bd. **III**, 450–461. Gosselchozizdat
    USSR, Kiev.
Ruggeri, D. (1966). *Phytopath. Mediter.* 5, 60–61.
Ruppel, E. G., Hills, F. J. and Mumford, D. L. (1975). *Pl. Dis. Reptr* **59**, 283–286.
Russell, G. E. (1965). *Pl. Path.* **15**, 97–100.
Russell, G. E. (1966). *Trans. Br. Mycol. Soc.* **49**, 621–628.
Russell, G. E. (1970). *N.A.A.S. Q. Rev.* **87**, 132–138.
Saad, A. T. and Minassian, V. (1971). *Phytopath. Mediter* **10**, 50–56.
Sandu-Ville, C. (1967). Ciupercile Erysiphaceae din Romania. Ed. Acad. RSR, Bukarest,
    1–358.
Savulescu, O. and Sandu-Ville, C. (1929). Die Erysiphaceen Rumäniens. I. c. 1, 82 S.
Seeliger, R. (1939). *Arb. biol. Reichsanst. Land- u. Forstw.* **22**, 453–478.
Skoyen, J. O., Lewellen, R. T. and McFarlane, J. S. (1975). *Pl. Dis. Reptr* **59**, 506–510.
Vanha, J. (1903). *Z. ZuckInd. Böhm.* **27**, 180.
Viennot-Bourgin, G. (1958). *Annls Épiphyt.* **9**, 97–210.
Walter, H. and Lieth, H. (1967). Klimadiagramm-Weltatlas. VEB Fischer Verlag, Jena.
Weltzien-Stenzel, M. (1959). *Höfchenbr.* **1**, 29–51.
Weltzien, H. C. (1963). *Phytopath. Z.* **47**, 123–128.
Weltzien, H. C. (1965). Der Echte Mehltau der Rüben. Mitt. BBA Land- und Forstwirt-
    schaft Berlin-Dahlem 115.
Weltzien, H. C. (1966). Neue Wege im Pflanzenschutz? Vorträge 20. Hochschultagg.
    Landw. Fak. Univ. Bonn, Landwirtschaft–Angewandte Wissenschaft, Landwirt-
    schaftsverlag, Hiltrup. 103–116.
Weltzien, H. C. (1967). *Z. PflKrankh.* **74**, 175–189.

Weltzien, H. C. (1968). *Zucker* **21**, 241–246.
Weltzien, H. C. (1970). *Arch. PflSchutz dienst* **6**, 217–224.
Weltzien, H. C., (1972). *A. Rev. Phytopath.* **10**, 277–298.
Weltzien, H. C. and Christias, C. (1963). Crop Prod. and Prot. Div., Fac. Agric. Sci., Amer. Univ. Beirut, Mimeo Pamphlet No. 24.
Weltzien, H. C. and Weltzien, M. (1962). *Z. PflKrankh.* **69**, 664–667.
Wenzl, H. (1957). *Sydowia* **1**, 342–352.
Wenzl., H. (1976). *Pflanzenarzt* **29**, 83–87.
Wiesner, K. (1960). *NachrBl. dt. PflSchutzdienst, N.F., Berl.,* **14**, 130–133.
Woodward, R. C. (1927). *Trans. Br. Mycol. Soc.* **12**, 173–204.
Yarwood, C. E. (1937). *I.P.D.R.* **21**, 180–182.
Yarwood, C. E. (1950). *Am. J. Bot.* **37**, 636–639.
Zaracovitis, C. (1964a). *Ann. appl. Biol.* **54**, 361–374.
Zaracovitis, C. (1964b). *Ann. Inst. Phytopath. Benaki* **6**, 73–106.
Zhukova, L. M., 1962. *Sakh. Svekla* **7**, 29–30.

Chapter 13

Powdery Mildews of Bush and Soft Fruits

A. T. K. CORKE and V. W. L. JORDAN

Long Ashton Research Station, Bristol

I. Introduction

Yields of soft fruit are closely related to the health of the plants from which they are harvested, not only in the cropping year but also in the preceding year, consequently, losses due to disease are always difficult to assess. Clearly, given freedom from loss of crop through other agencies, good control of fungal diseases will increase the yield, though the true cropping potential of a completely healthy plant, growing under optimum conditions, cannot easily be determined. Some indication of possible reductions in yield due to mildew infection may be deduced from disease control experiments, which incorporate assessments of infection levels and records of crop weights. Even when the level of infection is very low, considerable loss occurs: for example, blackcurrant bushes apparently only lightly infected with American gooseberry mildew had 68% fewer flowers and bore 83% less crop (by weight) than bushes showing little or no infection in the previous year (Corke, 1965). By the time infection by mildew becomes readily visible to the eye, an extensive mycelium with many haustoria has become established and considerable and irreparable damage has been done to the surface of the plant.

Few powdery mildew species are found on soft fruit plants in the temperate zones: gooseberry and black- and redcurrant bushes are all attacked by American gooseberry mildew (*Sphaerotheca mors-uvae* (Schw.) Berk.) and less commonly by the European gooseberry mildew (*Microsphaera grossulariae* (Wallr.) Lev.). Strawberry and raspberry plants are attacked by *Sphaerotheca macularis* (Wall. ex Fries) Jaczewski but whereas infections on the strawberry are widespread and can cause considerable damage, the disease is relatively uncommon on the raspberry. The synonymy of *S. macularis* with *S. humuli* was discussed by Junell (1965).

As a result of selective breeding, blackcurrant and gooseberry cultivars differ widely in their susceptibility to mildew. Bauer (1955), however, found none of the many cultivars which he tested to be resistant to *S. mors-uvae*, and no specificity was observed when inoculum was transferred from naturally-infected blackcurrant leaves (cv. Baldwin) to leaves of gooseberry (cv. Leveller) or from the gooseberry to leaves of ten blackcurrant cultivars (Jordan, 1968). On the other hand, although 93 species of wild and cultivated plants have been reported as hosts to *S. macularis*, physiological specialization prevented cross-inoculation with inoculum from 25 species (including raspberry), reputed to be hosts of *S. macularis*, and cultivated strawberry plants (Peries, 1961). Peries therefore considered that the mildew on strawberry is a *forma specialis* of *S. macularis*.

II. American Gooseberry Mildew

Since it was first reported by Massee (1900) on the leaves of gooseberry bushes in Ireland, American gooseberry mildew has been mainly confined to gooseberry bushes in this country. Although Salmon (1908) had reported that gooseberry mildew also occurred on cultivated blackcurrant bushes in England, Wilson *et al.* (1964) found that, during the summers of 1962 and 1963, the incidence of mildew on a wide range of unprotected blackcurrant seedlings was still sporadic and generally at a low level, even late in the season. In 1964, however, infection of blackcurrant plantations by *S. mors-uvae* became widespread; heavy infections which occurred early in the season led to restricted growth and reduced crops and provided moribund tissue highly susceptible to subsequent infection by *Botrytis cinerea*.

As with other diseases common to more than one *Ribes* species, the behaviour of mildew varies according to the host species. On blackcurrant bushes, the young leaves and the tip of the extension growth become most heavily infected; the thick mat of mycelium causes the leaves to become distorted and they remain small. Shoot extension is restricted and the formation of lateral buds is drastically reduced when infection is severe (Fulton,

1960), but although infection of the green berries by the fungus does occur, it is not a major cause of loss of crop (Lovett and Barss, 1923).

The number of flowers produced is strongly influenced by the activity of neighbouring leaves and internodes (Rudloff and Lenz, 1960) and it was shown by Corke and Wilson (1964) that not only is there a progressive decrease in the number of flowers formed per truss but also a decrease in the number of flower trusses per shoot, with loss of leaf tissue from blackcurrant shoots. Further, the detrimental effects of loss of leaf become progressively less as defoliation is deferred. As might be expected, the effect on the plant of mildew infection of the leaves is very similar to that which results from defoliation. Deliberate inoculation of uninfected shoots, at intervals of several weeks from late June to mid-August, resulted in levels of infection that were significantly different from each other (generally at 0.1% level) at the end of the season, earlier inoculation being associated with more severe damage (Jordan, 1967). Large numbers of naturally-infected shoots placed in one of five categories according to the severity of infection during the summer were removed from the bushes and examined in detail in November by dissecting buds on the dormant shoots (Corke and Jordan, 1965). As mildew infection increased in severity, the numbers of flower primordia and fruit buds on a shoot fell, so that on heavily infected shoots there were only 7% of those on uninfected shoots. Counts of flowers and berries on similar shoots, made in the following year, confirmed that the numbers of flowers and fruit were smallest on shoots which had been most heavily infected in the previous summer: crop loss on shoots in the highest category of infection was accentuated by the large number of shoots which carried no crop at all.

The effect of mildew infection on the growth of blackcurrant shoots was also examined experimentally (Jordan, 1967). Elongation of infected shoots became progressively slower than that of uninoculated shoots and approximately three weeks after inoculation, growth ceased. The shoot apices were by then thickly covered with mycelium, and the shoots were only half the length of those remaining uninfected.

It is clear that the damage done to the leaves and shoots of blackcurrant bushes by mildew infection depends on weather conditions suitable for rapid colonization of young tissues and the time which elapses between the establishment of infection and the end of the growing season. From the evidence available, it seems almost certain that new infections are initiated each year on blackcurrant leaves by the discharge of ascospores from cleistocarps which, on blackcurrants, remain embedded in the hyphal felt which persists throughout the winter on the shoots and fallen leaves.

According to Merriman and Wheeler (1968) the formation of cleistocarps on blackcurrant leaves in England has begun by August and is completed by October, while those which overwinter on blackcurrants in Bulgaria were found by Zakharieva and Stayanov (1972) to develop on severely infected leaves in mid-summer. Merriman and Wheeler found few ascospores in cleistocarps on leaves in the spring and no spores were discharged under natural conditions but Jordan (1967) found that in cleistocarps formed on infected shoots the differentiation of asci was well advanced by the end of February, maturation of ascospores reaching a peak in about the middle of May, at the time of full blossom. The apparently conflicting evidence as to whether cleistocarps on leaves or shoots provide the inoculum for primary infections in England was resolved to some extent by Jackson and Wheeler (1974), who showed that it is cleistocarps formed late in the season which, although few in number, are nevertheless most important in providing spores for spring infections. Not only are more ascospores discharged, relative to the number of cleistocarps, than from those formed earlier in the season but spores are discharged from these cleistocarps at a time when young currant leaves are available for infection. Also important in the provision of inoculum in spring is the survival of cleistocarps during the winter. Low temperatures lead to degeneration of the ascospores (Jordan, 1967) and soil micro-organisms destroy cleistocarps on leaves overwintering on the ground (Jackson and Wheeler, 1974). The suggestion has also been made by Jackson and Wheeler that differentiation within cleistocarps stops after leaf-fall, so that it is perhaps the continuing attachment to the mycelium on living host tissues, and elevation above the soil where attack by micro-organisms is less intense, which contribute to the survival and subsequent maturation of cleistocarps on shoots. Certainly, discharged ascospores carried in air currents were caught in increasing abundance in early May, a few days before the first infections became visible on nearby bushes (Jordan, 1967) and it seems likely that a large proportion of the primary inoculum comes from cleistocarps overwintered on shoots.

Daebeler et al. (1969) have suggested that gooseberry mildew overwinters as mycelium in the buds in East Germany (as does Podosphaera leucotricha on apple) as well as by cleistocarps. No evidence of the presence of mycelium in currant buds was obtained by bud dissection (Merriman and Wheeler, 1968), although hyphae having dimensions similar to those of S. mors-uvae were detected in dormant buds by Jordan (1967). No infection developed in either case on shoots previously heavily infected, surface-sterilized and subsequently kept in isolation. Again, no infection resulted from the transfer of overwintered mycelium to the surface of young currant leaves kept under conditions in which spores readily gave rise to infections (Jordan, 1967).

Thus, primary infections on blackcurrant are almost certainly initiated only by ascospores discharged from ruptured cleistocarps formed on stems and fallen leaves. The cleistocarps require moisture for differentiation and discharge of the ascospores (Merriman and Wheeler, 1968) and an atmospheric relative humidity of at least 97% is needed for germination of ascospores on leaf surfaces (Jordan, 1967). Hence, dry periods following rain or heavy dew, with temperatures between 5° and 27°C (Salmon, 1914a), are probably those in which discharge is most likely to occur. Ascospores, with a mean size of 13·1 × 27·6 μm, are borne in convection currents to the leaf surfaces, where germination occurs (optimum temperature, 18°C) and hyphae spread rapidly over the surface. Spore production begins after about 5 days and the infection generally becomes visible after 10 days, when chains of conidia have been formed. The extent of the damage already done to the leaf at this early stage in infection is revealed by a doubling of the rate of respiration, and a marked reduction in the rate of CO_2 uptake, largely due in all probability, to reflection of light by mycelium and spores; greater changes in the rates of both processes follow a further increase in infection (Jordan, 1967).

The hyaline conidia, which measure 22·4 × 14·9 μm, are abstricted from conidiophores to form chains up to 26 spores in length. Once detached from the conidiophore, and dispersed by wind to new infection sites, the conidia germinate and cause new infections. As with ascospores, a temperature of 18°C is the optimum for infection, which takes about 12 hours. While no more than 10·5% of conidia germinate on glass at relative humidities below 97%, infection occurs readily when conidia are present on leaves maintained in an atmospheric relative humidity of 60%, conditions approaching those normally found in blackcurrant plantations (Jordan, 1967).

As the mycelium becomes old or as autumn approaches, the mass of hyphae becomes brown and cleistocarps develop, embedded in the felt, particularly at the base of the lamina or on the petiole, or on the terminal internodes of the shoot. The cleistocarps (about 92 μm in diameter) are typically colourless when young but become brown to black on maturity and remain attached to the mycelium by several unbranched, tortuous appendages, light brown in colour and of variable length. The cleistocarp contains a single obovoid ascus (57 × 83 μm) in which the eight ellipsoid ascospores are eventually formed.

On gooseberry bushes, the usual host of this mildew, infection of the young leaves and shoots generally occurs earlier and is more severe than on blackcurrant bushes. Gooseberry fruits also become infected to a much greater extent than currants and, if infected early, the fruits may become completely enveloped by the mycelium. Not only do infected fruits remain small but their market value is much reduced by the blemishes on the skin, caused by the infection. The earlier and more intense infection of young gooseberry leaves and shoots is probably the result of perennation as mycelium within the bud. This is widely held to be the only way in which S. mors-uvae overwinters on gooseberry, since bud infection occurs often enough to provide inoculum for the annual spread of the disease (Merriman and Wheeler, 1968) and viable ascospores are rarely discharged from overwintered cleistocarps in gooseberry plantations (Jackson, 1974). Also, mature cleistocarps are readily dislodged from the mycelium (Salmon, 1914b) to fall on to the ground where they come under the destructive influence of the soil microflora.

The effect of soil micro-organisms in reducing the number of overwintered cleistocarps able to discharge mature ascospores in the following spring has been demonstrated by Jackson (1974). Rapid degeneration of asci and ascospores resulting from treatment with urea of leaf tissue bearing cleistocarps was attributed primarily to a direct effect on the asci, with possible secondary effects resulting from changes in the microflora. While not of immediate practical value as a method of reducing infection by S. mors-uvae, ways of manipulating the leaf microflora may become better understood, leading to the incorporation of biological methods in mildew control programmes. Non-chemical control methods previously used have included the removal of all infected material from plants and soil, the removal of infected shoot tips and pruning aimed at ensuring good ventilation and reducing soft growth. Annual pruning alone, without the use of chemicals, was shown to be sufficient to keep infection on gooseberry bushes at an acceptably low level over a period of twelve years (Dillon Weston and Taylor, 1945) and although Salmon (1914b) recommended pruning in August so that infected tissues were re moved before mature cleistocarps had formed, pruning carried out too early can lead to regrowth, with increased susceptibility.

Other cultural practices affecting plant vigour may also influence the susceptibility of the host to mildew infection (Corke, 1969). The use of herbicides, for example, not only reduces competition for nutrients but the herbicides themselves may contribute nutritionally to the soil, increasing growth and susceptibility (Upstone and Davies, 1967). Incorrect fertilizer applications and moist, stagnant air may also encourage mildew infection, although it was Yarwood's opinion (1957) that susceptibility to mildew infection was not positively correlated with increased host vigour, as is commonly believed: too many exceptions had been cited to allow such generalizations to be made. Thus, whereas Vukovits (1961) found increased infection related to high levels of nitrogen, Jordan (1968) found no such effect.

Control of infection by the use of chemicals has been approached in two ways: the eradication of overwintering inoculum by spraying the bushes and the surrounding soil during the dormant season, or the protection of leaves and stems from secondary infections initiated by conidia formed on the primary lesions. The role of chemicals in the control of mildew during the dormant season was discussed by Corke (1965) and Jordan (1969), who found that reductions of more than 78% in the numbers of ascospores discharged from heavily infected shoots (with numerous cleistocarps) resulted from a single treatment with tar/petroleum oil or phenolic materials. Control of mildew during the summer months requires repeated spray applications, to give the degree of disease reduction which will ensure that no reduction in crop occurs. The timing of spray applications is of the utmost importance, the first being necessary not later than "first green fruit", a stage in the development of the blackcurrant bush which coincided in two successive years with the peak ascospore discharge period (Jordan, 1969), and at least one further application after three weeks. The need for an extended spray programme is largely dictated by the prevailing weather conditions.

On gooseberry bushes, chemical control has been centred round the use of sulphur. In spite of the damage done to certain sulphur-sensitive cultivars, and although a number of fungicides are now available which will efficiently control *S. mors-uvae* on both currant and gooseberry bushes, it has been pointed out (Smith, 1965) that sulphur may be superior to organo-chlorine insecticides in protecting currant bushes against the spread of the reversion virus. As well as restricting shoot growth, thereby reducing mildew severity, feeding by the gall mite (*Cecidophyopsis ribis* Nal.) is inhibited by the presence of sulphur on the plant surface, and the spread of the virus is reduced. Speculation on the part played in mildew susceptibility by the restriction on shoot growth, brought about by the use of sulphur, led Smith and Corke (1966) to examine the effects, on diseases and mites, of treating bushes with a growth retardant, such as chlormequat chloride (CCC). The level of mildew infection was measurably less on young, non-fruiting bushes treated with CCC than on untreated bushes: differences were more marked when more sprays or higher concentrations were used. Less mildew developed on fruiting bushes sprayed in spring, when shoot growth is fastest but infection increased rapidly when the rate of growth subsequently returned to normal (Corke, 1969). Although CCC had no direct effect on mildew spores, significantly less infection ($P = 0.001$) occurred on blackcurrant leaves sprayed with CCC both before and after inoculation (Jordan, 1967). Reductions obtained by treatment before inoculation were equivalent to those achieved with mildew fungicides.

In addition to various formulations of sulphur, the list of chemicals currently recommended for control of American gooseberry mildew on currant or gooseberry bushes includes benomyl, carbendazim, dinocap, quinomethionate and thiophanate-methyl, with the addition of drazoxolon and thiabendazole for treatment of currant bushes only (Anon., 1976).

III. European Gooseberry Mildew

This mildew, caused by *Microsphaera grossulariae*, attacks gooseberry bushes less commonly than *S. mors-uvae* and is rarely found on currant bushes in the United Kingdom and Europe. The disease causes little damage to the bush, attacking only the upper surface of the leaves and causing some premature leaf fall. Cleistocarps (65–130 μm in diameter) formed in the ageing mycelium have numerous, much-branched appendages 5–22 μm in length, by which they are held to the dead leaves on the soil during the winter. When mature, cleistocarps contain 4–10 broadly ovate or oblong asci (46–62 × 28–38 μm) each containing 4–6 ascospores (20–28 × 12–16 μm) (Salmon, 1900). Chemical control follows closely that recommended for the control of American gooseberry mildew. Benomyl and triarimol, for example, were found to be effective against *M. grossulariae* in field trials carried out in Poland (Borecki *et al.*, 1971).

IV. Powdery Mildew of Blueberry

The cultivated highbush blueberry of North America (derived from *Vaccinium australe* and *V. corumbosum*) is commonly attacked by *Microsphaera penicillata* var. *vaccinii* (Schw.) Cooke, but the mycelium grows predominantly on the leaves and develops late in the season (Stretch, 1968).

Symptoms of infection vary according to the cultivar but the leaves of susceptible cultivars may become white with mycelium and spores growing on either surface. Chlorotic and necrotic areas appearing on the upper surface of infected leaves are bounded by reddish tissue and marked on the lower surface by water-soaked areas. As the infection ages, the production of cleistocarps begins on the leaf surfaces. Since the presence of infection does not become evident until after harvest and the effect on yield which may result from leaf infection and premature defoliation is not known, little attention has been paid to the control of this disease. The use of sulphur has been reported as being effective and to give a very economical control when added to insecticide applications made from the air (Stretch, 1968). Nutrition appears to be important in predisposing blueberry bushes to infection: Trevitt *et al.* (1969) found that leaves on severely infected lowbush blueberry bushes had significantly less nitrogen and phosphorus and a higher N:P ratio than those on uninfected bushes.

V. Powdery Mildew of Raspberry

For a long time the mildew attacking raspberry and strawberry plants was thought to be the hop mildew *Sphaerotheca humuli* (DC) Burr. However, a wide range of host species is attacked by the morphological species *S. macularis*, many being unaffected by inoculum from the same fungus (morphologically) colonizing the tissues of a different host (Blumer, 1933). Since Peries (1961) found that the strawberry mildew fungus is physiologically restricted to the strawberry to the exclusion of the raspberry, the latter is presumably attacked by another *forma specialis* of *S. macularis*.

Mildew infection is not common in the United Kingdom, but it was reported to have reached epidemic proportions on the cultivar Latham in Minnesota in the late 1920s (Peterson and Johnson, 1928). The fungus was also found on the leaves of black raspberry, blackberry and dewberry plants, but never abundantly. Where infection does occur it is generally confined to the lower surfaces of the leaves, which are prevented from expanding normally, and to the young fruits, which are generally damaged. In heavy attacks, the infection may develop very early in the growth of the young canes, restricting elongation and leading to the formation of shorter canes bearing smaller leaves than is normal: the leaves may also become mottled. Production of conidia is a less conspicuous feature of infection than is the case with many mildews and no cleistocarps have been seen to be formed on infected plants.

In the absence of cleistocarps, the fungus must be assumed to overwinter only in the infected buds of old canes, the development of conidia on the newly expanding leaves in spring serving as the primary source of infection. Control of infection by the application of ten sprays was found by Peterson and Johnson (1928) to be ineffective, the fungus being in a position to produce abundant inoculum at a time when susceptible new growth is plentiful, thus making protection by chemicals extremely difficult. Unless infection throughout the growing season can be kept at a very low level, the only solution to the problem of reducing the inoculum overwintering in buds is to remove the canes. This method, together with the removal or burial of all other infected tissue, was shown to be very effective in controlling this disease in the field (Peterson and Johnson, 1928).

VI. Powdery Mildew of Strawberry

S. macularis (Wall. ex Fries) Jaczewski forma specialis *fragariae* (Peries, 1961) was first recorded in the United Kingdom by Berkeley (1854) and has since been found throughout the world, wherever strawberries are grown. Perithecia are not commonly formed in the open, but Arthur (1888) in the USA was first to report their presence. Gilles (1961) stated that cleistocarps appear to play no part in perennation in France and suggested that mycelium on old leaves could provide inoculum in the spring. Peries (1961), who observed the production of perithecia only on heavily infected leaves on plants growing under sheltered, moist conditions, reported that in the spring overwintered cleistocarps contained only degenerated asci and ascospores. However, under the temperate conditions of south-west England, green leaves were present in the open throughout the winter and conidia were readily formed on infected leaves when a rise in temperature occurred.

Given optimum conditions, conidia on strawberry leaves germinate after

4 hours, and after 12 hours appressoria have usually been formed. Penetration of the host tissue is completed within the first 24 hours; the formation of conidiophores follows after a further three days and the production of conidia begins one day later (Peries, 1962). All parts of the plant above ground are attacked but it is the damage done to the flowers and fruit at all stages of development which causes the heavy losses which result from infection. The mycelium is mainly confined to the protected, lower surfaces of the leaves, forming a dense white mat and causing the leaves to curl at the edges as the attack progresses. Reddish-purple necrotic spots which appear on the lamina continue to enlarge with an increase in infection. Flowers which are attacked become completely enveloped in mycelium and are either deformed or killed. Green fruit becomes hard as a result of infection, and fails to ripen normally, while ripe fruit which is attacked remains soft and pulpy (Peries, 1961).

The economic importance of this disease is difficult to determine because of the rapid distribution and sale of the fruit. Since Freeman and Pepin (1969) showed that mildew infection of the plant after harvest had no effect on fruit yields in the following year, losses must be wholly due to attacks on the flowers and fruit. Field observations and fungicide trials suggested that significant increases in yield obtained in Finland were attributable to the control of mildew achieved with benomyl (Tapio, 1971), and a 60% loss of crop in the United States was reported to be entirely due to mildew infection of the fruit (Horn *et al.*, 1972). Even lightly infected fruit is affected after harvest; the shelf life and storage quality are impaired, losses of 6–11% in weight during storage for four days being recorded (Jordan and Hunter, 1973).

Chemical control of this mildew during the dormant season has not attracted much interest from research workers because of the absence of perithecia. Peries (1961), however, showed that defoliation by the use of a herbicide in early winter (1·1 kg ha^{-1} of Diquat) was sufficiently effective to be worthy of further investigation. Killing the overwintering leaves, on which mycelium is able to survive in the United Kingdom, should provide an effective control measure. As with other mildews, control trials during the spring and summer have centred round formulations of sulphur and its successors: these have been applied as protectants for the flowers and fruit and to reduce infection towards the end of the season. Systemic fungicides have recently been shown to give improved control following applications made at intervals of 10–14 days after the first flowers open.

VII. Discussion

Evidence that a great deal of damage is still often caused to bush and soft fruit by powdery mildews is strangely at variance with the fact that the causal fungi have a characteristically superficial habit which, it might be thought, would make them particularly vulnerable to the action of chemicals throughout the growing season, and the fact that many fungicides are available today, which have been developed specifically for the control of the mildews.

For a long time, emphasis has been placed on the need to spray during the summer months in order to obtain control of the mildews by simultaneous protection and eradication: failure to control mildew must, therefore, be largely due to poor application or to incorrect timing of the application. Although mildew mycelium is notoriously difficult to wet, and considerable problems are involved in achieving an adequate cover of fungicide early enough in the season, which is the period of rapid growth and high susceptibility, it is nevertheless necessary to maintain a high level of control until late in the growing season, when the invasion of the buds and the formation of cleistocarps occur.

The level of overwintering inoculum can be reduced to some extent by the removal, during annual pruning, of wood which has carried the current year's crop on bush and cane fruit plants, or by soil cultivation between rows, which may bury much of the infected plant debris. Effective eradication of inoculum would contribute greatly to lowering the level of infection early in the season and would improve the control achieved by spraying, while reducing the number of applications needed. In Yarwood's opinion (1957), winter sprays have no value but marked reductions in mildew infection during the summer were achieved by the use of winter washes applied before bud burst (Corke and Jordan, 1966). More recently, powdery mildew in dormant apple buds has been eradicated by the use of surface-active agents (Hislop and Clifford, 1976), adding weight to the view that the possibility of breaking the mildew cycle during the dormant season by the use of chemicals warrants serious consideration in the future, as an additional measure to the removal of infected tissues.

References

Anon. (1976). *MAFF, Agric. Chem. Approvals Scheme*, HMSO, 191 pp.
Arthur, J. C. (1888). *Rep. N.Y. St. agric. Exp. Stn* **1887**, 343–371.
Bauer, R. (1955). *Proc. 14th Int. hort. Congr., Scheveningen*, **1**, 685–696.
Berkeley, M. J. (1854). *Gdnrs' Chron.* 236.
Blumer, S. (1933). *Beitr. KryptogFlora. Schweiz* **7**, 483 pp.
Borecki, Z., Millikan, D. F., Puchala, Z. and Bystydzienska, K. (1971). *Pl. Dris. Reptr* 55 932–933.
Corke, A. T. K. (1965). *Proc. 3rd Br. Insectic. Fungic. Conf., Brighton*, 336–339.
Corke, A. T. K. (1969). *J. Sci. Fd Agric.* **20**, 401–402.
Corke, A. T. K. and Jordan, V. W. L. (1965). *A Rep. agric. hort. Res. Stn. Univ. Bristol* **1964**. 142–144.

Corke, A. T. K. and Jordan, V. W. L. (1966). *A Rep. agric. hort. Res. Stn. Univ. Bristol.* **1965,** 184–192.

Corke, A. T. K. and Wilson, D. (1964). *A Rep. agric. hort. Res. Stn. Univ. Bristol* **1963,** 71–73.

Daebeler, F., Giessmann, H-J. and Hingst, M. (1969). *NachrBl. dt. PflSchutzdienst, Berl.,* N.F. **23,** 246–248.

Dillon Weston, W.A.R. and Taylor, E. (1945). *Trans. Br. mycol. Soc.* **27,** 119–120.

Freeman, J. A. and Pepin, H. S. (1969). *Can. Pl. Dis. Surv.* **49,** 139.

Fulton, R. H. (1960). *Ext. Bull. Mich. St. Univ.* **370,** 18 pp.

Gilles, G. (1961). *Proc. 12th Symp. Phytopharma, Ghent,* **1960,**

Hislop, E. C. and Clifford, D. R. (1976). *Ann. appl. Biol.* **82,** 557–568.

Horn, N. L., Burnside, K. R. and Carver, R. B. (1972). *Pl. Dis. Reptr* **56,** 368.

Jackson, G. V. H. (1974). *Trans. Br. mycol. Soc.* **62,** 253–265.

Jackson, G. V. H. and Wheeler, B. E. J. (1974). *Trans. Br. mycol. Soc.* **62,** 73–87.

Jordan, V. W. L. (1967). *M.Sc. Thesis, Univ. London,* 125 pp.

Jordan, V. W. L. (1968). *Ann. appl. Biol.* **61,** 399–406.

Jordan, V. W. L. (1969). *Proc. 5th Br. Insectic. Fungic. Conf., Brighton,* 127–129.

Jordan, V. W. L. and Hunter, T. (1973). *A Rep. agric. hort. Res. Stn. Univ. Bristol* **1972,** 128

Junell, L. (1965). *Trans. Br. mycol. Soc.* **48,** 539–548.

Lovett, A. L. and Barss, H. P. (1923). *Bull. Ore. agric. Coll. Exp. Stn* 42.

Massee, G. (1900). *Gdnrs' Chron.* **28,** 143.

Merriman, P. R. and Wheeler, B. E. J. (1968). *Ann. appl. Biol.* **61,** 387–397.

Peries, O. S. (1961). *Ph.D. Thesis, Univ. Bristol,* 166 pp.

Peries, O. S. (1962). *Ann. appl. Biol.* **50,** 211–224.

Peterson, P. D. and Johnson, H. W. (1928). *Phytopathology* **18,** 787–796.

Rudloff, C. F. and Lenz, F. (1960). *Erwerbobstbau* **2,** 214–217.

Salmon, E. S. (1900). *Mem. Torrey bot. Club* **9,** 292 pp.

Salmon, E. S. (1908). *Gdnrs' Chron.* **44,** 203.

Salmon, E. S. (1914a). *Ann. appl. Biol.* **1,** 177–182.

Salmon, E. S. (1914b). *Gdnrs' Chron.* **55,** 325–326.

Smith, B. D. (1965). *Proc. 3rd Br. Insectic. Fungic. Conf., Brighton,* 286–292.

Smith, B. D. and Corke, A. T. K. (1966). *Nature, Lond.* **212,** 643–644.

Stretch, A. W. (1968). *1st Symp. Wkg Gp. Blueberry Culture in Europe. Int. Soc. hort. Sci.* **1967,** 133–143.

Tapio, E. (1971). *Ann. Agric. Fenn.,* **11,** 79–84.

Trevitt, M. F., Hilborn, M. T. and Durgin, R. E. (1969). *Res. Life Sci.* **17,** 10–14.

Upstone, M. E. and Davies, J. C. (1967). *Pl. Path.* **16,** 68–69.

Vukovits, G. (1961). *Pflanzenarzt* **14,** 105–106.

Wilson, D., Corke, A. T. K. and Jordan, V. W. L. (1964). *A Rep. agric. hort. Res. Stn. Univ. Bristol* **1963,** 74–78.

Yarwood, C. E. (1957). *Bot. Rev.* **23,** 235–301.

Zakharieva, T. and Stayenov, S. (1972). *Ovoshtarstvo.* **51,** 33–37.

Chapter 14

Powdery Mildews of Cucurbits

WAYNE R. SITTERLY

Clemson University Truck Experiment Station, Charleston, South Carolina, USA

I. Introduction

Members of the Cucurbitaceae are important to man as sources of food and fibre. Cucurbits as food crops are less important on a world basis than cereals and legumes, but in widespread regions from the tropics to the milder portions of the temperate zone, they serve as a source of carbohydrates, as dessert and salad ingredients, and as pickles. Some cucurbits are used as baskets, as insulation, as oil filters and as alternatives to pottery utensils. The cultivated cucurbits have needed man to ensure their survival since bona fide specimens of their wild counterparts have apparently never been collected.

There are about 90 genera and 750 species in the family Cucurbitaceae, but only 6 genera and 12 species are cultivated by man. These cultivated cucurbits are as follows:

watermelon *Citrullus lanatus* (Thum.) Mansf.
cucumber *Cucumis sativus* L.
muskmelon *Cucumis melo* L.
gherkin *Cucumis anguria* L.
dish rag gourd *Luffa cylindrica* Roem.
white-flowered gourd *Lageneria siceraria* (Standl.) Mol.
squashes and marrow *Cucurbita pepo* L., *Cucurbita maxima* Duch., and
    Cucurbita moschata Poir.
pumpkin *Cucurbita mixta* Pang.
figleaf gourd *Cucurbita ficifolia* Bouche
chayote *Sechium edule* Sev.

All the species are frost sensitive. Within each species there is a wide assortment of sizes, shapes, colour variants, flesh textures, flavours, etc. (Whitaker and Davis, 1962).

The major cucurbit crops are thought to have originated in several regions of the world. Watermelon was shown by DeCandolle (1882) to be indigenous to the dry open areas on both sides of the equator in tropical Africa, with perhaps a strong secondary centre of diversification in India. It has been cultivated for centuries by people bordering the Mediterranean Sea. De-Candolle (1882) showed cucumber to be native to India, where it has been cultivated for over 3000 years. Cucumbers were spread eastward to China and westward where they were enjoyed by the Romans and Greeks. A chromosome count of 14 and morphological features such as angular stems separate it from other members of the genus *Cucumis* which have 24 chromosomes. The place of origin of cantaloupe has never been fully resolved (Whitaker and Davis, 1962) but appears to be tropical and sub-tropical Africa. From there it spread into Asia and, when placed in this congenial environment, a number of sub-species evolved. The squashes and pumpkins are indigenous to North America, with the source apparently in central or southern Mexico or the northern part of Central America. As northern migration occurred they developed both xerophytic and humid adaptive types, which the native Indians of this area utilized for many of their needs (Whitaker and Davis, 1962).

Cucurbit powdery mildews occur throughout the world, in both glasshouses and the field. Powdery mildews generally are of minor importance on field cucurbits except in limited geographical areas with cool dry seasons. Much world food and cucurbit production does occur in such areas and powdery mildews definitely do exert an influence there. This is particularly true in specific tropical areas where it is sometimes so common that there is a tendency to accept it as a normal part of the plant foliage. Powdery mildews are the principal disease on cucurbits in glasshouse culture. Other than in glasshouse culture, powdery mildew losses are not too spectacular on a world basis, probably because of lack of systemic infection, lack of root infection, relatively slow increase during the growing season, infrequent host death, ease of control and a requirement for dry weather. The most striking aspect of disease loss is a reduction of quality rather than of yield, an aspect particularly devastating to the quality of cantaloupe.

Although negative aspects of plant diseases are most commonly emphasized, cucurbit powdery mildew has positive characteristics which should be mentioned because they allow easy separation of the pathogen and the disease. The disease is common and easily produced and the pathogen is an obligate parasite. These characteristics make this powdery mildew an ideal plant disease for classroom plant pathological usage and for utilization in the development of chemical fungicides to control plant diseases.

II. Taxonomy and Morphology of the Cucurbit Powdery Mildew Pathogens

There have been three genera and six species of powdery mildews recorded on the major species of the Cucurbitaceae. These include *Erysiphe cichoracearum* DC ex Mecat; *Erysiphe communis* (Wallr.) Link; *Erysiphe polygoni* (DC) St.-Am.; *Erysiphe polyphaga* Hammarlund; *Leveillula taurica* (Lev.) Arnaud; and *Sphaerotheca fuliginea* (Schlecht. ex Fr.) Poll. The genus *Leveillula* is usually considered as a synonym of *Erysiphe*. Some reports merely list cucurbit mildews as caused by *Oidium* sp. so that it is impossible in many cases to determine which species are involved. The two most commonly recorded species are *E. cichoracearum* and *S. fuliginea* and the major portion of this chapter will pertain to these two organisms. More than one species may occur in the same locality and on the same plant. (Ballantyne, 1975). The cucurbit powdery mildew pathogens are classified as being in the class *Ascomycetes*, the order *Perisporiales*, the family *Erysiphaceae*, and the subfamily *Erysipheae*, and the species and genus either *Erysiphe cichoracearum* DC or *Sphaerotheca fuliginea* (Schlecht. ex Fr.) Poll. The evolution of cucurbit powdery mildew is not well understood with the major problem being to decide which characters are primitive and which are not. Additionally, much confusion has existed as to which of two species is the primary causal organism of cucurbit powdery mildew.

As this disease is best known to growers when only the conidial stage is present, the problem is how to identify species when the cleistothecial stage is absent, as is usually the case with cucurbit powdery mildews.

Classification to genus has been based on the perfect stage characteristics of mycelium location, types of appendages on the cleistothecia, and number of asci in the cleistothecia. The mycelium grows closely appressed to the host, tending to follow the depressions at the contact of two epidermal cells. Hyphal cells are thin-walled, uninucleate and vacuolate with large nuclei. Along the hyphae are lateral swellings or appressoria, one on each alternate hyphal cell. Appressoria are lobed. Haustoria rise from the centre of attachment and penetrate host cells by a very narrow penetration tube. Most haustoria are uninucleate and globular. (Yarwood, 1957). The cleistothecia of *E. cichoracearum*, which occur very infrequently, are 80–140 μm in size, and contain 10–15 asci. Each ascus is narrowly to broadly ovate, more or less stalked, and 30–35 × 58–90 μm in size. Cleistothecia of both *E. cichoracearum* and *S. fuliginea* are borne on the host surface and are dark closed bodies with flexuous, indeterminate, slightly-coloured appendages.

The conidia of both species are 4–5 × 5–7 μm in size, continuous, elliptic, hyaline, and are borne in chains on short unbranched conidiophores. Conidiophores occur at right angles to the host surface. There is a stipe of one or more cells attached to the vegetative hypha or generative cell. (Walker, 1952).

Ballantyne (1975), in an excellent investigation concerning the identity of the powdery mildew on Cucurbitaceae, searched the available literature and developed a set of consistently reproducible criteria for identifying the cucurbit powdery mildews from the conidial stage. The three criteria are:

1. Type of conidiophore—both *Erysiphe* and *Sphaerotheca* have *Oidium*-type conidiophores with the long conidial chains and external mycelium.
2. Presence or absence of well-developed fibrosin bodies, as recognized by Clare (1958). *Erysiphe* has a granular form, and *Sphaerotheca* has a cylindrical or coneshaped form. (Homma, 1937).
3. Mode of conidial germination, as recognized by Hirata (1955) and Zaracovitis (1965). The germ tubes of *Erysiphe cichoracearum* are single with inconspicuous appressoria, while some of the germ tubes from *S. fuliginea* conidia are forked.

The other fungi causing cucurbit powdery mildews also fit into this scheme as devised by Ballantyne but their characteristics will not be discussed here.

Much confusion exists because the name of the perfect stage of a powdery mildew fungus has been so frequently applied to the imperfect stage without valid confirmation. Before 1958 *E. cichoracearum* was assumed to be the most widespread of the cucurbit powdery mildew fungi, with *S. fuliginea* thought not too important. Indeed, Walker (1952) stated that *E. cichoracearum* was apparently more common in the USA, and *S. fuliginea* more common in China, Japan, and Russia. In the USA this was enhanced by the report of Randall and Menzies (1956), who observed the cleistothecial stage of *E. cichoracearum* on cucumbers. They stated that cleistothecia matured with an average of 11 asci, each with two spores. Most cleistothecial records do indicate that either *E. cichoracearum* (Khan and Khan, 1970; Roder, 1937; Deckenbach and Koreneff, 1927) or *S. fuliginea* (Tarr, 1955; Rayss, 1947; Sohi and Nayar, 1969; Dingley, 1959; Nagy, 1970; Deckenbach, 1924) is the most commonly occurring powdery mildew on cucurbits throughout the world. Ballantyne (1975) showed that a mildew having major features of the imperfect stage of *S. fuliginea* is the predominant mildew in some countries and probably the only species in others. In only two instances has *E. cichoracearum* been positively identified from the imperfect stage outside of the continental United States; one in Hawaii (Raabe, 1966) and one in Hungary (Nagy, 1970).

Using the criteria of Ballantyne several researchers have identified *S. fuliginea* in the USA in Georgia (Sowell and Clark, 1971), Iowa (Clark, 1975), New York (Kable and Ballantyne, 1963), and Wisconsin (Shanmugasundaram, et al., 1971); and in New South Wales, Australia (Ballantyne, 1963). Investigators in other areas of the world should apply the above criteria to the cucurbit powdery mildew fungus existing in their geographical areas, and positively identify the organism for the purpose of eliminating past taxonomic confusion.

Taxonomic identification of cucurbit powdery mildew in the tropics has not been extensive and has certainly been conservative. Most reports of powdery mildews from the tropics, where the perfect stage rarely if ever occurs, indicate only that the fungus belongs to the genus *Oidium*. There is less inclination here to classify the fungus as either *E. cichoracearum* or *S. fuliginea* than there is with temperate zone investigators. There are possibly numerous powdery mildews in our tropics that have not been described and that do not occur in temperate zones. Wellman (1972) states that in tropical America, one or more species of *Oidium* is found on cucurbits. When mildews are subjected to positive identification and the organism determined, it is frequently designated by both the imperfect designation of *Oidium ambrosiae* or *Oidium communis*, and the perfect stage designation of *E. cichoracearum* or *S. fuliginea*. Not much attention has been given to details of many of the tropical hosts parasitized by *E. cichoracearum* or *S. fuliginea*.

III. The Disease

A. Symptomatology

Tiny, white, round, superficial spots appear on stems and leaves of cantaloupe, cucumber and squash, becoming powdery as they enlarge. These white lesions increase in number and coalesce, and eventually may cover the stems and both surfaces of leaves. Infection on young leaves may result in general chlorosis and eventually death of the leaves. Severely affected leaves become brown and shrivelled. Under ideal conditions premature defoliation may occur as the fungus covers the leaf surfaces. Black pinpoint bodies rarely occur but are conspicuous when they do occur. Roots are not attacked, and fungal growth on herbaceous plants usually stops above the ground line. Fruit of cucumber, cantaloupe and squash are free of visible infection, even when the foliage is heavily infested though fruits may ripen prematurely and lack flavour. This reduced quality is the chief damage in cantaloupe. Yield reduction occurs in proportion to time and severity of disease development, with late fruits often failing to mature and being small and misshapen. (Walker, 1952). Symptoms are identical in the tropics.

B. Epidemiology

Cucurbit powdery mildew is generally favoured by dry atmospheric and soil conditions, moderate temperatures, reduced light intensity, fertile soil and succulent plant growth. (Yarwood, 1957).

The relative importance of cucurbit powdery mildew in different regions is correlated with rainfall in those areas. Powdery mildew, for example, may be very severe in the summer in the arid portions of California, in western United States, but not very important in the eastern portion of the United States where abundant rainfall normally occurs. Incidence of powdery mildew thus increases as rainfall decreases. Spores may germinate in the absence of water and in relative humidities below 20%. This fact implies that conidia have a high water content and an extremely efficient water conservation system. Water is needed for germination of spores, but is internal in vacuoles instead of requiring an external source (Yarwood, 1957). This process was confirmed by Duvdevani, et al. (1946) who prevented dew formation on cucurbits by covering certain foliage areas with canvas, a procedure which did not reduce powdery mildew infection.

Tolerance of cucurbit powdery mildew to heat is usually lower than that of the host. Conidia germinate from 22–31°C, with a peak at 28° (Hashioka, 1937). They live for only a few hours at 26·7° or higher, survive much longer at 4·4°, and are again damaged at 1·1° or less. Cucurbit powdery mildew is able to thrive in hot climates because vines shade the ground and mycelium develops on the under side of the leaves (Walker, 1952). Powdery mildew develops better in shade than in full light, thus is more severe in close plant spacings and under a high carbohydrate level with its subsequent luxuriant growth.

Powdery mildew is more severe in glasshouses than in the field because of reduced air circulation, reduced light intensities, higher temperatures of glasshouses and continuous cropping. Foliage infection in glasshouses can be so serious that chlorotic spots are formed which later become necrotic.

Tropical conditions for powdery mildew development are similar to temperate zone conditions. In the tropics, cucurbit powdery mildew is especially severe in crops grown on hillsides, mountain table lands and shallow valleys at fairly high, dry and cool elevations. In both the hottest and wettest portions of the tropics there is the least damage from powdery mildew except for areas where a prolonged dry season occurs. After the rainy season begins in some of the drier areas, it is often only a few days before powdery mildew is practically gone (Wellman, 1972).

C. Etiology

The life cycle is initiated by either conidia or ascospores. If the spores of the pathogen make contact with the host under conditions of reduced light intensity, a temperature of 22–31°C and absence of moisture, germination commences within two hours. The first germ tube is usually short, forming a convoluted appressorium. A penetration tube grows into the centre of the cell lumen, a process which Hashioka (1937) demonstrated was also favoured by a temperature of 28°. While the first haustorium is being established, additional germ tubes are formed from other points on the same spore and hyphae are sent from the primary appressorium along the leaf surface. On all hyphae after the first germ tube, appressoria are formed laterally, and branching is acute and regular. The mother spore remains a living part of the thallus and does not collapse after the fungus has established nutritive relations with the host. About four days after infection, conidiophores begin to form. Colonies from single spores are rarely over 2 cm in diameter, and it appears that senescence of host tissue slows parasite growth. The actual method of conidial abstriction is not known. Dissemination of conidia is almost exclusively by wind. Conidia fall in the air at the rate of $1 \cdot 2$ cm sec^{-1}. A life cycle takes five or six days. (Yarwood, 1957).

Cleistothecia, if formed, occur several weeks after conidial formation. Nutritive conditions of the host and sexuality determine whether cleistothecia will be produced. Homma (1937) presents a detailed description of cleistothecial formation for *Sphaerotheca*. Uninucleate antheridia and ascogonia form in the centre of a colony and coil around each other. After secondary cleistothecial cell walls develop, the antheridium withers and disappears. When four or five cell layers of the wall are developed, appendages begin to arise near the base of the cleistothecium. When two layers of the outer cell wall are developed, the male and female nuclei in the ascogonium conjugate and the ascogonium divides to form the ascus. The ascus nucleus divides to form the ascospores. Heterothallism seems common. When ascospores mature they are forcibly discharged as the cleistothecium is ruptured by the swelling asci and ascospores.

E. cichoracearum produces cleistothecia and ascospores on some plants which could permit overwintering in cold climates. Cleistothecia are not observed in the tropics. Here the powdery mildew fungi overwinter as active mycelial and conidial stages, in sheltered situations, or on a variety of volunteer and cultivated cucurbits (Bessey, 1943; Ainsworth and Bisby, 1950). Different strains of the same species may vary greatly in their host range. These strains are not identical but constitute a number of physiological forms that differ in ability to attack particular host species. Cucurbit powdery mildew is most severe on cantaloupe and of less intensity on pumpkin, squash and cucumber. Although not immune, watermelon is not often affected. In addition, this powdery mildew attacks zinnia, phlox, aster, lettuce and sunflower (Walker, 1952). In New South Wales the cucurbitaceous weeds prickly paddy melon (*Cucumis myriocarpus* Naud.) and wild watermelon (*Citrullus lanatus* var *lanatus*) may be affected (Ballantyne, 1975). Schmitt (1955) found the strain of *E. cichoracearum* on zinnia has a greater host range than the strains on *Inula, Cerinthe, Helianthus*, phlox, or cucurbits. The greater variability of this organism was further emphasized by finding some strains on cucumber and sunflower that will cross infect (Reed, 1908) and other forms that will not (Schmitt, 1955). Isolates from sunflower, squash, hollyhock, dahlia and *Nicotiana glycines* have all infected cucumber but not equally (Reed, 1908). In actual crop production, however, one plant genus is usually not considered a source of powdery mildew inoculum for other genera.

IV. Genetic Control

A. Nature of disease resistance

One of the most successful procedures for controlling a plant disease is through the development of resistant varieties. Powdery mildew is no exception, and much success has been achieved for cucumbers and cantaloupes. Development of powdery mildew resistant squashes and watermelons has not been successful to date.

Development of powdery mildew resistant cantaloupes provides a classic example of plant breeding to combat a specific disease. Jagger and Scott (1937) screened cantaloupe plant material from all over the world and found powdery mildew resistance in a seed lot from India (PI 78374). Utilizing a backcross programme combined with field selection, they found that resistance to race 1 was controlled by a single dominant gene, referred to as Pm^1. This programme resulted in the introduction of PMR 45 in 1936. Jagger, *et al.* (1938) found that PMR 45 and selections of PMR 45 were susceptible to powdery mildew, indicating a new race. Genes for resistance to this race were also available from PI 78374. Utilizing PMR 45 as one parent, cultivars with resistance to race 2 were developed—PMR 5, PMR 6, and PMR 7. Race 2 resistance was thought to be controlled by two or three genes, without complete dominance (Whitaker and Pryor, 1942). Using these race 2 resistant varieties, Bohn and Whitaker (1964) demonstrated that race 2 resistance was controlled by a partly dominant gene, designated Pm^2. They stated that two modifier genes differentiated extreme resistance to race 2, and were epistatic to PM^2 but hypostatic to pm^2. Harwood and Markarian (1966), using Seminole as the resistant stock, demonstrated that resistance in their germ plasm was controlled by two gene pairs whose effect were unequal and partly additive. The major gene showed incomplete dominance, while the minor gene showed complete dominance. This resistance was different from that for Pm^2. Seminole, developed from PI 124112, was highly resistant in the eastern United States but only moderately resistant in the western United States. Thus Harwood and Markarian (1968) proposed the following genetic designations:

Pm^1 PMR 45, single dominant
Pm^1, PM^2 PI 79376, PI 78374, partially dominant with several modifiers
Pm^3 PI 124111, single dominant gene
Pm^4 Seminole, partially dominant
Pm^5 Seminole, complete dominance

Cucumber powdery mildew resistance has been intensively investigated and is still somewhat confused. Smith (1948) found that Puerto Rico 37 cucumber was resistant to powdery mildew and, upon crossing it with Abundance, demonstrated the F_1 to be susceptible but found a very small number of resistant individuals in the F_2. This suggested polygenic resistance. Barnes (1961), utilizing PI 197087, stated that perhaps three genes were involved but plants with all three genes may have been discarded due to close linkage with a gene, or genes, that resulted in the plants being unable to utilize manganese, as had previously been demonstrated by Robinson (1960). Hujieda and Akiya (1962) demonstrated that resistance was controlled by a single recessive gene in their resistant variety Natsufushinari. Kooistra (1968) developed cucumber lines in which resistance to powdery mildew was governed by three recessive genes—two from Natsufushinari and one from PI 200818. His resistance was of the hypersensitive type because Kooistra (1968) was working with supersensitive genes from Japanese plant material, while Barnes (1961) was working with a different set of three separate genes that produced a cumulative effect. This effect was aptly demonstrated by Barnes in his powdery mildew resistant varieties Ashley (one resistant gene) and Poinsett (three resistant genes). Wilson, et al. (1956) found the Burmese variety Yomaki (PI 200818) contained two pairs of duplicate recessive genes. Perhaps the nature of genetic resistance to powdery mildew in cucumber has been

best explained in a fine investigation by Shanmugasundaram, *et al.* (1971) who
hypothesized that by the addition of one recessive gene to the above data, the
apparently divergent inheritance patterns could be unified. These investigators
said resistance was controlled by a major recessive gene, *s*, which determines
hypocotyl resistance (intermediate) and is also essential for leaf resistance
(complete). Leaf resistance is controlled by a dominant gene, *R*, which only
expresses itself in the presence of *s*. Gene *I* is an inhibitory gene which prevents
expression of complete resistance, but does not affect gene *s*. Genes for
resistance in PI 212233, PI 234517 and Natsufushinari appear to be the same.
Genes for cucumber powdery mildew resistance were independently inherited from genes for scab resistance, spine colour and cotyledonary bitterness.

Obtaining powdery mildew resistance in squash cultivars has not been as
successful as for cantaloupe and cucumber. Whitaker (1959) found *Cucurbita
lundelliana* Bailey (Peter's Gourd) to be resistant to powdery mildew. Rhodes
(1959, 1964), utilizing the interspecific cross technique with *C. lundelliana* as
a bridging parent, developed a common gene pool of divergent germ plasm in
a series of crosses between *C. lundelliana*, *C. pepo*, *C. moschata*, *C. mixta*,
and *C. maxima*. By selecting resistant plants from this gene pool he demonstrated that resistance in squash is controlled by a single dominant gene
that had been transferred from *C. lundelliana* to *C. moschata*. Sitterly (1971)
obtained germ plasm from Rhodes and transferred resistance into the bush
type *C. pepo*. No resistant squash varieties have been released.

B. Sources and range of resistance

The best place to find resistance to any plant disease is in those areas where
the host and the pathogen have been developing over the centuries, allowing
only the resistant hosts to survive. Almost without exception the best powdery
mildew resistant collections of *Cucumis* (cantaloupe and cucumber) have
come from the primary gene centre of India and adjacent Asian areas.
Although many sources have now been identified, selected sources of cantaloupe powdery mildew resistance are USPI 79374, USPI 124111 and USPI
134198, all from India. A range of cantaloupe resistance embodying American
varieties would be: susceptible—Hales Best and Honey Dew; moderately
resistant—Georgia 47; race 1 resistant—PMR 45; race 2 resistant—PMR 7
and Seminole; highly resistant to race 1 and race 2—Planters Jumbo. Other
selected resistant varieties are: Perlita, Gulfstream, Floridew, Campo, Cum
Laude and Southland from the United States; Yokniam 54 and Ananas PMR
from Israel; Iyo 1 from Japan; and Yanco Treat and Yanco Delight from
Australia.

Selected sources of cucumber powdery mildew resistance are USPI 197087,

USPI 233646 and USPI 200815. A range of resistance utilizing American varieties would include: susceptible—Model; one resistant gene—Palmetto and Ashley; two resistant genes—Poinsett and Cherokee; three resistant genes—USPI 197087. Other specific resistant cultivars are: Polaris, Stono, Pixie and Ashley from the United States; and Natsufushinari from Japan, Leppik (1966) reports that several wild species of cucumber were either resistant or immune, but could not be crossed with cultivated varieties.

A source of Cucurbita resistance to powdery mildew is *Cucurbita lundelliana*. Although active breeding programmes exist in the United States, no resistant varieties have been released to date.

There are no published reports of varietal resistance to powdery mildew in watermelon.

V. Inoculation Procedures With Cucurbit Powdery Mildew Pathogens

A. Collecting and preparing inoculum

If infected leaves are available one has only to collect them and either wash off conidia and adjust spore concentration, or allow the leaves to remain on the plants and use indirect transfer. Powdery mildew cannot be grown in artificial culture but susceptible hosts may be grown in the greenhouse and the pathogen propagated by continuous living host transfer. For storage, infected leaves may be dried at room temperature and kept for a year in paper bags without elaborate storage equipment or requirements.

B. Inoculation in greenhouse and field

In the greenhouse, plants are placed in either shaded glass sash chambers or cheesecloth tents. Plants are inoculated at least twice—at cotyledon expansion, and again at first true leaf expansion. Conidia from heavily infected leaves are applied either by blowing over the test plants via human breath or by a small electric blower, by dusting with dry conidia, or by leaf contact. Leaf contact is least reliable. It is not necessary to wet the plants, as a 30–50% r.h. is probably most desirable. Under very dry conditions, however, walls of the inoculation chamber should be moistened to raise the relative humidity to a moderate level. Remove any coverings after 24 h. Severity of infection can be increased by a more abundant conidia level, increased plant population and protection from free water. There are three modifications to the above procedure which might better suit an individual investigator. Markarian and

Harwood (1967) grew host seedlings at 24°C in benches of sandy loam which
had been leached to reduce available nutrients. A susceptible variety was
planted every 30 rows to ensure uniform inoculation. Spores were dusted over
seedlings twice daily for one week beginning when first true leaf was ½ in
long. After inoculation the temperature was dropped to 16°C to aid conidial
germination, then raised to 24°C after two weeks. Clark (1975) kept greenhouse temperature at 19–20°C and utilized natural light intensity. Shanmugasundaram, et al. (1971) inoculated by dusting with conidia for three
consecutive days when plants were in the cotyledonary stage. Growth of
powdery mildew was noted in three to four days. Under all inoculation systems,
powdery mildew ratings may be taken two to three weeks after inoculation.

For field inoculation, conidia may be placed in a water suspension in a
knapsack sprayer and applied at 10 psi. Mid-morning would be the preferred
application time to allow more rapid and certain drying of the water carrier.
Three applications at daily intervals should be made to ensure infection.
Field inoculation, subject to environmental fluctuation, is not so consistently
reliable as greenhouse inoculation. Greenhouse testing is better for selecting
highly resistant items, and field performance may be predicted from results of
rapid greenhouse seedling evaluation. However, field testing is better for
selecting degrees of resistance and observing powdery mildew reactions under
natural conditions. Field testing is an absolute necessity for selecting fruit
and plant characteristics.

C. Rating systems

There are almost as many rating systems of the host–parasite interaction as
there are investigators reporting. A major difficulty is lack of a satisfactory
quantitative method adapted to statistical analysis, because, when rated
visually, scales are not mathematically linear nor even continuous. Generally
the susceptible reaction is expressed by all the leaf and stem area being covered
with abundant sporulating powdery mildew colonies. The resistant reaction is
manifested by one or two weak colonies on the entire plant. The intermediate
reaction may be characterized by mild or severe leaf and cotyledonary
infection, with nothing on remaining plant parts. Most investigators have
expressed these reactions on a 0–5 rating scale based on plant location and
abundance of powdery mildew colonies and mycelium, with abundance
frequently denoted by various percentage scales. Additionally any necrosis of
leaf tissue may be designated alphabetically, such as "A" (Bohn and Whitaker,
1964; Harwood and Markarian, 1968; Clark, 1975; Kooistra, 1968; Pryor,
et al., 1946; and Ballantyne, 1975). More than a single rating should be made
to be certain plant reactions do not vary.

VI. Breeding Procedures with Cucurbit Hosts

Cucurbits exhibit both advantages and disadvantages concerning plant breeding. The disadvantages are: plants require much space which makes large populations expensive; hand pollination is generally necessary for controlled genetic investigation; with the exception of squash, pollination must be done before selection; chromosomes are not easily differentiated from cytoplasm in pollen mother cells and chromosomes are small and not well separated from each other. Advantages are: flowers are large and easily handled; plants are easily grown by simple methods; plants are indeterminate with flowers available over a long period of time; fruits are fairly durable; and most fruits yield many seeds.

Cucurbit flowers are of a yellow or orange colour and the number of staminate flowers always exceeds the number of pistillate flowers. Staminate flowers have three stamens, with free filaments, united by their anthers. Pistillate flowers are epigynous with a pistil having one to five carpels and a thick short style terminated by three papillate stigma. There may also be sterile rudimentary stamens. Cucurbit fruit is an inferior berry or pepo (fruit indehiscent with fleshy floral tube adnate to pericarp). Seaton and Kremer (1939) showed the most important climatological influence on anthesis and anther dehiscence is temperature. Minimum temperature for pumpkin and squash is $10°C$; watermelons and cucumbers is $16°C$ and cantaloupes is $18°C$. Mann and Robinson (1950) showed pollen grains germinate in a few hours and the pollen tubes reach the ovary after 24 h. Tiedjens (1928) demonstrated that light influenced both fertilization of ovules and shape and growth of fruit by influencing sugar availability. He also showed light influenced sex expression with an abundance increasing the number of staminate flowers. Hubbard (1940) demonstrated the pronounced inhibitory effect of growing fruit on subsequent setting of pistillate flowers.

When pollinating cucurbit flowers, buds must be closed the day prior to natural opening to keep insects from applying undesired pollen. Paper tape reinforced by wire centres, paper fasteners, or string may be used. For watermelon, $1\frac{1}{4}$ in sections of $\frac{9}{16}$ in diameter plastic tubing sealed at one end with adhesive tape are preferred. A location marker placed at a $45°$ angle aids in locating the chosen pistillate flowers. Pollinations should be made within four hours after pollen dehiscence. Where hermaphroditic flowers are involved, as in some C. melo, crosses may be made by emasculating chosen flowers the afternoon before pollination. After pollination, place the location marker at a $90°$ angle to the soil. Each pollinated flower should be tagged to denote parentage and the tag should be attached directly to the base of the ovary. All crosses should have an additional duplicate tag on the location marker. After tagging, the pollinated flower should be covered to prevent pollen contamination. Cucumbers and cantaloupes may be covered with miniature glassine bags, or with gelatin capsules if located in a dry area. Watermelon, squash, and pumpkin are covered with No. 1 Kraft paper bags. Removal of the flower covers after a few days is desirable as flower tissues make good media for the growth of fruit rotting organisms when high temperature and high humidity prevail inside. Where soil-borne fruit rotting organisms are a problem, fruit selected for saving may be placed on hardware cloth, asphalt shingles, etc. to erect a barrier between the fruit and the soil.

In cucumber, since the percentage set of flowers pollinated varies from 50–90% it is advisable to pollinate at least two flowers per plant in a segregating population. Each fruit will produce between 100 and 200 seeds. For gynoecious plants the seed set from selfing is usually much smaller because most of the staminate flowers contain reduced quantities of pollen. Thus three to four flowers per plant should be used. For squash and pumpkin, fruit set is 90% of the flowers pollinated and pollinating one flower per plant is adequate. Each fruit will produce 100 to 200 seed. For watermelon and cantaloupe, percentage seed set by hand pollination is about 50% of what bee pollination would accomplish. Most intervarietal crossing of melons is done in a greenhouse using five plants of each parent. Seed will average 200 to 300 per fruit for watermelon and 300 to 400 for cantaloupe; hence one hybrid fruit is sufficient for the original cross but several are needed for F_2 seed.

A normal crossing and backcrossing programme combined with rigid greenhouse and field selection should suffice. To avoid discarding of beneficial modifying genes, etc. one should carry forward a number of lines from the original F_1 material. The most promising lines should be intercrossed in the F_4 or F_5 generation so as to pool as many desirable genes as possible into a single entity. Standard breeding procedures may then be followed. This "pooling" may be done as often as necessary to obtain a satisfactory combination of powdery mildew resistant genes and desirable horticultural characteristics. Concurrently an investigator should make pathogen collections from a wide range of localities and test his advanced lines against them for biotype reactions. All good breeding programmes include collections of plant material from areas in which the host is endemic that will adequately represent the variation found within the species.

The success of a cross is measured by the usefulness of the segregants selected out of it and is usually revealed in early generations. The larger the population the better the chances of finding the desired individual in a segregating population. For cucumbers, two-thirds of the seed of an individual fruit is adequate in the breeding generations. If selection proves an item desirable, most of the reserve seed should be planted in the next season for effective utilization. Some seed should be saved to compensate for occasional crop loss. For squash and pumpkin, 60 seed are considered adequate for the breeding generations, with the remainder of the procedure similar to cucumber. For melons, breeding generations have been most successfully handled by growing in isolation plots and letting the bees perform random interpollination resulting in extensive gene transfer. In this mass selection, population has seldom exceeded 600 plants with as few as 100 for watermelon and 240 for cantaloupe. In these minimum isolation plots (30 feet long), watermelons alternate with cantaloupes and are separated by a 30 foot barrier of cucumbers. There is only 1–2% foreign pollination contamination and the bees will also produce about 50% self-pollination. Seed is extracted from the 20–50 melons with the highest score in any single plot and this seed is then blended for the next generation. This process occurs for about eight generations. In early generations, periods of one to two generations of enforced sib-crossing are interspersed with mass selection.

Greenhouses may be used in the off-season period to evaluate disease resistance under a controlled environment; to increase small quantities of selfed seed for field production of larger quantities; to produce original crosses and to start a field crop indoors to avoid a waste of seed in short supply. If fruit is to be produced, plants grown in 10 inch pots perform better than those in ground beds.

Most pedigree systems provide for a basic accession list with number identification for each parent variety, each genetic stock, and each original cross. The system needs to show generation of selection, year of origin and how pollinated. Accession books should be set up as a permanent record.

In harvesting seed, the ripe mature fruit of cucumber and melons are used. Seed, pulp, and juice are allowed to ferment until seeds settle to the bottom (two days at 18°C). Seed are then washed, treated with bichloride of mercury to eliminate seed-borne organisms and spread to dry. The bichloride of mercury treatment involves soaking of seed in a 1:1000 solution for five minutes and then rinsing. If mercury isn't available, soak seed in water at 38°C. Hutton (1941) accelerates the seed separation process by adding 3 fl oz of hydrochloric acid to 25 lbs of freshly threshed material, followed by washing the seed after 15–30 minutes. Squash seed do not necessarily need fermentation as seed may be removed from fruit, washed, and dried. Freshly harvested cucumber seed are normally dormant, and excessively low temperatures and humidities may prolong dormancy. No special dormancy problem occurs in melons and squash, although freshly harvested seed will not germinate as well as comparable seed stored a few months. Triploid and tetraploid watermelon seed present a special problem because they do not adequately withstand the fermentation process and they have poor embryo development. A storage temperature of 20°C and a humidity below 40% is adequate for a five year storage period. Breeding seed must also be protected from mice in storage and in the field. All the cucurbits are treated with a seed protectant fungicide before planting.

Disease resistance should not be looked upon as a goal independent of other objectives. Genotypic balance is a proper objective (the true goal of evolution) and leads to the single character all successful varieties must have–good appearance. All cucumber varieties must have resistance to drought and temperature extremes. Fruit should have the desired conformation, a small seed cavity surrounded by crisp flesh, slowly developing seed, waxy cover and non-bitter flavour. Pickle cucumbers should have a definite length: diameter ratio and moderate warty protuberances, they should be free of carpel separation with hollow placenta and they must be suitable for preserving in salt brine. Slicer cucumbers should be free of warty protuberances and have a dark green colour. All melon, squash and pumpkin varieties must be productive, transportable and edible, to be useful in commerce.

Breeding cucurbits for disease resistance has resulted in several spectacular accomplishments in a relatively short period of "plant time". The most obvious benefit of developing resistance to powdery mildews is commercial production of cucurbit crops in areas where economic production was not feasible because of the presence of powdery mildews, or where production was decreased. Another beneficial result is the decrease in cost of crop production as a result of requiring few or no fungicide applications. This also decreases the danger of air, soil and water pollution with pesticidal chemicals, the hazard of injury to man associated with the application of pesticides and the contamination of foods with toxic residues. As individual plant diseases are conquered by breeding it is practical to combine the genes for resistance to individual diseases into multiple-disease resistant varieties. Examples of multiple-disease resistant cucumber varieties which contain resistance to powdery mildew are: pickle variety—Chipper (downy mildew, powdery mildew, angular leafspot, anthracnose and cucumber mosaic virus); slicer variety—Poinsett (powdery mildew, downy mildew, angular leafspot and anthracnose). Multiple-disease resistant cantaloupe varieties are: Seminole (powdery mildew race 2 and downy mildew); and Edisto 47 (powdery mildew and downy mildew). After development of multiple-disease resistant cucumber varieties the next logical step was the synthesis of hybrids to obtain earlier production and even higher yields. The procedure has been most successful with cucumbers, cantaloupes, and watermelons.

VII. Chemical Control

Chemical control of cucurbit powdery mildews has not been as thoroughly investigated as has chemical control of many other plant diseases. Cucurbit powdery mildew is not difficult to control and use of early control methods would be of value in developing countries today.

The powdery mildew fungus is vulnerable to the action of sulphur through most of its life cycle, except for the cleistothecial stages (Yarwood, 1957). Sulphur acts by selective toxicity—more toxic to the parasite than to the host. An effective sulphur application rate in the field would be 4·5 kg elemental sulphur ha^{-1} applied when the fungus is first observed, followed by a repeat application two weeks later. Do not apply sulphur to cantaloupes because they are sulphur-sensitive. Cucumbers are somewhat sensitive to sulphur; gourds, pumpkins, squashes and watermelons are sulphur-tolerant. Indeed, the principal objection to the use of sulphur in the field is the plant injury which may result. Damage may be reduced by use of low-grade sulphur, addition of inert fillers to dusts and wettable powders and by application to the soil under the plants. Dusts are less injurious than sprays but also are less effective than sprays because they do not result in as effective leaf coverage.

In the greenhouse plants are more susceptible to fungicide injury than in the field because of tender foliage and higher temperatures. Sulphur offers long lasting protection and is effective in the absence of moisture. The higher the temperature, the greater the vaporization of the sulphur. Thus small doses are more desirable in the greenhouse—preferably in the 15–20% concentration area. Because sulphur acts as a vapour it may be effective for several feet by merely painting sulphur on greenhouse heating pipes during colder growing periods. During warmer growing periods in greenhouses, sprays, dusts or aerosol bombs may be used. Although vaporization from sprays or aerosol bombs is more uniform, dusting is more economical. Dusts are also more effective on cloudy days because ventilators can be kept closed longer thus allowing better disposition of sulphur on plant foliage. In cool damp weather, one may have to use a 1:1:50 Bordeaux mixture (be careful on cucumbers). Do *not* mix copper and hydrocyanic acid gas, nor sulphur and Volck insecticide (Guba, 1928).

Harrigan (1941) attempted a unique control of powdery mildew on cantaloupes by developing sulphur-tolerant varieties. These varieties were not horticulturally suitable to the commercial trade and were not accepted. These varieties could, however, be used in areas where this characteristic was decisive in determining the presence or absence of production.

Around 1952, the first of the effective organic fungicides for control of powdery mildew on sulphur-sensitive cucurbits was developed. This material was dinitro capryl phenyl crotonate, frequently sold under the trade name of Karathane. As so frequently has been the case with other compounds since 1952, this compound is an effective miticide as well as an effective mildewicide. Karathane acts as an eradicant fungicide, killing both mycelium and spores (Godfrey, 1952). A 1% dust at 39·1–44·7 kg ha^{-1} or a spray using 0·5 kg ha^{-1} will give adequate control if the lower leaf surfaces are covered. Timing of application is important. Karathane should be applied when fruits are just beginning to set and should be used at seven-day intervals until just before harvest. Some areas recommend additional applications by the second day after 0·25 in of rainfall. Godfrey (1952) also demonstrated that for vine crops, aerial application of Karathane at 16·8 kg ha^{-1} was inadequate due to lack of thorough coverage.

In 1966 the first trials with benomyl (Benlate) for the control of cucurbit powdery mildew were conducted by numerous investigators. Delp and Klopping (1968) first reported systemic activity by benomyl and also resistance by the organism when applied to the foliage. Schroeder and Provvidenti (1968) also reported these same effects when benomyl was used as a soil drench or as a seed treatment to control powdery mildew. The seed treatment was to use benomyl as a slurry carried in a 4% methocel solution at the rate of 2% of the seed weight. One ml of this solution was used per 10 g of seed. For soil drench treatment these authors used 1·5 mg per 4 inch pot of soil, initially applied to young plants in the two leaf stage and with repeat applications every other week. When these treatments were applied to cucurbits the primary development of scattered colonies on plants previously treated with effective dosages, instead of a restricted development pattern, indicated a form of the fungus resistant to the systemic effects of this fungicide.

In South Carolina, USA, benomyl is used at the rate of 0·5 kg of active ingredient ha^{-1} and applied every other week for control of cucurbit powdery mildew. When benomyl is applied every week, the organism acquires resistance. Application every third week is just as effective but commercial growers may forget their schedule, or it may be raining, etc. Application every other week permits one week with which to physically manoeuvre, yet prevents the organism acquiring resistance.

Although benomyl has been extensively investigated for the control of cucurbit powdery mildews and other diseases since 1966, particularly in the United States, the tendency for pathogens to develop resistance in response to benomyl has stimulated investigation of other systemic chemicals as potential mildewicides. Perhaps due to differences in national toxicological requirements, investigation has progressed more rapidly in Europe than elsewhere. Triforine (CELA W524) was reported by Fuchs et al. (1971) to be completely effective in controlling cucurbit powdery mildew when applied as a preventative spray treatment. Fuchs et al. (1973) also demonstrated that 5–6 days after root application, the concentration of triforine in cucumber shoots far exceeded the amount necessary for control and was converted to non-phytotoxic piperazine rather quickly. Other systemic chemicals tested effectively were: dimethirimol (used commercially as Milcurb) and ethirimol (Byrde, 1971); triarimol when applied at 10 ppm on a 14-day schedule (Banihashemi et al., 1976) and, in greenhouses, the barbiturate secobarbital (Gorter and Nel, 1974). Limitations to utilization of systemic fungicides to control cucurbit powdery mildews have been suggested by several investigators. Systemics are limited by the probability of residues in the harvested crops, thus necessitating a balance between too rapid loss of activity and too slow degradation (Byrde, 1971). The selective effect of systemics could cause a shift in environmental balance as fungi adapt more readily than they would to conventional fungicides (Dekker, 1973). Restriction of use of systemic powdery mildewicides to specific situations and not as protectants, would fully utilize their curative characteristics (Evans, 1973). Cucurbit powdery mildew can be effectively controlled if attention is given to correct dosages, proper time of application and adequate leaf coverage.

A novel control procedure for the home gardener or small commercial grower with an abundant water supply has been suggested by Yarwood (1939), who found that high humidity favours growth of certain fungus parasites of powdery mildew and that rainfall, through mechanical action, materially decreases disease incidence. Thus home gardeners, or small growers could control powdery mildew by simply spraying the leaves with an ordinary garden hose in the afternoon.

VIII. Goals to be Achieved

Cucurbit plant breeders are continually developing improved varieties that are better adapted to a wider range of growing-conditions, have multiple disease resistance and meet the fluctuating demands of economics and consumers. With all cucurbits, work must be directed toward the development of small, compact vines for mechanical harvest. For all curcurbits, the nature of the mechanism that makes mites and cucurbit powdery mildew susceptible to the same pesticides should be elucidated. For cucumbers, better resistance to powdery mildew is needed in the greenhouse and in the tropics. For cantaloupe, a better understanding is needed of the relationship between cantaloupe and powdery mildew. For squash and pumpkin, the development of any powdery mildew resistant variety is necessary. For watermelon, we should know why this crop is apparently resistant to powdery mildew in relation to the susceptibility levels of the other cucurbits.

References

Ainsworth, G. C. and Bisby, G. R. (1950). "A Dictionary of the Fungi". 3rd, ed. Commonwealth Mycological Institute, Kew, Surrey.
Ballantyne, B. (1963). *Aust. J. Sci.* **25**, 360–361.
Ballantyne, B. (1975). *Proc. Linn. Soc.* N.S.W. **99**, 100–120.
Banihashemi, Z., Rafii, Z. and Azad, H. (1976). *Iran J. agric. Res.* **4**, 17–24.
Barnes, W. C. (1961). *Proc. Am. Soc. hort. Sci.* **77**, 417–423.

378 WAYNE R. SITTERLY

Bessey, E. A. (1943). *Pap. Mich. Acad. Sci.* **28**, 3–8.
Bohn, G. W. and Whitaker, T. W. (1964). *Phytopathology* **54**, 587–892.
Byrde, R. J. W. (1971). *Trop. Sci.* **12**, 105–111.
Clare, B. G. (1958). *Aust. J. Sci.* **20**, 273.
Clark, R. L. (1975). *Pl. Dis. Reptr* **59**, 1024–1028.
DeCandolle, A. (1882). "Origine des Plantes Cultives". Germes Bailliere, Paris.
Deckenbach, K. N. (1924). *Rev. appl. Mycol.* **5**, 70–71.
Deckenbach, K. N. and Koreneff, M. S. (1927). *Rev. appl. Mycol.* **2**, 273.
Dekker, J. (1973). *Organ. Eur. Mediter. Prot. Plant.* **10**, 47–57.
Delp, C. J. and Klopping, H. L. (1968). *Pl. Dis. Reptr* **52**, 95–99.
Dingley, J. M. (1959). *N. Z. Jl. agric. Res.* **2**, 380–386.
Duvdevani, S., Reichert, S. and Palti, J. (1946). *Palest. J. Bot.* **5**, 127–151.
Evans, E. (1973). *Organ. Eur. Mediter. Prot. Plant.* **10**, 59–67.
Fuchs, A., Doma, S. and Voros, J. (1971). *Neth. J. Pl. Path.* **71**, 42–54.
Fuchs, A., Viet Versweij, M. and DeVries, F. (1973). *Phytopath. Z.* **75**, 111–123.
Godfrey, C. H. (1952). *Phytopathology* **42**, 335–337.
Gorter, G. J. M. A. and Nel, D. C. (1974). *Phytophylactica* **7**, 21–24.
Guba, E. F. (1928). *Phytopathology* **18**, 847–860.
Harrigan, B. A. (1941). *Produce News* **13**, 6.
Harwood, R. R. and Markarian, D. (1966). *Proc. XVII Int. hort. Congr.* **1**, 454.
Harwood, R. R. and Markarian, D. (1968). *J. Hered.* **59**, 126–130.
Hashioka, Y. (1937). *Rev. appl. Mycol.* **17**, 93.
Hirata, K. (1955). *Bull. Fac. Agric. Niigata Univ.* **7**, 24–36.
Homma, Y. (1937). *J. Fac. Agric. Hokkaido Univ.* **38**, 183–461.
Hubbard, A. D. (1940). *Proc. Am. Soc. hort. Sci.* **37**, 825–836.
Hujieda, K. and Akiga, R. (1962). Jap. Soc. hort. Sci. **31**, 30–32.
Hutton, E. M. (1941). *J. Coun. scient. ind. Res. Aust.* **16**, 97–103.
Jagger, I. C. and Scott, G. W. (1937). "Development of Powdery Mildew Resistant Cantaloupe No. 45". *Circ. US Dept. Agric.* 441.
Jagger, I. C., Whitaker, T. W. and Porter, D. R. (1938). *Pl. Dis. Reptr* 22, 275–276.
Kable, P. F. and Ballantyne, B. (1963). *Pl. Dis. Reptr* **47**, 482.
Kahn, M. W. and Kahn, A. M. (1970). *Indian Phytopath.* **23**, 497–502.
Kooistra, E. (1968). *Euphytica* **17**, 236–244.
Leppik, E. C. (1966). *Euphytica* **15**, 323–328.
Mann, L. K. and Robinson, J. (1950). *Am. J. Bot.* **37**, 685–687.
Markarian, D. and Harwood, R. R. (1967). *Q. Bull. Mich. St. Univ. agric. Exp. Stn* **49**, 404–411.
Nagy, G. S. (1970). *Acta Phytopath, hung.* **5**, 231–248.
Pryor, D. E., Whitaker, T. W. and Davis, G. N. (1946). *Proc. Am. Soc. hort. Sci.* **47**, 347–356.
Raabe, R. D. (1966). *Pl. Dis. Reptr.* **50**, 411–414.
Randall, T. E. and Menzies, J. D. (1956). *Pl. Dis. Reptr* **40**, 255.
Rayss, T. (1947). *Palest. J. Bot.* **4**, 59–76.
Reed, G. M. (1908). *Trans. Wis. Acad. Sci. Arts Lett.* **15**, 527–547.
Rhodes, A. M. (1959). *Proc. Am. Soc. hort. Sci.* **74**, 546–551.
Rhodes, A. M. (1964). *Pl. Dis. Reptr* **48**, 54–55.
Robinson, R. U. (1960). *Proc. Am. Soc. hort. Sci.* **57**, 21.
Roder, K. (1937). *Rev. appl. Mycol.* **16**, 653.
Schmitt, J. A. (1955). *Mycology* **47**, 688–701.
Schroeder, W. T. and Provvidenti, R. (1968). *Pl. Dis. Reptr* **53**, 630–632.
Seaton, H. L. and Kremer, J. C. (1939). *Proc. Am. Soc. hort. Sci.* **36**, 627–631.

Shanmugasundaram, S., Williams, P. H. and Peterson, C. E. (1971). *Phytopathology* **61**, 1218–1221.

Sitterly, W. R. (1971). *In* "Breeding Plants for Disease Resistance—Concepts and Applications". (R. R. Nelson, Ed.), pp. 287–306. Penn. State Univ. Press, University Park and London.

Smith, P. G. (1948). *Phytopathology* **38**, 1027–1028.

Sohi, H. S. and Nayar, S. K. (1969). *Indian Phytopath.* **22**, 410–412.

Sowell, G., Jr., and Clark, R. L. (1971). *Bull. Ga. Acad. Sci.* **29**, 99–100.

Tarr, S. A. J. (1955). "The Fungi and Plant Diseases of the Sudan". Commonwealth Mycological Institute, Kew, Surrey.

Tiedjens, V. A. (1928). *J. agric. Res.* **36**, 731–736.

Walker, J. C. (1952). "Diseases of Vegetable Crops". McGraw-Hill Book Company, Inc., New York.

Wellman, F. L. (1972). "Tropical American Plant Diseases". Scarecrow Press, Inc., Metuchen, N. J.

Whitaker, T. W. (1959). *Madrono* **15**, 4–13.

Whitaker, T. W. and Davis, G. N. (1962). "Cucurbits". Wilty and Sons Ltd., (Interscience Publishers,) Inc., New York.

Whitaker, T. W. and Pryor, D. E. (1942). *Proc. Am. Soc. hort. Sci.* **41**, 270–272.

Wilson, J. D., John, C. A., Wohler, H. E. and Hoover, M. M. (1956). *Pl. Dis. Reptr* **40**, 437–438.

Yarwood, C. E. (1939). *Phytopathology* **29**, 288–290.

Yarwood, C. E. (1957). *Bot. Rev.* **23**, 235–300.

Zaracovitis, C. (1965). *Trans. Br. mycol. Soc.* **48**, 553–558.

Chapter 15

Powdery Mildew of the Hop

D. J. ROYLE

*Department of Hop Research, Wye College (University of London),
Near Ashford, Kent, England*

I. Historical Background

Powdery mildew or "mould", caused by *Sphaerotheca humuli* DC. (Burr.) is the oldest known fungal disease of the cultivated hop (*Humulus lupulus* L.). Originally known as "fen", it was the most serious hazard to hop growers in the east- and mid-Kent areas of England at least as long ago as 1700 when the mould growth seen on diseased hops was considered to be the effect rather than the cause. The disease was then attributed to atmospheric conditions or to an unhealthy sap associated with the luxuriant foliage produced in over-manured hop gardens. Despite no understanding of infectious micro-organisms, early descriptions of infected hops suggest that both white, powdery pustules of the asexual stage and dark-coloured cleistocarps of the sexual stage of *S. humuli* were observed (Marshall, 1798). When, later, this fungus was recognized as the cause of the hop disease it was attributed by Cooke (cited by Salmon, 1900) to the aggregate species *Sphaerotheca castagnei* Lév. Some confusion over its nomenclature then arose partly because of a report by Magnus (1899) that the fungus from hop could infect certain weed hosts and also because of its similarity to the strawberry mildew pathogen, *S. macularis* (Wallr. ex Fries) Jaczewski. However, Salmon (1900) firmly assigned the hop pathogen to *S. humuli* and, since it can be distinguished from *S. macularis* by the structure of the cleistocarp appendages (Liyanage, 1973), and seems highly specialized to hop (Liyanage and Royle, 1976), there can be little doubt that *S. humuli* is the correct binomial.

Although control by sulphur dusting was introduced around 1850, the disease had spread to all other hop-growing areas in England by the late 1800s (Whitehead, 1890). At this time it was also destructive in Western Europe and in Russia (Bondarzew, 1908) and mild attacks had been recorded in several regions of the USA (Blodgett, 1913). Powdery mildew subsequently became less severe on the continent but by 1909 had reached epidemic proportions in New York State causing total failure of many crops and an estimated loss in revenue of $\$\tfrac{1}{3}$ million in 1912, the worst year (Blodgett, 1915). These severe attacks were mainly responsible for the gradual migration of hop-growing from the eastern USA to the Pacific North-west where the disease has apparently never arisen.

Until downy mildew (*Pseudoperonospora humuli* (Miy. and Tak.) Wilson) first broke out in Europe in 1920, powdery mildew was the only fungal disease of consequence on hops. Aspects were investigated by Salmon (1907, 1921) at Wye College who formulated recommendations for control based on cultural and hygienic measures and sulphur dusts. Similar investigations were undertaken by Blodgett (1913, 1915) in the USA. During his hop-breeding programme, Salmon discovered resistance to the disease and examined in some detail the inheritance of various forms of resistance (Salmon, 1917a,b, 1919a,b, 1920, 1921, 1927). Research on powdery mildew was then discontinued and for the next 35 years or so its incidence was less frequently recorded. On the continent and in the USA it ceased to be a problem, while in England concern about the disease became overshadowed by the acute problems caused by downy mildew and later, by wilt (*Verticillium albo-atrum* Reinke and Berth.). Nevertheless, the disease remained troublesome and occasionally resulted in crop failures though in general it was kept under control probably because more frequent fungicide applications were made with those needed for the control of downy mildew.

In England increasing difficulty in control began to be experienced in the 1960s, particularly in the highly susceptible, widely-grown cultivar Northern Brewer. This culminated in widespread attacks in 1971 which compelled some growers to abandon large proportions of their crops. The disease then became serious for the first time in West Germany in 1972 since when it has annually required systematic control measures (Wiedemann, 1972; Kohlmann and Kastner, 1975). The renewed importance of the disease in England coincided with substantial changes in systems of hop management, notably that of replacing cultivations with herbicides for weed control (Royle, 1975). Continued escalation of the disease has recently been checked with new, more efficient fungicides and the introduction of resistant cultivars, the outcome of a breeding programme initiated in 1962 (Neve, 1964).

Fundamental investigations of hop powdery mildew recommenced only as recently as 1970, so our present knowledge of the disease and the pathogen is derived from a mixture of long experience and limited experimental work. This chapter attempts to blend together this knowledge in relation to changing practices in hop management.

II. Sources of Infection

The hop plant is a perennial rootstock producing a herbaceous shoot system which dies down to ground level each winter. Fresh shoots ("bines") emerge in spring and climb, twining clockwise, up strings or wires attached each year to a permanent framework of poles and wire. This framework may be up to 7·5 m high and hops reach the top within 10–12 weeks of starting growth.

Powdery mildew is usually first noticed before the shoots begin to climb in April/May as one of two forms—(1) as discrete, white, sporulating colonies of the fungus on leaves near to the ground, or (2) as a dense, sporulating mycelium smothering young basal shoots whose growth is usually retarded (Fig. 1). The first form has been common in hop gardens for many years and is caused by ascospores discharged from cleistocarps which have over-wintered on fragments of cones and leaves strewn over the hop garden (Salmon, 1907; Blodgett, 1913, 1915). Observations on cleistocarp popula-tions over three years (Liyanage and Royle, 1976) have shown that during the winter cleistocarps mature in two periods, ending in November and in March, when many of them differentiate eight, well-defined ascospores. In this respect *S. humuli* resembles *S. pannosa* (Wallr.) Lév. var. *rosae* Wor. on rose (Price, 1970). However, we have not been able to encourage cleistocarps to discharge ascospores before April even after exposing them to a variety of environmental conditions. After this apparent obligate dormancy, ascospore discharge coincides with the growth of new foliage, thereby helping to ensure pathogen survival. Commonly, 30–50% of cleistocarps have dehisced by the end of April, the exact proportion probably being determined by their time

Fig. 1.   Shoot of a hop plant in April colonized by *S. humuli* mycelium which had over-wintered in a dormant bud. Secondary fungal colonies are apparent on adjacent foliage of otherwise healthy shoots.

of initiation, as with *S. mors-uvae* (Schw.) Berk. on blackcurrant (Salmon, 1914; Jackson and Wheeler, 1974).

The precise conditions favouring ascospore discharge and infection are unknown. Cleistocarps will dehisce only when wetted so presumably rain or dew is required for ascospore release in the field. Although ascospores have been clearly established as the origin of mildew colonies on field plants, none have been detected in suction spore traps even when a concentrated source has been provided. It is possible that they are dispersed only by water splash but this aspect needs more study.

The origin of the primary disease on shoots is mycelium which perennates in dormant, aerial buds on hop rootstocks (Liyanage and Royle, 1976). This form of the disease is usually encountered on less than 3 % of the plants in a hop garden (i.e. 75 plants in 2700 ha^{-1}) but there may be several infected shoots on a plant. In some seasons, such as 1976, a lack of wet conditions precludes ascospore release and primary shoot infections start the secondary outbreaks. The precise time during the previous year when buds become infected is not known. Buds appear from July but never seem to have open scales which may be receptive to infection, as in apple with mildew (*Podosphaera leucotricha* (Ell. and Ev.) Salm.), (Burchill, 1958). The manner of bud infection is also uncertain but mycelium, with associated haustoria, is well-established between the bud scales by November (Fig. 2) and can persist and

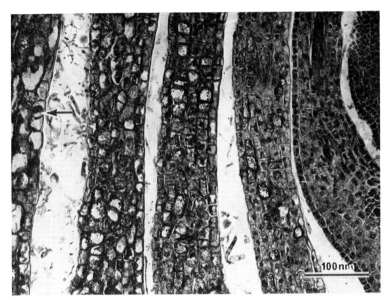

Fig. 2.   Portion of a longitudinal section of an infected dormant bud in October showing mycelium of *S. humuli* between the scales and haustoria (arrowed) in epidermal cells. The outermost bud scale is on the extreme left.

then develop with the growth of shoots in spring. Cleistocarps are sometimes associated with the bud mycelium (Figs 3 and 4) but are thought to function only if borne externally. Either mycelium or diseased buds may be killed according to the severity of the winter, as in apple (Covey, 1969) and gooseberry mildews (Preece, 1965). Most severe primary shoot infection would therefore be expected to follow mild winters and the experience of hop growers in recent years supports this view.

Perennation in buds is a relatively new element of the disease cycle and has compelled growers to start control measures much earlier in the season than they did previously. It was unknown in commercial gardens until the early 1960s and its increasing frequency has coincided with the gradual replacement of tillage with herbicides for control of weeds. Thus, rootstock buds, once removed in pruning operations or buried during cultivations, are now exposed aerially to infection during the summer. Modern cultural practices also encourage ascospore infection since hop debris containing cleistocarps remains on the surface of hop gardens, lodged in the crowns of dormant plants and is not buried beneath the surface. The greater abundance of initial inoculum caused by the changeover to non-cultivation is therefore considered to be a major factor contributing to the increased disease in recent years.

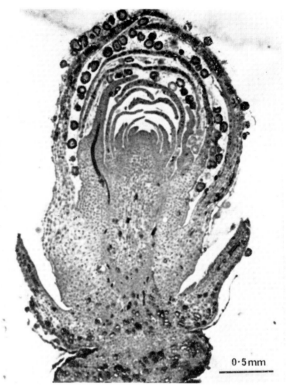

Fig. 3. Longitudinal section of an infected dormant bud in October showing the distribution of cleistocarps.

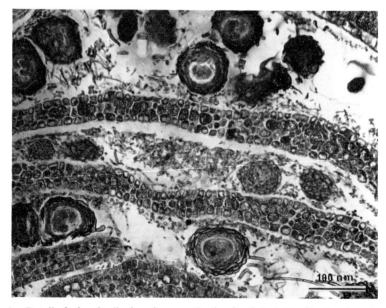

Fig. 4. Detail of a longitudinal section near the tip of an infected bud in October, showing cleistocarps in a dense matrix of mycelium between the bud scales. Ascospores are differentiating in several cleistocarps. The bud apex is to the right.

III. Main Features of the Disease

Within a wide range of conditions in spring the inoculum generated by the primary disease foci disperses on to neighbouring foliage and gives rise to secondary leaf and stem infections. At this stage most leaves on a plant are young, slow to attain full size and highly susceptible. Primarily infected shoots initiate a multitude of infections in localized concentrations (Fig. 1), whereas ascospores produce fungal colonies which are usually scattered throughout a hop garden. As the disease multiplies and gains height on the climbing shoots, conidia disperse more widely and the earlier differences in the distribution patterns of disease from the two sources are less marked. By June, powdery mildew may have spread generally unless effective control measures have been taken.

In response to infection, young susceptible leaves initially develop discrete humps or raised blisters, devoid of obvious fungal growth, which later bear

Fig. 5. Young, susceptible leaf of a glasshouse plant with raised blisters produced in response to infection. Colonies of the fungus then develop over the blisters.

the white, sporulating colonies typical of a powdery mildew (Fig. 5). Blisters were noted in the earliest reports of the disease (Marshall, 1798) and were described by Salmon (1917b). They appear to be an unusual reaction among powdery mildews and are suggestive of a localized hormonal imbalance. Blisters, sometimes chlorotic, are a common early indication of leaf infection in the field and are even more prominent on experimental plants grown in a glasshouse. When infection is severe, sporulating blisters coalesce and cause a puckering or distortion of leaves. No blisters form in response to infection of susceptible, older leaves nor of cones.

Susceptibility of hop leaves to infection declines as they age. Most infections occur on partly furled leaves and may appear as colonies on leaves of increasing age. However, there is some evidence that differences in susceptibility between leaves at successive nodes become less with increasing concentration of inoculum. Both leaf surfaces are intrinsically equally susceptible, although more mildew often develops on the lower surfaces of main bine leaves within 1·5 m of the ground in May/June because this surface is relatively protected from weather and fungicides. Such differences disappear higher up the plant where leaves are more exposed.

Since the hop plant produces an enormous leaf area, the commonly experienced levels of leaf infection are unlikely to affect the ultimate yield of cones. The sole importance in controlling leaf infection is therefore to eliminate conidial inoculum which, from July onwards, will otherwise invade the inflorescences or "burr" and then the young cones. Hop plants are dioecious, the cones (condensed inflorescences) of the female plant being the commercial product. On the continent and in most of the USA male plants are eradicated but in England it is customary to grow one male plant amongst every 150 females. As a result of pollination the female flowering period is shortened and the seeded cones grow out more quickly and become larger than seedless cones. Infection of the burr either prevents further development and results in hard, sterile knobs or it causes cones to grow out deformed (Fig. 6). This is usually considered to be a more serious effect on the crop than later attacks on maturing cones which can nevertheless still suffer localized restriction of growth. There have been moves in recent years to introduce seedless cropping in England to conform to regulations of the European Economic Community on seed content of hops. This would result in a longer burr period and conceivably increase the time during which mildew invasion could cause the most severe damage (Howard, 1905).

So far we have considered the so-called "white mould" form of the disease, the asexual stage of the pathogen, which can continue to multiply and directly damage the developing cones until they are harvested in September. Fruiting bodies (cleistocarps) of the sexual stage usually appear interspersed among hyphae and conidiophores of fungal colonies from July onwards. They develop most profusely on infected cones but also on leaves late in the

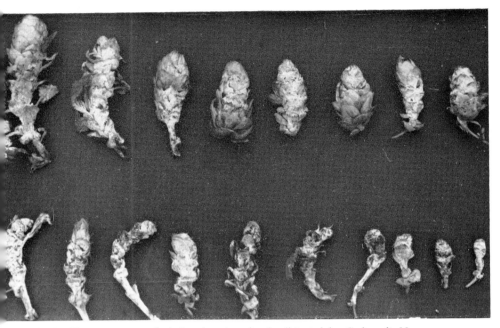

Fig. 6.   Hop cones severely infected and variously distorted by *S. humuli*. Numerous cleistocarps appear as black areas on some cones.

season. Cleistocarps form during virtually every outbreak of the disease and can develop so profusely as to turn infected cones black (Fig. 6). Infections bearing cleistocarps appear less powdery, due to suppression of conidial production, and are often the cause of a reddish colouration of cones, hence the growers' term "red mould" for this stage.

An insidious form of red mould can develop late in the season when cones may suddenly become discoloured and appear prematurely ripe. This rapid deterioration in their condition was first associated with powdery mildew by Hammond (1900) and its importance was re-emphasized by Coley-Smith (1964). Affected cones bear numerous cleistocarps but sparse fungal hyphae with few conidial chains. There is no cone distortion and it is supposed that infection occurs only at a late stage of cone development. The only opportunity for dealing with this problem is to prevent it by strict attention to control measures when cones develop.

The sexual stage can therefore contribute directly to damage of the crop but it is probably more important for carrying over the disease to the next season. Mildew-infected cones easily fragment and can transport countless cleistocarps to the ground during harvest. Severely infected hops, and mildly infected ones which may be surplus to requirements, are often left unpicked and unattended until after the cones have broken up naturally and fallen to the ground.

IV. Resistance

Salmon (1917a) first recorded resistance in the hop to powdery mildew and in a series of hybridization experiments investigated its inheritance and stability in glasshouse and field plants. Two types of resistance were recognized—(1) an immunity, in which some wild Italian male plants and a female cultivar called "Golden Hop" developed no signs of leaf infection after exposure to heavy inoculum in a glasshouse, and (2) a "semi-immunity" also in some plants from Italy, in which leaves of plants in a glasshouse responded to infection by producing blisters, as in a susceptible reaction (Salmon, 1917b), except that only a feeble, faintly white mycelium developed which then died to leave small brown patches of dead epidermal cells (Salmon, 1919a). Occasionally, some immune plants kept in a glasshouse became susceptible, later reverting to the immune condition, and others became susceptible in varying degrees in the field late in the season but Salmon could find no evidence that differential races of the fungus were responsible (Salmon 1919a,b, 1920, 1921, 1927).

Subsequently, during the present breeding programme at Wye, resistance reactions similar to those described by Salmon have been found and ascribed to the action of specific, dominant major genes, initially designated I_1, I_2, and B (Liyanage et al., 1973). Two of them (I_1 and I_2) confer immunity and may also have been responsible for the immune reactions noted by Salmon. The other (B) is closely linked to the inheritance of the sesquiterpene hydrocarbon, selinene, in the essential hop oil and gives a reaction (now called "resistant blister") apparently identical to that recorded by Salmon as "semi-immunity". It has not been possible to trace the origins of these reactions back to Salmon's material.

The resistant blister reaction (Fig. 7) is characteristic of glasshouse culture but it is also exhibited on field plants in response to high inoculum levels; otherwise B-genotypes appear immune in the field. On some of the plants exhibiting immunity, leaves appear as if they had not been challenged by the pathogen, but on others small chlorotic flecks of a hypersensitive reaction appear which were not recorded by Salmon.

Recent examination of the responses of a range of hop cultivars to infection by several isolates of S. humuli, shortly to be published, have suggested the presence of additional major genes for immunity. A genetic explanation for resistance based on a gene–for–gene relationship between the fungus and host is proposed in which five genes for resistance in the cultivars (R-genes) correspond to five genes for virulence in the pathogen. The R-genes relate to those postulated previously as follows:

Revised nomenclature: R_B R_1 R_2 R_3 R_4
Original nomenclature: B I_1 I_2 — —

Either gene R_B or R_2 is the basis of resistance to powdery mildew of four cultivars released into commerce in 1972–75 (see Section IXB, Control by resistant cultivars). Two further major resistance genes, R_5 and R_6, have been postulated in 1976.

Fig. 7.   Blisters associated with sparse fungal growth, characteristic of the response of plants of the R_B-genotype to infection.

V. Physiologic Specialization in *S. humuli*

Considerable variation in virulence appears to exist within natural populations of *S. humuli*. With the exception of a few of Salmon's cultivars which were not widely grown, all commercial cultivars grown before 1971 were susceptible. Selection pressure within pathogen populations would not therefore be expected to be great. However, marked differences were noted in the behaviour on various hop genotypes of eight isolates randomly collected in 1969 and 1970 from commercial hops growing in the major hop regions of England (South and West Midlands), from glasshouse plants at Wye and East Malling Research Station and from wild hedgerow hops in Rutland. Four physiologic races could be distinguished among the isolates,

Figs. 8–11. Stages in the development of a mildew pustule on a susceptible leaf.

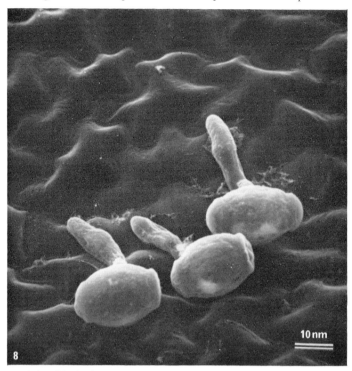

Fig. 8. Conidia with single germ tubes, 6 h after inoculation.

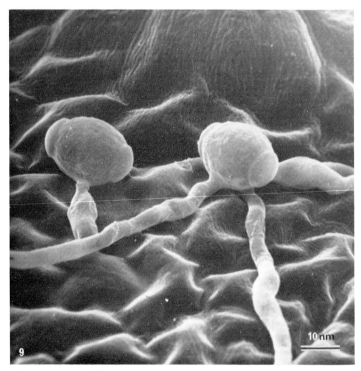

Fig. 9. Germinating conidia, one with three germ tubes, and early hyphal growth, 48 h after inoculation.

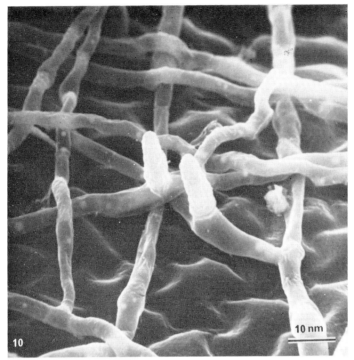

Fig. 10.   Young colony with conidiophore initials, 4 days after inoculation.

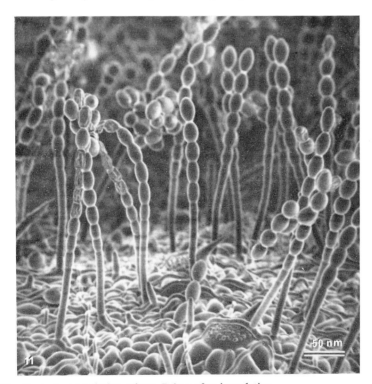

Fig. 11.   Part of a sporulating colony, 7 days after inoculation.

including one which allowed resistance in R_1-genotypes to be expressed for the first time at Wye since 1967 when it had been overcome after selection within the population of *S. humuli* prevalent in the area. None of the resistance genes was known to occur in either commercial or wild hops yet, interestingly, the isolate from Rutland was able to infect hosts with R_1 and R_4 in the experiments, suggesting that these genes may occur in wild hops.

Two further races of *S. humuli* were identified in 1973 and 1974 after they had overcome the resistance of genes R_2 and R_B, respectively. The physiologic races recognized so far are shown in Table 1.

TABLE 1

Postulated physiologic races of *S. humuli*

Race no.[a]
1, 3, 4
either O *or* B
1, 4
4
B, 4
1, 2, 3, 4

[a] Nomenclature relates to that of *R*-genes for resistance in the host.

VI. Pathogen Interactions with Host Genotypes

Some comparative studies on relationships of *S. humuli* with the host surface and its tissues in compatible and incompatible combinations have been carried out by Liyanage (1973), using several pathogen races and host genotypes. In this work, all observations were of the fungus on leaf discs whose growth was favoured by keeping them at $17 \pm 2°C$, 100% r.h. and under fluorescent lamps (4500 lux) to provide a 16 h photoperiod.

A. Interactions on the host surface

In compatible interactions, the first germ tube emerges from a conidium 6 h after inoculation (Fig. 8), produces no obvious appressorium and penetrates the host and establishes a haustorium within 12–15 h. Infection stimulates further development which, in terms of hyphal mass, then accelerates. After 48 h, the initial germ tube becomes a branching hypha and up to three more germ tubes arise (Fig. 9). The colony begins to abstrict conidiophore initials after 96 h (Fig. 10), and the first conidia are apparent shortly afterwards; sporulation intensifies from 5–7 days after inoculation (Fig. 11). The sequence proceeds without undue modification of the host surface.

In all incompatible combinations, i.e. when the chosen isolate allows expression of *R*-gene resistance, the pattern of fungal development until attempted host penetration 12–15 h after inoculation, is as on congenial hosts; conidial germination (emergence of the first germ tube) proceeds equally well. Subsequently, fungal interactions with host cytoplasm then differentiate compatible from incompatible reactions.

Gene R_B (resistant blister) allows mycelial growth which by 30 h is noticeably more retarded than on a susceptible host. Most conidia produce only one or two germ tubes although four occasionally arise. Sporulation is relatively weak—conidiophores are more sparse and the conidial chains are shorter. The host surface shows collapse of epidermal cells, even in advance of the hyphae (Fig. 12).

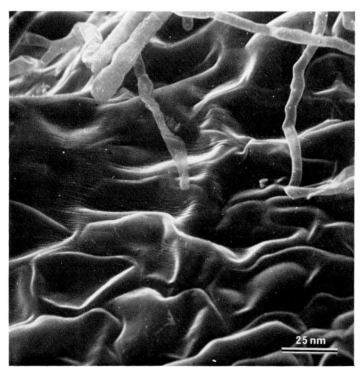

Fig. 12. Edge of a weak fungal colony on a leaf of the R_B-genotype with depressions in the leaf surface which are not artifacts but caused by collapse of epidermal cells in advance of the colony, 7 days after inoculation.

The R_2 gene conditions a more extreme response. The pathogen fails to develop beyond one germ tube. No other germ tubes arise and after 96 h the germinated conidium has collapsed (Fig. 13). The host surface undergoes no obvious change. Surface responses to the other immunity genes have not been examined in detail.

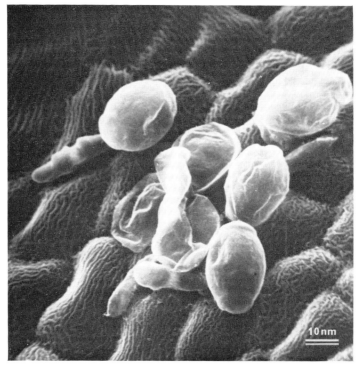

Fig. 13.   Germinated conidia which have collapsed after attempted penetration of cells of the R_2-genotype, 4 days after inoculation.

B. Interactions with host tissues

As many as three haustoria are produced in epidermal cells of congenial hosts, although one is most common. Such haustoria comprise a nucleated haustorial body contained in a matrix bounded by the host plasmalemma (Fig. 14). Electron micrographs show that the haustorial body produces lobed outgrowths which extend its surface within the sheath matrix, as in *S. pannosa* (Perera and Gay, 1976). The cytoplasm of the infected epidermal

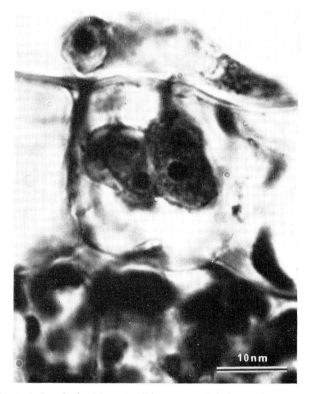

Fig. 14.   Haustoria in a leaf epidermal cell in a compatible interaction.

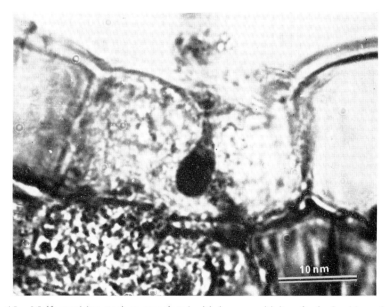

Fig. 15.   Malformed haustorium associated with hypersensitivity of a leaf epidermal cell of the R_B-genotype. Note also disorganization of neighbouring mesophyll cells.

cell remains organized as do the contents of the adjacent cells of the palisade mesophyll.

In contrast, haustoria in cells of uncongenial hosts either differentiate and then lyse, differentiate abnormally or are not produced at all. The exact development pattern depends upon the particular pathogen race:host genotype combination. Generally, the R_B gene allows the first haustoria, produced 12–15 h after inoculation, to form normally. By 21 h these have begun to lyse but not before supporting some hyphal development. Subsequent haustoria either behave similarly or are malformed at the outset, without clearly differentiated organelles. Dying or malformed haustoria are associated with disrupted epidermal and palisade cells which have reacted hypersensitively to the pathogen (Fig. 15). In immune hosts, haustoria are always malformed or else, with the R_2 genotype, are never produced and the epidermal cells react violently to attempted penetration. The size and frequency in the host cells of haustoria varies in proportion to these reactions.

VII. Epidemiology

Fundamental relationships between climatic factors (e.g. temperature, atmospheric moisture) and stages in the disease cycle (e.g. sporulation, colonization) have not been investigated for hop powdery mildew in controlled experiments. A few early workers (e.g. Steiner, 1908) studied interactions between other host plants and supposed forms of *S. humuli* now considered to be different species.

All kinds of weather have been associated with disease outbreaks in the past though only Blodgett (1913, 1915) made an earnest attempt to explain the influence of weather on the disease. From charts depicting components of weather and occasions when mildew infections were first seen in seven locality/year combinations, he inferred that rainy periods were responsible for outbreaks of mildew which arose 10–14 days afterwards. He considered that this explained why the worst epidemics appeared to develop in wet seasons. Re-examination of Blodgett's data suggests that they can be interpreted in several ways. The frequent occurrence of rainfall and of mildew infections in most of the years he studied makes it difficult to assign a positive role to rain. Blodgett's conclusions were probably influenced by the then common belief that both ascospores and conidia of *S. humuli* require water for germination. We know now that this may be true for ascospores but not for conidia whose germ tubes will emerge and develop on dry leaf surfaces in various atmospheric humidities. Furthermore, recent experience of severe mildew in dry weather conflicts with Blodgett's conclusions and it seems we must look elsewhere to explain behaviour of this disease.

Investigations at Wye College since 1973 have also taken a field approach but have aimed at identifying and measuring the factors responsible for disease fluctuations in hop gardens. Such information is considered im-portant for further rationalization of control measures. A few general remarks can be made at this early stage in the work.

The upper graph in Fig. 16 shows the number of leaf infections which developed, after incubation in standardized conditions, on batches of healthy

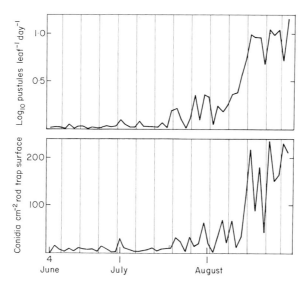

Fig. 16. Daily leaf infections on potted plants serially exposed in a hop garden (upper portion), and corresponding numbers of conidia caught by rod traps in 1974. The vertical lines represent periods of 3 days without data.

trap plants exposed for 24 h periods in an unsprayed hop garden in 1974. This pattern of disease is typical of that in the field in seasons in which the weather favours powdery mildew. Whether the disease originates from ascospores or from primarily infected shoots, there is a lengthy period in the early season when secondary infections remain at a relatively low level. Marked increases in infection begin around mid-July but these fluctuate appreciably before reaching high levels by mid-August. Like other powdery mildews (see Butt, this volume, Chapter 3) the hop pathogen establishes some infection on most days during the summer. Occasions when infection is zero are rare and they coincide with an absence of conditions which promote inoculum production. There is a close correspondence between the patterns of daily infection and of conidia caught in spore traps (Fig. 16). Preliminary multiple regression analysis has shown that inoculum explains more of the variation in infection than any other environmental variable even though the efficiency with which spores produce colonies varies considerably with climatic conditions.

In common with powdery mildews in general, sporulation in *S. humuli* is a rhythmic process and conditions a diurnal fluctuation in conidial release which on dry days begins at 0800 h GMT, reaches a maximum at 1300 h and then declines to a low level by 1900 h which is usually maintained through the night (Royle, 1967). Transient increases in the number of airborne spores are brought about by the onset of rain which, if it continues, removes all conidia from the air. Occasionally spore flights are detected in dry weather during darkness but these are unexplained. The factors governing variation in daily spore production are under investigation. So far, there is evidence that dew-wetness in the night previous to release of conidia favours their production and that high vapour pressure deficits are deleterious.

VIII. Disease Appraisal

The appearance of powdery mildew in a hop garden has always alarmed growers because of its property of rendering a crop totally unsaleable. As noted already, there have been many occasions in the past when crops have become worthless because of the disease. The scale of these losses has never been measured. Similarly, we have no information on quantitative effects on crop yield and quality and it is becoming increasingly important to know the levels of disease which can be tolerated without incurring a crop reduction. However, effects of powdery mildew on hop production are difficult to assess, not least because the criteria for judging economic loss relate to different stages in the preparation of the crop for market. These criteria—the weight of crop, its content of alpha-acid (see below) and the valuation of the final product—are each influenced by powdery mildew, though to what extent is unknown.

Indirect effects of the disease, such as that of leaf infection, are practically impossible to estimate and considerable problems also attend evaluation of direct effects. One difficulty is in acquiring a meaningful measurement of disease in a plant so large as the fully-grown hop. Also, the organization of plots which are representative of different disease levels is complicated by great variation in disease intensity between plants and across commercial gardens. Although area diagrams quantifying the disease on leaves and cones have been prepared, they require improvement since so far they are of value only for recording high levels of disease, as in unsprayed, susceptible crops. Moreover, the procedures of sampling plant units for disease/yield studies have not been investigated.

Despite little attempt to relate measurements of disease to the weight and quality of the crop, it has always been assumed, quite reasonably, that early attacks of the disease on the burr and young cones cause most appreciable reductions in crop weight. Some additional loss may occur because cones, infected later in their development, become brittle and easily shatter during machine-harvesting. Many of the cone fragments and small shrivelled cones are thereby rejected by the machine and so do not contribute to the yield.

The alpha-acid is formed in glands borne on the cone. It is the most important hop constituent in the brewing process where it is converted to iso-alpha-acid which gives beer its bitterness. The effects of powdery mildew on alpha-acid content (usually 5–10% by weight of cones) is also unknown but is likely to be most adverse in early attacks which prevent the formation of the glands.

The price paid to the grower for his hops is determined by their content of alpha-acid and by their appearance after being dried and pressed. The appearance is judged empirically and is mainly based on colour, which is an indication of ripeness, the conditions in which the hops were dried and their freedom from damage by pests and diseases. This valuation can be adverse if powdery mildew is visible as white mycelium and as cleistocarps and if the hops appear overripe due to the discolouration caused by the disease.

From these observations it might be thought that to demonstrate some sort of relationship of disease intensity with productivity of hops would be straightforward. However, the results from a fungicide trial, conducted in 1972 (Royle and Griffin, 1973) shows that this is not so. Powdery mildew was recorded in random samples of cones from 12 plots and ranged from 2·3–17·0% cones infected according to the success of fungicide treatment. No correlation could be found between level of disease and measurements of weight and alpha-acid content (Table 2). In contrast downy mildew, which

TABLE 2

Correlation coefficients between infection and yield and quality of hop cones in a fungicide trial, 1972 (Royle and Griffin, 1973)

	Cone wtah^{-1}	wtha^{-1}	Alpha-acid content (%)
Percentage of cones with:			
Powdery mildew	−0·139	0·148	0·105
Downy mildew	−0·811^a	−0·726^b	0·354

Significantly different from 0: ^{a} at $< 1\%$ ^{b} at $< 5\%$ level (n = 12)

in the same trial infected 0–37·7% of the cones, significantly depressed yield. The lack of a relationship for powdery mildew was probably due to late fungal attack on the cones which at harvest were not unduly deformed nor discoloured. Interference from downy mildew was also undesirable and possibly impaired the chances of success.

IX. Control

A. Cultural and sanitation methods

Modern recommendations for assisting control of powdery mildew are largely unchanged from those advocated by Salmon (1921). Most measures are rationally based and well known but some are often not adopted because they cannot conveniently be accommodated within farm work schedules. It is therefore difficult to assess their contribution to control in practice but indications are that it is not great. However, increasing problems with the disease in recent years have re-emphasized the importance of cultural and hygienic measures, a few of which are carried out as a routine part of hop garden management.

The lower leaves and extraneous shoots are traditionally the first to become infected and their removal helps to reduce progressive infection up the plant. This process was once done by hand but defoliant chemicals are now widely used in late May or early June to remove all surplus growth up to a height of about 1 m. Since the disease tends to arise earlier in the season nowadays, chemical stripping is often performed too late to prevent substantial disease spread.

Maintaining good aeration by removal of basal growth and by attention to plant spacing and methods of training has also long been recommended though it is likely that any advantages of an open growth habit could be offset by greater opportunities for inoculum dispersal. Balanced fertilizing has been advocated for control since the earliest times (Marshall, 1798) and we now know that heavy applications of nitrogen encourage soft, sappy growth of foliage on which S. humuli thrives. Growers nowadays are generally more discriminate in their use of fertilizers mainly because of the proven association between nitrogen and the incidence of Verticillium wilt (Sewell and Wilson, 1967).

One of the most important sanitary measures to reduce seasonal carry-over of disease is to collect and destroy infected hops before the over-ripe cones break up and scatter over the hop garden. To abandon a diseased crop in the field is a highly dangerous practice, particularly since ploughing hop gardens was superseded by non-cultivation systems. Nevertheless, many growers regard the harvesting of an unwanted crop as an inconvenient operation and in the past, as today, many infected crops have been allowed "to blow". Careful attention to the removal of unpicked hops in the 1960s would undoubtedly have facilitated better disease control. The importance of this practice is well illustrated by the results of a survey carried out in April 1972 (Liyanage and Royle, 1976) which showed that severe infection of young growth in half an area of 300 plants coincided with dense cone debris on the ground from an unpicked crop the previous year. The debris contained many cleistocarps, a large proportion of them dehisced. The remaining portion of the garden had been harvested, there was little debris and infection was slight (Fig. 17).

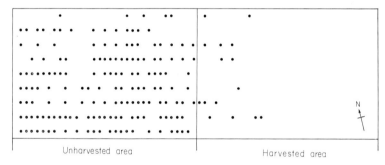

Fig. 17.   Distribution of plants cv. Bramling Cross (•) with infections which originated from ascospores in part of a commercial hop garden in April, 1972.

An additional hazard after harvest arises from infected shoots which can re-grow in some cultivars from the remaining bine bases. Conditions in the autumn seem to favour profuse fungal development on these shoots, often with abundant cleistocarps, and eradication of these high inoculum sources by fungicides or defoliant chemicals is important.

B. Resistant cultivars

With the exception of three of Salmon's cultivars, Sunshine, Early Choice and Copper Hop, which were grown on a small scale in England before 1963, all the hop cultivars grown in Europe until 1971 were susceptible to powdery mildew. In that year Wye Challenger was introduced as resistant to both downy and powdery mildews and it has since been widely planted in areas free from wilt. The following year Wye Target, resistant to powdery mildew

and wilt but susceptible to downy mildew, was made available. This cultivar
is now popular in the wilt-infested regions of the south of England. In
1975 two further powdery mildew-resistant cultivars, Wye Saxon and
Wye Viking, were released. Fig. 18 shows the area which resistant cultivars
have occupied since their introduction.

Fig. 18. Annual UK area planted with hop cultivars highly susceptible (—•—) and
resistant (—○—) to powdery mildew, 1965–76.

These cultivars all owe their powdery mildew resistance to major genes.
Resistance of Wye Challenger is based on gene R_B, which confers the resistant
blister reaction, though it also contains R_1. For three years on commercial
farms (and for eight years in the field at Wye College) this cultivar remained
free from both mildew diseases and required no fungicide. Then in 1974, it
became infected on a few farms and tests showed this to be due to a new
pathogenic race. This race $(B, 4)$ is now fairly widespread and has caused
some troublesome outbreaks which have necessitated control by fungicides.
Wye Challenger is still resistant to downy mildew. Wye Saxon and Wye
Viking also rely on gene R_B and so have become susceptible with Wye
Challenger in some areas.

Wye Target possesses the R_2 gene and has so far remained free from disease
in the field. Some potted plants became slightly infected in a glasshouse at
Wye in 1973 due to yet another race $(1, 2, 3, 4)$ which has not since re-appeared.

In the past, all susceptible cultivars have suffered severely from powdery
mildew and it is difficult to distinguish degrees of susceptibility between most
of them. However, Northern Brewer is notoriously the most sensitive and
since 1962 has been attacked more severely than has any other cultivar. In the
experience of growers, Wye Northdown (a descendant of Northern Brewer)

and Bramling Cross are also more susceptible than most other cultivars though distinctly less so than Northern Brewer. A considerable increase in the area planted to Northern Brewer, and possibly also to Bramling Cross, (Fig. 18) was certainly a factor in the increase of disease over the last 15 years.

C. Fungicides

1. Fungicide programmes

Ever since they were first advocated, fungicidal programmes have aimed at controlling the disease as soon as it became evident in hop gardens. With improved knowledge of the disease and its changing behaviour in response to modified cultural practices the start of such control schedules has been advocated progressively more early in the season.

The first programme, advocated 120 years ago using sulphur dusts (White-head, 1870), was surprisingly logical. It stipulated an initial application in June, whether or not the disease had appeared, followed by two more after a 10-day and then a 21-day interval. From his studies on the role of ascospores, Salmon (1921) called for an earlier start to sulphuring, in May, and suggested a more flexible later programme with applications at about 14-day intervals until the bines had reached the top wire, followed by further dusts according to disease severity. Both Salmon and Blodgett (1913, 1915) recognized the importance of applying sulphur to protect the inflorescences and to continue if necessary on the maturing cones.

Until the 1960s, recommendations were based firmly on Salmon's guide-lines using sulphur. Some past suggestions for timing of fungicide applica-tions are now recognized to be inadequate. For example, it was commonly considered that the first application should await the appearance of mildew spots, usually in mid-May. Nowadays we recognize the importance of applying fungicides much earlier to highly susceptible cultivars and in gardens with a recent history of severe mildew. A particularly early start, in April, is needed when the disease appears as primarily infected shoots. Another frequent suggestion, now recognized to be dangerous, was to cease fungicide applications if there appeared to be no further disease after burr. However, experience now shows that severe late attacks on cones can develop rapidly from slight inoculum foci and resumption of chemical control can then be to no avail.

Except for some specific scheduling of fungicides at the beginning and end of the vegetative growing period, the remaining treatments have always been applied in a relatively arbitrary fashion and not related to particular features of crop growth nor of the disease cycle. Broadly, fungicides are applied at 10–14-day intervals in order to provide continuous protective cover on a plant

406 D. J. ROYLE

whose extension growth can exceed 25 cm each day. When a grower has not
been sufficiently diligent with earlier control measures and has failed to check
the disease, this frequency may need to be increased to 7–10 days. Difficulties
in chemical control now most commonly arise from too long a delay in
starting control or from an interruption of the routine programme.

2. Fungicide usage.

Figure 19 shows the proportion of the English hop area treated, between 1965
and 1974, with various fungicides in relation to the control of powdery mildew.

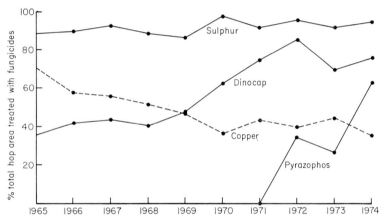

Fig. 19. Annual UK usage of fungicides of interest in hop powdery mildew control,
1965–74 (incorporating data of Umpelby and Sly, 1973).

Sulphur has been the mainstay of control since early times and has consistently
been applied in some degree to 90 % of the area as a dust, a wettable powder or
as colloidal formulations. Until 1972 sulphur and dinocap were the only fungicides available. As now, these chemicals often alternated as sprays in control
programmes or were applied together, usually each at half strength. Although
unpopular because of its unpleasantness to field workers, sulphur as a dust is
still the only proven fungicide to offer reasonable control of powdery mildew
within developing cones. Whether it should be applied in sunshine, when
sulphur gases are supposedly evolved, or during the night or early morning,
when dew assists deposition, has always been a vexed question which even
today is unresolved.

In growers' attempts to counteract the deteriorating disease situation,
dinocap usage increased steadily from an average of three applications on
36 % of the area in 1965 to four applications on 86 % of the area in 1972. Dur-

ing this period it substituted completely for sulphur in some growers' schedules and today is still widely used.

Pyrazophos, an organophosphorus fungicide with aphicidal side-effects was introduced to hop growing in 1972 and instantly applied on an average of two occasions to 30% of the acreage. This fungicide has systemic properties and has since given good control of the disease in difficult situations. By 1974 the area treated had risen to 63% and there can be no doubt that it has appreciably reduced the severity of the disease. Even wider usage is prevented by its high cost and tendency to cause some phytotoxicity early in the season when it is most beneficial. Pyrazophos appears to be less effective for control of the disease on cones than on leaves.

Although not used directly for control of powdery mildews, copper fungicides have been claimed to have beneficial side-effects (Aerts and Soenen, 1957; Moore, 1966). Conversely, dithiocarbamate fungicides have been considered to favour the disease (Horsfall and Lukens, 1966). A decline in the use of copper fungicides in favour of organic materials for control of hop downy mildew since the early 1960s (Fig. 19) has been held as an important factor favouring the deterioration of powdery mildew control over this period (Royle, 1975). The main evidence for this conclusion derives from the results of a commercial fungicide trial in 1972 where complete programmes of copper oxychloride, zineb or propineb were superimposed on to a routine dinocap schedule and compared for their effects on powdery mildew (Royle and Griffin, 1973). Copper oxychloride was significantly superior to zineb in restricting the disease on cones and was also somewhat better than propineb (Table 3). All fungicides controlled downy mildew equally well. How copper

TABLE 3

Percentage of cones infected with *S. humuli* at harvest in response to seasonal programmes of three downy mildew fungicides superimposed over dinocap (Royle and Griffin, 1973)

Plot No.	Zineb	Propineb	Copper Oxychloride
1.	7·3	3·0	5·3
2.	12·0	7·0	5·3
3.	17·0	—	5·7
4.	13·0	4·6	2·3
5.	7·0	—	—
Mean	11·3	4·9	4·7

exerts this effect is unknown. It is possible that it increases cuticle thickness (Bärner and Roder, 1964) and sugar content of epidermal cells (Horsfall and

Dimond, 1957) and it is widely believed by hop growers that whereas copper hardens leaves and cones, organic fungicides encourage soft, sappy foliage which is more susceptible to powdery mildew. None of the systemic fungicides, including benomyl, developed in the 1960s have been widely used against hop powdery mildew since they were either ineffective or were uneconomic to apply. Of the more recently introduced materials triforine is now widely advertised for control of cone disease in West Germany. This compound and triadimefon are showing promise in trials and their future depends on the residue levels found in treated hops and on their effects on yield and quality. These aspects are under investigation.

References

Aerts, R. and Soenen, A. (1957). Höfchen Br. Bayer PflSchutz-Nachr. **10**, 109–172.
Bärner, J. and Roder, K. (1964). Z. PflKrankh. **71**, 210–215.
Blodgett, F. M. (1913). Bull. Cornell Univ. agric. Exp. Stn **328**, 278–310.
Blodgett, F. M. (1915). Bull. New York agric. Exp. Stn **395**, 29–80.
Bondarzew, A. S. (1908). Bolez. Rast. **2**, 16–28.
Burchill, R. T. (1958). Rep. agric. hort. Res. Stn, Univ. Bristol, 1957. 114–123.
Coley-Smith, J. R. (1964). Rep. Dep. Hop Res. Wye Coll. 1963, 30–31.
Covey, R. P. (1969). Pl. Dis. Reptr **53**, 710–711.
Hammond, W. H. (1900). Jl. S.-e. agric. Coll. Wye **9**, 19–20.
Horsfall, J. G. and Dimond, A. E. (1957). Z. PflKrankh. **64**, 415–421.
Horsfall, J. G. and Lukens, R. J. (1966). Bull. Conn. Agric. Exp. Stn **676**, 7–8.
Howard, A. (1905). Jl. S-e. agric. Coll. Wye **14**, 211–219.
Jackson, G. V. H. and Wheeler, B. E. J. (1974). Trans. Br. mycol. Soc. **62**, 73–87.
Kohlmann, H. and Kastner, A. (1975). Der Hopfen. Hopfen-Verlag, Wolnzach, Germany.
Liyanage, A.de S. (1973). Studies on resistance and overwintering in hop powdery mildew. Ph.D. thesis. University of London.
Liyanage, A.de S. and Royle, D. J. (1976). Ann. appl. Biol. **83**, 381–394.
Liyanage, A.de S., Neve, R. A. and Royle, D. J. (1973). Rep. Dep Hop Res. Wye Coll. 1972, 49–50.
Magnus, (1899). Bot Zbl. **77**, 9–10.
Marshall, H. W. (1798). "The Rural Economy of the Southern Counties". Vol. I, London.
Moore, M. H. (1966). Ann. appl. Biol. **57**, 451–463.
Neve, R. A. (1964). Rep. Dep. Hop Res. Wye Coll. 1963, 8.
Perera, R. and Gay, J. L. (1976). Physiol. Pl. Path. **9**, 57–65.
Preece, T. F. (1965). Pl. Path. **14**, 83–86.
Price, T. V. (1970). Ann. appl. Biol. **65**, 231–248.
Royle, D. J. (1967). Rep. Dep. Hop Res. Wye Coll. 1966, 49–56.
Royle, D. J. (1975). Ann. appl. Biol. **81**, 441–442.
Royle, D. J. and Griffin, M. J. (1973). Pl. Path. **22**, 129–133.
Salmon, E. S. (1900). Mem. Torrey bot. Club **9**, 1–292.
Salmon, E. S. (1907). J. agric. Sci. Camb. **2**, 327–332.
Salmon, E. S. (1914). J. agric. Sci. Camb. **6**, 187–193.
Salmon, E. S. (1917a). J. agric. Sci. Camb. **8**, 455–460.
Salmon, E. S. (1917b). Ann. appl. Biol. **3**, 93–96.
Salmon, E. S. (1919a). J. Genet **8**, 83–91.

Salmon, E. S. (1919b). *Ann. appl. Biol.* **5**, 252–260.
Salmon, E. S. (1920). *Ann. appl. Biol.* **6**, 293–310.
Salmon, E. S. (1921a). *J. Minist. Agric. Fish.* **28**, 1–11.
Salmon, E. S. (1921b). *Ann. appl. Biol.* **8**, 146–163.
Salmon, E. S. (1927). *Ann. appl. Biol.* **14**, 263–275.
Sewell, G. W. F. and Wilson, J. F. (1967). *Ann. appl. Biol.* **59**, 265–273.
Steiner, J. A. (1908). *Zentbl. Bakt. ParasitKde* **2**, 677–736.
Umpelby, R. A. and Sly, S. M. A. (1973). *Pesticide Usage: Survey Report 5. Hops.* Ministry of Agriculture, Fisheries and Food.
Whitehead, C. (1870). *J. R. agric. Soc.* II, **6**, 336–366.
Whitehead, C. (1890). *J. R. agric. Soc.* III, **1**, 321–348.
Wiedemann, R. (1972). *Hopf. Rdsch.* **21**, 387–389.

Chapter 16

Powdery Mildews of Ornamentals

B. E. J. WHEELER

Imperial College Field Station, Silwood Park, Berks, England

I. Introduction

Diseases such as powdery mildews which disfigure leaves and blemish flowers defeat the very purpose of ornamentals, as plants grown to adorn our gardens and to decorate our homes. In these situations the aesthetic loss is paramount. Except for ornamentals grown commercially for sale to the public, the economic loss from these diseases is negligible since the powdery mildew fungi seldom kill the ornamentals which they parasitize. The leaves of the annual *Myosotis* species may become whitened with the ectotrophic fungal growth but the plants survive, even if they do appear unthrifty and unsightly. Similarly, the perennials such as rose and phlox renew their growth in the next season, sometimes with no apparent indications of severe mildew in the preceding year. The ability of these fungi to coexist with their hosts throughout the growing season is obviously advantageous to them in view of the obligate nature of their parasitism, but lack of a dramatic death of the plant has, perhaps, led to a certain tolerance of these diseases by the amateur grower and a lack of interest in them by the professional plant pathologist. Thus of the many ornamental plants on which species of the Erysiphaceae have been recorded, relatively few figure in studies of powdery mildew diseases as reflected in the yearly indexes of the 49 volumes of the *Review of Applied Mycology* (1922–69) and the succeeding volumes of the *Review of Plant Pathology* (1970 onwards). Only where the plant is pre-eminent as an ornamental and is extensively cultivated e.g. the rose, or where it is grown commercially especially in glasshouses and thus can involve financial loss e.g. begonias and chrysanthemums, does there seem to have been the necessary impetus for a scientific investigation of the respective powdery mildew disease. Inevitably any account of powdery mildews of ornamentals must reflect this.

In this review these diseases are considered in three categories.

1. Powdery Mildews which appear regularly on a widely-grown ornamental and which have attracted considerable scientific interest especially concerning control. Only one disease merits inclusion—rose powdery mildew.

2. Powdery mildews of somewhat sporadic occurrence but which have then been sufficiently important to attract scientific interest in the fungus and its control. Those considered are the powdery mildews of antirrhinum, begonia, cherry laurel (*Prunus laurocerasus*), chrysanthemum, clematis, delphinium, hydrangea, japanese spindle tree (*Euonymus japonicus*), Kalanchöe (*K. blossfeldiana*), lilac (*Syringa vulgaris*), lupin, phlox and sweet pea (*Lathyrus odoratus*).

3. Powdery mildews which have been recorded but seldom investigated.

In dealing with these fungi on ornamentals the problems of nomenclature loom large. Plant pathologists work mainly with the conidial states of these fungi, particularly when dealing with ornamentals grown under glass. In many instances these have been referred unwisely to genera such as *Erysiphe* or *Sphaerotheca* with no real assurance from collections of cleistocarps on the same host. There is also the problem of host range. It seems likely from the experiments of Hammarlund (1945) and Blumer (1951b, 1952) that a range of cultivated plants can be infected successfully by conidia from one host, especially in glasshouses. Hammarlund suggested that there was a plurivorous species which he designated *Erysiphe polyphaga* but, unfortunately, did not publish validly under International Rules. Regretfully, the use of this name by some authorities but not by others has caused further confusion in the naming of the fungi on ornamentals. Problems such as these are dealt with elsewhere in this book but inevitably they intrude into the accounts which follow. In these the causal fungi are named primarily with reference to the monographs of Salmon (1900) and Blumer (1967).

II. Rose Mildew

The rose has long held a special place amongst ornamentals. It has been cultivated since the earliest civilizations and a certain mystery has surrounded its development as a decorative plant. Apparently the first account of mildew on the rose was that of Theophrastus around 300 B.C. (Coyier, 1961) but many centuries passed before this was rightly attributed to a fungal pathogen.

A. The pathogen

Taxonomy

The fungus was first described by Wallroth in 1819 as *Alphitomorpha pannosa*, was transferred to the genus *Erysiphe* as *E. pannosa* by Fries in 1829 and was then placed in the genus *Sphaerotheca* which Léveillé described in 1851 (see Junnell, 1967). It has remained as *S. pannosa* (Wall. ex Fr.) Lév. apart from a recognition by some authorities of the division of this species by Woronichine (1914) into two varieties: *rosae* infecting roses and *persicae* infecting peach and almond.

Some workers have supported the view that this is not the only species of the Erysiphaceae on roses. Salmon (1900) in his monograph considered most of the specimens from North America to be *Sphaerotheca humuli* and many mycologists subsequently followed Salmon in thinking that whilst in Europe *S. pannosa* was the cause of rose powdery mildew, in North America this could be attributed to *S. humuli*. This concept of two distinct species on roses has been retained by Blumer (1967) in his revision of the Erysiphaceae in Europe. He, however, identifies the second fungus as *Sphaerotheca macularis* (Wallr. ex Fr.) Magnus, a species not considered by Salmon to be distinct from *S. humuli*. A re-appraisal by Coyier (1961) in which he examined herbarium and fresh material together representing collections from 32 states within the USA and from 22 foreign countries cast doubts on the separation

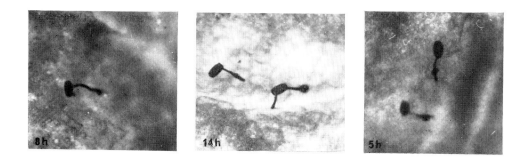

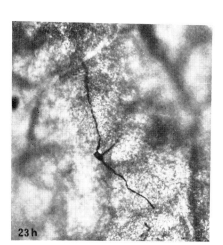

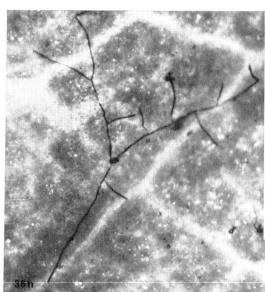

Fig. 1.   Development of *Sphaerotheca pannosa* on leaflets
of rose (cv. Frensham) at 20°. (From Price. 1969.)

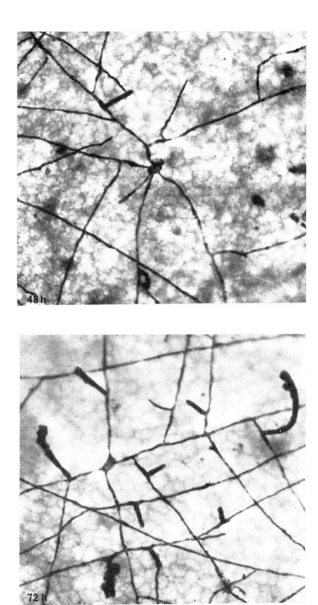

Fig. 1 *continuation*

of the two fungi *S. pannosa* and *S. humuli* sensu Salmon, and most workers there consider that in the USA all powdery mildew on the rose is caused by *S. pannosa* var. *rosae*.

Biological specialization

There are some indications of biological specialization within *S. pannosa*. For example, Yarwood (1952) found that infection of apricot leaves by *S. pannosa* from peach or *Rosa banksiae* resulted in large necrotic lesions but when the source was the rose cultivar Dorothy Perkins the lesions were small. In inoculations of detached shoots Coyier (1961) showed that conidia produced on *Rosa virginiana* were not capable of infecting the cultivars Red Garnette, Queen Elizabeth, Christopher Stone or Triomphe d'Orleans, though under the same conditions sporulating colonies developed on newly-inoculated *R. virginiana* in six days. Conversely, conidia from the rose cultivars Triomphe d'Orleans and Christopher Stone did not produce sporulating colonies on *R. virginiana*. In an extension of this work, Mence and Hidebrandt (1966) found that conidia from *R. virginiana* only infected detached leaflets of two rose cultivars (Fusilier and Better Times) from a total of seventeen cultivars which were tested, though they readily infected *R. virginiana* and *R. rugosa*.

Mycelial growth and conidial production

In common with many other powdery mildew fungi the mycelium of *S. pannosa* grows over the surface of the host with only limited intrusions by the haustoria, into the epidermal cells. The sequence of events shown in Fig. 1 for the development of *S. pannosa* on detached leaflets of the cultivar Frensham is typical of that found by other workers who have examined the growth of this fungus on leaf tissue (e.g. Corner, 1935; Longree, 1939; Coyier, 1961; Price, 1969, 1970; Perera, 1972). At 20°C and near 100% r.h., conidia start to germinate between 2h and 4h after deposition on the leaf, by producing a short tube, the primary germ tube, from one side of the conidium. This germ tube elongates and enlarges and by 6 h forms a club-shaped appressorial initial. A septum then develops at the proximal end of the appressorial initial and the initial enlarges into a mature appressorium. Between 8 h and 10 h a second germ tube is produced, usually on the opposite side but occasionally on the same side as the primary germ tube, and this coincides with the elongation of the primary germ tube from the distal end of the appressorium. Between 16 h and 20 h haustorial initials can be detected, with suitable clearing of the tissue, in the epidermal cells directly beneath appressoria. At this stage a third germ tube usually develops from a polar position. By 20–24 h mature haustoria can be seen in cleared cells and these appear as glistening spherical bodies, though E.M. studies now show these to be complex, lobed structures (see Chapter 9, this volume). After initiation of the third germ tube, mycelial growth is rapid with much branching of the germ tubes between 22 h and 36 h. By 48 h conidiophore initials form and by 72 h conidial chains are produced.

The development of the conidiophore of *S. pannosa* is described in detail by Foëx (1926). The conidiophore initial develops as a swelling on the hypha immediately above the nucleus. This initial elongates into a short tube which becomes separated from the hypha by a septum once the nucleus or a daughter nucleus formed by its division passes into it. The nucleus of the conidiophore initial then divides and a septum is formed to divide the initial into two cells. The basal cell becomes the pedicel of the conidiophore while the upper cell undergoes further division to form six to eight cells, the topmost of which become barrel-shaped as they mature into conidia. The size of the conidia vary slightly with the cultivar of the host on which they are growing (Coyier, 1961); the average length varies between 22·9 and 28·6 μm and the width between 13·6 and 15·8 μm (Bouwens, 1924). The fine structure of the conidia has been described by Akai et al., (1966).

The conidia of *S. pannosa* adhere to one another even after they appear mature but in common with most powdery mildew fungi they show a diurnal cycle of maturation and abstriction (Childs, 1940) which leads to diurnal periodicities of conidia in the air around rose bushes (Pady, 1972; Tammen, 1973). In a day free of rain, increasing numbers of conidia are released into the air as the r.h. decreases. Numbers reach a peak from midday to early afternoon and then decline as the conidiophores are depleted of mature conidia (Fig. 2).

Formation of cleistocarps

As *S. pannosa* develops on rose bushes it frequently forms a different mycelium from the thin evanescent type observed in the colonization of young leaves. This "secondary" mycelium consists of rather straight hyphae, c. 6 μm wide with much thickened walls and an almost-obliterated lumen. It usually persists in patches as a thick felt, the so-called "pannose mycelium" from which the species gets it name. It is predominantly within this mycelium that the ascocarps (variously termed cleistocarps, cleistothecia or perithecia) are found on some cultivars. These are globose to pyriform, 85–120 μm in diameter, usually c. 100 μm with a few, tortuous, mycelioid appendages which are usually short, pale brown and septate. The single ascus is broadly oblong to globose, 88–115 μm (average 100 × 60–75 μm) and contains eight ascospores, 20–27 × 12–15 μm (Salmon, 1900).

Cleistocarps form somewhat erratically. In some countries they appear to be relatively common; in other countries such as Japan, they appear not to

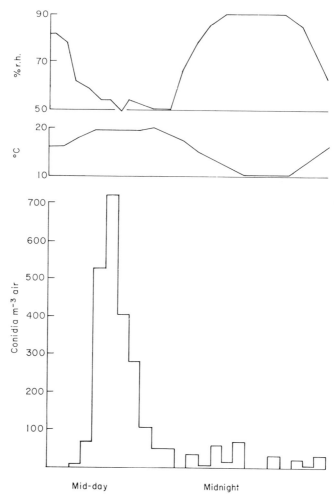

Fig. 2. Concentrations of conidia of *Sphaerotheca pannosa* near rose bushes (cv. Frensham), 24 Aug. 1971 at Imperial College Field Station, Silwood Park (as determined from catches in a Hirst volumetric spore trap).

occur at all. Within one locality they form on some cultivars but not on others; equally, at different localities they do not form consistently on one cultivar. In a survey of 741 examples of rose species and cultivars from 1966–68, Price (1970) found cleistocarps on only 32 cultivars. Generally they appeared to form more frequently on ramblers, climbers and old shrub roses where they were found embedded in the pannose mycelium usually round the thorns, stems and receptacles of the blooms. No cleistocarps were found on leaves. Coyier (1961), however, reports that cleistocarps formed on leaves of *R. virginiana* both in the greenhouse and in field plots in Wisconsin.

The irregularity with which *S. pannosa* forms its cleistocarps has apparently deterred investigators and there is no detailed account of their development.

Perennation

It has generally been supposed that in situations where cleistocarps are found they are the means by which the fungus perennates. There is little experimental evidence to support this. Price (1970) examined the perennation of cleistocarps during the winters of 1966–67, 1967–68 and 1968–69 in South-east England. Either lengths of stems or thorns with cleistocarps were placed in terylene net and this was pegged to soil. A similar pattern was obtained during each winter. From October to December there was a fall in the number of cleistocarps with asci and with ascospores; this was followed by a rise in January and almost immediately by a further decline. No cleisto-carp dehiscence was noted during these observations nor in more detailed experiments in which cleistocarps were kept outdoors in a chamber with slides for trapping ascospores. Examination of cleistocarps sampled directly from rose bushes during the winters of 1967–68 and 1968–69 showed that there were relatively few asci with ascospores and again no dehiscence was observed in any sample.

The only report of ascospore discharge appears to be that by Coyier (1961). He collected cleistocarps (in March) from the cultivar Triomphe d'Orleans grown in the greenhouse. After drying these for two weeks at room temperature (20–25°C) he mounted samples under a coverslip in distilled water. Within 3–5 min several asci ruptured their cleistocarps and the eight ascospores within each ascus clustered near the apex. Some 15–20 min following the rupture of the cleistocarp wall these ascospores were violently ejected from the asci.

On the available evidence it seems unlikely that cleistocarps are generally an effective means of overwintering and in those localities where mildew occurs abundantly but cleistocarps are not formed, the fungus must obviously perennate by other means. Several workers (Yarwood, 1939, 1944; Dillon Weston and Taylor, 1944; Ogawa *et al.*, 1967; Howden, 1968) suggested that *S. pannosa* perennates as mycelium in buds. This they deduced from field observations of young shoots in spring, some of which appeared to emerge from buds already covered with mildew. Perennation in buds was more convincingly demonstrated by Price (1969, 1970). He dissected dormant buds collected from the cultivar Frensham in January 1968 and examined the bud scales and leaves for mycelium like that of *S. pannosa*. Such mycelium was found in eight of 67 terminal buds and in five of 47 lateral buds examined. Within three of the terminal buds the mycelium bore developing conidio-phores clearly identifying it as that of *S. pannosa*. During the same winter he also collected lengths of stems with dormant buds, surface-sterilized these in hypochlorite and incubated them individually in plugged glass tubes. Under these conditions 393 buds were induced to burst; in twenty-three of these conidial chains of *S. pannosa* could be seen on the young growth even before the primary leaves emerged (Fig. 3).

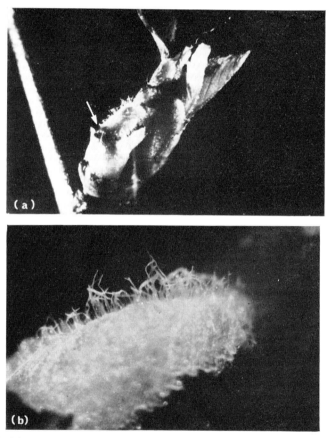

Fig. 3. Bud infection of rose (cv. Frensham) by *Sphaerotheca pannosa*. In (a) the bud scale (arrowed) has emerged, on a surface-sterilized shoot, already covered with mildew. A close-up (b) indicates the conidial chains of *S. pannosa*. (Form Price, 1969.)

The possibility that in mild winters the fungus survives as conidium-producing pustules, especially on the stems and thorns, has not been rigorously examined in experiments. Limited observations by Price (1969) during the winter of 1968–69 indicated that even as late as 9 December sporulating colonies could be found on rose cultivars but by 10 January 1969 all had been killed by frost.

B. The disease

Symptoms and effects

On very susceptible cultivars the white growth of the fungus consisting of mycelium and conidiophores first appears as discrete patches on the lower surfaces of young leaves. Given favourable conditions these may spread, covering the youngest leaves and often causing distortion, curling and premature leaf fall. Patches of fungal growth also develop initially on the young stems especially at the bases of the thorns. This growth tends to persist when the stems mature. The fungus may also attack the flowers. In most modern cultivars the petals of the opened flower are not often affected but on some, small circular patches of thin mycelial growth develop; these adversely affect the colour in the affected areas resulting in a spotting of the petal. However, on many cultivars the fungus grows abundantly on the pedicels, sepals and receptacles especially when the flower bud is unopened and this results in a bloom of poor quality. Overall, the development of mildew on the rose detracts from its value as an ornamental and in glasshouse-grown crops flower quality may be seriously affected. It is generally considered that severe mildew reduces leaf growth and may even decrease the number of flowers produced. Attempts by Wheeler (1975) to demonstrate statistically significant quantitative effects of the disease in field plots of the cultivar Frensham were unsuccessful, largely because of the considerable variation between plants in the production of susceptible leaf tissue. Generally, however, total leaf growth on marked shoots was least within plots where mildew was allowed to develop. There was also an association between high levels (25 and 50%) of mildew and small leaves but this may only reflect the abundance of mildew, especially late in the season, on the small leaves which are produced on shoots immediately below the flowers.

Host factors affecting disease development

Factors associated with the host profoundly affect the growth of S. pannosa. It grows well only on young tissues, it grows better on some parts of a given cultivar than on others and it grows abundantly on some cultivars but poorly on others.

Several studies indicate that the tissues of the rose become resistant to infection as they age. Those of Rogers (1959) and Mence and Hildebrandt (1966) show that in leaves this is associated with increasing thickness of the cuticle and outer epidermal wall (Table 1). It is not clear whether this increase in thickness directly prevents penetration or is itself associated with other changes inimical to the pathogen but abrasion of old rose leaves does not make them susceptible to S. pannosa. Whatever the mechanism the effect of

TABLE 1

Growth of *Sphaerotheca pannosa* on rose leaves (cv. Christopher Stone) in relation to age of leaf and thickness of cuticle + epidermal cell wall (Data of Mence and Hildebrandt, 1966)

Leaf age (days)	1	2	3	4	5	6	7	8	9	10
Thickness of cuticle + outer epidermal cell wall (μm)[a]	2·4	2·6	2·6	3·1	3·0	3·1	3·1	3·4	3·4	3·6
% established infections, 3 days after inoculation[b]	29	15	14	11	5	2	2	4	0	11
Infection rating[c], 2 weeks after inoculation	4·9	4·6	3·9	3·6	2·8	3·5	2·6	2·9	3·0	2·8

[a] Average of twenty measurements on each of two leaflets.
[b] 200 spores counted 3 days after inoculation.
[c] Average of eight leaflets, Infection rating scale: 1 = no infection; 2 = less than 25% cover of the leaflet; 3 = 25–50% cover; 4 = 50–75% cover; 5 = more than 75% cover

leaf age can be clearly seen in the field. Not only is the incidence of mildew associated with the main periods of vegetative growth but also its rate of increase is linked with the rate of host growth, high relative infection rates being associated with vigorous growth and low rates with periods of slight growth (Fig. 4). Typically, there is an increase in mildew as the shoots develop, an apparent decrease as these mature and terminate in a flower bud, followed by a further increase as lateral buds burst and new shoots arise. Where no control measures are used successive flushes of growth become increasingly affected.

Infected flower pedicels appear to be important sources of inoculum for the build-up of the fungus on the new growth (Price, 1970). Indeed, growth of the fungus on the pedicel and receptacle is often considerable even on cultivars whose leaves appear to be fairly resistant. The reasons for the apparent extreme susceptibility of these organs in many cultivars are not known but some possibilities have been examined by Price (1970) and Perera (1972). Price used the cv. Frensham. Even on this generally-susceptible cultivar the luxuriant growth of the fungus on the pedicels and receptacles contrasts with that on the leaves and stem and, somewhat unusual for a powdery mildew, results in a marked swelling of the pedicel (Fig. 5). This is mainly due to hypertrophy of the cortical cells, sometimes associated with meristematic activity, beneath the infected region (Perera, 1972). Price (1969) thought that the cylindrical form of the pedicel and its position might make it a particularly efficient spore trap but he could find no significant differences in numbers of conidia deposited per unit area of pedicels and of the leaves immediately below these within field plots. Perera (1972) examined the rates at which mildew developed on leaves and pedicels of comparable age of the susceptible miniature rose cv. Perla d' Alcanada. He found few

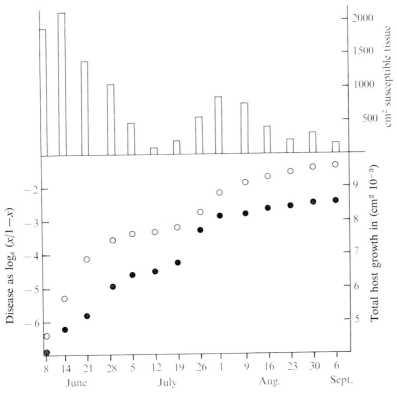

Fig. 4.   Relationship between mildew development and the growth of the rose (cv. Frensham). (From Wheeler, 1975.)

significant differences in mildew growth in experiments with both detached
and attached leaves and flowers, though there were some indications that,
initially, more conidia germinated and grew more quickly on pedicels than
on leaves. This suggests that the minimum time for infection might be less, so
that in the field, periods of favourable environmental conditions which are
too short to allow conidial germination and mildew establishment on leaves,
suffice for these processes on pedicels. On the other hand, the rate of mildew
growth *per se* need not necessarily be greater on pedicels than on leaves.
Other factors could result eventually in more extensive colonization of the
pedicels, one being a longer period of susceptibility. This was also investigated
by Perera (1972). He tagged newly unfolded leaves and pedicels of flowers
just emerged from their enclosing bracts on greenhouse-grown Perla d'
Alcanada and then at 4-day intervals he measured the cuticle thickness of
some leaves and pedicels and inoculated the remainder of the sample plants
with conidia of *S. pannosa*. His results indicated that pedicels lost their

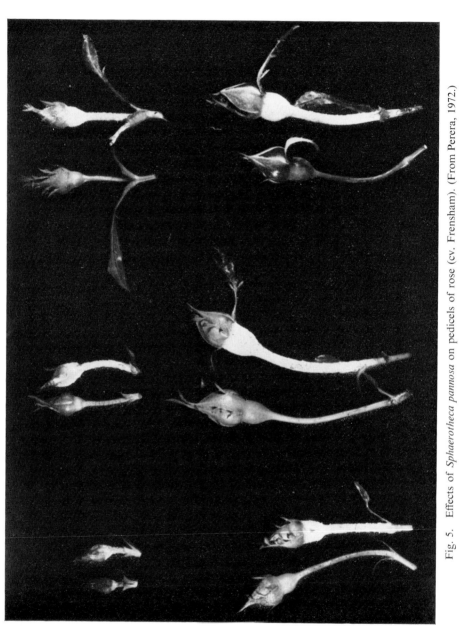

Fig. 5. Effects of *Sphaerotheca pannosa* on pedicels of rose (cv. Frensham). (From Perera, 1972.)

susceptibility with age more slowly than did leaves (Fig. 6). This appeared to be linked with a slower rate of cuticle formation, there being for both pedicel and leaf a marked inverse relationship between mildew growth and cuticle thickness.

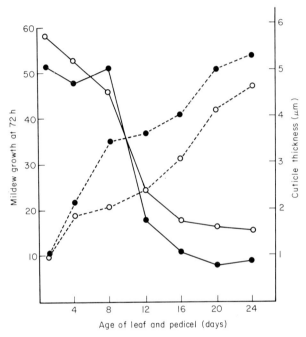

Fig. 6.   Susceptibility of leaves (●) and pedicels (○) of rose (cv. Perle d'Acanada) to infection by *Sphaerotheca pannosa* (———) in relation to age and cuticle thickness (– – –). Mildew growth was measured 72 h after inoculation with conidia on a scale 1 (single germ tube) to 4 (production of conidiophore initials) based on the stages shown in Fig. 1 (from Perera, 1972).

Although there are several recent reports in the literature of extensive field trials to examine the susceptibility of a range of rose cultivars (Docea *et al.*, 1970; Togliani, 1971; Simonyan, 1973; Rumberg, 1974) there have been relatively few studies to determine the reasons for the differences in susceptibility to *S. pannosa* which are commonly observed. Mence and Hildebrandt (1966) compared several stages of mildew development on the susceptible hybrid tea, Christopher Stone and the moderately-resistant grandiflora, Queen Elizabeth. No consistent difference was found between the two cultivars in the germination of conidia on detached leaflets nor in the initial penetration but on young leaves fewer haustoria developed normally in the

426 B. E. J. WHEELER

epidermal cells of Queen Elizabeth than on those of Christopher Stone. They concluded that resistance of rose leaves is determined by internal, post-penetration factors. In this respect the experiments of Mandre (1968, 1970, 1971) are noteworthy. He showed (Mandre, 1968) that mildewed rose leaves contained less anthocyanins and anthocyanidins than did healthy leaves because, as a result of oxidation in infected tissues, these compounds gradually give rise to colourless phenolic substances. Perhaps significantly, in view of the established antifungal nature of many phenolics, there were more anthocyanins in the leaves of resistant cultivars than in leaves of susceptible cultivars. There were also lower amounts of soluble carbohydrates in leaves of resistant cultivars than in those of susceptible cultivars (Mandre, 1970) but the significance of this remains in doubt, since in infected areas of leaves there were increases in the amounts of soluble carbohydrates associated with a decrease in saccharose (Mandre, 1971). There were also indications in this and in work by Bartlett (1964) of a relationship between amino acid content of leaves and their resistance; for example Bartlett found that both old leaves and those of resistant cultivars contained higher levels of cysteic acid.

Environmental factors affecting disease development

While the susceptibility of the host tissue is all important, the growth of the fungus on this tissue is considerably influenced by temperature, relative humidity and the presence of free water. Some effects of these factors have been studied by Longree (1939), Rogers (1959), Price (1970) and Perera and Wheeler (1975).

Temperature seems to be most important, affecting several stages in the development of *S. pannosa* on leaves. In experiments carried out at high relative humidity Longree (1939) found that the minimum temperature for germination on detached leaves of the cultivars Excelsa and Pernet was 3–5°C, the optimum 21°C and the maximum 33°C. At 3–5°C and 30–31°C haustoria were occasionally found but there was no evidence of further growth; no formation of haustoria could be detected at 33–34°C. Mycelial growth was poor in the range 6–10°C, good between 11–28°C and optimal between 18–25°C. The temperature limits for sporulation were somewhat narrower than those for mycelial development. There was no sporulation below 9–10°C or above 27°C. Conidiophores bore most conidia where leaves had been incubated from 21–27·5°C but generally towards the latter end of this range fewer conidiophores were formed.

Price's experiments were designed specifically to examine the effects of sub-optimal temperatures on the development of *S. pannosa* since he considered low temperatures to be largely responsible for the slow rate of spread from infected buds early in the season (Price, 1970). He inoculated and subsequently sampled detached leaves of the cultivar Frensham kept in damp

TABLE 2

Effects of sub-optimal temperatures on the development of *Sphaerotheca pannosa* on leaflets of rose, cv. Frensham (data of Price, 1970)[a]

Growth phase	Temp. (°C)	Incubation period (days)								
		0·5	1	2	4	7	11	14	21	28
Germination	15	78·4	93·0	96·8	97·7	99·0	—	—	—	—
	10	—	73·3	—	77·8	88·6	93·2	—	—	—
	3	—	—	—	—	93·4	—	97·0	92·5	94·9
Two germ tubes	15	—	21·1	7·1	1·0	—	—	—	—	—
	10	—	—	—	36·2	30·1	—	—	—	—
	3	—	—	—	—	36·4	—	21·6	54·5	3·5
Three germ tubes	15	—	—	50·7	51·5	66·2	—	—	—	—
	10	—	—	—	—	—	59·5	—	—	—
	3	—	—	—	—	0·9	—	13·6	49·4	70·9
Sporulation	15	—	—	—	—	66·2	—	—	—	—
	10	—	—	—	—	—	59·5	—	—	—
	3	—	—	—	—	—	—	—	—	46·0

[a] Figures as percentages of conidia present.

chambers at 3°, 10° or 15°C (Table 2). The rate of germination decreased with a fall in temperature but the final level of germination was similar at the three temperatures. The rate of colony development also decreased with decreasing temperature but contrary to Longree's findings the fungus developed to sporulation, albeit slowly, even at 3°C. In an extension of this work, Price subjected inoculated leaves to a regime of fluctuating temperatures which simulated those in the spring of 1968 and demonstrated that part, though not all, of the delay in appearance of the mildew in the field could be attributed to low temperatures.

Both Longree's and Price's experiments were carried out at high relative humidities since these seem to be best for mildew development. The germination of conidia on glass slides is much affected by humidity. Longree (1939) found that it was optimal between 99–97% r.h. but very low at or below 95% r.h. On inoculated rose shoots germination was reduced as the atmospheric humidity was decreased (see also Pathak and Chorin, 1968) but considerably less so than on glass slides. Indeed, on leaves, conidia appeared to germinate in what appeared to be very dry atmospheres. The mycelium of the fungus also continued to grow under these conditions, though more sparsely and fewer conidiophores formed. These results were confirmed by Rogers (1959) in somewhat more sophisticated experiments. He found that mildew development was not greatly affected by conditions designed to give a range of humidities at or near the upper leaf surface between 50–100% r.h.

In contrast, mildew development is adversely affected by the presence of water especially continuous films, the severity of the effect being directly related to the length of the wet period (Perera and Wheeler, 1975). The development of *S. pannosa* seems to be affected most when leaves are wetted immediately after the deposition of conidia (Fig. 7). Once the infection is

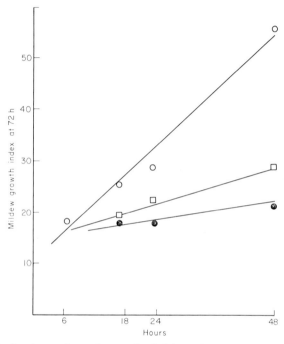

Fig. 7. Effect of water on the early growth of *Sphaerotheca pannosa* on rose leaves (cv. Perla d'Alcanada). Plants were subjected to an 8 h wet period immediately after inoculation (●) or 6 h later (□) or were not wetted (control, ○). Mildew growth was measured at 72 h after inoculation with conidia on a scale 1 (single germ tube) to 4 (production of conidiophore initials) based on the stages shown in Fig. 1 (from Wheeler, 1973, based on data of Perera, 1972).

established the effects are less marked though prolonged wetting causes the collapse of the conidial chains (Rogers, 1959).

Attempts have been made by Tammen (1973) and Wheeler (1973, 1975) to integrate some of these factors and evaluate conditions suitable for mildew growth in the field. The most favourable conditions appear to be a combination of about 15·5°C and 90–99 % r.h. at night which allows optimum conidial formation, germination of conidia and infection, with 26·7°C and 40–70 % r.h. during the day which favours the release of conidia (Tammen, 1973).

Control

New rose cultivars are continually being produced and in trials many of these show considerable resistance to mildew. Few appear to retain this high level once they are generally distributed, presumably because new strains of *S. pannosa* develop which can overcome this resistance. For example, the cultivar Super Star, noted for its resistance when it was first introduced in Britain, now becomes substantially mildewed in most seasons. While growers may hope to produce a cultivar with long-lasting resistance, control so far has relied mainly on protective sprays. The use of chemicals for the control of rose mildew was probably started in England by Radclyffe (1861) who advised spraying with copper sulphate but then withdrew the recommendation because of the phytotoxicity experienced. The subsequent pattern of chemical control of rose mildew has followed that for foliar diseases in general. An early period in which various forms of sulphur were used (Yarwood, 1957) was followed by one in which the use of more refined copper compounds, dithiocarbamates, some antibiotics and the fungicide dinocap was prominent (Palmer *et al.*, 1959; Fisher *et al.*, 1960; Deep and Bartlett, 1961; Jones and Swartout, 1961a,b; Brown, 1962; McCain *et al.*, 1962; Miller, 1962, 1966; Taylor, 1962; Cherkasskii *et al.*, 1964; Drozdovskaya, 1964; Smith, 1965; Misiga, 1966; Misiga *et al.*, 1967; Kavanagh, 1967; Immikhuizen, 1968; Baranowski, 1970; Pfister and Muller, 1970; Semeniuk and Palmer, 1970; Nichols and Nelson, 1970). In recent years there has been an increasing interest in the use of systemic fungicides such as benomyl and dodemorph (Beuzenberg and Scholten, 1969; Quarnström, 1970; Raabe and Hurlimann, 1970; Bourdin, 1971; Frost and Pattisson, 1971; Paulus *et al.*, 1971, 1974; Baresi and Roberti, 1973; Ponti and Bencivelli, 1973; Ambrosi, 1973; Burth and Zastrow, 1973; Vir and Raychaudhuri, 1973; Evans, 1974; Wheeler, 1975) though one report suggests that with repeated use of benomyl resistant strains of *S. pannosa* are induced (Yoshii and Morales, 1975). However, none of the materials presently available is translocated to any great extent in woody plants like the rose. Their chief advantage probably relies on their absorption into leaves where they are less affected by weathering than are the conventional protectant fungicides. The rapid production of susceptible tissue on developing rose shoots necessitates repeated application even with these materials (Wheeler, 1975). Until a fungicide truly systemic within the rose bush is found, the main consideration in spraying remains the timing of the application. More accurate forecasts of mildew development could help in this respect.

III. Powdery Mildews of Selected Ornamentals

A. Antirrhinum

Powdery mildew is relatively common on cultivars of *Antirrhinum majus* grown in glasshouses and is ascribed by both Salmon (1900) and Blumer (1967) to the species *Erysiphe cichoracearum* though their concepts of this species differ. Blumer (1952) also reported that *A. majus* could be infected in experiments using conidia from cucumber (*Cucumis sativus*), considered to be those of *E. polyphaga* sensu Hammarlund. However, as Junnell (1967) points out, this does not necessarily mean that this is the species which generally occurs on the inoculated host. Also, the differences between *E. polyphaga* and *E. cichoracearum* are not always well marked.

There are relatively few reports of the disease before 1950. Mayor (1936) in his mycological notes records it as affecting *A. majus* in the Neuchatel canton of Switzerland and Moore (1947), as "usually slight, under glass, and probably not uncommon" in England. He lists several collections, the earliest being in 1928 and describes the conidia as hyaline, almost rectangular, with slightly rounded ends or occasionally barrel-shaped, 24–45 × 12–17 μm (average 30 × 14·5 μm). Both authors list the causal fungus as *Oidium* sp. and there appear to be no reports of cleistocarps forming on cultivated antirrhinums. From 1950 the disease seems to have become of increasing importance on stocks of young plants raised in glasshouses especially in the USA and sufficiently so to warrant recommendations for its control in New York State (Dimock, 1953; Williamson, 1953). About this time the disease spread to the western USA and is first recorded in Pullman (Washington) in 1948 and then from California in 1949 by Baker and Maclean (1950). These authors note that the conidiophores resemble those of *E. cichoracearum*. Recently, control of the disease has been reported with some systemic fungicides. Raabe *et al.*, (1970) successfully applied benomyl as a spray or as a soil drench at 3-week intervals or mixed in the soil before planting. Riordain (1974) obtained good control with sprays of thiophanatemethyl and triarimol.

B. Begonia

A powdery mildew on cultivated begonias has been known for over 50 years and the causal fungus was originally described by Puttemans (1911) as *Oidium begoniae*. The disease first became important in Europe in the mid-1930s (Pape, 1939). It was found in Germany in 1934 (Gante, 1935), in Holland in 1935 (Van Poeteren, 1936), in England in 1938 (Moore, 1959) and in Portugal in 1939 (Da Camara and Da Luz, 1939). Since 1950 it has become more severe in some territories e.g. England (Moore, 1959) and has also extended its range to Australasia (Anon., 1954; Magee, 1955). In most instances only the conidial state is found, the conidia being barrel-shaped, 20–36 × 17 μm. A form of the fungus with larger conidia (34–72 × 9–22·8 μm) was later described by De Mendonca and De Sequeira (1962) from Portugal as *O. begoniae* var. *macrosporum*.

Hammarlund (1945) considered *O. begoniae* to be the polyphagous species *E. polyphaga* and he found cleistocarps on *Begonia* "Gloire de Lorraine" but gives few details other than that they contained 10–20 asci. Since then there have been three reports of cleistocarps from cultivated begonias. The best documented is that of a collection from Salisbury, England in February 1970 by Brooks. This was described by Sivanesan (1971) as *Microsphaera begoniae*. The conidia associated with this collection corresponded with those described as *O. begoniae* var. *macrosporum* and in Sivanesan's opinion it is this fungus which is the conidial state of *M. begoniae* and not *O. begoniae* Puttemans. This larger conidial form now appears to be the more common in collections from England (Audrey Brooks, personal communication) and also apparently from New Zealand (Boeswinkel, 1976). A *Microsphaera* from begonias has also been described from Rumania by Eliade (1972) as *M. tarnavschii* which she considers to be the perfect state of *O. begoniae* Puttemans. It is of interest that in inoculation experiments with conidia of this fungus she was unable to infect *Kalenchöe*, *Cyclamen*, *Primula* or *Saintpaulia*. There is also a report from Colombia of a *Microsphaera* on begonias (Molina, 1973) but the species is not described.

There appears to be a good deal of variation in susceptibility amongst begonia cultivars (Zobrist, 1946) and some are described as immune (Strider, 1974) but most control relies on the application of fungicides though this is difficult to obtain without phytotoxic effects (Martin, 1969). Thompson (1961) reported good control with Karathane (dinocap) sprays applied three times at 3-day or weekly intervals or with a dinocap—actidione dust, though with some treatments there was a slight bleaching of the petals. Strider (1974) also obtained good control with dinocap and with sulphur dust and thio-phanate-methyl but again there was some slight phytotoxicity. The fungicides triforine and parinol were slightly less effective in some respects but had the merit of not damaging the plants in any way.

C. Cherry laurel (*Prunus laurocerasus*)

Powdery mildew on cherry laurel is something of a rarity but occasional reports of it excite interest. The first substantial paper on the disease was that of Salmon (1906). He found the disease in July 1905 in Kew Gardens on leaves of young shoots and occasionally on stems, and using conidia from this material successfully infected young healthy leaves in inoculations of plants growing at Reigate, Surrey. According to Salmon the disease was first noted in Italy by Bertolini in 1879 who described the fungus as *Oidium passerinii*. In the following year what appeared to be the same fungus was discovered by Roumeguère at Tarbes in France in the conidial state but he later collected cleistocarps from this host and described these as *Erysiphe bertolini*. Salmon examined some of Roumeguère's dried material of the conidial state (which had been labelled *O. lauro-cerasi*) and considered it identical with his own from Kew. Salmon was impressed by the similarity in the pannose mycelium formed later on cherry laurel and that commonly found on rose and he considered that the fungus on cherry laurel was a form of *Sphaerotheca pannosa*. The conidia, too, from cherry laurel (23–28 × 12–15 μm) were not dissimilar to those from rose. He doubted, therefore, whether the conidial and perfect states of Roumeguère were really linked.

There were further reports of the disease from Italy in 1929 and 1953. Apparently it devasted a recently-pruned row of cherry laurels at Perugia in July 1929, causing deformation and distortion of the young leaves and twigs (Montemartini, 1930). The fungus involved was considered to be *S. pannosa* var. *persicae* but this opinion was based only on conidial size and the type of injury involved. The 1953 outbreak was similarly ascribed to *S. pannosa* (Sibilia, 1955). How far these claims for the identity of the fungus are justified remains a matter of some speculation because in 1919 Fischer found both conidia and cleistocarps on cherry laurel in the Botanic Gardens at Berne which corresponded to those of *Podosphaera oxycanthae* var. *tridactyla*.

In England, further collections were obtained from cherry laurel in 1937, 1938 and 1939 (Moore and Moore, 1949) and again in 1951 (Moore, 1952). The conidia in these collections were slightly longer (30–41 × 12–17 μm) than those found by Salmon and the presence of cleistocarps indicated that the material, like Fischer's, could be assigned to *P. oxyacanthae* var. *tridactyla*.

Fischer (1919) considered that the disease on cherry laurel was rare because mildew usually develops on other *Prunus* spp. after the cherry laurel leaves have become fully developed and are not then susceptible. Only in unusual situations where unseasonable pruning induces new growth is there susceptible tissue. The circumstances in which the disease has been recorded generally lend weight to Fischer's view.

A recent report from France indicates that, where necessary, control can be achieved effectively with benomyl sprays (Viennot-Bourgin, 1973).

D. Chrysanthemum

Powdery mildew is common on cultivated chrysanthemums both outdoors and under glass. On the commercial cultivars derived from *C. morifolium* (or *C. indicum*) the fungus is either called *Oidium chrysanthemi* a name proposed by Rabenhorst in 1853 (Blumer, 1967) or is considered to be the conidial state of *E. cichoracearum*. The perfect state is not found so the distinction depends on rather slight differences in conidial morphology which are of doubtful validity. Indeed, Hammarlund (1945) considered that the fungus on chrysanthemum was simply a form of the species he designated as *E. polyphaga*. Most of the plant pathological literature on this disease since 1922 is concerned either with susceptibility of different cultivars (Bohmig, 1937; Anon., 1952) or with control by fungicides. Amongst the materials most frequently recommended are sulphur sprays (Pape and Marggraf, 1936), formulations of copper + oil or thiram + petroleum, and dinocap used either as a spray or as a smoke (Hey, 1957; Martin, 1969). Recently good control with benomyl has been reported (Burth and Zastrow, 1973).

E. Clematis

Powdery mildew is reported to cause damage occasionally to clematis in the southern counties of England (Martin, 1969) and a recent report from the USSR suggests that it may be sufficiently common there to warrant trials of various hybrids (Beskaravainaya and Mitrofanova, 1973), those of the *Clematis heracleifolia* group proving the most resistant. Salmon (1900) considered the name of the causal fungus to be *Erysiphe polygoni* but in Blumer's treatment of the Erysiphaceae Salmon's species is subdivided so that two species are recorded on *Clematis*, *Erysiphe ranunculi* and *Erysiphe aquilegiae*. The latter appears less common since it is recorded only on *Clematis recta* and *C. vitalba*. The disease can be controlled by sprays of dinocap (Martin, 1969).

F. Delphinium

Powdery mildew is relatively common on cultivated delphinium and may affect not only leaves but also the flower spikes. It is often severe after the first flowering and reduces vigour, which is important in relation to the next season's growth. The species most commonly encountered is *E. polygoni* sensu Salmon (*E. ranunculi* in Blumer's treatment of the Erysiphaceae) but a second species, *S. humuli* var. *fuliginea* (or *S. delphinii* in Blumer) is reported on *Delphinium grandiflorum* and *D. amabile* and appears to be relatively common in California (Baker, 1943). There is also a report of *Oidiopsis taurica* on delphinium in India (Sharma, 1961).

   E. polygoni on delphinium has been the subject of several investigations concerned with the general biology of the Erysiphaceae. Thus, Yarwood

434   B. E. J. WHEELER

(1934) showed that this fungus had a diurnal cycle of conidial formation and
maturation and later (Yarwood, 1936) that the conidia would germinate at
low relative humidities. Allen (1936) gave a detailed account of the cytology
of cleistocarp formation and suggested that the fungus was heterothallic.
Smith and Wheeler (1969) presented evidence that in England cleistocarps of
E. polygoni overwinter on cultivated delphiniums and that ascospores initiate
infections in the spring.

Some species of *Delphinium* and their cultivars appear relatively resistant
to mildew (Pape, 1928; Baker, 1947) but the disease is mainly controlled by
the application of fungicides such as lime sulphur, copper oxychloride and
dinocap (Martin, 1969). So far there are no published reports of control using
benomyl or other systemic fungicides.

G. Hydrangea

Hydrangea powdery mildew was first noted as being of some importance in
1925 when it was recorded in a survey of Norwegian Erysiphaceae by Jørstad.
He named the fungus *Oidium hortensiae* (Jørstad, 1925) and this name is
retained by Blumer (1967). Further reports of the disease soon followed from
Holland (Van Poeteren, 1926), Switzerland (Blumer, 1926), Austria (Wahl,
1926), Denmark (Gram, Jorgensen and Rostrup,1927), France (Foëx, 1927)
and Germany (Pape, 1927). There was even a report from Columbus, Ohio,
of powdery mildew on hydrangea plants imported from Germany (Engel,
1928) though the disease may well have been present in the USA on some
stocks much earlier than that (Martin, 1928). The first English record of the
disease was in 1930 on hydrangeas in Warwickshire (Anon., 1934). In all these
collections only the conidial state was present. The conidial measurements
given by Pape (1927) were 37–45 × 16–21 μm and by Blumer (1928) as
32–40 × 17–19 μm (average 36·2 × 18·1 μm). The Dutch considered the
fungus to be *E. polygoni* (Bouwens, 1927) and this view was accepted by
several workers since the development of the germ tubes and appressoria
resembled that of *E. polygoni*. The conidial state is sometimes supposed to
belong to *E. polyphaga* but as Junnell (1967) points out, the conidia are longer
and more ellipsoidal than that species and neither Hammarlund (1945) nor
Blumer (1951b) found hydrangea susceptible to *E. polyphaga*.

Cleistocarps of the mildew on hydrangea were first found in 1930 in
Poland and were later described by Siemaszko (1933) as *Microsphaera
polonica*. The cleistocarps were globose, black, 110–120 μm diameter with
10–20 hyaline or brownish appendages, straight, occasionally dichotomously
branched, and about as long as the diameter of the cleistocarp. The asci were
ellipsoidal or oval, 38–44 × 30–36 μm and contained 4–6 ascospores, 20–22 ×
8–10 μm.

The fungus normally attacks the leaves causing brown spots on which the mycelium appears as a white coating but there is a report from Italy of an outbreak of the disease in glasshouses where only the inflorescences were affected (Goidanich, 1941).

Although from the early accounts there appeared to be considerable variation in susceptibility of hydrangeas, claims that some were immune (Ludwigs, 1927) were not supported by subsequent inoculation experiments (Blumer, 1928; Brassler, 1930). Control of the disease on susceptible cultivars relies on the use of chemicals (Grouet, 1963). Colloidal sulphur, phaltan and dinocap are the main conventional fungicides that are recommended (Valaskova, 1965; Martin, 1969) but benomyl is also reported to be effective (Peterson and Davis, 1970).

H. Japanese spindle tree (*Euonymus japonicus*)

Powdery mildew is often conspicuous and destructive on *Euonymus* spp. particularly on *Euonymus japonicus*. On this host the first published record is by Arcangeli who collected it at Livorno in Italy in 1900, though he says that it was also collected by a Dr Baroni of Florence at the end of 1899. The fungus was described by Saccardo in 1903 as *Oidium euonymi-japonicae* and this name is still the most commonly used. Other early records are those by H. and P. Sydow in Austria and by Salmon in England, both in 1903 (Salmon, 1905).

To Salmon must go the credit for first examining the overwintering of the fungus. He collected infected leaves bearing patches of mycelium during the winter of 1904 and incubated them in a damp atmosphere in a warm greenhouse (*c.* 65°F). After some days most patches of the mildew bore numerous conidiophores, with chains of conidia over their surface. Salmon considered that the fungus thus hibernated as these mycelial patches and this view was supported by the later work of Tokushige (1953) in Japan. A somewhat unusual means of overwintering was reported by Nannizzi (1927). Infection of leaves resulted in the formation of a suberized layer in which mycelium and chlamydospores hibernated and caused infection of new growth in the following spring. This remains the only report of this type and must be considered of doubtful validity especially as chlamydospores are not a feature of the Erysiphaceae.

There have been few reports of cleistocarps. Tokushige (1953) states that in Japan the mildew on *Euonymus japonicus* was named *Uncinula euonymi-japonicae* by Hara (1921) but it is not clear whether this was based on a collection of the perfect state, and other workers, for example Hino and Kato (1929), could find no cleistocarps. The first report of cleistocarps was by Viennot-Bourgin (1965). These were of a *Microsphaera* which he considered distinct from any other type and named *M. euonymi-japonicae* but without a latin diagnosis so it remains a *nomen nudum*. The cleistocarps were globose, 100–144 μm diameter bearing simple appendages 250–430 μm long with 1–3 cross walls. Measurements of the conidia given by Viennot-Bourgin are 21–36 × 7–13 μm; those recorded for *O. euonymi-japonicae* by Salmon (1905) are 30–38 × 13–14 μm and by Hino and Kato (1929), 28·9–37·3 × 8·9–13·3 μm. These are produced in a diurnal cycle (Childs, 1940).

Little has been published on this disease since 1965 but a paper by Nadernejad (1966) is interesting in that it records the disease as being serious in Iran since 1955 and reports successful control with sprays of wettable sulphur and with dinocap.

The mildew on the related *Euonymus europaeus* appears to be a distinct species, *Microsphaera euonymi*, a fungus originally described by De Candolle (as *Erysiphe euonymi*) in 1815. Salmon (1905) was unable to infect this species with conidia from *Erysiphe japonicus*. Blumer (1967) also records *Phyllactinia guttata on Euonymus europaeus* but there is little supporting evidence from the literature.

I. Kalanchöe (especially *K. blossfeldiana*)

A powdery mildew was first recorded on *Kalanchöe* in Germany by Lustner (1935) who suggested that it should be called *Oidium kalanchöeae*. The host was later identified as *K. blossfeldiana* (Sommer, 1935). The fungus formed greyish-white patches on the leaves, eventually killing the epidermis which could then be readily detached. In severe infections the mesophyll beneath the fungal growth also died and became desiccated. The disease appears to have been relatively common subsequently in Europe and Hammarlund (1945) used conidia from this host in extensive cross-inoculation experiments to show that the species involved was a polyphagous one which he called *Erysiphe polyphaga*. His findings were supported by the further experiments of Blumer (1952).

The disease was first recorded in England by Moore (1952) on plants of *K. blossfeldiana* cv Poelln grown under long-day conditions for experimental purposes at Rothamsted. This record is also notable because cleistocarps formed; these were 87–120 μm diam. (average 102 μm) and contained six to eight asci with immature ascospores. These features agree reasonably well with the somewhat incomplete description of cleistocarps of *E. polyphaga* in the literature (see Junnell, 1967). Apart from an earlier report from Denmark of cleistocarps on *K. globulifera* var. *coccinea* this remains the only report of cleistocarps on *Kalanchöe*.

Recently good control of powdery mildew on *K. blossfeldiana* has been reported by Manning *et al.*, (1972) using sprays of dinocap or benomyl and by Strider (1976) with benomyl, thiophanate methyl and triforine. Strider

(1976) also gives useful information concerning the resistance of some cultivars commonly grown in the USA.

J. Lilac (*Syringa*)

Powdery mildew on lilac is something of a curiosity. It is mentioned sporadically in the plant pathological literature from 1922 to about 1945, it appeared in epidemic form in Europe from 1948 to 1950 and attracted much interest, only to revert once more to a disease of occasional occurrence.

An early indication that lilac mildew might be sufficiently troublesome to warrant serious study is the paper by Crowell (1937) in which is listed the susceptibilities of about 300 species and cultivars, based on observations in the USA from 1933 to 1936. Crowell follows Salmon (1900) in calling the causal fungus *Microsphaera alni* (= *M. lonicerae* in Blumer's account of this genus, see Blumer, 1967). This was followed by a report (Lepik, 1943) that lilac in Estonia was fairly-heavily infected with *Phyllactina suffulta* (= *P. guttata* in Blumer, 1967). It also seems likely that it was fairly common in Canada at that time since in his study of conidial germination at low humidities Brodie (1945) included *M. alni* from lilac.

Cleistocarps on lilac in Europe were first reported near Paris by Viennot-Bourgin (1944). These were spherical, 150–300 μm diameter and bore 5–18 hyaline, rigid, acuminate appendages swollen at the base. Viennot-Bourgin considered the fungus to be *Phyllactinia corylea* (a synonym of *P. guttata* in Blumer, 1967). He noted that the infection first appeared in September and then became increasingly severe in early October, persisting until leaf fall.

During 1948–50 the disease was common in Europe. Blumer (1951a) reported that in Switzerland it was spreading in epidemic form from west to east. It was first recorded in England in 1948 and there were further collections in 1949 and 1950 (Moore, 1952). There was also a severe outbreak around Vienna in the autumn of 1951 (Schmidt, 1952). No cleistocarps were found and Blumer (1951a) proposed that the fungus should provisionally be called *Oidium syringae*. However, the conidia were like those in collections of lilac mildew from the USA assigned to *M. alni* and Blumer considered that the fungus had probably been imported into Europe on plants from America.

There have been no serious outbreaks of the disease in Europe since 1951 and the reason for its sudden appearance in the late 1940s remains a mystery.

K. Lupin

Interest in powdery mildew of lupins has centred not so much on the ornamental lupin (*Lupinus polyphyllus*) but on those cultivated as fodder or seed crops such as *L. luteus*, *L. angustifolius* and *L. albus*. The principal fungus involved comes within the broad concept of the species adopted by Salmon (1900) for *E. polygoni*. In the classification of Blumer (1967) this is subdivided and two species are considered to occur on lupins generally, *Erysiphe martii* and *Erysiphe pisi*, though on garden lupins (*L. polyphyllus*) probably only *E. martii*. In the USA another, quite different species is also recorded on lupins, *Microsphaera diffusa* (Luttrell and Samples, 1954).

While much is known about the biology of the fungus *E. polygoni* sensu Salmon, relatively little of this relates directly to the form on *L. polyphyllus*. However, Smith and Wheeler (1969) showed that in England cleistocarps overwintered on this host and that these discharged viable and infective ascospores in the spring. Smith (1970) also showed that the form on ornamental lupins was heterothallic.

Powdery mildew is seldom serious enough on garden lupins to warrant spraying but sulphur and dinocap have been recommended for this purpose (Beaumont, 1954; Pirone, 1970). On crops of the other species the possibility is being explored of using systemic fungicides as seed dressings provided these do not interfere with inoculations of rhizobia (Baldwin, 1976).

L. Phlox

Mildew is frequently severe on phlox towards the end of the growing season, some leaves appear completely white because they are covered with fungal mycelium. In most temperate areas, and apparently also in Egypt (Firky, 1936), the causal fungus is usually named as *E. cichoracearum*. This is seldom based on collections of cleistocarps but rather that the conidial state (named *Oidium drummondii* by von Thumen) is similar. However, in parts of the USSR (Gorlenko, 1974) and in India (Kamat and Patel, 1948; Sharma, 1961) the conidial state is an *Oidiopsis* and the fungus is then considered to be *Leveillula taurica* (or *L. polemoniacearum* in the revision of the genus *Leveillula* by Golovin, 1956). Both fungi appear specialized to the host attacking only cultivated phlox and not other ornamentals such as zinnia which have been tested (Schmitt, 1955; Schuepp and Blumer, 1963; Kamat and Patel, 1948). For this reason the fungus is often designated a special form (f. *phlogis*) of either *E. cichoracearum* or *L. taurica*. Blumer (1967) also lists *Sphaerotheca fuliginea* as occurring on *Phlox acuminata* based on a report by Mayor but otherwise there appear to be no further records of this fungus on phlox.

Varieties and cultivars of the garden phlox vary considerably in their susceptibility to powdery mildew (Mains, 1942). The most recent investigation is that of Thompson and Svejda (1965) in Canada who tested the resistance of 49 cultivars of *Phlox paniculata*. None was resistant, although the cv. Pyramid White initially showed some field resistance. Plants appeared to be more resistant in their first year than when three years old or more. Spraying with fungicides, e.g. dinocap (Martin, 1969) is the usual recommendation for controlling the disease but there are no recent published reports of field trials.

M. Sweet pea (*Lathyrus odoratus*)

Powdery mildew is occasionally troublesome on sweet peas grown in the glasshouse and is often found on outdoor plants towards the end of the summer. Beaumont (1951), for example, listed it as one of the commonest diseases of this host in the UK.

The causal fungus is classified as *E. polygoni* by Salmon (1900) and *E. martii* by Blumer (1967). There is, however, some degree of host specialization. Bouwens (1927) found that isolates from *L. odoratus* would infect garden pea (*Pisum sativum*) but none of the other hosts ascribed to *E. polygoni* sensu Salmon which were tested. Smith (1969) also successfully infected pea using conidia from *L. odoratus* but could not produce sporulating colonies on *L. odoratus* using conidia from pea as inoculum.

Although in England the disease appears late in the growing season there is some evidence that inoculum may be present much earlier since Smith and Wheeler (1969) showed that overwintered cleistocarps on dead leaves of *L. odoratus* contained viable and infective ascospores.

As with most powdery mildews of ornamentals, control has relied mainly on the use of fungicide sprays and there are recent reports that systemic fungicides such as benomyl are effective (Hammett, 1968; Umgelter, 1973).

IV. Other Powdery Mildews of Ornamentals

Most powdery mildews on ornamentals remain uninvestigated except by mycologists interested in the particular fungus and the taxonomy of the Erysiphaceae. The publications of Salmon (1900), Hirata (1966) and Blumer (1967) indicate the species recorded and those of Moore (1959), Pape (1955), Forsberg (1963), Pirone (1970), Wager (1970) and Stahl and Umgelter (1976), note the powdery mildews on ornamentals which are most common. Table 3 summarizes some recent reports of powdery mildews on ornamentals which supplement those to be found in the above-mentioned publications.

TABLE 3

Some recent reports of powdery mildews on ornamentals (1966–1976)

Host	Fungus[a]	Authors and publication	Notes
Stokesia laevis	Erysiphe cichoracearum	Kilpatrick, R. A. et al. (1975). Pl. Dis. Reptr. 59, 795	New disease for this host in USA Cleistocarps found.
Lagerstroemia indica	Erysiphe lagerstroemiae	McGuire, J. M. et al. (1975). Arkans. Fm. Res. 24, 14.	Control of the disease with sprays of benomyl.
Anthurium scherzerianum	Erysiphe communis	Schneider, R. and Kiewnick, L. (1974). Phytopath. Z. 79, 364–367.	Report of disease in a nursery in 1973.
Dimorphotheca sinuata	Sphaerotheca fuliginea and Leveillula taurica	Mathur, R. L. et al. (1971). Indian Phytopath. 24, 798–800	New host record for these mildews in India.
Dianthus barbatus	Oidium sp.	Ialongo, M. T. (1971). Annali. Ist. sper. Patol. Veg. 2, 137–141 (Rev. Pl. Path. 52, 3336).	Apparently new records [but note that Oidium dianthi Jacz. known on Dianthus caryophyllus and D. sinensis—see Blumer (1967)].
Celosia argentea	Oidium sp.		
Althaea rosea	Leveillula taurica	Sankhla, H. C. et al. (1971). Indian Phytopath. 24, 171.	New record on host for India.
Centaurea moschata	Erysiphe cichoracearum	Jain, J. P. and Singh, R. D. (1969). Indian Phytopath. 22, 251–252.	New host records for India.
Helipterum roseum	Erysiphe cichoracearum		
Helipterum album			
Helipterum roseum	Leveillula taurica		
Centaurea imperialis Centaurea cyanus	Oidiopsis taurica	Mathur, B. L. (1966). Curr. Sci. 35, 447.	New host records for India.
Adonis vernalis	Sphaerotheca fuliginea	Kowalski, J. (1966). Acta agrobot. 19, 5–16. [Rev. appl. mycol. 46, 1604]	New host for this fungus in Poland. Blumer (1967) records Erysiphe ranunculi on this host.

V. Discussion

With one or two notable exceptions, the striking feature of powdery mildews of ornamentals is the lack of information concerning the biology of the fungi involved. This mainly results from economic considerations. It would be difficult to justify an investigation of many of them on purely economic grounds since they occur only sporadically.

Ornamentals are selected for their attractiveness and in many instances a particular cultivar will gain favour for certain features and will be retained despite a susceptibility to powdery mildew, it being considered that this can be controlled reasonably well with fungicides. There are, indeed, fewer problems with the use of chemicals on these plants than on most. There are, for example, no problems of toxic residues in plant parts to be consumed by animals or humans and this gives greater flexibility of choice in the selection of materials and the time at which they can be applied. The main limitations are the dangers of handling certain materials and their phytotoxicity on delicate flowers. To these perhaps should be added the dangers of indiscriminate spraying with materials whose effect on plant and animal life remain largely unknown.

It is a pity, however, that so far these attitudes have somewhat stultified research on this group of diseases. We know little about the mechanism of resistance to rose powdery mildew or indeed, of the development of new strains of *S. pannosa*. There is still much confusion concerning the host range of some powdery mildew fungi, especially on glasshouse-grown crops and we know comparatively little about the survival of these fungi in our gardens. These are but some of the problems which require investigation.

References

Anon. (1934). *Bull. Minist. Agric. Fish. Fd., Lond.* no. 79, 117pp.
Anon. (1952). *Bull. Minist. Agric. Fish. Fd., Lond.* no. 92, 50pp.
Anon. (1954). *Agric. Gaz. N.S.W.* **65**, 100–103.
Akai, S., Fukutomi, M. and Kunoh, H. (1966). *Mycopath. Mycol. appl.* **29**, 211–216.
Allen, R. F. (1936). *J. agric. Res.* **53**, 801–818.
Ambrosi, M. (1973). *Notiz. Mal. Piante* no. 88–89, 171–175.
Baker, K. F. (1943). *Phytopathology* **33**, 832–834.
Baker, K. F. (1947). *Yb. Am. Delphinium Soc.*, 15–30.
Baker, K. F. and Maclean, N. A. (1950). *Pl. Dis. Reptr* **34**, 183–185.
Baldwin, G. (1976). Crop protection in grain lupins. M.Sc. thesis, University of Reading.
Baranowski, T. (1970). *Biul. Inst. Ochr. Rosśl., Poznań* no. 47, 233–245.
Baresi, F. and Roberti, L. (1973). *Notiz. Mal. Piante* no. 88–89, 153–162.
Bartlett, A. B. (1964). *Diss. Abstr.* **24**, 3045–3046.
Beaumont, A. (1951). *Gdnrs' Chron.* 3, **129**, 132.
Beaumont, A. (1954). *Gdnrs' Chron.* 3, **135**, 181.

442   B. E. J. WHEELER

Beskaravaĭnaya, M. A. and Mitrofanova, O. V. (1973). *Byull. glavn. bot. Sada, Moskva* no. 89, 94–97.
Beuzenberg, M. P. and Scholten, G. (1969). *Jvrsl. Proefstn. bloemist. Aalsmeer, 1969* 105–107.
Blumer, S. (1926). *Z. PflKrankh. PflPath. PflSchutz* **36**, 232–236.
Blumer, S. (1928). *Z. PflKrankh. PflPath. PflSchutz.* **38**, 78–83.
Blumer, S. (1951a). *Phytopath. Z.* **17**, 478–488.
Blumer, S. (1951b). *Phytopath. Z.* **18**, 101–110.
Blumer, S. (1952). *Ber. schweiz. bot. Ges.* **62**, 384–401.
Blumer, S. (1967). "Echte Mehltaupilze (Erysiphaceae)". Gustav Fischer, Jena.
Boeswinkel, H. J. (1976). *Trans. Br. mycol. Soc.* **67**, 152–155.
Bohmig, F. (1937). *Blumen-u. PflBau* **41**, 510.
Bourdin, J. (1971). *Phytiat.-Phytopharm.* **20**, 111–116.
Bouwens, H. (1924). *Meded. phytopath. Lab. Willie Commelin Scholten* **8**, 3–28.
Bouwens, H. (1927). *Meded. phytopath. Lab. Willie Commelin Scholten* **10**, 3–31.
Brassler, K. (1930). *Blumen-u. PflBau* **45**, 111–112.
Brodie, H. J. (1945). *Can. J. Res. C.* **23**, 198–211.
Brooks, A. V. (1970). *Jl R. hort. Soc.* **95**, 234–236.
Brown, I. F. (1962). *Phytopathology* **52**, 1220.
Burth, U. and Zastrow, J. (1973). *NachrBl. dt. PflSchutzdienst, DDR* **27**, 161–165.
Cherkasskiĭ, E. S., Selochnik, N. N. and Sheikman, A. K. (1964). *Dokl. Akad. Nauk SSSR* **156**, 1197–1200.
Childs, J. F. L. (1940). *Phytopathology* **30**, 65–73.
Corner, A. J. H. (1935). *New Phytol.* **34**, 180–200.
Coyier, D. L. (1961). Biology and control of rose powdery mildew. Ph.D. thesis, University of Wisconsin.
Crowell, I. H. (1937). *Pl. Dis. Reptr* **21**, 134–138.
Da Camara, E. de S. and Da Luz C. G. (1939). *Agronomia lusit.* **2**, 41–63.
Deep, I. W. and Bartlett, A. (1961). *Pl. Dis. Reptr* **45**, 628–631.
De Mendonça and De Sequeira, M. (1962). *Agronomia lusit.* **24**, 109.
Dillon Weston, W. A. R. and Taylor, R. E. (1944). *Trans. Br. mycol. Soc.* **27**, 119–120.
Dimock, A. W. (1953). *N.Y. St. Flow. Grow. Bull.* no. 93, 3.
Docea, E., Vulpe, O., Motiu, A. and Marin, J. (1970). *Revtă hort. vitic.* **19**, 80–86.
Drozdovskaya, L. S. (1964). *Nauchn. trud. Akad. kommun. kh-va 1962*, 115–116.
Eliade, E. (1972). *Lucr. Grăd. bot. Buc. 1970–1971*, 391–399.
Engel, E. (1928). *Gartenwelt* **32**, 314.
Evans, E. J. (1974). *Gdnrs' Chron.* **176**, 39, 41.
Firky, A. (1936). *Leafl. Minist. Agric. Egypt* no. 76, 6pp.
Fischer, E. (1919). *Schweiz. GartnZtg* **21**, 314–315.
Fisher, R. W., Chamberlain, G. C. and Kemp, W. G. (1960). *Pl. Dis. Reptr* **44**, 273–275.
Foëx, E. (1926). *Bull. Soc. mycol. Fr.* **41**, 417–438.
Foëx, E. (1927). *Revue Path. vég. Ent. agric. Fr.* **14**, 217–223.
Forsberg, J. L. (1963). "Diseases of Ornamental Plants". Special Publication no. 33. University of Illinois.
Frost, A. J. P. and Pattisson, N. (1971). *Proc. 6th Br. Insectic. Fungic. Conf.* **2**, 349–354.
Gante, T. (1935). *NachrBl. dt. PflSchutzdienst, Berl.* **15**, 14–15.
Goidanich, G. (1941). *Boll. Staz. Patol. veg. Roma* NS **21**, 161–174.
Golovin, P. N. (1956). *Trans. V. L. Komarov Bot. Inst. USSR Acad. Sci.* II, **10**, 195–308.
Gorlenko, M. V. (1974). *Mikol. i Fitopatol.* **8**, 497–501.
Gram, E., Jørgensen, C. A. and Rostrup, S. (1927). *Tidsskr. PlAvl* **33**, 781–841.
Grouet, D. (1963). "Maladies cryptogamiques de l'Hortensia". INRA, Versailles.
Hammarlund, C. (1945). *Bot. Notiser, 1945*, 101–108.

Hammett, K. R. W. (1968). *Pl. Dis. Reptr* **52**, 754–758.
Hey, G. L. (1957). *Comml Grow.* no. 3206, p. 1182.
Hino, I. and Kato, H. (1929). *Bull. Miyazaki Coll. Agric. For.* **1**, 91–100.
Hirata, K. (1966). "Host Range and Geographical Distribution of the Powdery Mildews". (mimeo). Niigata Univ., Niigata, Japan.
Howden, J. C. W. (1968). *Natn. Rose Soc. Rose A.* 131–136.
Immikhuizen, E. (1968). *Versl. Meded. plziektenk. Dienst Wageningen* no. 143, 95–105.
Jones, B. M. and Swartout, A. G. (1961a). *Pl. Dis. Reptr* **45**, 366–367.
Jones, B. M. and Swartout, A. G. (1961b). *Pl. Dis. Reptr* **45**, 794–795.
Jørstad, I. (1925). *Norske Videnskaps-Akad., Matem.-Naturvid. Kl. Skr.* **10**, 116pp.
Junnell, L. (1967). *Symb. bot. upsal.* **14**, 1–117.
Kamat, M. N. and Patel, M. K. (1948). *Indian Phytopath.* **1**, 153–158.
Kavanagh, T. (1967). *Res. Rep. Hort. For. Div. agric. Inst. Dublin 1966*, 120–132.
Lepik, E. (1943). *Zentbl. Bakt. ParasitKde*, 2, **106**, 89–93.
Longree, K. (1939). *Mem. Cornell Univ. agric. Exp. Stn* no. 223, 43pp.
Ludwigs, K. (1927). *Blumen-u. PflBau* **42**, 295–296.
Lüstner, G. (1935). *NachrBl. dt. PflSchutzdienst, Berl.* **15**, 41.
Luttrell, E. S. and Samples, J. W. (1954). *Pl. Dis. Reptr* **38**, 719–720.
Magee, C. J. (1955). *In* "Report of the plant diseases conference, Hawkesbury Agricultural College, N.S.W.".
Mains, E. B. (1942). *Phytopathology* **32**, 414–418.
Mandre, M. (1968). *Eesti NSV Tead. Akad. Toim.* Biol. seer. **17**, 229–235.
Mandre, M. (1970). *Eesti Loodus* **19**, 89–93.
Mandre, M. (1971). *Eesti NSV Tead. Akad. Toim.* Biol. seer. **20**, 255–261.
Manning, W. J., Vardado, P. M. and Connor, M. D. (1972). *Pl. Dis. Reptr* **56**, 405.
Martin, G. H. (1928). *Pl. Dis. Reptr. Suppl.* **65**, 400–437.
Martin, H. (Ed). (1969). "Insecticide and Fungicide Handbook". Blackwell Scientific Publications, Oxford.
Mayor, E. (1936). *Bull. Soc. neuchâtel. Sci. nat.* **61**, 105–123.
McCain, A. H., Byrne, T. G. and Bell, M. R. (1962). *Calif. Agric.* **6**, 11.
Mence, M. J. and Hildebrandt, A. C. (1966). *Ann. appl. Biol.* **58**, 309–320.
Miller, H. N. (1962). *Proc. Fla St. hort. Soc.* **74**, 400–404.
Miller, H. N. (1966). *Rep. Fla agric. Exp. Stn* **1965**, 163.
Misiga, S. (1966). *Polnohospodárstvo* **12**, 268–281.
Misiga, S., Pašmiková, A., Lietava, M. and Rakús, D. (1967). *Polnohospodárstvo* **13**, 332–343.
Molina, V. L. A. (1973). *Revtă Cienc. agric.* **5**, 51–61.
Montemartini, L. (1930). *Riv. Patol. veg., Padova* **20**, 201–206.
Moore, F. J. (1952). *Pl. Path.* **1**, 52–55.
Moore, W. C. (1947). *Trans. Br. mycol. Soc.* **31**, 86.
Moore, W. C. (1959). "British Parasitic Fungi". University Press, Cambridge.
Moore, W. C. and Moore, F. J. (1949). *Trans. Br. mycol. Soc.* **32**, 273–279.
Nadernejad, N. (1966). *Iran J. Pl. Path.* **3**, 21–24.
Nannizzi, A. (1927). *Atti R. Accad Fisiocritici, Siena*, 10, **2**, 399–403.
Nichols, L. P. and Nelson, P. E. (1970). *Scient. Agric.* **17**, 13.
Ogawa, J. M., Hall, D. H. and Koepsell, P. A. (1967). *In* "Airborne Microbes". (P. H. Gregory and J. L. Monteith, Eds) 247–267. University Press, Cambridge.
Pady, S. M. (1972). *Phytopathology* **62**, 1099–1100.
Palmer, J. G., Henneberry, T. J. and Taylor, E. A. (1959). *Pl. Dis. Reptr.* **43**, 494–495.
Pape, H. (1927). *Gartenwelt* **31**, 732–733.
Pape, H. (1928). *Gartenwelt*, **32**, 496–497.
Pape, H. (1939). *Blumen-u. PflBau* **43**, 522.

Pape, H. (1955). "Krankheiten und Schadlinge der Zierpflanzen und ihre Bekampfung". Paul Parey, Berlin.

Pape, H. and Marggraf, M. (1966). *Blumen-u PflBau* **40**, 202–203.

Pathak, S. and Chorin, M. (1968). *Phytopath. Mediter.* **7**, 123–128.

Paulus, A. O., Nelson, J., Shibuya, F., Miller, M. and Maire, R. G. (1971). *Calif. Agric.* **25**, 10–11.

Paulus, A. O., Nelson, J., Harvey, O., Maire, D. and Shibuya, F. (1974). *Calif. Agric.* **28**, 4–5.

Perera, R. G. (1972). Studies of conidial germination in some powdery mildew fungi. Ph.D. thesis, University of London.

Perera, R. G. and Wheeler, B. E. J. (1975). *Trans. Br. mycol. Soc.* **64**, 313–319.

Peterson, J. L. and Davis, S. H. (1970). *Pl. Dis. Reptr* **54**, 606–607.

Pfister, E. and Müller, H. J. (1970). *NachBl. dt. PflSchutzdienst, Berl.* N. F. **24**, 185–189.

Pirone, P. P. (1970). "Diseases and Pests of Ornamental Plants". Ronald Press Co., New York.

Ponti, I. and Bencivelli, A. (1973). *Notiz. Mal. Piante* no. 88–89, 163–170.

Price, T. V. (1969). Studies of the overwintering, epidemiology and control of *Sphaerotheca pannosa* on rose. Ph.D. thesis University of London.

Price, T. V. (1970). *Ann. appl. Biol.* **65**, 231–248.

Puttemans, A. (1911). *Bull. Soc. r. Bot. Belg.* no. 48, 238.

Qvarnström, K. (1970). *Vaxtskyddsnotiser* **34**, 43–45.

Raabe, R. D. and Hurlimann, J. H. (1970). *Calif. Agric.* **24**, 9–10.

Raabe, R. D., Hurlimann, J. H. and Sciaroni, R. H. (1970). *Calif. Agric.* **24**, 8.

Radclyffe, W. F. (1861). *Gdnrs' Chron.* **21**, 967.

Riordain, F. O. (1974). *Pl. Dis. Reptr* **58**, 12–13.

Rogers, M. N. (1959). *Mem. Cornell Univ. agric. Exp. Stn* no. 363, 38pp.

Rumberg, V. (1974). *In* "Bolezneustoichivost' rastenii". (Semenova, V. Ed.) 68–120. Academy of Sciences, Estonian SSR.

Salmon, E. S. (1900). A monograph of the Erysiphaceae. *Mem. Torrey bo. Club* no. 9, 292 pp.

Salmon, E. S. (1905). *Annls mycol.* **3**, 1–15.

Salmon, E. S. (1906). *Jl. R. hort. Soc.* **31**, 142–146.

Schmidt, T. (1952). *Pflanzenschutzberichte* **8**, 22.

Schmitt, J. A. (1955). *Mycologia* **47**, 688–701.

Schüepp, H. and Blumer, S. (1963). *Phytopath. Z.* **48**, 329–336.

Semeniuk, P. and Palmer, J. G. (1970). *Pl. Dis. Reptr* **54**, 598–602.

Sharma, O. P. (1961). *Sci. Cult.* **27**, 39.

Sibilia, C. (1955). *Boll. Staz. Patol. veg. Roma*, 3, no. 12, 221–239.

Siemaszko, W. (1933). *Revue Path. vég. Ent. agric. Fr.* **20**, 139–147.

Simonyan, S. A. (1973). *Biol. Zh. Armenii* **26**, 62–73.

Sivanesan, A. (1971). *Trans. Br. mycol. Soc.* **56**, 304–306.

Smith, C. G. (1969). *Trans. Br. mycol. Soc.* **53**, 69–76.

Smith, C. G. (1970). *Trans. Br. mycol. Soc.* **55**, 355–365.

Smith, C. G. and Wheeler, B. E. J. (1969). *Trans. Br. mycol. Soc.* **52**, 437–445.

Smith, P. (1965). *Rep. Glasshouse Crops Res. Inst. 1964*, 98.

Sommer, H. (1935). *Blumen-u. PflBau* **39**, 240.

Stahl, M. and Umgelter, H. (1976). "Pflanzenschutz im Zierpflanzenbau" Eugen Ulmer, Stuttgart.

Strider, D. L. (1974). *Pl. Dis. Reptr* **58**, 875–878.

Strider, D. L. (1976). *Pl. Dis. Reptr* **60**, 45–49.

Tammen, J. F. (1973). *Sci. Agric.* **20**, 10.

Taylor, J. C. (1962). *Scient. Hort.* **16**, 31–34.

Thompson, H. S. (1961). *Can. J. Pl. Sci.* **41**, 227–230.
Thompson, H. S. and Svejda, F. J. (1965). *Can. J. Pl. Sci.* **45**, 258–263.
Togliani, F. (1971). *Italia agric.* **108**, 83–92.
Tokushige, Y. (1953). *Ann. phytopath. Soc. Japan* **17**, 61–64.
Umgelter, H. (1973). *Erwerbsgartner* **27**, 569–570.
Valášková, E. (1965). *Acta prühon.*, **1965**, 51–62.
Van Poeteren, N. (1926). *Versl. Meded. plziektenk. Dienst Wageningen* no. 44.
Van Poeteren, N. (1936). *Versl. Meded. plziektenk. Dienst Wageningen* no. 83, 88pp.
Viennot-Bourgin, G. (1944). *Revue Mycol.* N.S. **9**, 75–77.
Viennot-Bourgin, G. (1965). *C.r. hebd. Séanc. Acad. Sci., Paris* **261**, 4222–4223.
Viennot-Bourgin, G. (1973). *Phytiat.-Phytopharm.* **22**, 272–279.
Vir, D. and Raychaudhuri, S. P. (1973). *Pesticides* **7**, 28.
Wager, V. A. (1970). "Flower garden diseases and pests". Purnell, Cape Town.
Wahl, B. (1926). *Verlag der Bundesanst für Pflanzenschutz, Wein* **II** 28pp.
Wheeler, B. E. J. (1973). *Jl R. hort. Soc.* **98**, 225–230.
Wheeler, B. E. J. (1975). *Ann. appl. Biol.* **79**, 177–188.
Williamson, C. E. (1953). *N.Y. St. Flow. Gr. Bull.* no. 93, pp. 2, 4.
Woronichine, N. (1914). *Bull. Soc. mycol. Fr.* **30**, 391–401.
Yarwood, C. E. (1934). *Phytopathology* **24**, 20–21.
Yarwood, C. E. (1936). *Phytopathology* **26**, 845–849.
Yarwood, C. E. (1939). *Phytopathology* **29**, 282–284.
Yarwood, C. E. (1944). *Phytopathology* **34**, 937.
Yarwood, C. E. (1952). *Bull. Calif. Dep. Agric.* no. 41, 19–25.
Yarwood, C. E. (1957). *Bot. Rev.* **23**, 235–300.
Yoshii, K. and Morales, O. (1975). *Fitopatologia, Bogota* **10**, 28–31.
Zobrist, L. (1946). *Gartnermeister* **49**, 17–19.

Chapter 17

Powdery Mildew of Tobacco

J. S. COLE

Tobacco Research Board of Rhodesia, Salisbury, Rhodesia

I. World Distribution and Economic Effects

Powdery mildew of tobacco (*Erysiphe cichoracearum* DC) has spread to most countries where tobacco is grown and is often a disease of major importance where resistant cultivars are not available. Perhaps the most notable exception is the United States of America, where, although it was reported on glass-house grown burley tobacco in Kentucky (Chapman, 1953), it has not spread to the tobacco crop in that area. Chapman noticed a severe infection on an interspecific cross between *Nicotiana bonariensis* × *Nicotiana alata*. The pathogen also infected nearby burley seedlings but growth was very slight and it seems unlikely that this was the same strain of the pathogen that readily attacks tobacco, including burley cultivars, in other countries. The disease was observed by Lucas (1975) on flue-cured tobacco in Jamaica in 1970 and in Nicaragua the following year on cigar tobacco, and is present in Brazil and Guatemala. One wonders why it has not established itself in the tobacco crops of the USA and Canada. In Europe it has been present for at least 100 years (Scaramuzzi, 1948), particularly in Italy, Greece, Portugal and other Mediterranean countries. It is also to be found in Russia, Indonesia, Taiwan, the African continent and Madagascar, and was recently identified in New Guinea (Shaw and Layton, 1975).

According to Renaud (1959), the disease can sometimes be very serious, or of secondary importance, according to the country, the season and the cultivar grown. Its economic importance in any particular situation depends on a number of independent and related factors. The weather can have a large effect on disease development, but more particularly by influencing the growth of the crop rather than acting directly on the pathogen. The cultivar, the time of primary infection, plant growth, water stress, cultural treatments such as removal of the flower (topping), method of harvesting the leaves, whether individually or by cutting the stalk, all can affect either the susceptibility of the leaves or the time available for the pathogen to spread. In the most severe conditions where the disease starts on the lowest leaves, which become fully susceptible and are reaped individually as they ripen, considerable losses can be caused. Cole (1963a) reported a yield depression of 19% in infected tobacco in 1957, compared with plants that had been protected with dinocap. The untreated tobacco had a mean percentage of approximately 40% of its leaf area infected. In other experiments there have been only slight losses from the disease and correlations between infection and crop loss are not very easy to detect. Ternovsky (1961) states that the disease is often so severe that leaves are not reapable but very little precise information on actual losses are available. In South Africa annual losses of 20–30% were estimated by Rossouw (1963). Khetsuriani (1966) reported that the disease reduced leaf yields by 16% monosaccharides by 63% and nicotine by 46%, but in contrast Cole (1963a) found only slightly less nicotine and sugars in severely infected leaves. Currently, in most countries, either resistant cultivars are grown or foliar sprays or dusts are used to protect crops, and losses are usually much less than those experienced previously.

II. Causal Pathogen

The exact identity of the fungus has been the subject of some controversy, particularly in countries where the perfect stage has not been found. Renaud (1959), in his comprehensive review, comments that although the oidium stage on tobacco is usually reported as *E. cichoracearum* it is possible that other species can parasitize and form conidia on tobacco and that this might explain why the perfect stage has not been found in certain countries, in Africa for example. The absence of compatible heterothallic strains or unsuitable environmental conditions are other possible reasons. Morrison (1960) showed that *E. cichoracearum* DC ex Merat is heterothallic. Clonal isolates from sunflower (*Xanthium pennsylvanicum*) and *Zinnia elegans* did not produce cleistothecia on leaf discs but intercrosses did and ascospores were produced in 6–7 weeks. Not all crosses were fertile, indicating that there were two allelomorphic series controlling host susceptibility. All the clones grew on sunflower. If the tobacco strains are similar, the absence of compatible ones could account for the lack of cleistothecia in some localities.

Cleistothecia have been identified on tobacco in several countries including Italy (Scaramuzzi, 1948), Turkey (Ducomet, 1928) and Brazil (Averna-Sacca, 1922). Marcelli (1949) noted that several infected tobacco plants in a glasshouse developed cleistothecia whilst a number of other genera in the house remained free from the disease; *Cucurbita maxima* was an exception in that it became infected but its level of infection was slight. Cross-inoculations from tobacco to *Cucumis melo*, *C. maxima*, *C. anguria* and *C. sativus* caused infections only on *C. anguria* and no cleistothecia were formed. Hopkins (1948) transferred conidia from potatoes to small defined areas on tobacco leaves and from tobacco to potatoes. Mildew developed in the inoculated areas but the isolate from potato did not grow as well on tobacco as the normal tobacco isolate and he concluded that it was probably a different strain of *E. cichoracearum*. Deckenbach (1924) found that squash seedlings, vegetable marrow, cucumbers and melons were infected after inoculation with conidia of tobacco powdery mildew but cleistothecia were not formed on any of the hosts, nor had they been identified on tobacco in Russia at that time. However, cleistothecia of *E. cichoracearum* were identified on some cucurbits. There are a number of other similar reports, some claiming successful inter genera infections and others reporting failure. It is not always clear whether the pathogen from other genera grew successfully on tobacco and sporulated freely but if its growth was as restricted as that of the tobacco pathogen sometimes is on resistant tobacco cultivars (Raeber *et al.*, 1963) then tobacco and the other species might not normally serve as alternative hosts in natural conditions.

III. Epidemiology

Initial field infections are usually caused by windborne conidia which germinate within a few hours, form appressoria, penetrate epidermal cells, develop haustoria and secondary hyphae and thus establish successful infections.

Levykh (1940) found that conidia on dry slides germinated best at relative humidities of 60–100% but inoculated leaves exposed to saturated air and an optimum temperature (16–24°C) showed no signs of infection after six days. Symptoms occurred and conidia were formed when the relative humidity (r.h.) was lowered to 70–76% (6–8 mbar saturation deficit (s.d.) at 22°C.) Rossouw (1957) reported that 23·5°C was the optimum temperature for infection and, rather surprisingly, that no infection occurred at 25°C. Minev (1956) found that an r.h. of 60–86% was optimum and saturated air was detrimental. He quotes an optimum temperature of 23°C and a maximum of 32°.

I measured germination of conidia at 25°C in sealed containers in which saturated solutions of inorganic salts produced saturation deficits in the range of 22–0 mbar (17–100% r.h.) and detected no differences in percentage germination. However, in experiments with seedlings in controlled environments (Cole, 1966e), germination was greater at near-saturation than at 8–9 mbar. Leaf age also had an effect on germination and more conidia germinated on older leaves than on the younger ones. On the other hand, hyphal growth and the percentage of germinated conidia that formed a successful relation with the host, equivalent to the elongating secondary hyphae of Masri and Ellingboe (1966) were favoured by larger saturation deficits of 7–9 mbar (Fig. 1). Nair and Ellingboe (1965), working with

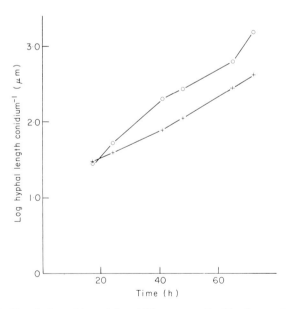

Fig. 1.   Effect of incubation at two air humidities on growth of hyphae on tobacco leaves.
Saturation deficits; +——+ 1·7 mbar; ○——○ 7·6 mbar.

Erysiphe graminis DC f. sp. *tritici*, found that the highest percentage germination was obtained in saturated air at 17°C for 1 h followed by 3 h at 22° and 9 mbar s.d. Saturated air was more beneficial for germination but it retarded the growth of secondary hyphae. Although the primary developmental processes of infection of tobacco by *E. cichoracearum* have not been studied in such detail as those of *E. graminis* on cereals, by Ellingboe and his colleagues, the same general conclusions seem to be valid that germination is greatest at near-saturation and hyphal growth is favoured by drier air. However, in common with most other powdery mildews, conidial germination, infection and sporulation are tolerant of a fairly wide range of environmental conditions.

A. Sporulation and dispersal of conidia

The conditions governing the formation of the perfect stage are not clearly defined but cleistothecia are found in most European countries where tobacco is grown and can act as an overwintering source of infection. Where they are not normally formed, conidia from overwintering plants initiate infection in the following crop. Conidiophores develop about 5–6 days after infection occurs; they grow away from the plant surface and conidia form in a basipetal chain by transverse septation of the generative cell. Hopkins (1956) considered that they were formed only in sunny weather and not in humid, cool and shady conditions. We now know that sporulation can occur in a wide range of temperatures, air humidities and light intensities but the environment can affect the processes involved in formation and dispersal. The principles established for powdery mildew of tobacco seem to apply to *E. cichoracearum* on other plants (Childs, 1940 and Pady *et al*, 1969).

Growth rates of conidia on attached tobacco leaves were measured by direct observation (Cole and Fernandes, 1970b) and by time-lapse photography (Cole and Geerligs, 1976). In constant temperature and continuous fluorescent light, conidia matured at a constant rate all day. Where, however, the light period was interrupted every 12 h by 12-h periods of darkness, the growth rate changed and more conidia were formed in the light than in the dark, although the 24-h total remained about the same as in continuous light. Similar periods of alternating temperature (18–24°C) in continuous light had no effect. The light–dark stimulus could also be produced by radiation in the near-ultraviolet (300–400 nm) range and it seems that it produces the main growth response. Stimulation of sporulation by near-ultraviolet light is fairly common in many fungi (Leach, 1962). Cole (1971) suggested that conidial maturation at different rates in light–dark cycles would synchronize populations of conidiophores and the majority of mature conidia would be ready for dispersal in the light period. This is almost certainly the main factor in the diurnal periodicity observed in natural conditions (Cole, 1966c). Conidia are passively dispersed and are normally dislodged by air or plant movements. In a wind of 0·8 m s^{-1} conidia were detached at a fairly constant rate in continuous light and constant temperature. Where darkness (12 h) alternated with fluorescent or near-ultraviolet light (12 h), many more conidia were detached in the light than in the dark and dispersal occurred on a greater number of occasions. There was, however, no clear pattern of release during the light period, probably because the sample was too small (Cole and Geerligs, 1976). Larger samples of the spore population were obtained in a controlled environment tunnel, in which spores were caught downwind from infected plants in a spore trap. A fairly comprehensive study of the combined and separate effects of light, temperature and air humidity at constant wind speed in the climate tunnel has clarified some of the factors in the sporulation and dispersal mechanisms. A daily periodicity similar to that found in field studies is produced in a 24 h programme of fluctuating temperature and humidity, modelled on natural conditions, and 13 h light and 11 h dark (Cole and Fernandes, 1970b). The catch has a normal distribution with a peak in the middle of the light period. Where temperature and humidity are constant and light either on or off, a constant number of conidia h^{-1} are dispersed throughout the day. Small changes in the s.d. of the ambient air (1–8 mbar) have no effect on the dispersal pattern and even large changes from 2–17 mbar (93–39% r.h.) produce only a slight increase in release in the dryer period (Cole, 1971). Temperature has a much greater effect than humidity and alternating 12 h periods of 16° and 23°C in continuous light or in continuous dark produce peak spore concentrations when the temperature changes to 23°C (Cole and Fernandes, 1970b). Changes in temperature and humidity (1–8 mbar s.d.) give responses similar to those of temperature alone. In constant temperature and humidity with fluorescent light alternating with dark (12 h periods), the catch follows a normal distribution with a peak about midway in the light period. The distribution is different from that of the temperature response, which is skewed towards the start of the warmer period. However, where programmes combine changes in temperature and light the response is similar to that of alternating light and dark alone and the temperature response is masked (Cole and Fernandes, 1970b). Alternating near-ultraviolet light gave the same response as that of fluorescent light (Fig. 2). Tobacco leaf discs in near-ultraviolet light lost 16% of their dry mass in three days whereas in fluorescent light they gained 52%, so the light acts directly on the pathogen and changes in photosynthesis do not cause the diurnal response. The response is also endogenous and although there is no lag in establishing it from continuous light or dark, it persists for about 24 h after the return to continuous light or continuous dark. This tends to support the hypothesis that changes in growth rate of conidia synchronize the population of conidiophores. The temperature response in continuous light is also

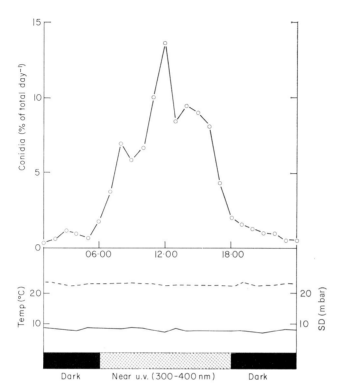

Fig. 2.   Effect of alternating near-ultraviolet light and dark on diurnal dispersal of conidia from infected plants. Geometric mean of 10 cycles ○; temp. – – –; saturation deficit (SD) ———.

established without a lag but does not persist after the change back to continuous light, as might be expected if no changes in growth rate are involved.

In natural conditions the daily light–dark cycle obviously determines the diurnal rhythm of spore production and dispersal. As temperature and humidity have little effect on this, significant correlations of hourly catches from the air around infected tobacco, with air temperature and humidity (Cole, 1966c) could be attributed to parallel rather than causally related changes.

The conidial chain in *E. cichoracearum* may be regarded as a cantilever whose moment acts on the joint between the youngest mature conidium and the rest of the conidiophore (Cole, J. S., 1976). As the chain lengthens with the addition of mature conidia, the force required to dislodge it as a unit becomes progressively less. In a synchronized population in light winds, the total daily production might be released at the same time each day giving the observed normal distribution with a fairly well defined peak, depending on the inherent variability. Large changes in wind speed, turbulence or physical disturbance could upset the pattern and give day to day variation. Bainbridge and Legg (1976) reported that barley leaves flapping at 0.6 m s^{-1} in a wind tunnel developed accelerations sufficient to dislodge conidia of *E. graminis* although a wind of this speed usually is not sufficient to drag conidia from a stationary leaf.

B. Leaf age and susceptibility

Although seedlings can be infected readily the disease is generally unimportant in seedbeds. The oldest leaves of field plants are usually the first to be infected. Cole (1963b) followed the progress of natural infection in several field experiments in Rhodesia during two seasons by assessing at 7-day intervals the percentages of leaf surface covered by the pathogen. Alternate leaves of plants were studied from emergence until they were senescent and removed. Increases in leaf area were measured and meteorological and soil data were collected. First infection was seen on the oldest leaves no earlier than six weeks after plants were transplanted from seedbeds. This observation has frequently been confirmed since. Small white colonies of sporulating mycelium develop on the abaxial and adaxial epidermes, usually on the proximal parts of the leaves first, and as the colonies spread and new ones appear the leaf surface is gradually covered. Leaf area measurements showed that infection became visible only as the leaf approached the end of its period of maximum expansion. This was investigated more fully later (Cole, 1966b) in similar field experiments. A minimum period of 35 days elapsed from the time of the emergence of each leaf from its bud to the time when it became infected (Table 1), irrespective of its position on the plant. Leaves emerging in the first week in January were not infected until the middle of February, although large numbers of conidia were in the air within the crop on each day. Leaves contained more sugars and usually more insoluble carbohydrate during their susceptible stage, than while they were expanding rapidly; amino nitrogen content was variable. There was no clear evidence that these constituents had a direct effect on changes from resistance to susceptibility in the leaves and Cole (1966d) suggested that it might be connected with the slowing down in the rate of protein synthesis that occurs at this time (Hellebust and Bidwell, 1964).

C. Differential susceptibility of a leaf

Stephen (1955) and Cole (1963b) observed that powdery mildew often starts on the proximal part of leaves of field plants. Cole (1966d) later measured

TABLE 1

Dates of emergence and natural infection with powdery mildew of alternate leaves
of intact tobacco plants

Leaf number	Days between emergence and infection	
	Not irrigated	Irrigated
2	39	41
4	35	37
6	35	40
8	41	42
10	45	40
12	50	38
14	45	40
16	44	42
18	48	44

disease onset and progress on the two halves of leaves. Proximal parts were usually infected first and secondary hyphae grew more quickly on discs from these parts, which also contained less amino nitrogen and sugars, than on discs from distal parts of leaves. Generally, hyphal growth on leaf discs from intact and topped plants was inversely correlated with the amino nitrogen concentration and water content of the tissue and directly correlated with the concentration of insoluble carbohydrate. Whether these are causal relations is open to question.

Apart from the possible differences in susceptibility of proximal and distal parts of leaves, climatic factors may also have more direct effect on the pathogen in field crops where plants are growing close together. Proximal parts of lower and middle leaves are more shaded than distal ends, their maximum temperatures tend to be lower (Cole, 1967), they are warmer at night and are less affected by external conditions such as cloud cover. Consequently dew is formed less frequently than on distal ends and conidia may have a greater chance of developing a successful relation with the leaf. This could be especially important when few conidia are present in the early stages of infection. Decreased light intensity may also provide more suitable conditions for fungal growth on the proximal parts, particularly of lower leaves. I know of no direct evidence for this but the prevalence of powdery mildew in glasshouses and on cigar wrapper tobacco in shade tents has often been reported. The sterilizing effect of ultraviolet light could also be detrimental on the exposed distal parts and on upper leaves of topped plants. This aspect involves the interaction of many factors and may only be resolved by a more complete study than has yet been reported.

D. Topping and leaf resistance

Cole (1963b) noted that there was very little infection on upper leaves and that the top 20% had hardly any. Plants were topped and it was thought that the absence of infection in upper leaves might be attributable to metabolic changes induced by the removal of the flower head. This was confirmed (Cole, 1966b,d) by field measurements with intact and topped plants. Only leaves that are still expanding when the flower is removed are affected and the rate of increase in whole plant infection is similar in topped and in intact plants at first (Fig. 3), then it decreases very sharply in those that have been

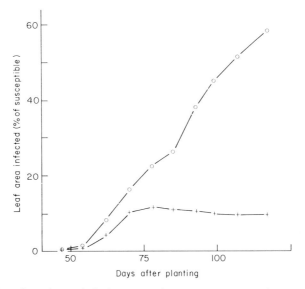

Fig. 3.   Effect of topping on infection. Topped +———+; not topped ○———○.

topped. Topping had no effect on the amount of infection of the lower six leaves (Table 2), which had completed 90% of their potential expansion when the plants were topped. It significantly reduced infection in leaf 8, which was about 80% expanded, and in the leaves above. Leaf 16 was less than half expanded at topping and had very little infection at all. Hyphal growth on leaf discs from topped field plants (Table 3) and on attached leaves of glass-house grown plants was much less than on comparable leaf tissue from plants whose flower head had not been removed (Cole and Fernandes, 1970a). Topping had little effect on amino nitrogen or carbohydrate content of the tissue.

TABLE 2

Percentage area infected with powdery mildew of leaves of irrigated field tobacco

Leaf No.	Not topped	Topped	S.E.	Lsd P=0·05
2	15·7	8·6	±15·24	32·9
4	57·5	40·0	± 9·21	19·8
6	65·6	42·5	±11·89	25·5
8	58·1	19·3	±13·42	29·0
10	67·5	5·6		
12	70·6	1·4		
14	54·4	1·2		
16	45·8	0·1		

Where conditions are ideal for plant growth during the period of maximum expansion, all leaves of intact plants can probably become equally susceptible. However, environmental factors often limit their full potential. One of these, as we have seen, is the removal of the flower head, which amongst other hormonal changes removes a metabolic sink into which nutrients are moving. These are then diverted into side shoot growing points in the axils of leaves when they start elongating, into the leaves that are still expanding and into the stems and roots. In tobacco culture, the side shoots are removed or suppressed, mechanically or chemically, so leaves receive additional nutrients and they continue to expand and to increase in mass for longer than those on intact plants. The changed leaf growth pattern and its consequent effects on metabolism that appear to interfere with some stage of the primary infection processes provide a potentially profitable field of study to clarify the host–pathogen interaction in this disease.

E. Soil moisture stress and leaf growth

Soil moisture stress also affects leaf growth and influences the potential susceptibility of individual leaves. Ternovsky (1961) observed that powdery mildew spreads more in tobacco crops that have supplementary irrigation. Cole (1963b) subsequently showed that tobacco grown without supplementary irrigation and subjected to occasional periods of water stress had less disease than comparable plants that were irrigated during the dry spells up to the time that the leaves were infected (Fig. 4).

Irrigation, therefore, presumably affected leaf growth or composition and not the pathogen directly. In other work with intact plants (Cole, 1966b), the pathogen grew more quickly on fully expanded leaves of irrigated plants than on those that had not been watered during the dry spell. However, the leaves

TABLE 3

Germination and mean hyphal length/conidia on discs from leaves 12 and 16, with their free amino nitrogen and carbohydrate contents at the start of incubation

| Leaf | Parts of leaf | germination (%) | | Mean hyphal length (μm) | | Amino nitrogen (μgcm^{-2}) | | Carbohydrate (μgcm^{-2}) | | | |
| | | | | | | | | Soluble | | Insoluble | |
		Intact	Topped	Intact	Topped	Intact	Topped	Intact	Topped	Intact	Topped
	S.E. (±)	2·70		8·58		0·31		28·0		142·3	
12	Proximal	23·2	19·4	74·2	34·0	2·1	2·3	454	385	2096	1943
	Distal	28·3	22·7	33·1	25·7	3·7	3·3	608	603	1895	1914
16	Proximal	20·6	18·0	51·0	27·8	2·4	2·9	434	360	1929	1800
	Distal	20·6	18·6	40·2	25·0	4·0	4·2	659	630	2220	2263
Main effects											
S.E. (±)		1·91		6·07		0·22		19·8		100·6	
Topping		—3·5		—21·5		0·1		—44		—55	
Proximal—distal		—2·3		15·8		—1·4		—217		—131	
Leaf 12–16		3·9		5·7		—0·5		—8		—91	

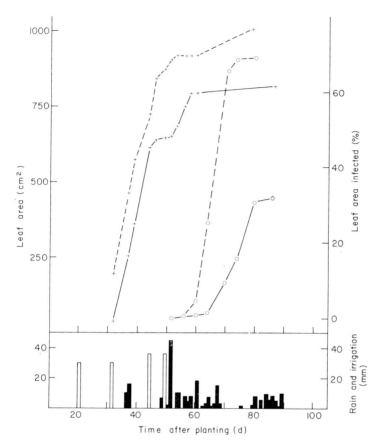

Fig. 4.   Infection of leaf 8 in relation to growth and irrigation. Infection ○——○ ; leaf area +——+ ; irrigated – – – ; not irrigated ——— ; irrigation □ ; rain ■.

of irrigated plants were also larger and, as discussed in the following section, in some circumstances leaf infection is positively correlated with leaf size, which may have been a separate factor here. Further up the stalk, leaves that were still expanding during the dry weather had a temporary growth check and then grew larger than those of irrigated plants. They were infected after the dry weather was over and had significantly less disease at harvest. The susceptible stage of a leaf can be delayed and its potential susceptibility decreased by water stress during leaf expansion and, to a certain extent, during the infection process. The mechanisms involved in these pathogen–plant interactions are not at all clear. Field observations indicate that an excess of rain and waterlogged soil, where it restricts leaf growth, can also adversely affect leaf susceptibility. These two extreme conditions can both produce physiological drought and may induce similar biochemical changes. Alternatively the poor leaf growth associated with an excess of rain in sandy soils could also involve nitrogen and other mineral deficiencies and these may be more directly involved in the disease resistance.

Drought can often disturb the hormonal balance in tobacco (Mizrahi and Richmond, 1972) and levels of cytokinins, abscisic acid and gibberellins are sometimes altered and may not return to the original state. Protein metabolism is interrupted and despite the ultimate recovery of turgidity irreversible changes may occur. There is a considerable amount of evidence (Levitt, 1972) that proline accumulates in water stressed leaves. It is either synthesized directly or results from disturbed protein metabolism. Weybrew et al. (1974) report that the cessation of nitrate reductase activity and the beginning of starch accumulation which accompanies the change from leaf growth to maturation is delayed by drought and excessive rainfall.

F. Plant growth and mineral nutrition

Humid air at the base of plants was thought to provide ideal conditions for powdery mildew (Moore, 1926; Thung, 1937; Hopkins, 1956; Renaud, 1959) and farmers were consequently advised to avoid poorly ventilated and badly drained sites for tobacco. Close spacing has also been cited as the cause of poor air circulation (Renaud, 1959) and aligning rows of tobacco SE–NW (Putterill, 1923) and E–W (Hopkins, 1956) to trap maximum sunshine were also recommended. Priming, the removal of lower leaves, was suggested by Hopkins as a means of decreasing humidity. In five experiments in Madagascar there were no correlations between plant spacing and disease severity (Renaud, 1959). In Rhodesia, widening the spacing between rows and to a much lesser extent between plants, increased plant susceptibility (Table 4). Increasing the supply of nitrogen and phosphorus had a similar effect

TABLE 4
Effect of plant spacing on powdery mildew of tobacco

Between rows (m)	Between plants (m)			Mean
	0·30	0·61	0·76	
S.E.		±1·84		±1·06
1·07	7·3	7·7	9·9	8·3
1·22	12·7	12·7	12·8	12·8
S.E.		±1·30		
Mean	10·0	10·2	11·4	

(Table 5). In these experiments there was most disease on plants from treatments that produced the biggest yield and there were highly significant correlations between leaf yield per plant and percentage leaf area infected, after allowing for the effects of treatments.

TABLE 5

Effect of nitrogen, phosphorus and potassium fertilization on powdery mildew infection of field tobacco

Nitrogen (kgha⁻¹)	P_2O_5 (kgha⁻¹)			K_2O (kg/a⁻¹)			Mean
	0	112	224	0	109	218	
S.E.		±2·9			±2·9		±1·7
0	19	21	28	25	21	22	23
31	26	30	34	28	31	32	30
62	29	37	39	35	30	40	35
K_2O							
S.E.		±2·9					±1·7
0	30	28	30				29
109	18	30	34				28
218	26	31	37				31
S.E.		±1·7					
Mean	25	30	34				

Plant nutrients can also influence disease susceptibility (Cole, 1964). In aerated nutrient solutions there was most plant growth and disease where nitrogen, phosphorus and potassium were not limiting. Powdery mildew infection had little effect on the dry mass, area or total nitrogen content but it decreased the amino nitrogen concentration. Where there was sufficiency of potassium and nitrogen, phosphorus deficiency decreased the disease slightly. At a low level of potassium with adequate nitrogen and irrespective of phosphorus, leaves were very resistant and the percentage of leaf area infected was only 4% of that of plants supplied with sufficient of the three main nutrients. Potassium deficient leaves had thin epidermal cell walls and large water contents. They contained significantly more amino nitrogen and also the largest number of individual amino acids, of which proline and asparagine were usually the most concentrated. The accumulation of proline in leaves affected by drought has already been mentioned and may indicate a common cause of resistance. Small amounts of glycine, asparagine and threonine were detected only in infected leaves supplied with adequate amounts of the nutrients and in healthy ones that were mineral deficient. Traces of phenylalanine, tryptophane and glutamine were found only in infected leaves.

Potassium deficiency had little effect on germination of conidia on leaf discs incubated on water or on 10% sucrose solution but hyphal growth was retarded. The amino nitrogen concentration in both potassium deficient and control discs increased during incubation on water so the concentration was always higher in those deficient in potassium. In discs floating on sucrose solution the amino acid concentration increased in the potassium sufficient discs and either remained unchanged or decreased in those which were potassium deficient. Although this decreased the difference between them it did not eliminate it. Obviously the high concentration of amino nitrogen in the very resistant tissue may be a coincidental effect of disturbed nitrogen metabolism and not the cause of the resistance but the presence of large amounts of particular amino acids such as proline and possibly asparagine may contribute to it. Samborski and Forsyth (1960) report that some amino acids and carbohydrates that stimulate growth of wheat rust may be toxic at high concentrations. Both proline and aspartic acid were fungitoxic and phytotoxic at certain concentrations so it is possible that metabolites in potassium deficient tobacco leaves, which produce the necrotic deficiency symptoms, may also be responsible for the resistance to powdery mildew.

Nitrogen deficiency, as well as disturbing the nitrogen metabolism of leaves, leading to an accumulation of amino acids and amides (Richards, 1956), also affects the carbohydrate metabolism. Discs taken from leaves of resistant, potassium deficient, plants initially contained more soluble carbohydrate than did the controls, but during incubation on water they lost it at a greater rate and later contained less than the controls (Cole, 1966a). Discs floated on sucrose solution gained sugar but there was no evidence that this affected hyphal growth. Insufficient carbohydrate supply obviously does not account for the resistance of potassium deficient leaves.

Where potassium was deficient, substituting sodium in the nutrient medium accentuated many of the characteristics of potassium deficiency such as leaf necrosis, increased sugar and amino nitrogen concentration and resistance to powdery mildew (Cole, 1966a).

G. Plant hormones and antimetabolites

The ability of E. cichoracearum to form a successful association with a tobacco leaf is influenced by metabolic changes in the leaf. Some aspect of protein synthesis may be a common factor between leaf resistance associated with active leaf growth during rapid expansion, the effects of removal of the flower head, soil moisture stress and potassium deficiency. Attempts were made to stimulate protein synthesis with cytokinins such as kinetin (furfurylaminopurine) and 6 benzyl aminopurine (Cole and Fernandes, 1970a) and to block it with chloramphenicol, puromycin and actinomycin D.

Chloramphenicol acts at the ribosome level by preventing the formation of peptide bonds; puromycin substitutes for tRNA and produces an incomplete polypeptide; actinomycin D prevents transcription of tRNA.

Cytokinins sprayed onto the surface of attached leaves once, or daily for seven days before infection, had no significant effect on the growth of hyphae on the opposite surface. But hyphal growth was retarded on leaf discs floated on a 10 μg ml^{-1} kinetin solution. There was also an interaction with light. Hyphal growth increased with increasing light intensity on water-incubated leaf discs, whereas on kinetin solution it decreased (Table 6). Dry

TABLE 6

Hyphal length/conidium (μm) of powdery mildew on tobacco leaf discs floated on water and on kinetin for 3 days

Fluorescent light (1x)	Water	Kinetin (10 μgml^{-1})
S.E.	±99·4	
4573	1098	331
1883	1013	486
<50	585	496

mass changes in uninfected leaf discs during incubation for 3 days on water or on kinetin were similar. It seems unlikely that kinetin acts directly on the pathogen because there was no effect on attached leaves; similar results were reported by Dekker (1963) with other powdery mildews. However, the fungicidal activity of benomyl is attributed by Skene (1972) to its kinetin-like properties.

Chloramphenicol, which blocks protein synthesis on chloroplast ribosomes but not on those in the cytoplasm in tobacco (Ellis, 1969), also retarded growth of the pathogen on discs floated on a 5 μg ml^{-1} solution (Cole and Fernandes, 1970a). A mixture of chloramphenicol and kinetin caused a significantly greater inhibition than chloramphenicol alone. The activity of kinetin, however, was only very slightly increased by the addition of chloramphenicol (Table 7).

Cytokinins have multiple effects in plant tissues, which are often dependent on interactions with other growth substances. They regulate and induce the production of nucleic acids and proteins and delay senescence and loss of chlorophyll in detached tissue. Various enzymes that are activated by protein breakdown in detached tobacco tissue, including peroxidases and ribonucleases, and those that are depressed, such as glycollic acid oxidase and catalase, are greatly inhibited by kinetin (Farkas, et al., 1963). However

TABLE 7

Log hyphal length/conidium (μm) of powdery mildew on tobacco leaf discs
floated on chloramphenicol and on kinetin for 3 days

	Water	Kinetin (10 μg/ml)
S.E.		± 0.067
Water	2·94	2·12
Chloramphenicol (500 μg/ml)	2·30	2·00

Parthier *et al.* (1964) show that although kinetin at 3 μg ml^{-1} (1·5 × 10^{-5} M).
induced an increased incorporation of amino acids into protein in *Nicotiana
rustica*, 32 μg ml^{-1} strongly inhibited it. There are insufficient data at present
to draw conclusions on how kinetin and chloramphenicol inhibited fungal
growth, especially as actinomycin D (2·5 μg ml^{-1}) and puromycin (5 μg ml^{-1})
had no effect.

H. Infection with other pathogens

Several instances of changes in resistance induced by infection with other
pathogens, viral or fungal, have also been reported. Chadha and Raychaudhuri (1968) found that tobacco leaves infected with chilli mosaic virus had a
smaller proportion of their surface infected with powdery mildew than had
healthy leaves and that chlorotic areas were less susceptible than was the
darker green tissue. Raeber *et al.* (1963) cited an example of an immune
tobacco cultivar being infected with *E. cichoracearum* following a field
infection with the tobacco rosette-bushy top complex of viruses, but this was
not observed again and was not confirmed in subsequent glasshouse tests.
Infection with *Peronospora tabacina* induced resistance to powdery mildew in
the cultivar Michal (Cohen *et al.*, 1975). The mechanisms involved in these
changes in resistance are, so far, unknown.

IV. Heritable Resistance

Tobacco cultivars resistant or immune to *E. cichoracearum* and currently
grown commercially in many countries, including Japan, Russia, South
Africa, Rhodesia and Taiwan, have been bred from interspecific crosses and
from within *N. tabacum*. Most *Nicotiana* species are very resistant and Ternovsky (1934) showed that species hybrids *N. tabacum* × *N. glutinosa* and

N. sanderae × *N. rustica* were also resistant. Since then (Ternovsky, 1961 and 1969) resistant tobacco cultivars have been bred from *N. digluta*, an amphidiploid between *N. tabacum* and *N. glutinosa*. The resistance is dominant and shows in all stages of the plant's growth. Cultivars Trebizond 161, Dubec 7 and Dubec 566 are immune to powdery mildew and tobacco mosaic virus (TMV). Germination of conidia and growth of hyphae are inhibited and few haustoria are formed on Dubec 566. The epidermal cells react rapidly to penetration and haustoria degenerate (Ternovsky *et al.*, 1973). These cultivars yield as much tobacco of good quality as do the susceptible cultivars. Rossouw *et al.* (1966) in South Africa and Nakamura *et al.* (1969) in Japan, also have bred resistant cultivars from *N. glutinosa*. Raeber *et al.* (1963), however, reported that they were unable to get true-breeding resistant cultivars from *N. digluta* and that resistance in their lines appeared to be controlled by many genetic factors. They speculated that the strain of *N. digluta* that they used, might have lost the major gene for immunity through meiotic irregularities but still contained a polygenic background that conferred a large degree of resistance. Kuznetsova and Terent'eva (1973) report marked disturbances in meiosis of amphidiploids of the *N. tabacum* × *N. glutinosa* cross including partial chromosome pairing.

The Japanese air-cured cultivar, Kuo-fan, has been the main breeding source for resistant cultivars within *N. tabacum*. Wan (1962) in Taiwan, crossed it with Hicks cultivar and concluded that resistance is controlled by duplicate recessive genes each of which has a slight cumulative effect. Chang *et al.* (1966) produced a new flue-cured cultivar in Taiwan called Vamfen-Hicks that is resistant to powdery mildew and TMV. It was bred from Kuo-fan and backcrossed to Hicks and a TMV resistant cultivar. Rossouw *et al.* (1966) in South Africa and Raeber (1967) in Rhodesia also released resistant cultivars bred from Kuo-fan; these are still resistant in Southern Africa but Corbaz (1975) recently reported that Kutsaga El cultivar, bred in Rhodesia, appears to have lost its field immunity in Bulgaria and is now only resistant. It has been grown in Bulgaria for a number of years and Corbaz concludes that a new strain of the pathogen has been selected. Wan and Chang (1967) concluded that genes for powdery mildew resistance in Vamfen-Hicks, *pm1* and *pm2*, are linked on both chromosomes with similar intensity to genes for female sterility, *st1* and *st2*. Later, Chen *et al.* (1971), by the use of suitable monosomic lines located genes *pm1* and *pm2* on chromosomes H and I, respectively. The twenty-four monosomic lines were identified and crossed to Vamfen-Hicks. The monosomic F_1 plants were re-crossed to Vamfen-Hicks and the progenies, which were tested for powdery mildew in the field, segregated in the ratio of three susceptible to one resistant, except for three lines. Two of these segregated in a 1:1 ratio and identified genes *pm1* and *pm2* on chromosomes H and I, respectively. The progeny of the W monosomic line were all, rather inexplicably, susceptible.

Recently (Smeeton, 1976), doubled-haploid cultivars with resistance to powdery mildew, wildfire (*Pseudomonas tabaci*), angular leaf spot (*P. angulata*) and tobacco mosaic virus have been developed by the use of the anther culture technique (Nitsch and Nitsch, 1969). Compared with normal breeding methods, this technique is claimed to save at least four years in the development of a new cultivar.

Partial resistance was demonstrated within *N. tabacum* by Raeber (1966) who compared powdery mildew infection in more than 20 cultivars during several seasons in Rhodesia. Canadel was consistently the most resistant but when it was included in similar comparisons in Europe by Corbaz (1975), it was as heavily infected as the most susceptible cultivars whereas Irabourbon 1, which was derived from an air-cured cultivar originating in South America, was much more resistant.

V. Fungicide Control

Renaud (1959) points out that the choice of fungicides for use against tobacco leaf pathogens is restricted by the need to keep residues to a minimum and to avoid chemicals that might alter the quality of the processed product. Sulphur, which was an obvious choice before the development of more specific fungicides, is completely unsuitable from all points of view other than that of disease control. D'Angremond (1923), in comparisons of lime sulphur, sulphur dust and Bordeaux mixture sprays, found sulphur dust to be very effective and much better than the other treatments. To overcome the disadvantage of leaf residues, he dusted the soil with sulphur at 500 kg ha^{-1}, and achieved some control. Later (D'Angremond, 1924) about 170 kg ha^{-1} was found to be adequate when used at the first appearance of disease, but if 3–6 leaves were already infected even 265 kg ha^{-1} did not stop the spread. Obviously it is difficult to apply sulphur dust to soil around growing plants without getting some on the leaves and the method was condemned in South Africa for this reason. Scaramuzzi (1949) used several formulations of sulphur, copper acetate and potassium permanganate in greenhouse tests; two forms of finely divided sulphur gave best control. Other work with various forms of sulphur have been reported by Hopkins (1928), Sempio and Lucacci (1953), D'Angremond (1926) and Lucacci and D'Armini (1953). Lithium salts have also been tested as systemic fungicides against tobacco powdery mildew. Lithium carbonate applied as three soil drenches at 6·85 and 13·7 g l^{-1} was absorbed from the soil and protected greenhouse seedlings for 30 days (Vidali and Ciferri, 1951). However, in the field, two drenches at 3·4 or 6·8 g l^{-1} did not control the disease, although the plants grew better than those which were not treated (Vidali, 1951). D'Armini (1953) re-tested lithium carbonate applied as a soil drench to potted plants and confirmed that it protected them against powdery mildew. He concluded that the optimum dose, which depends on soil type and amount treated, was easily either exceeded or not attained. No further work appears to have been done with this compound and its mode of action is, apparently, unknown.

Dinocap was the first fungicide to be used extensively and is probably still the most widely used. It was tested in South Africa (Rossouw, 1957), Italy (Marcelli, 1957), Madagascar (Renaud, 1959), Poland (Mickovski, 1957) and Rhodesia (Cole, 1963a) and results were generally satisfactory. Some phytotoxicity has been reported at high rates of application, especially in hot weather ($30°C$). Rates of 0.25 g litre^{-1} plus an additional wetting agent, and applied at 800 litre ha^{-1} controlled the disease. Where disease was severe, yield and quality of flue-cured tobacco were improved by the sprays (Cole, 1963a) and residues were usually less than 1 μg g^{-1}. Delaying the onset of infection for three weeks by means of two dinocap sprays at weekly intervals, reduced the total leaf area infected of topped plants by 37% at reaping. Four sprays delayed onset for four weeks and reduced infection by 69% (Fig. 5).

Salicylanilide was tested by Thung (1936) in Indonesia. Although it was effective he thought that its widespread use was not feasible. Marcelli (1952) reported that two sprays had a good preventative action, which was better than sulphur. Cole (1963a) compared it with dinocap but it gave comparable control only at 32 times the concentration. Baudin (1965) found that quinomethionate was more effective than dinocap, in Madagascar, and similar results have been found in Rhodesia, but it has not been tested there at severe infection pressure. Other compounds such as antibiotics, have been tested on seedlings but not on a field scale. Strijdom and Van Vuuren (1961) used one from *Bacillus subtilis* with some success, and Manucarjan (1964) applied trichothecin in four sprays of 15 ml plant^{-1} at 0.2 g litre^{-1} and obtained effective control without plant injury.

Some of the many systemic fungicides that are effective against powdery mildews have been tested on tobacco. Thiophanate ethyl, thiophanate methyl and benomyl are good protectants as foliar sprays in the glasshouse and field but soil applications are effective only with potted seedlings. Thiophanate (Yamaguchi and Yashida, 1972) applied to tobacco leaf surfaces was not translocated to untreated leaves, stem and roots, nor to the opposite surface or to untreated parts of treated leaves. Benomyl, thiophanate and tridemorph were tested in Rhodesia as three medium-low volume sprays (100 litre ha^{-1}) at 10–16 day intervals covering the susceptible period of topped plants. Benomyl and tridemorph sprays at 10 day intervals were more effective than thiophanate at equal concentrations. Overall, three sprays at 10 day intervals were more effective than the same number at intervals of 16 days. However, field control with benomyl has not always been successful. On one occasion when it was used on cigar wrapper in shade tents, initial protection was

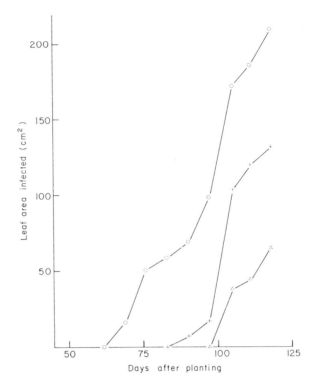

Fig. 5. Effect of delaying onset of infection of topped plants on leaf area infected. Not sprayed ○———○; 2 dinocap sprays (0·25 gl⁻¹) 49 and 56 days after planting+ ———+; 4 dinocap sprays (0·25 gl⁻¹) 49, 56, 63 and 70 days after planting △———△.

satisfactory but after the second spray the disease spread rapidly and strains of *E. cichoracearum* were isolated from leaves in which there was a carbendazim concentration of 35 μg g⁻¹. Growth of the standard strain was prevented at a concentration of 5 μg g⁻¹ (Cole, D. L., 1976). The benomyl-tolerant isolates, cultured on tobacco leaves containing benomyl, grew successfully on concentrations of 1 mg g⁻¹, but they grew more slowly than the wild type on leaf discs that did not contain benomyl. Spray experiments with benomyl and other fungicides in the same locality during the following year produced no evidence of the benomyl-tolerant strain and weekly sprays of benomyl were more effective than polyoxin B, dinocap, tridemorph or ethirimol. The wild strain would appear to be a better competitor than the benomyl tolerant one and should re-establish itself where the use of benomyl is not widespread.

Dimethirimol and ethirimol are absorbed by roots of potted seedlings in sandy soil and prevent growth of *E. cichoracearum* on the leaves. In the field, the fungicides were tested as soil drenches on cigar wrapper tobacco grown on heavy clay soil and irrigated. They were translocated to leaves but not in sufficient quantities to give economic control. Dosages of 0·25–3·0 g plant^{-1} (3·8–46 kg ha^{-1}) were applied once or twice, five weeks apart, around the stem. Two applications were more effective than the same amount applied once and the best treatment, dimethirimol at 46 kg ha^{-1}, reduced the mean percentage leaf area infected by 63%. The quality but not the yield of the cured leaves was negatively correlated with infection.

VI. Discussion

The interaction between tobacco and *E. cichoracearum* is intriguing because the pathogen is on the one hand very successful and yet there are many instances when the plant is resistant to it. The biochemistry of the leaf tissue that determines whether haustoria and secondary hyphae develop successfully seems to be very specific and fairly readily changed by external and internal factors. If the processes controlling resistance are all interrelated, it would be very interesting to know the mechanisms involved. There seem to be some common features in the results reviewed but much more work at the cell level would be necessary before a meaningful hypothesis could be formulated.

From the plantbreeders' standpoint, the situation seems to be particularly encouraging. Not only are there several sources of resistance, as shown by their method of inheritance, but most of the *Nicotiana* species contain one or other of them. As interspecific hybridization either directly or through amphidiploids is frequently not too difficult, the potential for alternative sources of resistance is great. Furthermore, the ability of the pathogen to overcome the resistance incorporated into commercial cultivars so far seems to be relatively limited so that many of the cultivars have maintained their resistance in commercial production for several years in different parts of the world. The one exception is Kutsaga E1, which contains double recessive genes for resistance but which recently became infected in the field in Bulgaria. In Africa, where it has been grown commercially for eight years, there may be less chance of a breakdown of resistance because of the absence of sexual reproduction in the pathogen.

Other interesting features are the comparative absence of resistance in *N. tabacum* and the lack of the disease in the USA. Many of the *Nicotiana* species that are considered to be closest to *N. tabacum*, such as *N. sylvestris*, *N. tomentosiformis* and *N. tomentosa*, and from which it was probably derived, are immune or have high levels of resistance. It might have been expected that the original tobacco cultivars would have inherited some of this resistance, particularly as it is usually dominant. The absence of powdery mildew from the tobacco crops in the USA is perplexing because of its presence in Central America and the reported infection of an interspecific *Nicotiana* cross and burley tobacco in Kentucky. Susceptible cultivars are grown in the USA and it seems unlikely that the environment is limiting when one considers the variety of climates in which the disease is endemic.

References

Averna-Sacca, R. (1922). *Bolm Agric.*, *S. Paulo* **23**, 201–268.
Bainbridge, A. and Legg, B. J. (1976). *Trans. Br. mycol. Soc.* **66**, 495–498.
Baudin, P. (1965). *Docums Inst. Rech. Agron.*, *Madagascar* 44.
Chadha, K. C. and Raychaudhuri, S. P. (1968). *Indian Phytopath.* **21**, 171–175.
Chang, E. Y., Wan, H. and Hsu, P. K. (1966). *J. agric. Ass. China* **56**, 26–34.
Chapman, R. A. (1953). *Pl. Dis. Reptr* **37**, 528.
Chen, C. H., Ueng, I. N. and Wu, J. K. (1971). *Rep. Tob. Res. Inst. Taiwan.* 1–4.
Childs, J. F. L. (1940). *Phytopathology* **30**, 65–73.
Cohen, Y., Reuveni, R. G. and Kenneth, R. G. (1975). *Phytopathology* **65**, 1313–1314.
Cole, D. L. (1976). D. Phil. Thesis, Univ. Rhodesia, Salisbury.
Cole, J. S. (1963a). *Rhod. J. agric. Res.* **1**, 65–70.
Cole, J. S. (1963b). *Proc. 3rd World Tob. Sci. Congr.* 29–37.
Cole, J. S. (1964). *Ann. appl. Biol.* **54**, 291–301.
Cole, J. S. (1966a). *Ann. appl. Biol.* **57**, 201–209.
Cole, J. S. (1966b). *Ann. appl. Biol.* **57**, 435–444.
Cole, J. S. (1966c). *Ann. appl. Biol.* **57**, 445–450.
Cole, J. S. (1966d). *Ann. appl. Biol.* **58**, 61–69.
Cole, J. S. (1966e). *Ann. appl. Biol.* **58**, 401–407.
Cole, J. S. (1967). *J. exp. Bot.* **18**, 254–268.
Cole, J. S. (1971). In "Ecology of Leaf Surface Microorganisms". (T. F. Preece and C. H. Dickinson, Eds), 323–337. Academic Press, New York and London.
Cole, J. S. (1976). In "Microbiology of Aerial Plant Surfaces". (C. H. Dickinson and T. F. Preece, Eds), 627–636. Academic Press, New York and London.
Cole, J. S. and Fernandes, D. L. (1970a). *Ann. appl. Biol.* **66**, 239–243.
Cole, J. S. and Fernandes, D. L. (1970b). *Trans. Br. mycol. Soc.* **55**, 345–353.
Cole, J. S. and Geerligs, J. W. G. (1976). *Trans. Br. mycol. Soc.* **67**, 339–342.
Corbaz, R. (1975). *Bull. Inf. CORESTA* 3–4, 20–22.
D'Angremond, A. (1923). *Meded. Proefstn vorstenl. Tabak Klaten* **49**, 7–25.
D'Angremond, A. (1924). *Meded. Proefstn vorstenl. Tabak Klaten* **52**, 5–23.
D'Angremond, A. (1926). *Meded. Proefstn vorstenl. Tabak Klaten* **56**, 5–48.
D'Armini, M. (1953). *Tabacco*, *Roma* **57**, 319–323.
Deckenbach, K. N. (1924). *Bolez. Rast. Leningrad* **13**, 98–102.
Dekker, J. (1963). *Nature. Lond.* **197**, 1027–1028.
Ducomet, V. (1928). *Revue Path. veg. Ent. agric. Fr.* **15**, 288.
Ellis, R. J. (1969). *Science, N.Y.* **163**, 477–478.
Farkas, G. L., Dézsi, L., Horváth, M., Kisbán, K. and Udvardy, J. (1963). *Phytopath. Z.* **49**, 343–354.
Hellebust, J. A. and Bidwell, R. G. S. (1964). *Can. J. Bot.* **42**, 1–12.
Hopkins, J. C. F. (1928). *Rhodesia agric. J.* **25**, 1342–1348.
Hopkins, J. C. F. (1948). *Rhodesia agric. J.* **45**, 330–342.
Hopkins, J. C. F. (1956). "Tobacco Diseases", Commonw. Myc. Inst. Kew, Surrey.

Khetsuriani, G. A. (1966). *Soobshch. Akad. Nauk. gruz. SSR* **42**, 701–706.
Kuznetsova, D. V. and Terent'eva, A. I. (1973). *Sb. nauchno-issled. Rab. vses. nauchno-issled. Inst. Tabaka Makhorki* **158**, 43–53.
Leach, C. M. (1962). *Can. J. Bot.* **40**, 151–161.
Levitt, J. (1972). "Responses of Plants to Environmental Stresses". Academic Press, New York and London.
Levykh, P. M. (1940). *A. I. Mikoyan pan-Soviet Sci. Res. Inst. Tob. Indian Tob. Ind. (VITIM), Rostoff-on-Don Publ.* **141**, 97–111.
Lucacci, G. and D'Armini, M. (1953). *Tabacco, Roma* **57**, 354–357.
Lucas, G. B. (1975). "Diseases of Tobacco". Biological Consulting Associates, Raleigh, USA.
Manucarjan, M. A. (1964). *Tabak, Mosk.* **25**, 25–27.
Marcelli, E. (1949). *Notiz. Mal. Piante.* **6**, 19–21.
Marcelli, E. (1952). *Tabacco, Roma* **56**, 370–374.
Marcelli, E. (1957). *Tabacco, Roma* **61**, 262–269.
Masri, S. S. and Ellingboe, A. H. (1966). *Phytopathology* **56**, 389–395.
Mickovski, J. (1957). *Duvan* **7**, 91–105.
Minev, K. (1956). *Ann. Fac. Phil. Skoplje Sect. Agric.* **10**, 5–57.
Mizrahi, Y. and Richmond, A. E. (1972). *Aust. J. biol. Sci.* **25**, 437–442.
Moore, E. S. (1926). *J. Dep. Agric. Un. S. Afr.* **12**, 428–455.
Morrison, R. M. (1960). *Mycologia* **52**, 786–794.
Nair, K. R. S. and Ellingboe, A. H. (1965). *Phytopathology* **55**, 365–368.
Nakamura, A., Matsuda, T. and Nakatogawa, H. (1969). *Bull. Jap. Tob. Exp. Stn. Iwata* **2**, 11–18.
Nitsch, J. P. and Nitsch, C. (1969). *Science, N.Y.* **163**, 85–87.
Pady, S. M., Kramer, C. L. and Clary, R. (1969). *Phytopathology* **59**, 844–848.
Parthier, B., Malaviya, B. and Mothes, K. (1964). *Pl. Cell Physiol., Tokyo* **5**, 401–411.
Putterill, V. A. (1923). *J. Dep. Agric. Un. S. Afr.* **7**, 131–134.
Raeber, J. G. (1966). *Proc. 4th Int. Tob. Sci. Congr.* 604–610.
Raeber, J. G. (1967). *Tob. Forum Rhod.* **24**, 26–30.
Raeber, J. G., Schweppenhauser, M. A. and Cole, J. S. (1963). *Proc. 3rd World Tob. Sci. Congr.* 230–236.
Renaud, R. (1959). *Revue int. Tabacs* **34**, 318–320.
Richards, F. J. (1956). *Potass. Symp.* 59–73.
Rossouw, D. J. (1957). *Fmg. S. Afr.* **32**, 27–28.
Rossouw, D. J. (1963). *Proc. 3rd World Tob. Sci. Congr.* 237–241.
Rossouw, D. J., Lamprecht, M. P. and Peens, J. F. (1966). *S. Afr. J. agric. Sci.* **9**, 419–428.
Samborski, D. J. and Forsyth, F. R. (1960). *Can. J. Bot.* **38**, 467–476.
Scaramuzzi, G. (1948). *Tabacco, Roma* **52**, 207–222.
Scaramuzzi, G. (1949). *Notiz. Mal. Piante.* **1**, 37–41.
Sempio, C. and Lucacci, G. (1953). *Tabacco, Roma* **57**, 45–55.
Shaw, D. E. and Layton, W. A. (1975). *Res. Bull. Dep. Agric. Stk. Fish., Port Moresby* **13**, 27–32.
Skene, K. G. M. (1972). *J. Hort. Sci.* **47**, 179–182.
Smeeton, B. W. (1976). *Tob. Forum Rhod.* **2**, 4–8.
Stephen, R. C. (1955). *Interim Rep. Tob. Res. Bd. Sth. Rhod.* **4**.
Strijdom, B. W. and Van Vuuren, P. J. (1961). *S. Afr. J. agric. Sci.* **4**, 255–260.
Ternovsky, M. F. (1934). *Züchter* **6**, 140–144.
Ternovsky, M. F. (1961). *Revue int. Tabacs* **36**, 27–29.
Ternovsky, M. F. (1969). *Sel. Khoz. Biol.* **4**, 813–821.
Ternovsky, M. F., Shinkarev, V. P., and Kuznetsova, D. V. (1973). *Mikol. Fitopatol.* **7**, 333–336.

Thung, T. H. (1936). *Meded. Proefstn vorstenl. Tabak Klaten* **82**, 27–35.
Thung, T. H. (1937). *Meded. Proefstn vorstenl. Tabak. Klaten* **84**, 25–42.
Vidali, A. (1951). *Notiz. Mal. Piante.* **16**, 35–39.
Vidali, A. and Ciferri, R. (1951). *Tabacco, Roma*, **55**, 95–102.
Wan, H. (1962). *Tob. Sci.* **6**, 180–183.
Wan, H. and Chang, E. Y. (1967). *Crop Sci.* **7**, 158–160.
Weybrew, J. A., Long, R. C., Dunn, C. A. and Woltz, W. G. (1974). *Bull. R. Soc. N.Z.* (1974). **12**, 843–847.
Yamaguchi, Y. and Yashida, D. (1972). *Bull. Hatano Tob. Exp. Stn* **71**, 149–152.

Chapter 18

Powdery Mildews of Tree Crops

R. T. BURCHILL*

East Malling Research Station, Maidstone, England

I. Introduction

Powdery mildews are present on most deciduous trees, but it is only on those cultivated for their fruit particularly *Prunus* spp. (*e.g.* apple, apricot and peach) and to a lesser extent some timber trees (e.g. oak and teak), that serious losses occur.

Usually one species of mildew is restricted to a single host species but occasionally as with the peach and the rose, the same fungus, *Sphaerotheca pannosa* (Wallr. ex Fr.) Lév. is common to both plants.

The severity of infection on any particular host plant varies between countries and within localities in the same country. Such variations may be due to the range in susceptibility of the different cultivars, the prevailing climatic conditions and occasionally a geographical isolation which is sufficient to exclude infection from extraneous sources.

In most years powdery mildew infections cause slight damage in contrast to diseases such as apple scab (*Venturia inaequalis* (Cke.) Wint.) and potato blight (*Phytophthora infestans* (Mont.) de Bary) which in dry years present no problems but in wet favourable seasons may cause severe losses. On some plants however, if uncontrolled, powdery mildew increases in severity with successive seasons, resulting in a stunting and distortion of leaves and a general decline in the health of the host plant. In a seven-year study of the incidence of apple powdery mildew (*Podosphaera leucotricha* (Ell. and Ev.)

* Present address: National Vegetable Research Station, Welltsbourne, Warwick England.

Salm.), Bennett and Moore (1963) demonstrated that the crop yield from severely mildewed trees was significantly reduced; furthermore the disease caused a russeting on the fruit which reduced the market value.

In common with powdery mildews on other crops, conidia or ascospores are responsible for the initiation of new infections at the commencement of each growing season. Once established, the fungus produces conidiophores and conidia which serve to intensify the level of infection on susceptible hosts.

During the winter most powdery mildews are inactive. Overwintering by means of cleistothecia does occur but in many species this stage appears to be functionless. Usually perennation is by mycelium present in the dormant buds of the host plant but occasionally overwintering as ordinary mycelium, as pannose mycelium, as chlamydospores or, as in warm southern regions of the world, by active infection has been recorded. Cleistothecia are formed by the apple mildew fungus but appear to play no part in overwintering so that perennation is entirely by mycelium present within dormant fruit and vegetative buds (Burchill, 1958). Overwintering by this method has been reported for powdery mildew of oak (Neger, 1915), peach (Smith, 1894), hawthorn (Dillon-Weston and Taylor, 1944) and plum (Yarwood, 1944).

The persistence of cleistothecia over many years in contrast to the short life of a few days for conidia suggests that the former may function as a means of survival over long unfavourable periods. Alternatively, the ascospores originating from the cleistothecia may be the source of new pathogenic races of mildew.

The powdery mildews recorded on tree crops contain species present in the eight genera of the Erysiphales listed by Brooks (1953) in his book "Plant diseases".

Erysiphe Hedwig f. ex Fries

Cleistothecia brownish-black, containing two to many stalked asci each with 2–8 spores; appendages short or floccose and interwoven with the mycelium.

Leveillula Arnaud

Cleistothecia brownish-black with numerous asci; appendages irregularly branched. Mycelium both intercellular and superficial; conidiophores simple or branched usually forming a single large conidium of irregular form at the apex.

Microsphaera Léveillé

Cleistothecia brownish-black with several asci; appendages repeatedly forked at the tip.

Phyllactinia Léveillé

Cleistothecia brownish-black, containing many asci; appendages equatorial, rigid, pointed with a bulbous base; the upper part of the cleistothecium with a mass of short branched outgrowths.

Podosphaera Kunze

Cleistothecia brownish-black spherical or sub-spherical, containing one 8-spored ascus; appendages basal and apical, the latter usually dichotomously branched at the apex.

Sphaerotheca Léveillé

Cleistothecia brownish-black, containing one 8-spored ascus; appendages basal, not regularly branched, often interwoven with the mycelium.

Uncinula Léveillé

Cleistothecia with several asci; appendages simple or dichotomously branched, uncinate at the apex.

Oidium Link

Cleistothecia not recorded; conidia in chains.

II. Host Plants

A number of powdery mildew species may be found attacking the member species of a particular host genus. In describing the mildews present on tree crops, a description is given of the host genus followed by an account of the members of the Erysiphales parasitizing each species within that genus.

Acacia *spp.* (*"Wattle"; Gum arabic tree*)

Contains a large number of species native to Australia, Burma, India, the drier parts of Africa, tropical America and some Pacific Islands. Members of this genus provide catechu (used for chewing in India) and gum arabic.

Four species of powdery mildew have been recorded. All produce a leaf necrosis which is associated with a white fine powdery growth of sporulating mycelium on the leaf surface. Severe attacks may lead to defoliation.

Erysiphe acaciae Blumer recorded in India on *A. catechu* Willd.

Phyllactinia acaciae Syd. recorded in South Africa on *A. robusta*.

Oidium sp. recorded in Zambia on *A. campylacantha* Hochst. ex A. Rich.

Leveillula taurica (Lév.) Arnaud recorded in the Sudan on the Gum arabic tree (*A. seyal* Delile). This powdery mildew frequently occurs on trees and shrubs belonging to the Leguminosae. It is distinguished from other powdery mildews by its endoparasitic habit, with the mycelium present both on the epidermal layers and penetrating and ramifying within the mesophyll. Entry into the host leaf is through the stomata and not the cuticle as with other powdery mildews.

Cleistothecia when formed are scattered and often embedded in a dense superficial mycelium. They measure from 135–250 μm in diameter and possess numerous appendages which are often hypha-like, short and indistinctly branched. Each cleistothecium contains many asci (usually 20). The ascospores are large, cylindrical to pyriform, sometimes slightly curved and measure 25–40 × 12–22 μm. The conidia which are of two distinct shapes, cylindrical and navicular are borne simply on short hyphal branches. They vary in size, from 25–95 × 14–20 μm.

A full description of *Leveillula* is given in the C.M.I. Descriptions of Pathogenic Fungi and Bacteria, No. 182, 1968.

Acer *spp.* (*Maple; Sycamore*)

Contains a large number of species found mainly in the North temperate zone, Africa, Asia and occasionally other parts of the world.

Severe attacks of powdery mildew caused by *Uncinula bicornis* (Lév) [= *U. aceris* (DC.) Sacc.] may cause defoliation, but usually infections are not damaging except on nursery stock.

The disease appears during September as a white powdery infection mostly on the lower surface of leaves. In late summer or autumn, cleistothecia appear on the mycelial mat. These measure 95–240 μm in diameter (usually about 180 μm) and possess numerous appendages which may be forked. Each cleistothecium contains 4–12 asci measuring 55–95 × 35–55 μm. There are eight pyriform to cylindrical shaped ascospores (rarely six) per ascus, measuring 15–26 × 8–15 μm. The conidia are ellipsoidal to oval and measure 24–36 × 15–20 μm.

Hosts include *A. tataricum* L. (Sinadskii, 1965), *A. macrophyllum* Pursh., *A. negundo* L. and *A. platanoides* L. Although common in Great Britain on *A. campestre* L. and to a lesser extent *A. pseudoplatanus* L. the disease is unimportant (Jones, 1944).

A full description of *Uncinula bicornis* is given in the C.M.I. Descriptions of Pathogenic Fungi and Bacteria, No. 190, 1968.

Uncinula circinata Cooke and Peck has been recorded in Canada on the foliage of *A. rubrum* L. and *A. saccharum* Marsh (Anon 1961).

Phyllactinia guttata (Wallr. ex Fr.) Lév. has also been recorded in Canada on *A. rubrum* L.

Alnus *spp.* (*Alder*)

There are few reports of powdery mildews parasitizing the genus *Alnus*, however Boyce (1961) noted the presence of *Erysiphe aggregata* (Peck.) Farl. on the female catkins of Alder.

Phyllactinia guttata (Wallr. ex. Fr.) Lév. and *Microsphaera alni* (Wallr.) Wint. have been recorded on *A. incana* Moench. (Spaulding, 1958, 1961.)

Berberis *spp*.

Powdery mildews are uncommon on this genus and when present cause little damage. Mayor (1967) noted the presence of *Microsphaera berberidis* (DC.) Mérat Lév. on foliage of *Berberis* sp.

Betula *spp*. (*Birch*)

Although many *Betula* species are attacked by powdery mildew, the disease rarely causes damage. Most infections are caused by *Phyllactinia guttata* (Wallr. ex Fr.) Lév. [= *P. suffulta* (Rabenh.) Sacc.] [= *P. corylea* (Pers.) Karst. em. Salm.]

The disease occurs in Asia (China, Formosa, India, Japan, Korea, Turkey, USSR) and is widely distributed in Europe, Canada and the USA (Hirata, 1966). Various biologic forms are known including f. sp. *coryli*; f. sp. *betulae*; f. sp. *fagi* and f. sp. *alni* (Blumer, 1967).

The mycelium present on leaves rarely forms definite patches. Conidia are borne singly and measure 50–90 × 10–20 μm. The cleistothecia are usually scattered, measuring 160–230 μm in diameter. The cleistothecia have 6–15 appendages which are equatorially inserted. Within the cleistothecium there are 8–25 subcylindric asci measuring 70–100 × 25–40 μm. Each ascus contains 2–3 ascospores which vary in size and shape from 25–45 × 15–25 μm.

A full description of *Phyllactinia guttata* is given in C.M.I. Descriptions of Pathogenic Fungi and Bacteria, No. 157, 1967.

P. guttata has been recorded on numerous woody plants chiefly belonging to the Betulaceae, e.g. *B. alleghaniensis* Michx. and *B. papyrifera* Marsh in Canada and on *B. pendula* Roth. in Canada and Britain, Fagaceae and Juglandaceae. It has also been recorded on species of *Acer, Alnus, Populus, Prunus, Salix* and *Ulmus* (Blumer, 1967; Hirata, 1966). Smitskaya (1955) recorded the presence of *P. suffulta* f. sp. *betulae* on leaves of *B. pubescens* Ehrh. Occasionally *Microsphaera alni* (Wallr.) Wint. has been recorded on *B. pendula* and *B. pubescens* (Spaulding, 1958, 1961).

Carica *spp* (*Papaya*)

Oidium caricae Noack., powdery mildew of papaya (*Carica papaya* L.) is found throughout the tropics (Raabe, 1966). The disease occurs on both surfaces of the leaf, on fruits and on the pedicels of male flowers. The disease causes little damage to the leaves of mature plants but on seedlings where the stems and upper leaf surfaces are infected, losses may be severe.

Chiddarwar (1955) described a powdery mildew causing severe damage to young papaw seedlings growing in Poona, India. Small circular lesions developed on both leaf surfaces, particularly on the upper surface of young leaves and these became turgid and were shed. In the advanced stages of infection, the growing shoots developed severe die-back. The pathogen was identified as *Oidium indicum* n. sp. It differed from *O. caricae* in that the conidia were barrel shaped and not elliptical.

Oidium caricicola has been recorded by Yen and Wang (1973) on papaw.

Carya *spp.* (*Pecan*)

A genus of fast growing trees allied to the Juglandaceae and mainly confined to North America.

Occasionally infection of pecans (*Carya olivaeformis* Nutt.) by *Microsphaera alni* (Wallr.) Wint. is sufficiently serious to justify control measures (Worley and Harmon, 1969).

Cassia *spp.*

The genus Cassia contains approximately 400 species. Powdery mildews although present on some species are relatively unimportant and do not cause serious losses. *Oidium erysiphoides* Fr. f. sp. *cassia* Sacc. and *Sphaerotheca cassiae* Pandotra and Ganguly have been recorded on *C. uniflora* in the Dominican Republic and *C. occidentalis* L. in India respectively. Unidentified *Oidium* spp. have been noted on *Cassia* spp. in Jamaica, India, Pakistan, Kenya, Tanzania, Malawi, Zambia, Nigeria and Venezuela.

Castanea *spp.* (*Chestnut*)

Phyllactinia guttata (Wallr. ex Fr.) Lév. has been reported by Hanlin (1969) as occurring on American chestnut (*C. dentata* Borkh.) and European chestnut (*C. sativa* Mill.). The fungus forms a powdery white superficial mycelium on leaves. Although widely distributed in the north temperate zone, it seldom causes severe infection.

Microsphaera alni (Wallr.) Wint. has been recorded on *C. sativa* Mill (Spaulding, 1958, 1961).

Celtis *spp.* (*Hackberry*)

Uncinula polychaeta (Berk. and Curt) Ellis is a parasite of *Celtis* and related genera. The fungus infects the leaves producing a white to grey powdery mildew on which are scattered minute, globular cleistothecia, (Spaulding, 1961). It has been recorded in north and south America and in India and Pakistan where Bagchee and V. Singh (1954) and later Huque (1964) noted its presence on *C. australis* L.

According to Snetsinger and Himelick (1957) a combination of a powdery mildew (*Sphaerotheca phytoptophila* Kell. and Swingle) and an eriophyid mite belonging to the genus *Aceria* is responsible for the disease witches' broom of common hackberry (*Celtis occidentalis* L.).

Trees infected with the disease bear brooms or knots composed of numerous abortive twigs each of which bears many buds. These buds are generally larger and have more open bud scales than healthy buds and tend to die back after one year's growth.

Ceratonia *spp.* (*Carob*)

The commonest disease of the carob tree (*Ceratonia siliqua* L.) is caused by *Oidium ceratoniae* Comes. The fungus attacks young leaves causing premature defoliation. Occasionally the flowers and young fruits are infected. The most favourable period for infection is September to December with the mildewed leaves being shed in February.

The disease is widely distributed in Mediterranean countries.

Graniti (1958) working in Sicily showed that male trees appear to be more susceptible than either female or hermaphrodite trees.

Citrus *spp.*

Tamayo and Pordesino (1959) working in the Philippine Islands reported the presence of powdery mildew (*Oidium tingitanium* Carter) on the mandarin orange cultivar Laddu.

Powdery mildew infections on orange (*Citrus aurantium* L.) are not common and when they do occur it is often in association with poor nutrition apparently predisposing the orange to infection.

Corylus *spp.* (*Filbert*)

Corylus spp. are only occasionally attacked by powdery mildew. Miller (1945) noted the presence of *Phyllactinia corylea* (Pers.) Karst. em Salm. on filbert growing in the Pacific Northwest USA and Smitskaya (1955) recorded *Phyllactinia suffulta* (Rabenh.) Sacc. f. *coryli avellanea* on hazel (*C. avellana* L.).

Cydonia *spp.* (*Quince*)

Podosphaera leucotricha (Ell. and Ev.) Salm. a powdery mildew common on apple (*Malus domestica* Burkh.) has been recorded on quince (*Cydonia oblonga* Mill.).

Podosphaera oxyacanthae (DC.) de Bary has also been reported on the foliage of quince.

Dalbergia *spp.* (*Sheesham*)

Dalbergia sissoo Roxb. ("Sheesham") is an important Indian timber tree. With the advent of the cold season Mukerji (1969) noted the appearance of powdery mildew (*Phyilactinia dalbergiae* Pirozynski) on the lower surface of some of the leaves. Initially the mycelium is white, but after it has produced abundant conidiophores and conidia, the colour changes from greyish white to pale yellow. The long conidiophores are bent inwards and spirally coiled at the base. The conidia which are variable in shape are borne singly at the tips of the conidiophores. In late December, cleistothecia form on the under surface of the infected leaves. Initially the cleistothecia are orange-brown but later they become black. They measure 180–280 μm in diameter with 6–15 appendages inserted equatorially. Each cleistothecium contains 10–30 asci (45–80 × 20–38 μm) with 2–3 ascospores (20–40 × 14–22 μm). The fungus causes severe defoliation but never kills the tree. New leaves develop in the spring.

The disease has also been recorded on *D. lanceolaria* L. and *D. volubilis* Roxb. by Patil (1962) and Patwardhan (1962) respectively.

Eucalyptus *spp.* (*Gum tree*)

Oidium eucalypti Rostr. and other *Oidium* spp. are known to occur in Great Britain, Portugal, Italy, Denmark, Poland, Argentina, Brazil and Australia.

The disease appears on leaves as small white patches which later spread and are associated with a leaf necrosis, leading to defoliation.

Yarwood and Gardner (1972) reported the occurrence of *Erysiphe cichoracearum* DC. ex Merat on eleven *Eucalyptus* spp. growing in the USA. Infections were most frequent in the cultivated crops and relatively uncommon in the undisturbed plant communities.

Brandenburger (1961) provisionally identified *Sphaerotheca macularis* (Wallr. ex Fr.) Lind. as the powdery mildew present on 6-month-old seedlings of *E. gomphocephala* DC. and *E. diversicolor* F.v. M. raised in a greenhouse.

Euonymus *spp.*

Oidium (*Uncinula*) *euonymi-japonicae* (Arcang.) Sacc. has been recorded on the leaves of *Euonymus japonicus* L. Tokushige (1953) reported that in Japan, it did not appear to produce cleistothecia but overwintered as superficial mycelium with haustoria on leaves infected in the previous year.

Fagus *spp.* (*Beech*)

Powdery mildews are unimportant on beech, although their presence has been recorded on a number of occasions. Cotton (1919) and Woodward *et al.* (1929) noted the presence of the oak mildew (*Microsphaera alphitoides*, Griff. and Maubl.) on beech. Cotton's record was on coppice beech growing in the immediate vicinity of severely infected oak.

Phyllactinia corylea (Pers.) Karst. and *Phyllactinia guttata* (Wallr. ex Fr.) Lév. have been recorded on *Fagus grandifolia* Ehrh. in Canada and *F. sylvatica* L. in Great Britain.

Fraxinus *spp.* (*Ash*)

Powdery mildews are uncommon on *Fraxinus* spp. Yarwood and Gardner (1964) recorded the presence of *Erysiphe cichoracearum* DC. ex Mérat. on white ash (*Fraxinus americana* L.)

Gmelina arborea *Roxb.*

Patil (1961) reported the presence of *Phyllactinia suffulta* var. *gmelinae* Patil on leaves of the timber tree *Gmelina arborea* Roxb. The fungus caused little damage.

Hevea *spp.* (*Rubber*)

Infections of the leaves of the rubber tree (*Hevea brasiliensis* Müll. Arg.) by *Oidium heveae* (Steinm.) although common in Malaysia and to some extent in Sri Lanka and Indonesia occur less frequently in other rubber producing countries (Wastie, 1975).

The increase in the incidence of the disease in the last 20 years has resulted from the introduction of newer, high yielding clones of rubber which are often susceptible to *O. heveae*.

The disease is most serious at the time of refoliation after wintering of the trees, when it may cause premature defoliation commonly termed "secondary leaf fall". Serious attacks weaken the trees, with the result that repeated attempts at refoliation divert metabolites from latex production and the depression in yield, which normally occurs after wintering, is prolonged. Even a moderately severe attack results in a poor canopy thereby reducing tree vigour and, as a result of allowing more light penetration to the ground, increases weed growth.

The severity of secondary leaf fall alters as the tree ages. It is insignificant before about the fifth year when the wintering cycle becomes established and even for the first few years of "tapping" yields may be satisfactory in spite of a depleted canopy. Thereafter increased susceptibility and the effects of defoliation are greater.

Control measures are particularly important in the years 8–15 of the tree's life.

The severity of infections increases rapidly once favourable weather coincides with the presence of susceptible leaves. Only if continuous wet conditions prevail can *O. heveae* be ignored.

Wastie and Mainstone (1969) reported that mildew may be found throughout the year on young plants grown under shade conditions, however the disease is never sufficiently serious to cause defoliation.

In Sri Lanka, *O. heveae* is present throughout the year on *Euphorbia pilulifera* L., a weed common in the drier parts of that country (Young, 1954).

Juglans *spp.* (*Walnut*)

Juglans regia L. produces highly prized timber and edible nuts, however, the quality of the nuts is often uncertain therefore trees grown for the latter purpose are usually grafted trees of named clones selected for the quality of their nuts and early fruiting.

Powdery mildew infections of the leaves of walnut cause little damage.

Cristinzio (1942) recorded *Microsphaera alni* (Wallr.) Wint. on *J. regia* in Italy. The disease was most frequent in thickly planted walnut groves, but caused little damage. *M. alni* has also been recorded on walnut by Spaulding (1958, 1961).

Phyllactinia guttata (Wallr. ex Fr.) Lév. has been recorded on members of the Juglandaceae. (C.M.I. Descriptions of Pathogenic Fungi and Bacteria, No. 157, 1967).

Malus *spp.* (*Apple*)

Apple powdery mildew (*Podosphaera leucotricha* (Ell. and Ev.) Salm. is found in all the apple growing districts of the world. In some countries it is of minor importance but in others (e.g. Great Britain) it is very serious and requires numerous fungicide applications to maintain an acceptable level of control.

New infections normally occur on the under surface of the leaf but later these spread to cover both surfaces. The conidia which contain distinct fibrosin bodies are produced in long chains and measure 22–30 × 15–20 μm. Cleistothecia are often formed on severely infected tissues. They possess two types of appendage, inserted apically and basally. The apical appendages are usually straight and undivided whereas the basal appendages are rudimentary, rarely well developed, short and tortuous. The single ascus measures 55–70 × 44–55 μm and contains eight ascospores measuring 22–26 × 12–15 μm.

A full description of *Podosphaera leucotricha* is given in C.M.I. Descriptions of Pathogenic Fungi and Bacteria, No. 158, 1967.

Apart from leaves, the fungus infects young shoots, fruits and the blossom trusses of most commercial varieties. A severe infection may lead to defoliation of the mildewed shoot with a subsequent curtailment of its growth and a reduction in the number of potential fruiting buds developed on the two-year-old wood. Infections of the blossom trusses result in the non-production of fruit, and successive attacks on the same tree will lead to a cumulative loss of crop, both directly and indirectly due to the diversion into vegetative growth of materials normally used in fruit production. On some varieties the fruit is attacked shortly after blossoming, but such infections seldom persist after the apple skin hardens in mid-summer. Evidence of a previous infection of young fruit is shown by the presence of a web-like russeting on the surface of the apple. This damage reduces the market value of the fruit.

During the spring and summer the large numbers of conidia produced in infected tissues are dispersed by the wind to healthy leaves where new infections ("secondary infections") develop.

Although cleistothecia may be formed, it is doubtful if they are of importance in the perennation of the disease.

The sole means of overwintering is by mycelium present within diseased fruit and vegetative buds (Burchill, 1958). The fungus having entered the buds during the growing season remains dormant during the winter. In the following season, when the buds break to produce either vegetative shoots or blossom trusses, the fungus resumes activity and rapidly covers the emerging tissues with mycelium and conidiophores; this constitutes the "primary infection" phase (Burchill, 1958).

Other species of powdery mildew recorded on apple include *Erysiphe heraclei* (DC.) St.-Am. (Puttoo, 1972) and *Podosphaera oxyacanthae* (DC.) de Bary.

Mangifera *spp.* (*Mango*)

The mango (*Mangifera indica* L.) is cultivated for its edible fruits. It is parasitized by the powdery mildew *Oidium mangiferae* Berthet. The fungus attacks the young tissues of the inflorescences, leaves and fruits, forming small patches of white powdery mycelium on the affected organs. The lesions eventually coalesce and infected tissues turn brown.

On the flower, the sepals are particularly susceptible with the petals more resistant. As the flower stalks and flowers become affected, the flowers cease to grow and are often shed. In susceptible varieties all the branches of the inflorescences may be completely covered by the disease and eventually turn black. On leaves infections are usually confined to the under surface but on some varieties both sides may be attacked. Affected leaves curl and become distorted.

Newly set fruits may be entirely covered by mycelium. As the fruit expands the epidermis in the infected area cracks and corky tissue is formed. Such fruits usually fall prematurely.

Cleistothecia have not been recorded. The disease persists on older leaves during seasons in which young susceptible tissues are unavailable or conditions are unfavourable for infection. When the climatic conditions become favourable, conidia are produced and are dispersed by the wind on to new flushes of growth or young flowers.

Palti *et al.* (1974) in Israel reported that the disease may cause serious losses in mango orchards. It occurs at latitudes of up to 40° North and South of the equator; in many countries south of the Sahara, in the Middle East, southern Asia and in America from the southern United States to Peru and Brazil. It has also been reported from New South Wales (Australia).

Morus *spp.* (*Mulberry*)

Bakshi and Singh (1961) in India recorded a powdery mildew, *Phyllactinia corylea* (Pers.) Karst. on leaves of the mulberry tree (*Morus nigra* L.).

Later Itoi *et al.* (1962) in Japan, identified the powdery mildew attacking mulberry in that country as *Phyllactinia moricola* (P. Henn.) Homma. The disease first appears in late May as a result of infection from ascospores released from overwintered cleistothecia present on mulberry "wattles".

Later Itoi, Nakayama and Kubomura (1962) in Japan, identified the powdery mildew attacking mulberry in that country as *Phyllactinia moricola* (P. Henn.) Homma. The disease first appears in late May as a result of infection from ascospores released from overwintered cleistothecia present on mulberry "wattles".

In Georgia, USSR the disease has been identified as *Phyllactinia suffulta* (Rabenh.) Sacc. f. sp. *moricola*. Chanturiya (1965) described the disease as overwintering as cleistothecia on withered leaves and peduncles on the soil surface and on branches in the central part of the shoot. The fungus was specific to mulberry.

Khairy *et al.* (1971) in the United Arab Republic identified the powdery mildew attacking mulberry as *Phyllactinia guttara* (Wallr. ex Fr.) Lév.

Nephelium *spp.* (*Rambutan*)

Nephelium lappaceum L. yields the edible rambutan fruit. In 1949, Hadiwidjaja (1950) working in Indonesia recorded the occurrence of *Oidium nepheli* nov. sp. on Rambutan. The disease attacks the flowers, fruits and the young leaves of the water shoots causing them to fall prematurely.

Persea *spp.* (*Avocado*)

The avocado (*Persea gratissima* Gaertn.) which produces a large pear-like fruit is seldom attacked by powdery mildew.

Franco (1946) has recorded an unidentified *Oidium* sp. on the foliage.

Platanus *spp.* (*Plane*)

A small genus of large, maple-like trees. Himelick and Neely (1959) in Urbana, USA recorded the presence of *Microsphaera alni* (Wallr.) Wint. on the foliage of London Plane (*P. hispanica* Muenchh. (× *acerifolia*) (× *hybrida*) and on the American sycamore (*P. occidentalis* L.)

Prunus *spp.*

P. domestica L. European plum or Plum

P. domestica L. is grown in large plantations for the production of the edible fruits. Powdery mildew attacks caused by the fungus *Podosphaera tridactyla* (Wallr.) de Bary (= *P. clandestina* (Wallr.) Lév. var. *tridactyla* (Cooke) are not common but on nursery plants infections may be severe causing damage to shoots and leaves (Radman, Ljubica, 1963).

Cleistothecia measuring 70–110 µm in diameter may be produced in abundance on infected shoots. These fruiting bodies usually possess 2–8 dichotomously branched appendages. They contain one ascus (60–80 × 60–70 µm) and eight ascospores (18–30 × 12–15 µm).

The conidia which are often sparse in number are formed apically in chains and measure 22–45 × 14–20 µm.

A full description of *Podosphaera tridactyla* is given in the C.M.I. Descriptions of Pathogenic Fungi and Bacteria, No. 187, 1968.

Podosphaera oxyacanthae (DC.) de Bary has also been recorded on plum.

P. amygdalus Stokes Almond

P. amygdalus Stokes is grown partly as an ornamental tree and, in many localities, for the production of the fruits. The latter consist of little more than a "stone" covered with a thick, dry, woolly skin. The contained kernel which is edible, is often used in confectionery and in cooking.

Weigle (1956) recorded the presence of a powdery mildew on the leaves and shoots of almonds growing in Wenatchee, USA. From the appearance of the cleistothecia the disease was identified as *Podosphaera tridactyla* (Wallr.) de Bary.

Other reports of powdery mildews include the occurrence of *Podosphaera leucotricha* (Ell. and Ev.) Salm. on almonds in the USA (Yarwood and Gardner, 1964) and *Sphaerotheca pannosa* (Wallr.) Lév. var. *persicae* Woron. on almonds in Spain (Tuset Barrachina, 1972).

P. armeniaca L. Apricot

In California, USA, Yarwood (1952) reports that the most common powdery mildew on apricot leaves is *Podosphaera clandestina* (Wallr. ex Fr.) Lév. var. *tridactyla* Cooke. The fungus overwinters by means of cleistothecia. He also recorded infection of apricot fruits and leaves by *Sphaerotheca pannosa* (Wallr. ex Fr.) Lév. These infections were associated with mildew on adjacent rose trees particularly on *Rosa banksiae* Ait. with disease incidence varying inversely with the distance between the roses and the apricots.

Delmas (1953) working in the eastern Pyrenees, France identified the powdery mildew attacking apricots in that area as *P. clandestina* var. *tridactyla*.

Infections developed on the young fruits sometimes as early as mid-April. These infections cause the epidermis at the site of the lesion to become necrotic, to harden and to crack. Leaves may be attacked on both surfaces and, if severely infected become contorted, fold over longitudinally and large areas of the blade die, causing premature defoliation. In some years cleistothecia are formed on the withered leaves, particularly towards the end of July. Generally the fungus overwinters by means of active mycelium on leaves remaining on the tree throughout the winter. In regions where low winter temperatures cause all the leaves to absciss, the fungus overwinters by means of cleistothecia.

El-Ghamrawy *et al.* (1967) working in the United Arab Republic identified the powdery mildew on apricot as *Podosphaera oxyacanthae* var. *tridactyla* Sal. In some seasons the disease caused severe losses.

A full description of *Podosphaera tridactyla* (= *P. clandestina* var. *tridactyla*) is given in the C.M.I. Descriptions of Pathogenic Fungi and Bacteria, No. 187, 1968.

A full description of *Podosphaera clandestina* (Wallr. ex Fr.) Lév. a powdery mildew common on hawthorn (*Crataegus monogyna* Jacq.) is given in the C.M.I. Descriptions of Pathogenic Fungi and Bacteria, No. 478, 1975.

P. avium L.	Sweet cherry
P. cerasus L.	Sour cherry
P. mahaleb L.	Mahaleb cherry
P. pumila L	Sand cherry
P. tomentosa Thunb.	Tomentose cherry

Cherries are widely cultivated for their edible fruits. Apart from the infections of nursery stock, powdery mildew is not a serious problem.

The disease is caused by *Podosphaera oxyacanthae* (DC.) de Bary. This species has also been recorded on plum (*Prunus domestica* L.), peach (*Prunus persica* Stokes), apricot (*Prunus armeniaca* L.), apple (*Malus domestica*

Borkh.), pear (*Pyrus communis* L.), quince (*Cydonia oblonga* Mill.) and haw-thorn (*Crataegus monogyna* Jacq.).

The fungus attacks young shoots and leaves causing the latter to curl. New infections occur throughout the season. At the end of the season, cleistothecia develop on the underside of the leaves. The cleistothecia overwinter on the fallen leaves, releasing ascospores in the spring to produce new infections. The disease may also overwinter as mycelium present within buds (see apple powdery mildew).

English (1947) in the USA recorded *P. oxyacanthae* infections on cherry fruits. The lesions were circular or more frequently irregular in shape due to the coalescence of two or more infections. Hyphae radiating from the centre of young lesions covered the infected portions giving the fruit a dull appearance. Older lesions were reddish brown with the skin tougher than that of healthy fruits.

Creelman (1967) in Canada has reported a powdery mildew on cherry which was identified as *Podosphaera clandestina* (Wallr.) Lév. The disease was occasionally severe in British Columbia and Ontario.

P. instita L. Damson

Podosphaera oxyacanthae var. *tridactyla* Sal. has been reported by Mayor (1967) on damson (*P. instita* L.).

P. persica Stokes Peach

Although not a disease of major importance, powdery mildew of peach like that of apricot may be a problem when weather conditions are favourable for infection. The disease is often serious in nurseries where seedling stocks infected early in the growing season often become stunted.

Two species of powdery mildew attack the peach. The most common and destructive is *Sphaerotheca pannosa* (Wallr. ex Fr.) Lév. usually known as rose mildew and the other less common species is the cherry powdery mildew (*Podosphaera oxyacanthae* (DC.) de Bary).

S. pannosa attacks the leaves, young shoots and fruits. The young leaves may become coated with a thick layer of mycelium causing them to curl and become stunted. On older leaves large white patches develop. On the fruit, the disease first appears in the form of white round spots which increase in size until a large portion or the whole fruit is covered. The fruit becomes pink and later turns brown, with the epicarp becoming leathery and often cracking. Such fruits are worthless.

Cleistothecia are rarely found on peach but are more common on other hosts such as the rose. The cleistothecia measure 80–120 μm in diameter and have few appendages. They contain one ascus (60–75 × 80–100 μm) with eight ascospores (20–30 × 12–17 μm). Once new infections are established the fungus produces long conidiophores bearing chains of conidia (20–35 × 14–20 μm).

The disease is found in most peach growing areas in the world.

Fikry (1937) reported it as being one of the most serious diseases of peach in United Arab Republic, attacking shoots, leaves and fruit.

Louw (1946) in South Africa reported serious outbreaks of mildew on peaches grown in the Cape Province. The fungus, which he identified as *S. pannosa* var. *persicae*, caused a dehydration and abscission of leaves, twig die-back and hard white patches on the fruit which often resulted in cracks. First infections occurred chiefly on the fruit with the inoculum originating from the diseased dormant shoots and buds of the previous season. Leaf and shoot infections occurred much later.

Earlier, Yarwood (1939) had claimed that *S. pannosa* was able to overwinter in infected peach buds. Weinhold (1961) demonstrated that *S. pannosa* overwintered as mycelium in dormant buds. Shoots growing from infected buds were covered with the fungus on emergence in the spring and these shoots provided the inoculum for the current season's infection.

Peach varieties differ in their susceptibility to infection. Ogawa and Charles (1956) reported that *S. pannosa* var. *persicae* affects non-glandular varieties of peach more seriously than glandular varieties.

A full description of *Sphaerotheca pannosa* is given in the C.M.I. Descriptions of Pathogenic Fungi and Bacteria, No. 189, 1968.

Pyrus *spp.* (*Pear*)

Powdery mildew infections on pear (*Pyrus communis* L.) are rarely serious. Occasionally the leaves may be infected and on some cultivars e.g. Doyenné du Comice the fungus grows on the fruit producing russeted areas.

Usually the fungus is identified as *Podosphaera leucotricha* (Ell. and Ev.) Salm. the species common on apple (*Malus domestica* Borkh.).

Other species reported to occur on pear include *Podosphaera oxyacanthae* (DC. ex Mérat) de Bary f. *piri* f. nov. (Golovin, 1956); *Phyllactinia suffulta* (Rabenh.) Sacc. (Jovićevic, 1955) and *P. suffulta* f. sp. *pyri* (Akhundov, 1963).

Quercus *spp.* (*Oak*)

Powdery mildew is the only serious foliage disease on oak (*Quercus* spp.). The disease first appears in the spring about mid-May on the leaves and shoots arising from buds infected in the previous year.

New infections appear as cinnamon-coloured spots which later turn white and spread to cover the surface of the leaf. As the infection develops the leaves become distorted and reduced in size. In severe attacks, growth of the shoot may cease prematurely and the leaves soon wither and fall.

Cleistothecia may be formed; these remain on the surfaces of the fallen leaves during the winter producing ascospores in the spring, but these spores are seldom seen to germinate. Perennation is mainly by mycelium present within dormant buds. Severe attacks result in reduced growth, reduced acorn crops, imperfect ripening of shoots and increased liability to attack by secondary parasites.

Damage occurs mainly on nursery and young material, regeneration material, on the second flush of leaves after initial defoliation and on coppice material. In the nursery the disease can be controlled with fungicide sprays but on large trees, control by this means is difficult and expensive.

Powdery mildew of oak occurs in most areas of the world where *Quercus* is grown, although the severity of infection is often dependent upon the particular species of mildew present and the species of oak attacked.

Quercus is parasitized by more species of Erysiphaceae than any other host plant, with some of the most important parasites belonging to the genus *Microsphaera*.

Microsphaera alphitoides Griff. and Maubl. (conidial state *Oidium quercinum* Thüm.).

Widely distributed in the north temperate zone, including Great Britain where it was formerly known as *Microsphaera quercina* (Schw.) Burr.

Recorded on *Q. alnifolia* Poech., *Q. cerris* L., *Q. petraea* Liebl. and *Q. robur* L. *Q. cerris* is reputed to be fairly resistant. *M. alphitoides*, which is sometimes considered to be identical with *Microsphaera extensa* (Cooke and Peck) is one of the most injurious pathogens of *Quercus* in Great Britain.

Microsphaera penicillata (Wallr. Fr.) Lév. (= *M. alni* (Wallr. Wint.)).

Widely distributed in the north temperate zone, including Canada. It is common and sometimes destructive in Europe. Reported on *Q. robur* L. and other *Quercus* spp. (Spaulding, 1958, 1961).

A full description of *Microsphaera penicillata* is given in the C.M.I. Descriptions of Pathogenic Fungi and Bacteria, No. 183, 1968.

Microsphaera extensa Cooke and Peck

Syn. *M. alni* (Wallr.) Wint. var. *extensa* (Cooke and Peck) Salm., sometimes considered to be identical with *M. alphitoides*.

Widely distributed in the north temperate zone, but unimportant in North America. Occurs on the foliage of *Quercus* spp. e.g. *Q. alba* L., *Q. cerris* L., *Q. petraea* Liebl., *Q. robur* L. and *Q. rubra* L. sec Du Roi. (Spaulding, 1956, 1961).

Sphaerotheca lanestris Harkn.

Syn. *Cystotheca lanestris* (Harkn.) Miyabe

Widely distributed in Asia and the USA. Hosts include *Q. ilex* L. in Pakistan and *Q. incana* Roxb. in India.

Infection of young trees may result in stunting, and sometimes in the development of witches' brooms.

Gardner, Yarwood and Duafala (1972) recorded four species of powdery mildews occurring on 28 species of oaks (*Quercus* and *Lithocarpus*) in the San Francisco Bay area of California. Each species of mildew was distinctive in morphology and to a lesser extent in ecology:—

Sphaerotheca lanestris Harkn. had the widest host range being recorded on 24 oak species.

Phyllactinia corylea (Pers.) Karst. occurred frequently on *Q. agrifolia* Nee. trees where it was limited to the lower leaf surfaces.

Erysiphe trina Harkn. was apparently limited to *Q. agrifolia*, *Q. chrysolepsis* Leibm. and *Lithocarpus densiflora* Rehd. and is distinctive because of its conspicuous mycelium, transparent cleistothecia and almost complete absence of conidia.

Microsphaera penicillata (= *M. alni*) occasionally occurred.

All four species occured on *Q. agrifolia* but usually there was only one species of mildew on any given host species.

Vlasov (1957) in the USSR identified a powdery mildew attacking *Quercus* spp. as *Microsphaera silvatica*. Roll-Hansen (1961) described *Microsphaera hypophylla* Nevodovskij. as a mildew new to Norway. Later he recorded it in Switzerland, Sweden and Austria. Unlike *M. alphitoides*, *M. hypophylla* attacks the spring leaves of *Q. robur* and produces cleistothecia in abundance.

Rodigin (1956) recorded the occurrence of *Phyllactinia roboris* (Gachet) Blum. on leaves of *Q. robur* L.

Salix *spp.* (*Willow*)

Smith (1971) noted the presence of *Uncinula salicis* DC. ex Wint. and *Microsphaera penicillata* (Wallr. Fr.) Lév. on willows (*Salix* spp.). Both fungi survived the winter as cleistothecia releasing viable ascospores in the spring. Both species were also recorded on aspen (*Populus tremula* L.).

Tectona *spp.* (*Teak*)

Teak (*Tectona grandis* L.) is one of the most important hardwood timbers. It is indigenous throughout parts of India, Burma, SE Asia and Java and is established as a major plantation species in those areas as well as being on trial in a number of countries in Africa, the West Indies, Sri Lanka and parts of America.

Two powdery mildews are known to attack teak, *Phyllactinia guttata* (Wallr. ex Fr.) Lév. and *Uncinula tectonae* Salm. Both species cause leaf necrosis and defoliation, which may be severe under warm, humid conditions.

Tilia *spp.* (*Lime*)

Basswood powdery mildew (*Uncinula clintonii* Peck) has been recorded in North America and eastern Asia on the foliage of *Tilia americana* L. (Boyce, 1961).

Ulmus *spp.* (*Elm*)

Elm powdery mildew (*Uncinula macrospora* Peck) has been found in North America on the foliage of white elm (*Ulmus americana* L. (Boyce, 1961). Calonge (1968) identified *Uncinula clandestina* Schroet. on elm trees in Spain. Bakshi and Singh (1961) recorded the same species on *Ulmus wallichiana* Planch.

Hanlin (1969) in Georgia, USA recorded a powdery mildew identified as *Phyllactinia angulata* (Salmon) Blumer on winged elm (*Ulmus alata* Michx.).

Zizyphus *spp.* (*Ber tree*)

Oidium erysiphoides Fr. is reported to form a mildew on the leaves of *Zizyphus mauritania* Lam. in India. (Bagchee and U. Singh, 1954; Butler and Bisby, 1960).

Acknowledgement

The assistance of Mrs. G. M. Tardivel is gratefully acknowledged.

References

Akhundov, T. M. (1963). *Izv. Akad. Nauk azerb. SSR Ser. biol. med. Nauk*. **5**, 15–18.
Anon, (1961). Canadian Forest Insect and Disease Survey Report. Canada Dept. of Agriculture. 1961.
Bagchee, K. D. and Ujagar Singh (1954). *Indian Forest Rec*. (*N.S.*) **1**, 199–348.
Bakshi, B. K. and Singh, S. (1961). *Indian Forester* **87**, 542–545.
Bennett, M. and Moore, M. H. (1963). *Rep. E. Malling Res. Sta.* **1962** 105–108.
Blumer, S. (1967). "Echte mehltau pilze (Erysiphaceae)". Gustav Fischer Jena.

Boyce, J. S. (1961). "Forest Pathology". McGraw Hill Book Company Inc., New York

Brandenburger, W. (1961). *Sydowia* **15**, 194–196.

Brooks, F. T. (1953). "Plant diseases". Oxford University Press, London.

Burchill, R. T. (1958). *Rep. agric. hort. Res. Stn. Univ.* Bristol 114–123.

Butler, E. J. and Bisby, G. R. (1960). "The fungi of India". Indian Council of Agricultural Research, New Delhi.

Calonge, F. D. (1968). *An. Inst. bot. A. J. Cavanilles* **26**, 15–35.

Chanturiya, N. N. (1965). *Soobshch. Akad. Nauk gruz. SSR.* **38**, 637–644.

Chiddarwar, P. P. (1955). *Curr. Sci.*, **24**, 239–240.

C. M. I. Descriptions of Pathogenic Fungi and Bacteria. Commonwealth Mycological Institute, Kew, Surrey.

Cotton, A. D. (1919). *Trans. Br. Mycol. Soc.* **6**, 198–200.

Creelman, D. W. (1967). *Can. Pl. Dis. Surv.* **47**, 31–68.

Cristinzio, M. (1942). *Ric. Ossni Divulg. fitopat.* Campan. Mezzogiorno **9**, 17–64.

Delmas, H. G. (1953). *Ann. Inst. Rech. agron.*, C, **4**, 59–89.

Dillon Weston, W. A. R. and Taylor, E. (1944). *Trans. Br. Mycol. Soc.* **27**, 119–120.

El-Ghamrawy, A. K., Sabet, W. A. and Fayed, G. E. (1967). *Agric. Res. Rev.*, Cairo **45**, 5–19.

English, H. (1947). *Phytopathology* **37**, 421–424.

Fikry, A. (1937). *Egypt. Min. Agr.*, *Tech. Sci.*, *Bul.* 183.

Franco, E. (1946). *Bolm. fitossanit.* **3**, 91–97 .

Gardner, M. W., Yarwood, C. E. and Duafala, T. (1972). *Pl. Dis. Reptr* **56**, 313–317.

Golovin, P. N. (1956). *Trudy Inst. Bot., Baku* **10**, 195–308.

Graniti, A. (1958). *Ann. Canad. Sci. For.* **7**, 309–328.

Hadiwidjaja, T. (1950). *Landbouw* **22**, 247–257.

Hanlin, R. T. (1969). *Pl. Dis. Reptr* **53**, 395–397.

Himelick, E. B. and Neely, D. (1959). *Phytopathology* **49**, 831–832.

Hirata, K. (1966). Host range and geographical distribution of the powdery mildews. Niigata Univ., Japan.

Huque, H. (1964). Plant quarantine situation in Pakistan. FAO/IUFRO Symposium on internationally dangerous forest diseases and insects. Oxford.

Itoi, S., Nakayama, H. and Kubomura, Y. (1962). *Bull. sericult. Exp. Sta. Tokyo* **17**, 321–445.

Jones, E. W. (1944). *J. Ecol.* **32**, 215–252.

Jovićević, B. (1955). *Zašt. Bilja* **30**, 37–39.

Khairy, E. A., Michail, S. H. and Abd-El-Rehim, M. (1971). *Phytopath. mediter.* **10**, 269–271.

Louw, A. J. (1946). *Fmg. S. Afr.* **21**, 93–99.

Mayor, E. (1967). *Ber. schweiz. bot. Ges.* **77**, 128–155.

Miller, P. W. (1945). *Stat. Bull. Ore. agric. Exp. Stat.* 24.

Mukerji, K. G. (1969). *Mycologia* **61**, 181–184.

Neger, F. W. (1915). *Naturw. Z. Forst. u. Landw.* **13**, 544–550.

Ogawa, J. M. and Charles, F. M. (1956). *Calif. Agric.* **10**, pp. 7 and 16.

Palti, J., Pinkas, Y. and Chorin, M. (1974). *Pl. Dis. Reptr* **48**, 45–49.

Patil, B. V. (1961). *Curr. Sci.* **30**, 155–156.

Patil, S. D. (1962). *J. Univ. Poona* **22**, 119–125.

Patwardhan, P. G. (1962). *Mycopath. Mycol. appl.* **18**, 149–150.

Puttoo, B. L. (1972). *Indian J. Mycol. Pl. Path.* **2**, p. 193.

Raabe, R. D. (1966). *Pl. Dis. Reptr* **50**, 519.

Radman, Ljubica (1963). *Zašt. Bilja* **14**, 251–254.

Rodigin, M. N. (1956). *Not. syst. Sect. crypt. Inst. bot. Acad. Sci. USSR* **11**, 103–104.

Roll-Hansen, F. (1961). *Medd. norske SkogsforsVes.* **17**, 37–61.

Sinadskii, Yu. V. (1965). *Byull. glavn. bot. Sada, Moskva* **59**, 83–88.

Smith, C. G. (1971). *Trans. Br. mycol. Soc.* **56**, 275–279.

Smith, E. F. (1894). *Jour. Mycol.* **7**, 90–91.

Smitskaya, M. F. (1955). *Bot. Zh., Kŷyiv* **12**, 87–92.

Snetsinger, R. and Himelick, E. B. (1957) *Pl. Dis. Reptr* **41**, 541–544.

Spaulding, P. (1956). *Agric. Handb. US Dep. Agric.* No. 100, pp. 144.

Spaulding, P. (1958). *Agric. Handb. US Dep. Agric.* No. 139, pp. 118.

Spaulding, P. (1961). *Agric. Handb. US Dep. Agric.* No. 197, pp. 361.

Tamayo, B. P. and Pordesino, A. N. (1959). *Philipp. Agric.* **43**, 236–239.

Tokushige, Y. (1953). *Ann. phytopath. Soc. Japan* **17**, 61–64.

Tuset Barrachina, J. J. (1972). *An. Inst. nac. Invest. agrar.* **2**, 27–33.

Vlasov, N. A. (1957). *Referat. Zh. Biol.* **20**, 177.

Wastie, R. L. and Mainstone, B. J. (1969). *J. Rubb. Res. Inst.* Malaya **21**, 64–72.

Wastie, R. L. (1975). *Pestic. Abstr.* **21**, 268–288.

Weigle, C. G. (1956). *Pl. Dis. Reptr* **40**, 584.

Weinhold, A. R. (1961). *Phytopathology* **51**, 478–481.

Woodward, R. C., Waldie, J. S. L. and Steven, H. M. (1929). *Forestry* **3**, 38–56.

Worley, R. E. and Harman, S. A. (1969). *Hort. Sci.* **4**, 127–128.

Yarwood, C. E. (1939). *Phytopathology* **51**, 478–481.

Yarwood, C. E. (1944). *Phytopathology* **34**, 937.

Yarwood, C. E. (1952). *Bull. Calif. Dep. Agric.* **41**, 19–25.

Yarwood, C. E. and Gardner, M. W. (1964). *Pl. Dis. Reptr* **48**, 310.

Yarwood, C. E. and Gardner, M. W. (1972). *Phytopathology*, **62**, 799.

Yen, J.-M. and Wang, C.-C. (1973). *Revue Mycol.* **37**, 125–153.

Young, H. E. (1954). *Q. Circ. Rubb. Res. Inst. Ceylon* **30**, 51–60.

Chapter 19

Powdery Mildews of Vegetable and Allied Crops

G. R. DIXON

National Institute of Agricultural Botany, Cambridge, England

I. Introduction

Powdery mildews are highly successful pathogens because they gradually cause the death of their hosts and infect a wide host range; Hirata (1966) estimated that between 13–38% of Angiosperms in a given geographical region could be infected by them. Most of the vegetable crops used in temperate and tropical regions are included within this host range, many of which provide staple foods for populations in developing countries. The significance of these pathogens to the yield of host crops has only recently begun to be appreciated; the few yield loss studies which have been made however, indicate that they cause very considerable reductions in crop production.

In this chapter diseases of five major crop families are considered, individual powdery mildews of particular crops are described together with host symptoms, geographical distribution, and where possible estimations of crop loss, studies of resistance and of host–parasite physiology. Many crops are attacked by several powdery mildews and often only the imperfect stage is seen. In some crops therefore it has been difficult to decide which particular pathogen is being reported by an author. For this chapter the generic classification proposed by Yarwood (1957) is used:

A. Perfect genera

1. Erysiphe

Mycelium superficial, cleistothecia with mycelial like appendages and several asci.

2. Microsphaera

Mycelium superficial, cleistothecia with appendages which are dichotomously branched at the tip.

3. Leveillula

Mycelium partly intercellular, cleistothecia with mycelium like appendages and several asci.

B. Imperfect genera

1. Oidium

Unbranched conidiophores and barrel shaped conidia.

2. Ovulariopsis

Clavate conidia.

3. Oidiopsis

Branched conidiophores emerging from the stomata.

At the specific level the nomenclature used by the Commonwealth Mycological Institute in their *Descriptions of Pathogenic Fungi and Bacteria* has been adopted. Host families are arranged in the ascending evolutionary order used by Clapham, Tutin and Warburg (1952).

II. Cruciferae

The major powdery mildew on crucifers is caused by *Erysiphe cruciferarum* a classification which resulted from the revision of *Erysiphe communis* by Junell (1967). Originally, Salmon (1900) used the collective name *Erysiphe polygoni* for this pathogen on swede and turnip. Some members of this species were removed by Blumer (1933) and raised to specific level. The remainder, classified as *E. communis*, were forms whose morphology was indistinctly known and where cleistothecia seldom developed.

Symptoms mainly consist of patches of thin mealy mycelium on the upper leaf surface; these gradually coalesce until the entire surface is covered by the pathogen; spread to the lower surface is common. Other organs such as stems and the swollen axillary buds of Brussels sprouts are frequently infected. Heavily infected organs become chlorotic but death of the host usually takes a long time. Disease distribution is world-wide. Transmission is thought to be mainly by wind-blown conidia and perennation may take place through mycelium which can survive as "subinfections" (Searle, 1914).

A. Forage brassicas (*B. napus and B. campestris*)

Powdery mildew in the United Kingdom has only recently been appreciated as a severe cause of yield loss. This is similar to the situation with cereal mildew (*Erysiphe graminis*) which was thought to have little effect on yield until Doling and Large (1962, 1963) proved that heavy losses were attributable to the disease. Thus statements that the disease is not "generally a serious problem on brassicas" (Channon and Maude, 1971) or that the effects "on yield were insufficient to warrant use of fungicides" (Ogilvie, 1969) have been made in ignorance of the true situation in field crops. Searle (1914) working in East Kent investigated the local farming theory that sowing swedes before June enhances the possibility of mildew infection. In these investigations the first signs of mildew were seen at the end of August and could indeed devastate the yield of susceptible stocks. It was also reported that experience in northern England showed that bronze swedes were more susceptible than purple and green types. In field trials in Wales, Whitehead (1940) found that resistance of some stocks varied with season. This reinforced Searle's (1920) observations that no varieties of swede and rape were immune and that swedes were more severely attacked than turnips. Notes of outbreaks are scattered through the annual reports of various research stations indicating years in which the disease has been more than usually severe, for example, in Yorkshire (1939) and in Norfolk (1941–43) (Thomson, 1947). Willey (1953) describes the complete defoliation of swedes at Cambridge and in Norfolk in 1945 and also points to "considerable yield reduction". In Wales (Griffiths, 1958) there were severe attacks and heavy losses in Hungry Gap Kale and Rape-Kale. In Scotland (Anon, 1934) swede trials were heavily infected and some lines were more quickly infected than others. It was also observed that juvenile foliage became more rapidly colonized following rain.

Systematic analysis of cultivar response to these diseases began with the trials of Doling and Willey (1969) who showed that late sowing to avoid mildew infection led to losses of 60% in the yield of swedes. Growers were therefore paying a heavy penalty for using susceptible cultivars which required this treatment. More recently Dixon and Furber (1971) reported on the resistance and susceptibility of a wide range of swede cultivars and on the usefulness of fungicides. Recommended resistant cultivars now include: Bangholm, Marian, Merrick, Seefelder, Ruta Øtofte, Vogesa, Wilhelmsburger Prima and Wilhelmsburger Sator Øtofte (Anon, 1976). This work has led to the development of keys for disease assessment on a field scale and to studies of cross infection under glasshouse conditions. The latter indicate that strains of *E. cruciferarum* on *B. napus* and *B. campestris* may be different from those on *B. oleracea* (Dixon, unpublished data). In fungicidal studies the eradicant material dinocap is highly effective but materials with systemic action such as benomyl and tridemorph are more efficient on crops like swedes than on Brussels sprouts. In sprouts the later incidence of infection and slow translocation rates of the host tend to favour materials like dinocap. Studies of forage rape cultivars indicate that Lair, Emerald, Blako and Late Dwarf tend to be most resistant (Anon, 1975).

Breeding work indicates that resistance is of a continuous nature under multiple gene control. Chignecto and selections of Danish material with high resistance to powdery mildew are being used in Wales to develop powdery mildew resistant swede cultivars (Johnston 1972). Work is also in progress to utilize the resistance of fodder radish to powdery mildew. This is done by producing Raphano-brassica hybrids which in limited field trials have shown high levels of resistance (McNaughton, 1974). There are however, strains of powdery mildew which are known to infect Raphano–brassica hybrids (Buhr, 1958).

Various studies have been made of the biological forms of brassica powdery mildew, (e.g. Searle, 1920). Inoculations from *B. campestris* to *B. oleracea* invariably led to "subinfections". On *B. oleracea* in the field these were thought to offer a means of overwintering since under favourable conditions they could develop into full infections. Conidia present in subinfections were shown to be viable. In some subinfections "minute black spots" developed which gave the appearance of localized cell death, i.e. hypersensitivity, (Smith, 1900).

Limited physiological studies have been made. Zinc frit added to the soil of potted plants reduced infection on turnips and the subsequent high levels of zinc in the foliage indicated rapid translocation of this heavy metal (Tomlinson and Webb, 1958). In studies of the infection process Purnell and Preece (1971) showed that washing leaves prior to inoculation inhibited primary and secondary hyphal production. If, however, there was a five-day interval between washing and inoculation then washing appeared to stimulate the infection process. There was no effect of washing on germination and appresorium development *per se* but it did reduce the numbers of conidia and length of primary hyphae. Washing removed wax from the leaf cuticle and little of this was replaced, also washed leaves were leached of carbohydrate which was subsequently replaced to excess.

Hirata (1966) reported species causing powdery mildew of Brassicas as *E. communis*, *Leveillula taurica*, *Erysiphe cichoracearum* and *Oidium spp.* Subsequently, Connors (1967) found *E. polygoni* on *B. campestris* and *B. napobrassica* in Canada; Al-Hassan (1973) described *E. communis* on *B. campestris* and *B. rapa* in Iraq; Baker (1972) has reported *E. cruciferarum* as widespread on swedes and rape crops in the UK; and Oran (1974) found *E. communis* on *B. napus* and *B. rapa* in Turkey.

B. Cole crops and other *B. oleracea* types

The major agricultural forage crop in this group is Marrow Stem Kale which is generally highly resistant to *E. cruciferarum*. The later developing Maris Kestrel Kale is however, more susceptible (Thompson 1963) although attempts are being made to introduce more resistant inbreds into hybrid seed production.

Horticultural brassicas for human consumption are highly susceptible to powdery mildew and Chupp and Sherf (1960) cite the following host list— cabbage, chinese cabbage, kohlrabi, broccoli, kale, mustard, collards, cauliflower, radish and in Europe horse-radish. The disease is found on all foliage especially the upper surface and when severe leads to leaf curling and abscission.

Breeding work has been carried out in the USA (Walker and Williams,

1965) to combat powdery mildew on cabbage. Resistance was thought to be due to a single dominant gene and that this was influenced by numbers of minor genes since in parental generations resistance was incompletely dominant. Thus under conditions of heavy inoculum a heterozygotic host may support limited fungal growth. Powdery mildew is found in semi-arid and humid areas of USA especially after protracted drought. It is especially significant on kraut cabbage where it may cause disfigurement and irregular development; the latter interferes with mechanical harvesting and processing. Symptoms range from the whole plant being covered in mycelium to a fine necrotic flecking which may indicate limited resistance (similar to Searle's observations, 1920).

Multiple disease resistant cabbages were produced by Williams et al. (1968), thus Hybelle (Badger Inbred 12 × Globelle) and Sannibel (Badger Inbred 13 × Globelle) were aimed at quality sauer-kraut production in the northern USA and at fresh market production in Florida during the winter. Globelle was derived from Resistant Glory which was in turn bred from the European cultivar Glory of Enkhuizen. The Badger Inbreds 12 and 13 came from Wisconsin All Seasons and Bugner. Globelle has been shown to be homozygous for resistance to powdery mildew (Walker and Williams, 1973).

In the UK powdery mildew is of increasing importance especially on Brussels sprouts. This is thought to be due to the intensification of production and use of F_1 hybrid cultivars (MAFF Closed Conference, 1975). Studies of cultivar resistance have been made over a number of seasons together with the development of assessment keys (Dixon, 1974); cultivars such as Gibsons No. 1 (Bedford Selection), Lancelot, Nelson, Rido and Wellington have proved to be particularly resistant. Disease develops from late August onwards, initially infecting the foliage and stems and then progressing to the axillary buds (buttons). Deep penetration of the buttons has been reported but it is unclear how a pathogen which generally invades the epidermis can achieve this. Early-maturing cultivars can be badly diseased, possibly they present mature tissue for colonization at a period of the year when rapid epidemic development can occur. Later-maturing cultivars tend to include more resistant material derived from parents selected for resistance in areas such as Bedfordshire and the Vale of Evesham for many generations. The disease has now spread to most sprout growing areas in the UK including Lincolnshire and Lancashire. Losses are due not only to an immediate reduction in yield but also to loss of quality. Infected produce can be downgraded in the open market or completely unacceptable for processing. It is difficult to obtain accurate information on the financial loss to growers through either of these causes but a heavy loss is indicated by their readiness to utilize fungicidal controls and resistant cultivars. In periods of cold weather the quality of infected buttons may be reduced still further by the melanization of mycelium on the button surface giving a black speckled appearance.

Invasion by *E. cruciferarum* can also give a portal for entry by secondary rotting organisms such as grey mould (*Botrytis cinerea*). There are differences in levels of infection up and down sprout plants and it has been shown that in some cultivars the leaves may be more resistant than the buttons (e.g. Ulysses) and vice versa, (e.g. Fasolt), indicating that, as with cabbages, resistance is under multigenic control. This will also be influenced by microclimatic and husbandry factors, wide spacing tends to encourage mildew infection as do certain forms of nitrogen fertilization. It is likely that mildew perennates on other *B. oleracea* crops which are generally freely available in the areas of sprout production. Growers suggest that periods of moisture stress make plants more susceptible to mildew infection but this may be more of an effect of temperature since irrigated crops suffer equally from the disease (Dixon, unpublished data). Some level of chemical control has been achieved with the eradicant dinocap but systemic materials are of limited value due to the slow rates of translocation in sprouts during the autumn.

In cabbage and cauliflower two forms of infection have been noted, firstly necrosis of the outer leaves with obvious powdery mildew lesions accompanied by much reduced curd size and secondly disease lesions on the outer leaves and necrosis of the inner wrapper leaves starting at the tips and progressing towards the curd which at the same time began to go rotten. Cultivars of summer cabbage, Vienna Green Wonder and Wiam Persista were least badly affected, while in winter hardy cauliflowers Armado May, Filgap 70, Improved June Market, Juno, Late Adonis, Markanta and Progress, were resistant (Dixon, unpublished data).

Powdery mildews of *Brassica* and *Raphanus* species are reported by Hirata (1966) as *E. communis*, *E. cichoracearum*, *L. taurica* and *Oidium* spp. More recently Connors (1967) found *E. polygoni* on *B. oleracea acephala*, *B. pekinensis* and *R. raphanistrum* in Canada, Oran (1974) found *E. communis* on *B. oleracea* in Turkey and Al-Hassan (1973) described *E. communis* on *B. oleracea* and *E. cichoracearum* on *R. sativus* in Iraq.

C. Oil crop brassicas and linseed

Infection with powdery mildew (*E. polygoni*) has been reported on rape and mustard seed crops in Europe and India. In France the disease was found on *B. napus* var. *oleifera* (Darpoux 1945). Cook (1975) however, suggests that in France and Germany the disease is seen late in the season and is not of economic significance. In the UK the disease is regularly seen on maturing crops, particularly on spring-sown material; differences in cultivar resistance, have been noted (Furber, unpublished data).

In India, these crops are a particularly valuable source of oil. *E. polygoni* infects all parts of the plant leading to either a complete inability to set seed or the production of small and shrivelled seed (Bhander *et al.*, 1963). Cross inoculation experiments showed the pathogen would infect *B. campestris* (yellow mustard/sarson); it was also possible for this pathogen to infect *B. juncea* (Sankla, *et al.*, 1967a). Infections occured in March-April with a dirty white hyaline mycelium producing granular barrel shaped conidia on the lower leaves. Resistance to *E. polygoni* was found by Narain and Siddiqui (1965) in *Eruca sativa, B. campestris* var. brown sarson, *B. chinensis, B. napus, B. alboglabra, B. japonica* and *B. alba.* Other powdery mildew pathogens reported by Hirata (1966) include *E. communis, E. cichoracearum. L. taurica* and *Oidium spp.* Since then *E. communis* has been found on *B. nigra* by Oran (1974) in Turkey.

Linseed is the most valuable Indian oil seed crop. Losses due to powdery mildew (*Oidium lini*) range from 2–10%, seed is sown in November and infections are easily seen by February (Kushwah and Chand, 1971), the effects being the reduction in both seed size and number (Sharma, *et al.*, 1972). In trials cvs Hira and Sabaur showed resistance (Singh and Kaurav, 1973).

Infections have also been seen in the UK late in the season (early August) which probably caused little damage to the crop (Furber-pers. comm.). Other reports of the disease have been made by Riley (1960) in Tanzania, Chorin and Palti (1962) in Israel and Bernaux (1965) in France.

III. Papilionaceae

A. Pea

Powdery mildew in peas is caused by *Erysiphe pisi* (often referred to as *E. polygoni*), this species is seen on *Pisum, Medicago, Vicia, Lupinus* and *Lens* and can be distinguished from *Erysiphe trifolii*, which is more commonly found on Papilionaceae, by the length and shape of the cleistothecial appendages, (Kapoor, 1967c). It has a world wide distribution. Hammerlund (1925) distinguished three biologic forms; f. sp. *pisi* on *P. sativum* and *P. arvense*; f. sp. *medicaginis sativae* on *Medicago sativa* and *M. falcata* and f. sp. *viciae sativae* on *Vicia sativa, V. sepium* and *V. sylvatica.*

Mycelium grows typically on the upper leaf surface as powdery, greyish-white lesions which eventually coalesce. Usually late sown crops and those at the dried pea stage of maturity are most liable to infection. Leaves, stems and pods may be infected causing death of the vine, withering of foliage and occasional plant death. In pod infection the fungus may penetrate to the seeds which may become grey-brown (Chupp and Sherf, 1960). The disease may be seed-borne as was illustrated by work in New Mexico by Crawford

(1927). Powdery mildew became a limiting factor in pea production in this area but use of hot water seed treatment and sulphur dust fungicides in the field led to a doubling of yield. Other early workers (van Hook, 1906 and Stevens, 1921) found evidence of seed transmission, one seed batch was found to contain 22·5% infected seeds.

Using Peruvian material, Harland (1948) showed resistance was controlled by a single recessive gene er. This was supported by Pierce (1948) who found resistance in cv. Strategem which he crossed with cvs Glacier and Shasta illustrating that resistance was controlled by a single gene pair with susceptibility dominant. Later this resistance broke down under field conditions (Schroeder and Providenti, 1965), the population which contained virulence against gene er had a wide host range in the field. Some 400 cultivars were later assessed by Cousin (1965) against natural field infection in France, some Mexican and Peruvian material was highly resistant; in the cvs Mexique and Resistant Strategem, resistance was controlled by similar single recessive factors. Later Heringa et al. (1969) in Holland listed a wide range of pea accessions from Peru, Hawai and Geneva USA; it was thought that resistance in the Peruvian material was confined to the leaves and due to a second recessive factor er_2. According to Yarnell (1962) the resistant Peruvian material comes from peas taken there by the Spaniards. Resistance of the er type is 35% linked to factor A which is responsible for purple flower colour and also linked to node number. Cousin (1965) also supports the idea that infection is correlated to tissue age; powdery mildew is an important disease in Central France progressing northwards in July as the crops mature and acting as a limiting factor to pea production.

Considerable breeding work is now in progress in the United States. This is justified agronomically since Gritton and Ebert (1975) report that mildew infection adversely affected total plant weight, weight of shelled peas, numbers of peas per pod, number of peas per plant, plant height and nodes per plant, moreover, infection raised tenderometer readings and thereby depressed quality. Earlier work by Gritton and Hagedorn (1971) led to the release of processing pea lines with powdery mildew resistance using progeny from Oregan State University 42 × New York 59–29. This was back crossed to Sprite, Dark Skinned Perfection, New Era and New Line Early Perfection to give suitable "quick freeze" cultivars. Powdery mildew is a troublesome disease under prolonged warm, dry daytime conditions and when the nights are cool enough for dew formation (Hagedorn, 1973). The disease is also a problem where peas are grown in low wet areas with high soil moisture which maintains the plants in the vegetative phase. At present these areas account for 5% of the US acreage but this is likely to increase with the release of late-maturing cultivars which are resistant to root rot (*Fusarium* spp) (Muehlbauer, pers. comm., 1976). Screening for resistance is done on the basis of natural field infections using a late sowing technique to encourage the disease. Sources of resistance used at present include: Oregon State University 42, Geneva 59–29 and PI Nos 142775, 142777, 180792, 201497, 203064 and Strategem. Some of these are similar to the lines reported by Ford and Baggett (1965) who in addition used PI numbers 222069 and 244155. Inoculations were made in the glasshouse using a "spore suspension" at temperatures of $72 \pm 5°F$. Resistance has also been noted in some of the leafless peas developed at the John Innes Institute in the UK and in breeding material from India and Malaya (Matthews, pers. comm., 1976).

In Canada the field pea (*P. arvense*) is being looked at as a potential breakcrop for the Prairie cereals but Morrall *et al.* (1972) and McKenzie and Morrall (1973, 1975) have already reported infections by powdery mildew. Also in Canada, Ali Khan *et al.* (1973) carried out trials of 1200 cultivars of which only 18 had some resistance. The disease poses little threat to processing crops either in Canada (Basu *et al.*, 1973) or in Europe (Dixon, personal observation) since the crops are harvested well before infection can develop. It is the increasing dried pea acreage which is likely to be infected by this disease because the leaf tissue is sufficiently mature for infection. When pulling peas were grown on a large scale in the UK they were regularly infected. Beaumont and Hodson (1928) reported that some badly infected crops were not worth harvesting.

Powdery mildew was known on peas in Rhodesia, where it became severe enough to cause the cessation of production (Mathur *et al.*, 1971). The disease is also found in South Africa (Boelema, 1963) and Australia where Wark (1950) assessed the resistance of 500 cultivars of *P. sativum*, *P. arvense* and various edible podded peas from the Middle East. An inoculation technique of floating leaf discs on 10% sucrose was used and leaves of all cultivars were found to be resistant at the juvenile stage. Differences between cultivars could be detected when older tissue was tested, in susceptible material there were increases of fungal growth and sporulation. Six cultivars reported by Harland (1948) as immune in Peru were susceptible in Australia, as were some species of *Lathyrus*.

In India and other developing countries the disease poses its most severe threat because peas are an essential protein source. Sharma and Shrivasta (1968) report very large crop losses and thought that tallness was associated with some levels of resistance. Similar reports were made by Hawre (1971), especially heavy losses occur in the winter crop sown in October, early-maturing cultivars are attacked more quickly. The pathogen can reduce the number of harvests from six or seven to one (Mathur *et al.*, 1971). Loss of yield can be judged by the results of fungicide trials, plots dusted with sulphur produced twice the yield (1121 kg ha^{-1}) of untreated controls. Pod weight was found to be a good index for disease assessment by Vasudeva (1962); later Munjal *et al.* (1963) showed there was a 20–30% reduction in pod number and about 25% reduction in pod weight, due to the disease. Powdery mildew is also important in Italy (Cirulli pers. comm., 1975) where in the South, sowings are made in December and disease epidemics commence in March; resistance is being sought using irradiated material from Sprinter. An extensive list of hosts and geographical locations is given by Hirata (1966), other areas where the disease is found include Afghanistan (Gattani, 1962), Iraq (Al-Hassan, 1973), Libya (Krantz, 1965), Malaya (Thompson and Johnston, 1953), Portugal (Sequeira da Silva, 1966), Russia (Kirik and Kitsno, 1974), Sudan (Tarr, 1955), Tanzania (Riley, 1960), Turkey (Oran, 1974) and Zambia (Griffee, 1969). Peas can also be attacked by *L. taurica* and this was noted by Tarr (1955) in the Sudan.

A few physiological studies have been made of *E. pisi* particularly that by Oku *et al.* (1975) who demonstrated phytoalexin production in an obligate host parasite relationship for the first time. Pisatin was first detected two days after leaf inoculation at $3\mu g\ g^{-1}$ fresh weight and increased logarithmically to $300\mu g\ g^{-1}$ by day 4. When spores of the non-pathogenic fungus *E. graminis* f. sp. *hordei* were used, pisatin was found only 15 hours after inoculation. Conidia of the pea pathogen were 13 times more tolerant to pisatin than those of the non-pathogenic form. Concentrations of pisatin were greater at the infection site than elsewhere in the leaf and are therefore thought to have a role in determining host specificity to powdery mildew. Pisatin was produced in advance of susceptible cell collapse or hypersensitive necrosis. Externally applied nitrogen has an effect on this host–parasite relationship, plants (cv. Pauli) given ammonium sulphate were free of disease on the leaves and only slightly infected on the pods whereas untreated plants were badly infected (Wijngaarden and Ellen, 1969).

Colonization and pathogen specificity were studied by Smith (1969). It was possible to discern ten developmental stages in the colonization of cv. Onward by *E. pisi*. Only isolates from *Lathyrus odoratus* would cross infect to *P. sativum*. This work also suggested that the condition of leaf tissue determines the degree of infection since colonization was more extensive on leaf discs than on whole plants.

Other crops which can be seriously attacked by *E. pisi* include gram (*C. arietinum*), pigeon pea (*Cajanus cajan*), black gram (*Phaseolus mungo*), green gram (*Ph. aureus*) and lentil (*Lens esculentus*) (Gupta, 1974). Infection of lentil leads to rapid chlorosis and defoliation (Sankhla *et al.*, 1967). In India infections are first seen in February as white foliar lesions on both leaf surfaces which then spread to the pods. Studies of infection on cowpeas (*Vigna sinensis*) were made by Paulech and Herrera (1969, 1970) who divided the infection process into six stages over the first 150 hours of colonization; germ tube formation started within an hour of inoculation and the whole leaf became covered with mycelium. Infection is localized to small dark spots in resistant plants. These crops were also at risk from *L. taurica* and this pathogen was reported on pigeon pea in: India (Kamat and Patel, 1948), Mauritius

(Orieux and Felix, 1968) and Sudan (Tarr, 1955); and on chick pea (Tarr, 1955). A full account of host range and distribution was given by Hirata (1966).

B. Soybean

Powdery mildew of soybean may be caused by one or more species of fungi, (Dunleavy *et al.*, 1966), early accounts attributed *E. polygoni* as the causal agent but more recently *Microsphaera diffusa* has been implicated. Under glasshouse conditions the fungus spreads rapidly on susceptible cultivars but in the field it is not adapted to prolonged survival and occurs only sporadically. Symptoms include the development of small colonies of thin, light grey or white mycelium on the upper leaf surface with reddening of the undersurface, lesions rarely coalesce to cover all the leaf surface. Many spores are produced and heavily infected leaves turn brown and drop off.

Earliest reports on this disease are those of von Wahl (1921) and Lehman (1931) who attributed the disease to *E. polygoni*. Salmon (1900), however, reported *M. diffusa* on *Glycyrrhiza lepidota* a relative of soybean. Demski and Phillips (1974) tested cultivars for resistance to *M. diffusa*, in Georgia, USA, those with no field infection included: Bragg, Cobb, Dowis, Essex, Hardee, Hutton, Jackson, McNair 800, while Hale –7, Hampton 266A and Ramson were field infected, Demski and Phillips also report colonization of the pods. An epidemic rapidly developed in Wisconsin (Arny *et al.*, 1975) in 1974, resistant cultivars were commonly found to have Lincoln-Richland in their parentage. The disease has now been reported in Delaware, Iowa, South Carolina and Texas (Anon., 1960), Illinois (Paxton and Rogers, 1974), Mississippi (Johnson and Jones, 1961). Grau and Lawrence (1975) report two forms of resistance to *M. diffusa*.

1. a resistant reaction where mycelial development and sporulation are sparse.
2. a highly resistant reaction where mycelium does not form on the leaves.

Reaction 1. is thought to be due to a single dominant gene. Disease occurrence has been associated with relatively low temperatures and could be a menace to crops in Georgia, Illinois and Minnesota but the potential effects of this disease have not been evaluated.

C. Beans

The world wide distribution of *E. polygoni* on beans was described by Zaumeyer and Thomas (1957). In the USA losses can exceed half the crop especially in the autumn. All types of bean are attacked: dwarf french, runner, snap (*Phaseolus vulgaris*) and lima (*Ph. lunatus*). Infection begins on the foliage as a slight darkening which resembles virus mosaic, but these areas soon become covered with mycelium (Chupp and Sherf, 1960). Lesions are inclined to follow the veins and young foliage when infected becomes dwarfed, shrivelled and distorted. Unlike pea powdery mildew the disease is not thought to be seed-borne despite a report of Miller (1952). In the Southern USA there is continuous infection while in the North infection occurs via wind-blown conidia. *Erysiphe* grows best at 68–75°F (Yarwood, 1936) and the conidia contain large quantities of water and can therefore germinate under very dry conditions.

The disease has been well documented since it was first recognized about 130 years ago. In Bermuda, Whetzel (1921), found *E. polygoni* to be a serious parasite of bean crops which were grown as cash crops and marketed in New York, the disease flourished all year round with pod infections leading to stunting and poor setting. A similar report was made by Ogilvie (1925). Conditions unfavourable to host growth i.e. low temperature and lack of soil moisture (Cook, 1931), favoured mildew development. This correlated with Yarwood (1949) who reported severe mildew when there were low soil moisture levels and high concentrations of soil nutrients.

Inheritance of resistance was studied by Dundas (1936, 1941) using detached leaflets supported on cotton wool soaked in 10% sucrose; cultivars, Hopi, Lady Washington, Hungarian, Yellow Long Kidney and Pinto were resistant in all stages of development while Pink was susceptible in young stages and resistant in older stages. Resistance was due to a single dominant (cf. peas) factor which was thought to be common to all resistant cultivars. Susceptibility may be influenced by virus infection; Maxfield (1968) showed that beans infected with Southern Bean Mosaic developed 90% fewer powdery mildew colonies and there was a 75% reduction in haustoria.

Various *Vicia* species are also infected by powdery mildew, this is usually attributed to *E. polygoni*. In California and Georgia, however, *Microsphaera* has been identified as a causal agent, although this may be limited geographically. In Canada McKenzie and Morrall (1975) found *Vicia faba* infected with *Microsphaera penicillata var ludens*. There was an original report by Yu (1946) in Yunnan Province, China, where powdery mildew was found on the leaves, petioles, stems and pods of broad bean plants although little economic damage was caused. It was one of the few reports of cross infection to peas— the pathogen was capable of overwintering as either conidia or perithecia on beans, peas and other *Vicia* species and infection usually started in the early spring at flowering time.

Other reports of *E. polygoni* on *Vicia* beans have been made by Tarr (1955) in Sudan where in irrigated crops north of Khartoum 50% yield losses were recorded due to leaf shedding and drying-up of plants. An extensive host list was given by Hirata (1966) and later reports include Al Hassan (1973) in

Iraq (pathogen reported as *E. cichoracearum*) and Oran (1974) in Turkey (pathogen reported as *E. communis*).

Both *Phaseolus* and *Vicia* species may be attacked by *L. taurica*; this has been reported in the Sudan (Boughey 1946) and Tanzania (Anon 1958, Riley 1960). Additionally *Oidium erysiphoides* was reported in Argentina (Feldman and Pontis 1960) on peas and beans.

D. Clover

Erysiphe trifolii is common on *Trifolium* and *Lathyrus* and is occasionally seen on other Leguminosae (Kapoor, 1967a). There is world wide distribution and five biologic forms on *Trifolium* and two on *Lathyrus* were distinguished by Hammerlund (1925). One of the most extensive accounts of the disease was by Horsfall (1930). It was first reported in USA in 1908 by Sheldon (1922) in West Virginia and spread to New York State by 1921, the spread westwards against the prevailing winds in a series of seasonal cycles is well documented in *Plant Disease Reporter*. Perennation is thought to be largely by mycelium persisting on foliage, rather than by sexual means. On mature leaves in the autumn haustoria were found deeply penetrating the palisade and even vascular tissue (Klika, 1922). The disease was found on Prince Edward Island by Hurst (1931); large numbers of *Trifolium* species were infected probably by separate biotypes of *E. trifolii* since Salmon (1903) showed that powdery mildew on *T. pratense* is a separate race restricted to this species and therefore highly specialized. Also Searle (1920) showed that conidia from *T. pratense* would not infect *T. hybridum*, *T. repens*, *Vicia faba* or *Onobrychis viciifolia*; conidia from *T. hybridum* would not cross-inoculate to *T. repens*.

The disease led to a failure to set seed and loss of foliage (Hurst 1930), host clones from Quebec were found to be resistant as were European cultivars. Several reports (e.g. Mains 1923) have suggested the European red clover is more resistant than North American. The probable explanation for this is that in North America the pathogen population does not possess the necessary virulences to overcome any resistance in European material. The reverse situation may apply in Europe since in England some American cultivars possess a high level of resistance (Dixon, unpublished data).

Recently Carr (1971) has given the following excellent description of symptoms and colonization

"symptoms: yellow mottling following initial penetration of the germ tube, penetration is direct by an appressorium and infection peg leading to establishment of functional haustoria which are bulbous and lack digitate processes. A prostrate whitish grey weft of superficial mycelium develops which forms secondary appressorial swellings and haustoria at frequent intervals. From the prostrate mycelium erect conidiophores arise abstricting elliptic uninucleate hyaline conidia only one of which is fully formed and mature at any one time on the conidiophore. Conidia can germinate within a few hours of being shed producing an haustorium in under 12 hours, while sporulation can commence within 5 days of infection. The pathogen is tolerant of a wide humidity range but sporulation is favoured by moderately high temperatures and low humidity and inhibited by rain. Host leaves lose their green colour some time after infection becoming chlorotic and finally dry and brown. At this time and generally late in the season cleistothecia appear. Often several hundreds per leaflet and visible to the naked eye as small globose structures initially light straw coloured turning dark and carbonaceous. A normal meiosis followed by mitosis occurs in the young ascus, 7 or 8 chromosome pairs being involved with each ascospore being delineated around one nucleus, only the unallocated nuclei remaining in the residual ascus cytoplasm."

Carr feels that cleistothecia, at least in the Aberystwyth area of Wales, are more common than suggested by Sampson and Western (1954) but are of doubtful significance in disease epidemiology, the main function being to provide a mechanism of genetic reassortment.

Asexual dispersal by conidia is more active in daylight especially from 0800 to midday than in darkness. Closure of clover leaves at night appreciably reduces chances of infection (Horsfall, 1930).

Some effects on yield have been studied, in the USA Yarwood (1931) found that green weight yield was inversely proportional to the incidence of powdery mildew, while seed yield was less badly affected. Estimates ranging from 2–10% fresh weight yield losses were made by Horsfall (1930), while in Canada, Berkenkamp (1971 and 1972) estimated losses at 2–3%. O'Rourke (1977) suggests the disease adversely affects yield and quality in Eire, Germany and Australia. The disease is especially prevalent in Eire from midsummer onwards particularly in periods of warm dry weather. This is also true in Great Britain where on susceptible cultivars extensive infections are noticed from July onwards (Dixon, unpublished). These outbreaks may be correlated with the optimum temperature of 24°C for conidial germination, mycelial growth and sporulation (Stavely and Hansen, 1966a). It is also thought that high levels of soil nitrogen and phosphorus favoured disease while potassium retarded it (Sturm, 1958).

Considerable biochemical and physiological studies have been made by Stavely, Hanson and co-workers. They showed that E. trifolii has physiological races differing in host range within and between species (Stavely and Hanson, 1967a); and found that isolates from red clover could infect 46 other legume hosts but not Lathyrus. The same authors identified 12 physiological races on six different clones of red clover. Intercrossing these clones showed resistance was dominant and mostly monogenically inherited. Resistance in the American cultivar Lakeland was, however, multigenic giving a high level of field resistance irrespective of the race involved. Resistance to races 1, 3, 7, 8 and 10 was monogenic and dominant (Stavely and Hanson, 1966c). In one case resistance was governed by two complimentary dominant genes, resistance to race 6 in both hetero- and homozygous resistance clones appeared to be a hypersensitive reaction in the epidermal cells. No differences in the penetration of resistant and susceptible clones of *T. pratense* by *E. polygoni* were found by Stavely *et al.* (1969). An amorphous collar quickly formed at the penetration site around the invading fungus which invaginated the host plasmalemma. The host nucleus was located close to the penetrating fungus. Resistant and susceptible reactions could be differentiated soon after the fungus had penetrated the host cell wall and passed through the collar. In resistant cells the plasmalemma and nuclear membranes were rapidly destroyed and the grana dissociated from their normal stacked arrangement. Mitochondria and most of the haustorial sheath remained intact in resistant cells, haustoria in the resistant cells were smaller, less branched and had less well developed endoplasmic reticula than haustoria in susceptible cells. The sheath membranes in resistant cells never developed the vesicular invaginations that formed in sheath membranes of susceptible cells. The pathogen apparently did not develop a nutritional relationship in resistant cells. In the susceptible combination, invasion occurs relatively quickly but with this compatible association the host cell is not rapidly killed.

Earlier Stavely and Hanson (1967b) could find no differences in protein and phosphatase, peroxidase, malate dehydrogenase or glucose-6-phosphate dehydrogenase between resistant and susceptible non-inoculated red clover clones. Twenty four hours after inoculation, the number of protein and malate dehydrogenase bands in polyacrylamide gel electrophoresis was the same as before inoculation. The numbers of peroxidase and glucose-6-phosphate dehydrogenase bands increased in susceptible leaves while remaining the same in resistant leaves. An additional acid phosphatase band was observed in both susceptible and resistant leaves following inoculation. No peroxidases were detected in conidial extracts. After infection both susceptible and resistant clover epidermal cells stain with Azure B. Uninfected cells do not take up the stain; susceptible infected cells go blue and then purple after 2–3 weeks, in resistant cells the periphery goes greenish after 1–2 days and then purple after 4–7 days (McKeen and Leong, 1972). Prior inoculation of red clover with Bean Yellow Mosaic Virus retarded powdery mildew infection (King *et al.*, 1964). Conidia germinated on virus infected leaves but disintegrated following necrosis of the leaf tissue in the immediate area of the germinating conidium.

It has been noticed that seed treatment with boric acid, ammonium molybdate, potassium permanganate and copper sulphate increased resistance by red clover to mildew (Kivi, 1963). It is likely however, that the main means of control will be by resistant cultivars. Stavely and Hanson (1966b) studied resistance in the American cultivars Dollard, Graham's Mammoth, Kenland, Lakeland and resistant selections from Wisconsin University. In Great Britain, Toynbee-Clarke (1973) found Sabtoron to be susceptible, with somewhat lower levels of infection on Essex Broad Red and Maris Gemini; most tetraploids were free of infection although Maris Leda was moderately infected. Assessments over several seasons have shown the following cultivars to be resistant to the present mildew population in Great Britain: Aberystwith S123, Hungaropoly, Kuhn, Lakeland, Ostaat Treu, Ottawa, Red Head, Robusta, Teroba and Triel (Dixon, unpublished data). No infections were found in trials of white (*T. repens*) and alsike clover (*T. hybridum*).

Trifolium species have been reported as infected by a range of *Erysiphe* species (*E. communis, E. pisi, E. martii* and *E. polygoni*). These reports are summarized by Hirata (1966), the more important have been Iraq (Al Hassan, 1973) including *T. fragiferum*, Israel (Chorin and Palti, 1962) on *T. alexandrinum*, Kenya (Nattras, 1961) and Turkey (Oran, 1974).

Other herbage legumes, such as lucerne (*Medicago*) and vetch (*Vicia*) have also been found to be infected with *Erysiphe* species (see Boughey, 1946; Connors, 1967; Hirata, 1966; Oran, 1974). On perennial clover species *Microsphaeria alni* has been identified, (Anon, 1960), while *L. taurica* can cause disease in all herbage legumes especially in Mediterranean, Middle and Far Eastern Regions, Afghanistan (Gatani, 1962); Iran (Ershad, 1971, who referred to the pathogen as *L. leguminosarum*.); Israel (Chorin and Patti, 1962); Sudan (Tarr, 1955, Zapromentoff, 1930).

IV. Umbelliferae

Most crops in this family are attacked by *Erysiphe heraclei*, also referred to as *Erysiphe umbelliferarum* (Kapoor, 1967b), hosts include carrot, fennel, parsely and other umbellifers. There is a world wide distribution (Hirata, 1966), six biologic forms were distinguished by Hammerlund (1925) while Marras (1961) recognized three host strains on carrot, fennel and parsley.

A. Carrot

Powdery mildew usually attacks Palestinian carrots when 10–12 weeks old and was especially severe on winter sown crops (Rayss, 1940). Various recommendations for chemical control were reported, notably the use of maneb and morestan mixtures and triphenyltin compounds. The disease is still a problem since Netzer and Katzer (1966) estimate powdery mildew (*E. umbelliferarum*) as one of the two major foliar diseases of Israeli carrots. In Mediterranean Regions carrots are infected, thus Matta (1962), describes outbreaks on three carrot cultivars: Nantes, Palesean and Nantes × Palesean in experimental plots at Turin, Italy, reference is also made to the perfect stage and the pathogen named as *E. polygoni*. In Great Britain occasional infections are seen on carrots, Hawkins and Phillips (1960) found outbreaks in the Black Fen Region of East Anglia which caused little obvious crop loss but encouraged premature foliar senescence. Severe infections were also seen on East Anglian crops following the dry summer of 1976 (Dixon, unpublished data). The same authors also reported the disease as having occurred in Jersey, and Bernaux (1965) reported the disease in France.

B. Parsnip

In recent years parsnip crops in Great Britain have been severely infected with powdery mildew (presumably *E. heraclei*): Davies, pers. comm. (1976) described an infection at Newmarket, use of fungicides gave no significant yield response but there was a trend towards increased total weight and larger root sizes in sprayed plots. In cultivar trials Avonresister remained relatively uninfected compared to most other commercially available cultivars (Bedford, pers. comm.).

C. Angelica, celery, dill, fennel and parsley

All five of these crops have been sporadically reported as badly infected with *E. heraclei* (Hirata, 1966). Matta and Garibaldi (1969) reported severe losses to crops of dill in 1967 in the Albenga region of Italy. Similar losses occurred to fennel crops (Noviello, 1961) which is an increasingly essential oil crop in Southern Italy, especially Sicily. Symptoms include the development of a white mycelial cobweb on the leaves, the outer leaves being attacked first, young plants are either killed or reduced in vigour. In common with many powdery mildew diseases rain causes a regression of disease development.

D. Coriander

Coriandrum sativum is an important crop in areas of India and Pakistan where it may be attacked by several powdery mildew pathogens. Infections of coriander in the Agra region at the end of March, by *E. polygoni* were described by Gupta and Dalela (1962). The pathogen spread rapidly on the leaves, stems and peduncles, development of the flowers and fruits was retarded and they remained immature. Late sown crops were especially affected, disease losses of 83% were recorded. Similar heavy losses were reported by Srivastava *et al.* (1972) in Rajasthan, India, there was a reduction in seed yield and also loss in the colour and seed quality. A grading system was given for disease evaluation and formulae for calculating percentage disease index and control; spraying with sulphur fungicides produced some measure of control.

L. taurica will also cause powdery mildew disease of coriander. It was reported by Khan and Kamal (1962) in Pakistan who noted white powdery patches on the abaxial leaf surface and yellowish discolouration on the upper surface. Lesions gradually turned brown and became obvious, ultimately the leaves died and fell off. Most umbelliferous crops are attacked by *L. taurica*, Boughey (1946) reported the pathogen on Sudanese carrots, fennel and parsley; while Chorin and Palti (1962) report *Oidiopsis taurica* on Israeli carrots, plots where moisture status was preserved by mulching had less mildew than unmulched plots with fluctuating moisture levels. Celery has been reported as infected by *L. taurica* in France, Italy, USSR and India (Hirata, 1966).

Records of infection by the imperfect *Oidial* stage occur for all the crops described and have been summarized by Hirata (1966). Seed transmission of *E. heraclei* on carrot, fennel, parsley and parsnip seed has been reported by Noviello (1961) Boerema *et al.* (1963), Noble and Richardson (1968) and Ferri (1969).

V. Solanaceae

Leveillula taurica is the major powdery mildew pathogen of this family, it is present in all dry areas around the Mediterranean and is especially serious in Israel and parts of Turkey, the oidial stage has been found in tropical countries such as Mexico (Chupp and Sherf, 1960). Disease affects mainly leaves and more rarely the stalks, flower parts or fruits. The most common symptom is a yellowish spot on the upper leaf surface and a white powdery covering on the lower surface, shedding of the foliage is a prominent symptom chiefly in pepper (*Capsicum annuum*). Wide variations, depending on host species exist in the degree of endophytism of *L. taurica* on leaves, in the size of conidiophores, conidia, cleistothecia, asci and ascospores and in the number of asci in each cleistothecium.

Geographical distribution of *L. taurica* varies greatly with host species, it is widest in both dry and humid zones on hosts such as *C. annuum* but more strictly limited to dry conditions for other hosts. The fungus is considered to have originated in warm dry regions and to adapt itself progressively to more humid, but not to cooler, conditions. Differences in cultivar resistance to *L. taurica* have been reported for tomato (*Lycopersicon esculentum*), egg plant (*Solanum melongena*) and the composite artichoke (*Cynara scolymus*) (Palti, 1971). Chilli pepper (*C. annuum*) in India was affected (Mathur *et al.*, 1972) leading to severe defoliation and reduction in size and numbers of fruits. There was a 25–30% increase in yield with the use of fungicides.

Although similar pathogenic properties have been found in isolates of *L. taurica* from tomatoes, peppers and eggplants, there seem to be distinct differences in disease development on *Lycopersicon* compared to *Solanum* species (Reuveni and Rotem, 1973). On peppers, disease developed better at high than low day humidities while shedding of infected leaves was more pronounced at low than at high day humidities. Epidemics on tomatoes developed more rapidly at low than at high day humidity but in neither case were infected leaves shed. In tomatoes a cycling 15–25°C regime was associated with a higher rate of epidemic development than a regime of 10–20°C. At a practical level Rotem and Cohen (1966) investigated the effects of irrigation on infection of tomatoes by *Oidiopsis taurica* (= *L. taurica*). Disease was 30–40% more severe on furrow-irrigated compared to sprinkler-irrigated crops. The latter may have increased the relative humidity and reduced air temperature. Differences in symptom expression between tomatoes and peppers were also noted by Palti (1971), peppers in Israel are grown in more humid areas and seasons than tomatoes. Palti postulates that peppers shed their leaves while tomatoes retain them in response to daytime humidities rather than those at night. Cross-inoculation tests were attempted by collecting infected leaves of pepper, tomato and eggplants from the field and placing them in, or on, the foliage of test plants, Palti also exposed potted plants in the field for 2–3 days. Water suspensions of conidia were a successful form of inoculum for tomato and eggplant but not for peppers. No host specificity was noted, the cross-inoculations took easily. The infection of peppers was greater under wet than dry conditions while the reverse was true for tomato. This suggests a host effect if there was no isolate specificity.

Other geographical areas where the disease was reported included tomatoes in Dominica (Castellani, 1958) and peppers in Tacua Province of Peru (de Segura, 1961). Symptoms caused by *Oidiopsis taurica* were described as a white powdery covering of the lower leaf surface while the corresponding upper surface turned yellow; infected tissue became necrotic and diseased leaves withered and dropped leading to severe defoliation. Zwirn (1943) thought there were at least 12 strains of *L. taurica* which could be distinguished due to host specificity–this conflicts with Palti's work. The pathogen was markedly xerophytic, conidia germinated best at 26°C on a dry slide with 52–75% r.h. some germination took place at as low as 30% r.h. Zwirn contended that infection does not take place through the stomata as Blumer (1933) suggested but by passage of hyphae between two epidermal cells inferring that the fungus dissolved the host cuticle and middle lamella. The pathogen disrupted host water metabolism—the transpiration rate rose but there was less water present in the diseased host, hence the leaves became shrivelled and were abscissed.

The geographical and host range of *L. taurica* has been listed by Hirata (1966). Oran (1974) commented that *L. taurica* had numbers of hosts in severe steppe climates: 72 in Persia and 162 in Kajkas region of Russia. Oran was one of the few to note *L. taurica* on potato (*Solanum tuberosum*), others were Orieux and Felix (1968) in Mauritius. Studies of host range have been made by Al-Hassan (1973) in Iraq, Allison (1952) who found *Oidiopsis taurica* on irrigated tomatoes in Iraq, Boughey (1946) in Sudan, Fahim (1969) reported *L. taurica* on tomato (cv. Moneymaker) in Egypt, Gatani (1962) in Afghanistan, Kranz (1965) in Libya, Nattras (1961) in Kenya, Tarr (1955) in Northern Sudan and Thompson and Johnston (1953) in Malaya. Unfortunately Ershad (1971) erected the name *Leveillula solanacearum* for powdery mildew on eggplant in Iran. Cultural control measures aim chiefly at planting in seasons when crops can escape infection, at reducing concentrations of inoculum by avoiding neighbouring older crops and at irrigation management.

Some exophytic powdery mildews are pathogens of the Solanaceae, in particular *E. cichoracearum* which has been widely reported as a parasite of potato (*S. tuberosum*). It has been reported by Ducomet (1920) and Marchal and Foex (1927) in France, Muller (1928) in Germany, Nattras (1934) in Cyprus, Anon (1934) in the UK, Szembel (1927) in Astrakhan, Palti and Moeller (1944) in Palestine, in North America by Kunkel (1936) in New Jersey and Valleau (1941) in Kentucky. Perithecia are seldom seen and this leads to difficulties in identification, Vanha (1903) named the pathogen *Erysiphe solani*, a name also used by Ducomet (1921) while Marchal and Foex (1927) and Valleau (1941) identified the pathogen as *E. cichoracearum*. Generally the disease is inconspicuous and confined to terminal foliage and the petioles, little commercial damage is caused. *Solanum nigrum* is more profusely colonized by what some authorities think may be a separate biotype of the pathogen since when grown together with commercial cultivars, *S. nigrum* is either attacked alone or more severely.

In the UK, Thomas (1946) reported infections on glasshouse grown seedlings at Cambridge. Some field infections were also found to the South-west of Cambridge on cultivars such as Majestic, Dunbar Rover, Abundance, Dunbar Standard, Arran Victory, President and King Edward hybrids. No infections were present on Kerr's Pink, Gladstone, Red Skin and other Solanaceous plants. Artificial infection was successful on *Nicotiana tabacum* cv. White Burley. Infections were heavier than those reported in the USA, generally disease was more severe on the upper leaf surface with round oval white patches 1–3 cm in diameter; no infections were found on the stems, and petioles. In 1959 extensive field infections were recorded at Rochester,

Kent on cv. Majestic (Baker, 1972) while in 1964 the disease was seen at Rothamsted on glasshouse grown Aran Victory.

Other reports come from Western USA (Threshow and Cannon, 1961), Chile (Bertossi, 1963), Mexico (Delgado Sanchez et al., 1969) and New Zealand (Dingley, 1960). This latter was a severe field infection with 50% of plants cv. Shurchup being infected in a ten hectare crop. Connors (1967) in Canada reported E. cichoracearum on Solanum nigrum and S. demissum in New Brunswick; S. tuberosum cultivars Green Mountain, Plymouth and Delus were also susceptible. The author referred to the pathogen as E. polyphaga after Hammerlund, but only the oidial stage was seen. Similarly, Feldman and Pontis (1960) reported the imperfect stage on mature potatoes in Argentina.

In India potato breeding stocks at Simla were infected (Dutt et al., 1973). Ulster Torch was found to be resistant whereas other commercial cultivars were susceptible. These authors postulate that resistance is of a polygenic nature on the basis of segregation in wild type seedlings.

Oran (1974) found E. cichoracearum on 32 hosts in Turkey, while species such as E. communis were recorded on tomato. Anon (1960) recorded E. cichoracearum on eggplant (S. melongena) and E. polygoni on tomato.

Roy (1965) in Assam described Oidium erysiphoides on eggplant in Jorhat. It was first seen in November as spots about 1 cm in diameter on both leaf surfaces which later coalesced to form large irregular lesions.

Collective species such as E. polyphaga have been reported on eggplant (S. melongena) in India and much earlier Zaprometoff (1930) found Oidium melongenae (or E. communis) on eggplant in Uzbekistan. Bernaux (1976, pers. comm.) reported E. polygoni on tomato at Montpellier, France, while Hirata (1966) mentioned Sphaerotheca fuliginea on S. melongena in Formosa and Japan.

VI. Compositae

The best documented powdery mildew pathogen of this family is E. cichoracearum. This is recognized by its two-spored asci and basally inserted appendages which are much longer than the diameter of the ascocarp. It resembles S. fuliginea in possessing conidia in long chains which has possibly led to confusion in some reports. Conidia of S. fuliginea produce forked germ tubes whereas those of E. cichoracearum form well differentiated appressoria (Zaracovitis, 1965). The pathogen has world wide distribution; Blumer (1967) distinguished 13 formae speciales based on a single species or a single section of a genus. A biologic form on safflower was named E. cichoracearum f. sp. carthami by Milovtzova (1937) while wild and cultivated lettuce appear to be attacked by separate strains (Schnathorst et al., 1958).

A. Lettuce

Since 1950 Schnathorst and colleagues have worked extensively on *E. cichoracearum* infecting lettuce in California, USA. The disease was thought to be confined largely to that area and many reports prior to 1951 have been shown to be erroneous, although that of Pryor (1940) was valid but based on infection of a commercial × wild lettuce (*Lactuca serriola*) hybrid (41858). The pathogen strain responsible for outbreaks on cultivated lettuce (*Lactuca sativa*) since 1951 is physiologically distinct from the Salinas Valley wild lettuce strain; it is pathogenic to cultivated lettuce cultivars and more sensitive to high temperatures. Presence of mixed infections of the two strains on wild lettuce in the field suggested that the cultivated lettuce powdery mildew was a mutant endemic to the Salinas Valley. Infections have occurred in most years since 1951, in some instances leading to severe damage. Blumer (1933) cited two outbreaks in Europe (1913 and 1914) but only the oidial stage was seen. In the USA the disease is now spreading north and south of the Salinas Valley and may have reached Arizona (Snyder *et al.*, 1952). Anon (1960) reports the disease in Michigan while Chorin and Palti (1962) reported the disease in Israel, and Hirata (1966) cited the disease in France, Greece, Switzerland and the USSR. The disease is usually first seen in August and reaches a peak in October causing great reductions in yield and quality (Miller, 1953). Infected plants have a dusty off-colour with yellowing and browning of diseased leaves and perceptible odour. If plants are infected near to maturity diseased leaves can be trimmed off but infection of young plants often leads to total crop loss. Initially only the asexual stage was found on cultivated lettuce with perithecia on wild lettuce but in 1952 perithecia were found in abundance on cv. Great Lakes. Susceptibility in cultivars was described by Schnathorst and Bardin (1958), none of the crisphead or non-heading leafy types were resistant and only butterheads such as Arctic King, Big Boston, Salad Bowl and Bath Cross (a red leaved cos type) possessed resistance. Differences in resistance may be due to a physico-chemical mechanism whereby high osmotic pressures in leaves of resistant material inhibit water uptake and growth of the pathogen (Schnathorst, 1959a). Conidial germination was highest and growth of germ tubes and mycelia most rapid at 18°C (Schnathorst 1960). The minimum temperature for infection was 6–10°C and the maximum 27°C. Moisture stress giving the best germination of conidia was in the range 95·6–98·2% r.h. between 25–30°C but a humid atmosphere (100% r.h.) inhibited germination. Development of powdery mildew was most affected by temperature but atmospheric moisture influenced the rapidity of disease development and severity. Light intensity was shown to affect resistance (Schnathorst, 1959b); the cv. Great Lakes at 13°C and 186 lux light intensity was resistant but if light intensity was reduced to 28–37 lux the plants were susceptible. Spread is primarily in the direction of the prevailing wind (Schnathorst, 1959a). Conidia could be disseminated for at least 120 miles, spores being liberated between 1200 and 1600 hours in groups of two or more. Perithecia formed equally on leaves of all ages; ascospores are also an important part of the life cycle but initial infections each season were due to perithecia formed in the previous season. Perithecia burst on absorption of water at 15–22°C while ascospores could germinate both inside and outside the ascus.

B. Other compositae

In addition to lettuce *E. cichoracearum* will attack artichokes, sunflower, chicory and endive (Chupp and Sherf, 1960). Disease can develop on seedling leaves and stems appearing on the upper surface of older but still green foliage where the white mycelial areas enlarge and will eventually cover the whole upper surface and occasionally the lower surface. Affected foliage loses lustre, curls, becomes chlorotic, then brown and dies. Apart from the USA the pathogen has been reported on sunflower in the Sudan (Tarr, 1954, 1955) as the oidial stage; on safflowers in the Jalalabad area of Afghanistan (Gattani, 1962). An extensive list of hosts and geographical locations is given by Hirata (1966).

Stokes aster (*Stokesia laevis*) which is usually thought of as an ornamental crop but has recently been looked at in USA as a potential oil crop (Gunn and White, 1974) was reported by Kilpatrick *et al.* (1975), as becoming infected by *E. cichoracearum* which caused the foliage to become chlorotic.

Considerable damage is caused to Compositae by *L. taurica* (also known as *Oidiopsis taurica* (Lev.) Salm.) and *Ovulariopsis cynarae* (Ferr. et Massa) Cicc.), chiefly in Mediterranean regions. On artichokes the pathogen causes white indeterminate flecks on the lower foliage but owing to the hairiness of the leaves such flecks are difficult to detect in the early phases of invasion. Later, infected tissue becomes yellow but without marked lesions (Ciccarone, 1953). On this host *L. taurica* will produce long germ tubes which are superficial and do not directly penetrate stomata (Tramier, 1963). Conidiophores develop singly from external mycelium remaining unbranched and grow to 290–500 μm in length. Ciccarone (1953) found 150–160 conidiophores mm^{-2} on artichoke leaves. No cross-infection from artichoke to tomato, olive or pepper was achieved by this author but spores from *Cynara annuum* would infect *C. cardunculus* and vice versa. Germination of *L. taurica* spores may be stimulated by substances present in artichoke leaves but absent in those of peppers (Laudanski, 1957). Optimal germination with *L. taurica* spores from *C. annuum* was achieved at 100% r.h. (Clerk and Ayesu–Offei 1967). At low humidities there was a decline in germination and a reduction in mean germ tube length. When submerged in water 6% of conidia germinated as compared to 37% of dry spores kept at 100% r.h. (Nour, 1958).

Disease epidemics on artichoke are associated with limited rainfall and decreasing autumn temperatures. Older cultivars which have almost entire leaf blades and no spines are more resistant than later cultivars with lobate leaves (Ciccarone, 1953). Despite the difficulty of obtaining artifical cross-infection it is likely that there is transmission of *L. taurica* between artichoke and pepper (Tramier, 1963). In Tunisia Laudanski (1957) recommends separating these crops and keeping first and second year plantings of artichokes apart; in addition infected basal leaves should be removed. Fungicide applications should be made when conidiophores appear on four or five lesions of the lower leaves. Apart from the Mediterranean region the disease is reported on globe artichoke and sunflower in Iraq (Natour *et al.*, 1971) and at Montpellier, France (Bernaux, 1976, pers. comm.), Hirata (1966) gives an extensive list of hosts and geographical locations.

VII. Minor Crops

A vast range of tropical and temperate crops are described by Hirata (1966) as hosts of powdery mildew fungi. Since then descriptions have been made of *E. polygoni* on *Arachnis hypogea* (Orieux and Felix, 1968) in Mauritius; *E. martii* on *Melitotus officinalis* and *E. pisi* on *Trigonella foenum-greekum* (Ershad, 1971) in Iran; *E. cichoracearum* on *Symphytum officinale* and *Oidium* sp. on *Salvia officinale* (Baker, 1972) in England; *E. labiatarum* on *Mentha spicata* (Bernaux, 1972) in France; *E. cichoracearum* on *Hibiscus esculentus*, *L. taurica* on *Physalis peruviana*, and *Oidium* sp. on *Sesamum indicum* (Al Hassan, 1973) in Iraq; *L. taurica* on *Hibiscus esculentus*, *E. pisi* on *Lupinus* sp. *E. martii* on *Melilotus officinalis* and *M. albus* (Oran, 1974) in Turkey; *Oidium* sp. on *T. foenum-greekum* (Bernaux 1976 pers. comm.) in France.

References

Al-Hassan, K. K. 1973. *Pl. Prot. Bull. FAO*. **21**, 88–91.
Ali-Khan, S. T., Zimmer, R. C. and Kenaschuk, E. O. (1973) *Can. Pl. Dis. Surv.* **53**, 155–156.
Allison, J. L. (1952). *Pl. Prot. Bull. FAO*. **1**, 9–11.
Arny, D. C., Hanson, E. W., Worf, G. L., Oplinger, E. S. and Hughes, W. H. (1975). *Pl. Dis. Reptr* **59**, 288–290.
Anon. (1934). Bulletin 79. Ministry of Agric and Fish HMSO.
Anon. (1934). *Rep. Scott. Pl. Breed. Stn.* **1933.**
Anon. (1958). *Rep. Dep. Agric. Tanganyika* **1958.** (*II*), 16–33.
Anon. (1960). *USDA Handbk No. 165.*

Anon. (1975). *Nat. Inst. Agric. Bot.* Farmers Leaflet No. 2.
Anon. (1976). *Nat. Inst. Agric. Bot.* Farmers Leaflet No. 6.
Baker, J. J. (1972). Tech. Bull. 25. Minist. Agric. Fish Fd. HMSO.
Basu, P. K., Crete, R., Donaldson, A. G. Gairley, C. O., Haas, J. H., Harper, F. R., Lawrence, C. H., Seaman, W. L., Toms, H. N. W., Wong, S. I. and Zimmer, R. C. (1973) *Can. Pl. Dis. Surv.* **53**, 49–57.
Beaumont, A. and Hodson, W. E. H. (1928). Pamphlet 25. *Rep. Dep. Pl. Path. Seale-Hayne agric. Coll.* **1927.**
Berkenkamp, B. (1971). *Can. Pl. Dis. Surv.* **51**, 96–100.
Berkenkamp, B. (1972). *Can. Pl. Dis. Surv.* **52**, 51–55.
Berneaux, P. (1965). *Bull. Soc. mycol. Fr.* **81**, 165–187.
Berneaux, P. (1972). *Bull. Soc. mycol. Fr.* **88**, 305–314.
Bertossi, E. O. (1963). *Rev. Univ. Santiago* **48**, 41–56.
Bhander, D. S., Thakur, R. N. and Husain A. (1963). *Pl. Dis. Reptr* **47**, 1039.
Blumer, S. (1933). *Beitr. Kryptog Flora.* **7**, 1–483.
Blumer, S. (1967). "Echte Mehltan pilze (Erysiphaceae)". Fischer, Jena.
Boelema, B. H. (1963). *Fmg. S. Afr.* **38**, 65–67.
Boerema, G. H., Dorenbosch, M. M. J. and Van Kesteren, H. A. (1963). *Versl. Meded. plziekfenk. Dienst Wageningen* **1962**, 138.
Boughey, A. S. (1946). *Mycol. Pap. 14*, C.M.I., Kew.
Buhr, H. (1958). *Arch. Feunde NatGesch. Mecklenb.* **4**, 9–88.
Carr, A. J. H. (1971). Herbage Legume Diseases *In* "Diseases of Crop Plants". (J. H. Western *Ed.*) Macmillan, London.
Castellini, E. (1958). *Pl. Prot. Bull. FAO.* **7**, 33–36.
Channon, A. G. and Maude, R. B. (1971). Vegetables. *In* "Diseases of Crop Plants". (J. H. Western, *Ed.*) Macmillan, London.
Chorin, M. and Palti, J. (1962). *Israel J. agric. Res.* **12**, 135–166.
Chupp, C. and Sherf, A. F. (1960). "Vegetable Diseases and their Control". Ronald Press Co. New York.
Ciccarone, A. (1953). *Pl. Prot. Bull. FAO.* **1**, 49–51.
Clapham, A. R., Tutin, T. G. and Warburg, E. F. (1952) "Flora of the British Isles". The University Press, Cambridge.
Clerk, G. C. and Ayesu-Offei, E. N. (1967). *Ann. Bot.* n.s. **31**, 749–754.
Connors, I. L. (1967). *Can. Dep. Agric. Pub.* 1251.
Cook, H. T. (1931). *Bull. Va. Truck. Exp. Stn.* **74**, 931–940.
Cook, R. J. (1975). "Diseases of Oil Rape in Europe". Minist. Agric. Fish and Fd.
Cousin, R. (1965). *Annls. Amél.* **15**, 93–97.
Crawford, R. F. (1927). *Bull. New Mex. agric. Exp. Stn* **163.**
Darpoux, H. (1945). *Annls. Epiphyt* n.s. **11**, 71–103.
Delgado–Sanchez, S., Fucikovsky, L. and Cadena-Hinojosa, M. (1969). *Pl. Dis. Reptr* **53**, 189–190.
Demski, J. W. and Phillips, D. V. (1974). *Pl. Dis. Reptr* **58**, 723–726.
Dingley, J. M. (1960). *N.Z. Jl. agric. Res.* **3**, 461–467.
Dixon, G. R. (1974). *Pl. Path.* **23**, 105–109.
Dixon, G. R. and Furber, M. J. (1971). *J. natn. Inst. agric Bot.* **12**, 308–313.
Doling, D. A. and Large, E. C. (1962). *Pl. Path.* **11**, 47–57.
Doling, D. A. and Large, E. C. (1963). *Pl. Path.* **12**, 128–130.
Doling, D. A. and Willey L. A. (1969). *Expl. Husb.* **18**, 87–90.
Ducomet, V. (1920). *Bull. Soc. Path. vég. Fr.* **7**, 57–58.
Ducomet, V. (1921). *Bull. Soc. Path. vég. Fr.* **8**, 153–154.
Dundas, B. (1936). *Hilgardia* **10**, 243–253.
Dundas, B. (1941). *Hilgardia* **13**, 551–565.

Dunleavy, J. M., Chamberlain, D. W. and Ross, J. P. (1966). *USDA Agric. Handbk.* 302.
Dutt, B. L., Rai, R. P. and Harikishore, S. (1973). *Indian J. agric. Sci.* **43**, 1063–1066.
Ershad, D. (1971). *Iran J. Plant Path.* **6**, 50–59.
Fahim, M. M. (1969). *Proc. 1st Cong. Medit. Phytopath. Union.* 134–135.
Feldman, J. M. and Pontis, R. E. (1960). *Pl. Prot. Bull. FAO* **8**, 117–119.
Ferri, S. (1969). *Phytopath. Mediter.* **8**, 56–58.
Ford, R. E. and Baggett, J. R. (1965). *Pl. Dis. Reptr* **49**, 787–789.
Gattani, M. L. (1962). *Pl. Prot. Bull. FAO* **10**, 30–35.
Grau, C. R. and Lawrence, J. A. (1975). *Pl. Dis. Reptr* **59**, 458.
Griffee, P. J. (1969). *Trans. Br. mycol. Soc.* **53**, 491–493.
Griffiths, D. J. (1958). *Rep. Welsh Pl. Breed. Stn* **1950–1956** 90–92.
Gritton, E. T. and Hagedorn, D. J. (1971). *Crop Sci.* **11**, 941.
Gritton, E. T. and Ebert, R. D. (1975). *J. Am. Soc. hort. Sci.* **100**, 137.
Gunn, C. R. and White, G. A. (1974). *Econ. Bot.* **28**, 130–135.
Gupta, P. K. S. (1974). (*Pestic. Abs. News Summ.*) **20**, 409–415.
Gupta, J. S. and Dalela, G. G. (1962). *Agra Univ. J. Res. Sci.* **11**, 123–127.
Hagedorn, D. J. (1973). Peas. *In* "Breeding Plants for Disease Resistance: Concepts and Applications". (R. R. Nelson Ed.) Pennsylvania State Univ. Press, University Park.
Hammerlund, C. (1925). *Hereditas* **6**, 1–126.
Hawkins, J. H. and Phillips, D. H. (1960). *Pl. Path.* **9**, 111–114.
Harland, S. C. (1948). *Heredity* **2**, 263–269.
Hawre, M. P. (1971). *Ind. J. Mycol. Pl. Path.* **1**, 82–84.
Herenga, R. J., van Norel, A. and Tazelaar, M. F. (1969). *Euphytica* **18**, 163–169.
Hirata, K. (1966). "Host range and Geographical Distribution of the Powdery Mildews". Niigata University, Japan.
Horsfall, J. G. (1930). *Mem. Cornell Univ. agric. Exp. Stn.* 130.
Hurst, R. R. (1931). *Rep. Dep. Agric. Can., 1931, Botany* 183–184.
Johnson, H. W. and Jones, J. P. (1961). *Pl. Dis. Reptr* **45**, 542–543.
Johnston, T. D. (1972). *Rep. Welsh Pl. Breed. Stn* 101–102.
Junnell, L. (1967). *Svensk. Bot. Tidskr.* **61**, 209–230.
Kamat, M. N. and Patel, M. K. (1948). *Indian Phytopath.* **1**, 153–158.
Kapoor, J. N. (1967a). CMI Descriptions of Pathogenic Fungi and Bacteria No. 156.
Kapoor, J. N. (1967b). CMI Descriptions of Pathogenic Fungi and Bacteria No. 154.
Kapoor, J. N. (1967c). CMI Descriptions of Pathogenic Fungi and Bacteria No. 155.
Khan, S. A. and Kamal, M. (1962). *Pakist. J. scient Res.* **5**, 41.
Kilpatrick, R. A., Vecker, F. A. and White, G. A. (1975). *Pl. Dis. Reptr* **59**, 795.
King, L. N. Hampton, R. E. and Diachuns. (1964). *Science* **146**, 1054–1055.
Kirik, N. N. and Kitsno, V. E. (1974). *Mikol. Fitopatol.* **8**, 353–355.
Kivi, K. (1963). *Eesti põllum Akad. tead. Tööde Kogunuk* **28**, 40–52.
Klika, R. S. (1922). *Ann. Mycol. Berl.* **20**, 74–80.
Kranz, J. (1965). Phytopathological Paper No. 6. CMI Kew.
Kunkel, L. O. (1936). *Phytopathology* **26**, 392–393.
Kushwah, U. S. and Chand, J. N. (1971). *Indian Phytopath.* **24**, 200–201.
Landanski, F. (1957). *Rapp. Trav. Rech. effect. Serv. Bot. agron. Tunis 1956* **2**, 239–255.
Lehman, S. G. (1931). *J. Elisha Mitchell scient. Soc.* **46**, 190–195.
Mains, E. B. (1923). *Proc. Indiana Acad. Sci.* **1922**, 307–313.
Marchal, P. and Foex E. (1927). *Annls Épiphyt.* **13**, 383–454.
Marras, F. (1961). *Studi Sassaresi Sez III* **9**, 12pp.
Mathur, R. L. Singh, G and Gupta, R. B. L. (1971). *Indian J. Mycol. Plant Pathol.* **1**, 95–98.
Mathur, R. L. Singh, G. and Gupta, R. B. L. (1972). *Indian J. Mycol. Plant. Pathol.* **2**, 182–183.
Matta, A. (1962). *Riv. Patol. veg Ser 3* **2**, 342–344.

Matta, A. and Garibaldi, A. (1969). *Pl. Prot. Bull. FAO* **17**, 115.
Maxfield, J. E. (1968). *Diss. Abstr.* **28** (7)B. 2686.
Mckeen, W. E. and Leong, P. C. (1972). *Proc. Can. Phytopath. Soc.* **39**, 36.
McKenzie, D. L. and Morrall, R. A. A. (1973). *Can. Pl. Dis. Surv.* **53**, 187–190.
McKenzie, D. L. and Morrall, R. A. A. (1975). *Can. Pl. Dis. Surv.* **55**, 97–100.
McNaughton, I. H. (1974). *Rep. Scott. Pl. Breed. Stn* **1974**, 16–18.
Miller, H. J. (1952). *Phytopathology* **42**, 114.
Miller, P. R. (1953). *Pl. Prot. Bull. FAO* **1**, 104–105.
Milovtzova, M. O. (1937). *Trudy Inst. Bot. Khar'kiv* **2**, 7–13.
Morrall, R. A. A., McKenzie, D. L., Duczek, L. J. and Verma P. R. (1972). *Can. Pl. Dis. Surv.* **52**, 143–148.
Muller, K. O. (1928). *Nachr Bl. dt. PflSchutzdienst. Berl.* **8**, 19–20.
Munjal, R. L., Chenulu, V. V. and Hora, T. S. (1963). Indian Phytopath. **16**, 260–267.
Narain, A. and Siddiqui, J. A. (1965). *Indian Oilseeds J.* **9**, 153–154.
Natour, R. M., El-Behaldi, A. H. and Majeed, M. G. (1971). *Pl. Dis. Reptr* **55**, 192.
Nattrass, R. M. (1934). *Rep. Dir. Agric. Cyprus 1933, Mycology* 48–57.
Nattrass, R. M. (1961). Mycological Paper No. 81. CMI Kew.
Netzer, D. and Katzir, R. (1966). *Pl. Dis. Reptr* **50**, 594–595.
Noble, M. and Richardson, M. J. (1968). Phytopathological Paper No. 8. CMI Kew.
Nour, M. A. (1958) *Trans Br mycol Soc.* **41**, 17–38.
Noviello, C. (1961). *Phytopath Z.* **42**, 167–174.
Ogilvie, L. (1925). *Rep. Dep. Agric. Bermuda, 1924, Pathology*, 32–43.
Ogilvie, L. (1969). Bull. Minist. Agric. Fish. Fd. 123.
Oku, H., Ouchi, S., Shiraishi, T. and Baba, T. (1975). *Phytopathology* **65**, 1263–1267.
Oran, Y. K. (1974). *J. Turkish Phytopath.* **3**, 1–27.
Orieux, L. and Felix, S. (1968). Phytopathological Paper No. 7. CMI Kew.
O'Rourke, C. J. (1977). "Diseases of Grasses and Forage Legumes in Ireland". An Fovas Taluntais, Carlow, Eire.
Palti, J. (1971). *Phytopath Mediter* **10**, 139–153.
Palti, J. and Moeller, S. (1944). *Palest. J. Bot. Rehovot* **4**, 148–156.
Paulech, C. and Herrera, S. (1969). *Biologia Bratisl.* A, **24**, 720–727.
Paulech, C. and Herrera, S. (1970). *Biologia Bratisl.* A, **25**, 3–10.
Paxton, J. D. and Rogers, D. P. (1974). *Mycologia* **66**, 894–896.
Pierce, W. H. (1948). *Phytopathology* **38**, 21.
Pryor, D. E. (1940). *Pl. Dis. Reptr* **25**, 74.
Purnell, T. J. and Preece, T. (1971). *Physiol. Plant Path.* **1**, 123–132.
Rayss, T. (1940). *Palest. J. Bot. Jerusalem* I, 313–315.
Reuveni, R. and Rotem, J. (1973). *Phytopath Z.* **76**, 153–157.
Riley, E. A. (1960). Mycological Paper No. 75 CMI. Kew.
Rotem, J. and Cohen, Y. (1966). *Pl. Dis. Reptr* **50**, 635–639.
Roy, A. K. (1965). *Pl. Prot. Bull FAO* **13**, 42.
Salmon, E. (1900). *Mem. Torrey bot Club* **9**, 1–292.
Salmon, E. (1903). *Beih. bot. Zbl.* **14**, 261–315.
Sampson, K. and Western, J. H. (1954). "Diseases of British Grasses and Herbage Legumes". Cambridge University Press, Cambridge.
Sankla, H. C., Singh, H. G., Dalela, G. G. and Mathur, R. L. (1967a) *Pl. Dis. Reptr* **51**, 800.
Sankla, H. C., Singh, H. G. and Sharma, L. C. (1967b). *Indian Phytopath.* **20**, 251.
Schnathorst, W. C. (1959a). *Phytopathology* **49**, 464–468.
Schnathorst, W. C. (1959b). *Phytopathology* **49**, 562–571.
Schnathorst, W. C. (1960). *Phytopathology* **50**, 304–308.
Schnathorst, W. C. and Bardin, R. (1958). *Pl. Dis. Reptr* **42**, 1273–1274.
Schnathorst, W. C., Grogan, R. G. and Bardin, R. (1958). *Phytopathology* **48**, 538–543.

Schroeder, W. T. and Providenti, R. (1965). *Phytopathology* **55**, 1075.

Searle, G. O. (1914). *Rep. con. Mycol. S.-E. agric. Coll. Wye, Kent* **1913–1914**.

Searle, G. O. (1920). *Trans. Br. mycol Soc.* **6**, 274–294.

Segura, de B. C. (1961). *Pl. Prot. Bull. FAO* **9**, 130–131.

Sequiera Da Silvade, M. P. (1966). *Agronomia lusit.* **28**, 145–168.

Sharma, H. C. and Shrivastava, S. F. (1968). *JNKVV Res. Jnl.* **2**, 69–70.

Sharma, L. C., Mathur, R. L. and Gupta, S. C. (1972). *Indian Phytopath.* **25**, 600–601.

Sheldon, S. (1922). *Pl. Dis. Bull.* **6**, 53.

Singh, B. P. and Kaurav, L. P. (1973). *Sci. Cult.* **39**, 228–229.

Smith, C. G. (1969). *Trans. Br. mycol. Soc.* **53**, 69–76.

Smith, G. (1900). *Bot. Gaz.* **29**, 153–184.

Snyder, W. C., Bardin, R. and Grogan, R. G. (1952). *Pl. Dis. Reptr* **36**, 321–322.

Srivastava, U. S., Rai, R. A. and Agraweit, J. M. (1972). *Indian Phytopath.* **24**, 437–440.

Stavely, J. R. and Hanson, E. W. (1966a). *Phytopathology* **56**, 940–943.

Stavely, J. R. and Hanson, E. W. (1966b). *Phytopathology* **56**, 309–318.

Stavely, J. R. and Hanson, E. W. (1966c). *Phytopathology* **56**, 795–798.

Stavely, J. R. and Hanson, E. W. (1967a). *Phytopathology* **57**, 193–197.

Stavely, J. R. and Hanson, E. W. (1967b). *Phytopathology* **57**, 482–485.

Stavely, J. R., Pillai, A. and Hanson, E. W. (1969). *Phytopathology* **59**, 1688–1693.

Stevens, F. L. and Hull, J. G. (1921). Diseases of Economic Plants. p. 213, Macmillan, New York.

Sturm, H. (1958). *Z. Acker-u PflBau* **107**, 203–240.

Szembel, S. J. (1927). *Comment Inst. Astrakhanensis a Defensioneum Plantarum.* I. 44–48.

Tarr, S. A. J. (1954). *Pl. Prot. Bull. FAO* **161**–165.

Tarr, S. A. J. (1955). "The fungi and plant diseases of the Sudan". CMI Kew.

Thomas, D. G. (1946). *Nature Lond* **158**, 417–418.

Thompson, E. G. (1947). *J. natn. Inst. agric. Bot.* **5**, 129–133.

Thompson, K. F. (1963). *Rep. Pl. Breed. Inst.* **1961–1962**. 51–53.

Thompson, A. and Johnston, A. (1953). Mycological Paper No. 52. CMI. Kew.

Threshow, M. and Cannon, O. S. (1961). *Pl. Dis. Reptr* **45**, 354–355.

Tomlinson, J. A. and Webb, M. J. W. (1958). *Nature Lond.* **181**, 1352–1353.

Toynbee-Clarke, G. (1973). *Rep. Pl. Breed. Inst.* **1972**. 114.

Tramier, R. (1963). *Annls. Épiphyt.* **14**, 355–369.

Valleau, W. D. (1941). *Phytopathology* **31**, 357–358.

Vanha, J. (1902). *Dt. ZuckInd.* **27**, 180–187.

Van Hook, J. M. (1906). *Bull. Ohio agric. Exp. Stn* 173.

Vasedeva, R. S. (1962). *Rep. Agric. Res. Inst. New Delhi* **1958–1959**. 131–147.

Wahl, von C. (1921). *Z. PflKrankh.* **31**, 194–196.

Walker, J. C. and Williams, P. H. (1965). *Pl. Dis. Reptr* **49**, 198–201.

Walker, J. C. and Williams, P. H. (1973). Crucifers. *In* "Breeding Plants for Disease Resistance: Concepts and Applications". (R. R. Nelson *Ed.*) Pennsylvania State University Press, University Park.

Wark, D. C. (1950). *J. Aust. Inst. agric. Sci* **16**, 32–38.

Whetzel, H. H. (1921). *Rep. Dep. Agric. Bermuda* **1920**, 30–64.

Whitehead, T. (1940). *Welsh J. Agric.* **16**, 99–110.

Willey, L. A. (1953). *J. natn. Inst. agric Bot.* **6**, 468–474.

Williams, P. H., Walker, J. C. and Pound, G. S. (1968). *Phytopathology* **58**, 791–796.

Wijngaard, Tj and Ellen, J. (1969). *Pl. Soil* **30**, 143–144.

Yarnell, S. H. (1962). *Bot. Rev.* **28**, 467–537.

Yarwood, C. E. (1931). "Powdery mildew of red clover". M.S. Thesis Purdue University.

Yarwood, C. E. (1936). *Phytopathology* **26**, 845–859.

Yarwood, C. E. (1949). *Phytopathology* **39**, 780–788.

Yarwood, C. E. (1957). *Bot. Rev.* **23,** 235–301.
Yu, T. F. (1946). *Phytopathology* **36,** 370–378.
Zaprometoff, N. G. (1930). *Cotton Industry Tashkent* **1,** 143–145.
Zaracovitis, C. (1965). *Trans. Br. mycol. Soc.* **48,** 553–558.
Zaumeyer, W. J. and Thomas, H. R. (1957). Tech. Bull. 868 USDA.
Zwirn, H. E. (1943). *Palest. J. Bot. Jerusalem* **3,** 52–53.

Chapter 20

Powdery Mildew of the Vine

J. BULIT and R. LAFON

INRA Bordeaux Station de Pathologie Végetale, France

I. Introduction

Although powdery mildew was the first really serious disease to be encountered in vines, an effective treatment was quickly discovered and this has scarcely received any modifications since. As a result of this there have not been many further original studies of the disease, after the many valuable early investigations. Instead there has been a tendency to reiterate this early data.

This report on grape powdery mildew is not the result of our own latest research but a summary of the information already available. It includes relatively old data from Viala (1893), Galloway (1895), Yossifovitch (1923), Marguerite Gaudineau (1946), Delp (1954) as well as recent material by

Boubals (1961), Toma (1974), Oku *et al.* (1975). The excellent reviews by Arnaud and Madeleine Arnaud (1931), Viennot-Bourgin (1949), J. Lafon *et al.* (1966), Galet (1973–1975) and Geoffrion (1976) were consulted for the disease, and those of Torgeson (1969) and Lhoste and Lambert (1970) in the case of the fungicides.

II. Historical Review

Although the fungus responsible for grape powdery mildew, *Uncinula necator* (Schw.) Burr., seems to have always existed in North America on different species of *Vitis*, it was first described in only 1834 by Schweinitz in his "Synopsis fungorum americ. bor.". Very few significant studies of the myco-logical flora of the United States had been published before this work. In addition the species of *Vitis* cultivated on this continent during the last century were not particularly susceptible to powdery mildew. Thus it was hardly surprising that the disease was not reported earlier in its country of origin.

Not all authors, however, agree on the geographical origin of *U. necator*. Hesler and Whetzel in 1917 considered that the fungus came from Japan (Arnaud and Arnaud, 1931). Be that as it may, it was the spread of powdery mildew into Europe and the vineyards of *Vitis vinifera* which ascribed real economic significance to this disease.

The story of the European invasion by grape powdery mildew is well known (Arnaud and Arnaud, 1931; Galet, 1973; Enjalbert *et al.*, 1974; Geoffrion, 1976). It started in 1845 in the glasshouse vines of Margate, situated near to the mouth of the River Thames, in England. It was here that the gardener Tucker observed the disease for the first time in Europe. The discovery was published in 1847 in the "Gardeners Chronicle", by the botanist Berkeley who described the fungus responsible under the name of *Oidium Tuckeri* (sic).

A feasible hypothesis is that the development of steam-boats and the increase in trade between Europe and the New World was responsible for this initial localization of the disease near to a great European sea port. Spread of the disease was then extremely rapid. By 1847, it has been reported in France, near to Paris, in the glasshouses of J. de Rothschild, and from 1850 powdery mildew was known in all the vine-producing regions of France and Europe. Later, vines all over the world came more or less rapidly under attack and at the present time the disease may be observed in all climates, including the tropics, wherever vines are grown.

The psychological and economic consequences of the appearance of powdery mildew in Europe were of great significance. It was in fact the first time that a serious disease had invaded vineyards and the vine growers, ignorant of the cause of the injury, were scared by the mysterious and seemingly inevitable nature of the disease. The economic consequences of grape powdery mildew were extremely serious. Whereas in 1847, the French wine harvest had been $54·8 \times 10^6$ hl this was reduced to only $10·8 \times 10^6$ hl in 1854 when powdery mildew became really severe. Certain ruined vine-growers chose to emigrate to North Africa or South America as a result.

The term "Vine Disease" was employed to describe grape powdery mildew thus recognizing it to be the major enemy of this plant. Later on, the discovery of sulphur treatment diminished its economic importance. The term "Vine Disease" was then successively accorded to *Phylloxera vastatrix* P. L. in 1868, and then to downy mildew (*Plasmopara viticola* (B. and C.) Berlese and de Toni) in 1878 when these diseases appeared with such severity in France.

It is somewhat remarkable to note that the sexual form of *Oidium Tuckeri* was only discovered in Europe by Couderc in 1892, 47 years after its intro-duction. As a result of this observation the European disease could be finally identified with the American one.

III. Appearance of the Disease and Vine Injury

Powdery mildew may be observed on all the green parts of the vine: leaves, branches, inflorescences and grapes. The extent of the damage varies con-siderably according to the organs affected and the period of attack. Powdery mildew is especially destructive to the grapes causing them to burst and sub-sequently to dry up or to rot.

1. Leaves

The leaves may be attacked at any age or time—essentially on the lamina, both surfaces of which are equally vulnerable (Fig. 1). In the spring, the first indications of the disease are highly discrete, scattered blemishes, dull dark green in colour, on the upper surface of the lamina. These indicate the presence of the parasite when it cannot yet be seen with either the naked eye or a lens. These very first symptoms are usually followed by varying degrees of blistering of the middle part of the lamina and crinkling of the margin. The crinkled regions exhibit small yellowish or brownish necrotic spots when viewed against the light.

The symptoms arise because the parasite only attacks the epidermal cells, checking their growth, whereas the underlying parenchymatous tissue con-tinues to develop. The resulting swelling and stretching of the attacked zones makes the epidermis smoother, appearing more shiny.

18A (8 pp.)

Fig. 1.   Appearance of contaminated leaf, (a) upper surface. (b) lower surface.

In time the symptoms become more characteristic, especially if climatic conditions favourable to the parasite set in. The laminal blemishes expand becoming clearly visible. They develop a covering of powdery felt which is whitish on the upper surface and grey on the lower one. This felt represents the parasites' external mycelium, conidiophores and conidia. The affected leaves curl up and become twisted.

Another phase may sometimes be observed on the leaves at the beginning of summer. The lamina exhibits colourless regions which may be several mm to one cm in diameter, resembling the "oil-spot" appearance associated with downy mildew (*Plasmopara viticola* (B. et C.) Berlese and de Toni) although the discolouration is less pronounced than in the latter disease (Arnaud and Arnaud, 1931).

All the authors agree that grape powdery mildew confined to the leaves is in itself of little importance and does not justify the application of special treatments. However foliage attack generally heralds the certain spread of the disease, notably to the racemes, thus presenting a serious menace to the harvest which must be taken into consideration (Galet, 1973).

2. Branches

New branches are attacked by the parasite before maturation of the wood occurs although traces of powdery mildew may remain on the shoots until the end of the following winter.

Sometimes at the beginning of spring young shoots may be observed which are completely whitened due to invasion by powdery mildew. These have been given the name of "flags", no doubt because their deformed leaves orientate themselves in the same direction in relation to the branch, and also because they attract attention from a distance (Viala, 1893). They are characteristic but tend to be few in number.

Prior to maturation of the wood, weakly-defined circular or unilateral blemishes, which stand out against the green background, can usually be seen on the branches. Although whitish to begin with, they quickly turn brown. Their extension is peripheral as the mycelium which causes them gradually dies off in the centre.

After maturation of the wood these blemishes (Fig. 2), which may be from one to several centimetres in size and either isolated or confluent, become dark brown or reddish in colour and their periphery fibrillar in appearance. The parasitic mycelium has disappeared by this stage leaving only the cells destroyed along its course which give the blemishes their characteristic appearance and colour.

As in the leaves, powdery mildew injury in the branches is also usually unimportant. However it should be noted that in the event of an early and

Fig. 2.   Symptoms on mature wood.

severe attack, the branches stop growing, do not mature completely and may
even dry out.

3. Inflorescences

The inflorescences may sometimes be parasitized by powdery mildew before
fertilization has occurred. Attack is usually partial, affecting only those parts
of the inflorescence that are very close together. The fungus envelops this part
in a light greyish white felt causing the flowers to dry up and drop off. The
fungus remains on the stalk which becomes covered with blemishes similar to
those on the branches.

Although destruction of the inflorescences at an early stage of their
development is rather exceptional in vineyards, it is more frequently reported
in glasshouse vines (Viennot-Bourgin, 1949).

4. Fruit

The grapes are susceptible to powdery mildew attack from the time of setting
to ripening and it is the injury caused to these organs which is responsible for
the great economic significance of the disease. Although the green berries are
no more receptive than other parts of the vine, the whole of their surface is
vulnerable to infection and their situation is highly conducive to attack.

After setting, when they are scarcely 2–3 mm in diameter, the young infected grapes take on an oily ash green colouration and soon become covered
with numerous conidial fructifications. The mass of conidia on the surface of
the skin gives the impression that the grapes are sprinkled with flour and this
impression is enhanced by the smell of mouldy flour released by the diseased
grapes. Many of the attacked berries dry up and fall. Those which do not drop
off remain tiny and their skin thickens considerably and hardens (Fig. 3).

Fig. 3.   Damage to grapes.

The development of powdery mildew has a very serious effect on the grapes
especially during the highly active swelling phase and before the closing of the
cluster. Because the epidermal cells are killed by the fungus, the epidermis can

no longer extend, by cell multiplication, at a sufficient rate to keep up with the swelling of the grape. Under the internal pressure of the expanding pulp it splits, the rupture extending deeply into the pulp sometimes as far as the pips. In general the grapes burst in two or sometimes three or four parts. Depending on the climatic conditions the split grape then either dries up or rots.

At the ripening stage the berries practically stop swelling and powdery mildew attack no longer causes bursting although it marks the fruit and reduces the quality of table grapes. By the end of the ripening phase the green matter has disappeared from the external layers of the skin and the grapes are no longer subject to attack.

To summarize, the importance of powdery mildew lies in the continuous parasitic pressure that it exerts on the aerial parts of the vine throughout the period of active vegetative growth. The constant presence of the parasite weakens the plant and the outcome of the disease, if this affects the grapes, is to destroy the harvest.

IV. The Parasite Responsible: *Uncinula necator* (Schw.) Burr.

A. Taxonomic classification and characteristics

The fungus causing powdery mildew in vines is a Perisporial Ascomycete belonging to the genus *Uncinula*, in the family Erysiphaceae. The characteristic features determining its taxonomic position may be summarized as follows (Viennot-Bourgin, 1949):

1. The vegetative apparatus

This consists of a mycelium of slender septate hyphae which extends over the surface of the green parts of the vine. This aerial mycelium, which is always external to the parasitized organ, forms a more or less dense felt, white to greyish in colour. Specialized suckers, the haustoria penetrate the epidermal cells thus ensuring simultaneous nutrition and anchorage to the organ.

2. The reproductive apparatus

This gives rise to conidia (or oidia) and ascospores. The mycelium produces the conidial, or Oidium-type, stage regularly and over a long period whereas the ascospore stage is formed less frequently.

The Oidial form consists of upright mycelial strands which develop into conidiophores, the basal segment being topped by a club-shaped element which divides to give short chains of simple hyaline conidia.

The perithecia are superficial closed receptacles containing several asci and an irregular number of ascospores. They are furnished on the outside with hyaline fulcra which are bent back in a simple or double crosier at their tips.

B. Microscopic description

1. The external mycelium

The mycelium appears in the form of slender hyaline filaments, 4–5 μm in diameter, slightly septate, with generally thin walls which may thicken, however, in places in the autumn. During mycelial growth the hyphae divide abundantly becoming tangled on the surface of the parasitized organs and forming a felt which is white in colour becoming greyish-white with age. The mycelial hyphae send out suckers at various places along their path which penetrate the epidermal cells. The suckers appear as mycelial diverticula consisting of an external, swollen, more or less lobate, *appressorium* which is applied to the epidermis and extends into the epidermal cell by means of an ovoid *haustorium*.

2. The conidiophores

From the mycelium develop the conidiophores. These are single filaments of 10–400 μm in length which are hardly differentiated from the mycelium (Yossifovitch, 1923). At their base they are of approximately the same diameter as the mycelial hyphae but are much more swollen at their distal end. They are flexuous at their base and may remain more or less prostrate before straightening up perpendicularly to the parasitized support. They remain continuous with the mycelium for quite a long time as the partition which isolates them from the hypha only develops when the conidial chain is almost complete (Viala, 1893).

3. The conidia

These split off progressively from the tips of the conidiophores. The first septum develops some distance from the tip and is followed by others, on average three of four, lower down. The last conidium is almost formed before all the septa are visible. The conidia remain linked in short chains of several, usually three to five, elements. During formation their contents are granular in appearance becoming highly vacuolate after liberation. They are cylindro-ovoid to ovoid in shape and measure 28–40 × 14–16 μm. The size of the conidia is dependent on the nature of the organ attacked. A special study carried out in Rumania (Toma, 1974) revealed that the conidia produced by powdery mildew on the leaves are larger in volume than those produced on the grapes:

Conidia on leaves
(mean derived from 7 varieties 32·42(±0·32) × 16·24(±0·18) μm from 1961 to 1967).
Conidia on grapes
(mean derived from 6 varieties 31·76(±0·33) × 15·36(±0·17) μm from 1961 to 1967).

However the relationship length:breadth does not vary at all, remaining between 2 and 2·1.

4. The perithecia

These are closed spherical conceptacles, the part in contact with the supporting organ being slightly flattened. They measure on average about 100 μm in diameter. Although colourless at the onset of their formation they then become yellowish before browning and finally appearing black when viewed with a lens or the naked eye. They remain superficial and are attached, at their flattened base, by numerous hyphae to the mycelial felt.

The perithecia bear fulcra of varying number, about ten on average, which are inserted slightly above their base. These fulcra appear as septate filaments being rectilinear and brown at their base becoming crosier-shaped and almost colourless at the tip. Their length varies from three to seven times the diameter of the perithecium. This bending of the fulcra into a crosier-shape is characteristic of the genus *Uncinula* (Viennot-Bourgin, 1949).

At maturity the perithecia contain six pear-shaped asci on average (sometimes five, rarely eight), each 48–55 × 36–44 μm in size. Each ascus contains four to eight ascospores, often six, which are ovoid and hyaline, measuring on average 15 × 10·5 μm.

To complete this description of the fungus *U. necator*, the relative frequency of appearance and the importance of the various structures should be mentioned. The mycelium and the conidial stage which directly follows it are both observed frequently. They justify the name of Oidium assigned to the disease and are responsible for the symptoms and damage observed on vines.

The ascospore stage is far less frequent. Regular and abundant formation of perithecia is still a rare occurrence in the majority of vineyards. The sexual phase is not essential to fungal development and because of this its role is minimized.

V. Development of Epidemics

Colonization and spread of powdery mildew over the vine depends essentially on the parasitic inoculum available at a given time, the climatic conditions likely to affect its development and also the receptivity of the host-plant.

A. Overwintering of the fungus

The perithecia are obviously the usual and most effective means of assuring conservation of the fungus throughout the winter. However their existence is not always certain. In Europe they were only discovered approximately 50 years after the introduction of powdery mildew whereas they had been known for a long time in North America. At the present time they are produced regularly and abundantly in all the vine-growing areas of Rumania. In France, they may be found almost every year in certain regions, and practically never in others. The same goes for most of the other vine-growing countries of Europe.

Although perithecia may be borne on all the parasitized parts of the vine, the conditions necessary for their formation have not yet been clearly determined. According to the observations available, which are unfortunately very few both in time and space, it appears that the formation of perithecia is induced by the setting-in of conditions that are unfavourable to the parasite. Such conditions may act directly on the parasite or through the intermediate of the plant host (Arnaud and Arnaud, 1931).

If and when perithecia are produced it is certainly possible that they play a part in the conservation of the fungus. They usually appear in July–August and release their ascospores through until April–May of the following year, that is, after the reinitiation of vegetative growth of the vine. Artificial infection studies in Rumania have demonstrated the role played the ascospores although it has not yet been possible to see if they actually cause the initial springtime infections (Toma, 1974).

Although it is not yet possible to accord general significance to the overwintering of powdery mildew in the form of perithecia, it is, on the other hand, well accepted that the fungus overwinters as mycelium in certain dormant buds on the vine branches. Mycelium may in fact be observed in the buds in both winter and spring. Even conidia are to be found in all parts of the buds although they are usually localized on the internal surface of the protective scales (Lafon et al., 1966). The mycelium remains inactive during bud dormancy but is reactivated in the spring at the onset of vegetative growth. It develops with the bud and spreads over the young shoot as it expands. The ultimate result of invasion of these young shoots by the mycelium harboured in the source buds is the appearance of the so-called "flags". This form of winter conservation is similar to that of apple powdery mildew (*Podosphaera leucotricha* (Ell. et Ev.) Salm.), also responsible for the first mildewed shoots observed in the spring.

It has also been shown that the buds at the base of the branches are most frequently infected by powdery mildew.

Bud whorl from the branch
  base upwards   1 2 3 4 5 6 7 to 16
Percentage buds infected on
  Carignan vine   0 0 66 50 50 35 0
                                                (from Boubals, 1961).

Other methods of overwintering of grape powdery mildew have also been proposed, notably where the conidia and mycelium remain alive on the green parts of the vine. Although this is quite feasible it is clearly only relevant to certain special situations such as the cultivation of vines in glasshouses or in tropical climates.

B. Dispersal and germination of spores

1. Dispersal and germination of the ascospores

This has not been the subject of many studies on account of the relative rarity of perithecia. Only rain is actually able to detach the perithecia from their mycelial substrate (Yossifovitch, 1923). It washes them away, disseminating them over the organ on which they were borne and thence on to other organs or on to the soil. The asci and perithecia swell up and burst simultaneously at maturity. Both open by large slits, the asci liberating their ascospores via an aperture of dehiscence at their apex (Galloway, 1895).

Recent research has shown that in Rumania although the perithecia are able to release their contents throughout the entire period of vegetative dormancy, the largest number reach maturity and liberate their ascospores in April and May. It has also been shown (Galloway, 1895), that in the spring the ascospores are able to germinate in water after four to five hours and that their germinating power may persist for nearly two years.

2. The dispersal and germination of conidia

These are better understood. Wind is the essential climatic agent responsible for the separation of mature oidia from the conidial chain and their subsequent transport. Being so light, the conidia may be carried far from their place of origin and can disseminate the disease throughout the vineyard.

Conidial germination depends on three major physical factors of unequal significance.

(a) Temperature

Germination is initiated around 4°C, is slow up until 10–12°C, accelerates from 15°C onwards and, according to the most recent data, is optimal around 25°C. Although germination does not occur above 35°C the conidia remain viable, but viability is lost at temperatures greater than 40°C (Yossifovitch, 1923; Delp, 1954; Oku *et al.*, 1975).

(b) Humidity

Water in liquid form is not essential for germination, atmospheric moisture being sufficient. A small proportion (12%) of the conidia are able to germinate in a relatively dry atmosphere of 20% r.h. (Oku *et al.*, 1975). From 40% onwards the relative humidity is no longer of significance, the level of germination remaining at a more or less constant average value of 65% at humidities of 40–100% (Toma, 1974).

(c) Light

Trials have shown that germination is checked in bright sunlight and favoured by diffuse light, 47% conidial germination occuring in diffuse light and only 16% in sunlight (Toma, 1974).

C. Conditions of development

When infection has occurred, mycelial growth is followed rapidly by the formation of the conidial stage. Subsequent spread of the disease depends on various climatic factors, the major ones being temperature, humidity and light although the individual action of each of these is not always easily discernible.

Early data is available on the development of grape powdery mildew:

1. As a function of temperature: although powdery mildew is able to appear in the spring when temperatures do not exceed 10°C, it does not develop below 15°C and only becomes serious around 20–25°C. High temperatures between 35°C and 40°C check its development at least momentarily. This behaviour of the disease is associated with the relative cold and heat resistance exhibited by the mycelium and the

conidia. The mycelium is only, in fact, destroyed by frost and temperatures higher than 45°C, whereas the conidia are able to resist temperatures of 4°C for almost a month but only two hours at 38°C.

2. As a function of humidity: the development of grape powdery mildew
   increases with atmospheric humidity. The lower limit seems to be
   20–30% r.h. and, if continuous, this can alter the germinative ability of
   the conidia. Rainfall is not necessary and may even be harmful especially
   when violent, carrying away the conidia and disrupting the mycelium
   that it hits directly.

   Humidity has a marked effect on conidium production and consequently on the relative production of inoculum throughout the
   vegetative phase of the vine. It has been shown that the average number
   of spores produced in 24 hours by the conidial chains ranges from
   2·07–2·95 and 4·67 respectively when the atmospheric humidity rises
   from 30–40% to 60–70% and 90–100% (Toma, 1974).

3. As a function of light: under natural conditions it is difficult to dissociate
   the action of light from that of temperature and relative humidity.
   Intense light, however, slows down the development and spread of
   powdery mildew and this is most clearly evident in vines shaded by trees
   bordering the vineyard or in the very heart of the foliage.

More specific data has been obtained recently in Rumania (Toma, 1974)
to show that the thermal optimum for the development of powdery mildew is
22°C. The duration of incubation is about seven days at this temperature,
fourteen days at 14·5°C and ten days at 26°C. An equation, whereby the
duration of incubation in days may be evaluated as a function of temperature,
has been proposed:

$$n = 7 \times 1/x$$

where the coefficient x is a function of the temperature and values for temperatures between 10°C and 29°C are presented ($x = 1$ at 22°C).

D. Plant receptivity

This particular powdery mildew appears only to parasitize the Vitaceae, and
in general, within this family, resistance is seen to increase as one recedes, in a
phylogenetic sense, from the genus *Vitis*. Thus in order of increasing resistance
we have *Vitis, Ampelopsis, Parthenocissus, Cissus* (Boubals, 1961).

Within the genus *Vitis* the European *V. vinifera* and the Asiatic species (for
example: *V. betulifolia, V. pubescens, V. Davidii, V. Pagnucii, V. Piazezkii*)
are the most susceptible whereas the American species are more resistant.
V. Berlandieri, V. Labrusca, V. riparia and *V. rupestris* are practically immune
and *V. cinerea* and *V. cordifolia* are little attacked.

Not all varieties of a given species exhibit the same degree of resistance. This is clearly the case in *V. vinifera* and also in *V. Labrusca* and *V. cinerea*.

The causes of resistance to grape powdery mildew are of two sorts: those which hinder sucker penetration and those which more or less limit the spread of the parasite over the plant (Boubals, 1961). In the first category, villosity, of the lower leaf surface essentially, may be cited but as the fungus can easily develop on the upper leaf surface, this cause of resistance is of only limited significance.

One of the causes of limitation of spread is due to necrosis of the *appressoria* which no longer form suckers; this is largely responsible for the resistance to powdery mildew observed in *Parthenocissus* and *Cissus*. Necrosis of the epidermal cells, which harbour the haustoria, may also occur and this phenomenon explains the field resistance observed in the American vines *V. riparia* and *V. rupestris* in particular.

As these mechanisms do not account for all the observed cases of resistance one is also forced to postulate the existence of a mechanism of a physiological nature which acts on the nutrition of powdery mildew, affecting this to a greater or lesser degree depending on the biochemical constitution of the plant host (Boubals, 1961).

VI. Control of Grape Powdery Mildew

A treatment for grape powdery mildew was very quickly developed in France in the years following the initial outbreak of the disease.

The importance of the damage and also the granting of subsidies, by the Société d'Encouragement pour l'Industrie Nationale, stimulated the inventive genius of many research workers at that time. Numerous methods were put to the test, even the most unlikely ones. However treatment with sulphur was the only one to yield positive results (Geoffrion, 1976).

Sulphur was first used in French vines in 1850 at Versailles in the trials of Hardy, Grison and Duchartre. They used flowers of sulphur (sublimated sulphur) and applied it to diseased vines under different conditions. Grison also concocted a mixture composed of sulphur and lime, termed "Bouillie sulfocalcique" or "Bouillie versaillaise", which was later adopted by the Americans under the name of lime-sulphur and widely used since to control other diseases, notably in fruit trees (Arnaud and Arnaud, 1931).

A difficult series of experiments was undertaken during the years that followed. Everything had to be discovered and it was necessary to tolerate the scepticism and sarcasm of retrograde thinkers. It was very quickly evident that the treatment of vines with sulphur was highly effective. Numerous methods of application were compared. Eventually the method of dry dusting with bellows was adopted. However this treatment required a lot of time and personnel; for example one worker would need a day to treat 600 to 1000 vines depending on the density of plantation. Nevertheless from 1847 onwards, thanks to the efforts of popularization made by progressive men such as Mares and De la Vergne, the use of sulphur became generally widespread in vineyards and the injury observed diminished considerably. French wine production rose from 10.8×10^6 hl in 1854 to 53.9×10^6 hl in 1858 thus returning to the same level as in 1847 (54.3×10^6 hl) when powdery mildew first made its appearance (Galet, 1973).

At the present time, sulphur is still used to protect vines. Despite the discovery of numerous synthetic fungicides for the control of grape powdery mildew it is still the best product for practical use because it is both effective and cheap.

A. Preventive measures

Apart from the chemical method no other is sufficiently effective to ensure the protection of vines against powdery mildew attack. Nevertheless certain beneficial preventive measures may be applied to check the disease. These include taking account of environmental factors and avoiding planting highly susceptible varieties especially in those situations which are favourable to spread of the disease.

Although environmental factors are difficult to control in a vineyard, the site of plantation, vine habit and winter pruning also affect, indirectly, the aeration and insolation of the grapes and thence spread of the disease.

Planting in shady sites, as, for example, beside trees in poorly ventilated valley bottoms, should be avoided.

Winter pruning should be hard so as to encourage the production of a small number of vigorous branches bearing well-aerated, sun-exposed grapes.

Where it is possible to choose a particular vine habit, less susceptible low forms are preferable to trellis and pergola types and other high forms.

The choice of vine variety is important. Although there are no varieties of *Vitis vinifera* that are truly resistant to powdery mildew, appreciable differences in susceptibility exist. The choice of variety obviously does not depend exclusively on susceptibility to powdery mildew. However when soil and the type of production permit, it is desirable to choose the least susceptible variety.

According to observations made at the Ecole Nationale Supérieure Agronomique de Montpellier (Galet, 1975), vine varieties may be classified into three categories and the following examples are cited:

   1. Highly susceptible varieties: Carignan, Cinsaut, Cabernet-Sauvignon, Cabernet franc, Muscadelle, Muscat de Frontignan, Chasselas, Frankenthal, Emperor.

2. Susceptible varieties: Alicante Bouschet, Clairette, Mourvèdre, Char-
   donnay, Pinot noir and gris, Meunier, Gamay noir, Riesling, Sylvaner,
   Müller-Thürgau, Gewürztraminer, Sauvignon, Chenin blanc, Colom-
   bard, Merlot noir, Portugais bleu, Tannat, Ugni-blanc.
3. Weakly susceptible varieties: Aramon, Cot, Folle blanche, Maccabeu,
   Baroque, Grenache, Syrah.

The practice of mixing varieties within a plot in the vineyard is not to be
recommended, especially when the most susceptible vines are the least
numerous. In reality, in this case, the treatments serve to protect the least
susceptible vines which, as a consequence, allows the disease to develop on
the most susceptible ones. In this way foci of infection are maintained which
are both dangerous to the neighbouring plants and very difficult to eliminate.

The secondary effects of treatments employed to control downy mildew
(*Plasmopara viticola* (B. and C.) Berlese and de Toni) should also be mentioned in relation to their effect on powdery mildew attack. It was observed,
in vineyards where downy mildew control necessitated the application of
several treatments, that the development of other diseases was also affected
by the fungicides employed. The secondary effect of such treatments on the
spread of powdery mildew is appreciable especially as the intense development of this fungus corresponds exactly to the period when it is also necessary
to protect the grapes against downy mildew.

Most of the dithiocarbamates, zineb in particular, encourage the development of powdery mildew. It is not known whether these compounds directly
favour growth of this fungus or whether they act indirectly via the intermediate of the host plant. As the latter hypothesis is most feasible in those
varieties which are highly susceptible to powdery mildew, it is preferable not
to treat downy mildew with a dithiocarbamate.

Of the various phthalimides used to control downy mildew captan favours
the growth of powdery mildew whereas folpel seems to inhibit it.

Copper and dichlofluanid act directly against powdery mildew. Although
these compounds cannot control the disease completely, they enhance the
action of sulphur to a useful degree.

B. Sulphur treatments

Sulphur is used in vineyards mainly as a dust (sublimed sulphur, triturated
sulphur) or as a wettable powder (wettable micronized sulphur) (Lhoste and
Lambert, 1970).

1. Various types of sulphur

(a) Sulphur dusts

These may be produced in two ways according to the type in question:

(*i*) *Sublimed sulphur:* is a very fine straw yellow powder obtained by distillation of crude sulphur. It exists in the form of hollow spherical utricles which may be isolated or strung together like beads and whose exterior consists of amorphous sulphur insoluble in carbon disulphide. The utricles are approximately 5 μm in diameter and may form strings containing five to ten utricles linked together. Sublimed sulphur has an apparent density of 0·4–0·5. This lightness confers it with maximal covering power at the time of application. The amorphous sulphur composing the external envelope of the utricles has a higher vapour pressure than that of crystalline sulphur which means that sublimed sulphur is also able to emit fungicidal vapours more readily than this latter

(*ii*) *Triturated sulphur:* is crude, ground sulphur whose particle size is represented by passage through 100 and 120 mesh sieves (125 and 150 μm). One form of this sulphur is obtained by selecting the finest particles by an air-blasting process: this blown triturated sulphur is able to pass through 150 to 300 mesh (100 to 50 μm). Triturated sulphur particles consist of orthorhombic crystals. Their apparent density is from 0·8–0·85.

To improve the application of sulphur dusts, industry has developed a number of processes which render the sulphur fluent, that is to say, able to flow as easily as a liquid through the tubing of the applicator and easier to distribute over the plants.

Other forms of powdered sulphur are available on the market (various sulphur mixes, black sulphur, etc.) but are little used.

In France, the most widely used sulphur dust is in the form of flowable sublimed sulphur.

(b) Wettable sulphur

Pure sulphur is immiscible with water and various adjuvants must be added to triturated sulphur to obtain miscibility. As the best results are obtained with the finest particle size, triturated sulphur is subjected to violent crushing which reduces 100% of the particles to a diameter of less than 40 mm of which 80% have a diameter less than 12 μm. This is termed wettable micronized sulphur (according to French standards).

Wettable sulphur is employed alone in water but more often in association with downy mildew fungicides. It adheres better to the vegetation than powdered sulphur although penetration of the spray droplets into the heart of the foliage is not so satisfactory. As the quantity of sulphur used is low, heavy attacks of powdery mildew may be difficult to restrain. For this reason wettable sulphur tends to be reserved for preventive treatments where a small quantity of sulphur, capable of checking infections as they start, is maintained on the vine.

Other sulphurs may be employed as sprays. These include the alkaline polysulphides, colloidal sulphur and lime-sulphur. These have never given such good results as elemental sulphur and also present disadvantages. Lime-sulphur, especially, is both phytotoxic to the vine, damages the machinery used for treatment and also reacts with copper compounds forming sulphides which are inactive against downy mildew.

2. Mode of action of sulphur

The internal mechanism of action of sulphur on the fungus has been and still is the subject of controversy (Torgeson, 1969).

At the present time the most likely hypothesis is that sulphur inhibits the production of energy by exerting a toxic effect on the mitochondria thus upsetting the dehydrogenases involved in the Krebs cycle with the result that the fungus is killed by blockage of the respiration process. This phenomenon occurs during the eight hours immediately following treatment with maximal activity around the third hour.

In practice sulphur exerts its effect through the vapours which are given off. Activity is conditioned mainly by: (a) the distribution of sulphur over the parts of the vine to be protected, and (b) factors favourable to the production of vapours.

3. Distribution

Owing to its toxic vapours, sulphur is able to act at a distance from its point of application. Nevertheless, as activity is linked to the quantity and nature of the compound used and to the temperature it is important to ensure the distribution of sulphur over the entire surface of the parts to be protected to obtain rapid and effective action. Dusting with flowable sublimed sulphur provides the most satisfactory distribution both on the exterior and interior of the foliage cover and on the grapes.

With this treatment a considerable quantity of sulphur is suddenly applied which has the advantage of producing an abundance of active fumes, though, the persistence of the dust is of only brief duration.

In comparison, use of wettable micronized sulphur involves only a small quantity of the compound but the very fine particles ensure good distribution. Immediate action is not as powerful as with sulphur dusts but the persistence is greater.

(a) Factors favouring the emission of fumes

(*i*) *Temperature*. Optimal vapour production occurs at too high a temperature to be supported by the vine so that the limits of plant susceptibility to sulphur phytotoxicity have to be respected. In the vineyard, sulphur exhibits good activity from 18°C with an optimum at 25°–30°C. Above this temperature although the fungicidal action increases still further, the risk of burning comes into play.

(*ii*) *Luminosity*. At a given temperature the production of sulphur fumes may be five times greater in clear as opposed to dull weather.

(*iii*) *Humidity*. At a given temperature, the activity of sulphur is always weaker in the presence of water, or in damp as opposed to dry air.

(*iv*) *Wind*. Whereas in calm weather the sulphur vapours remain in the vicinity of the vine and are able to act, under windy conditions they become diluted in the air and their action is reduced. This phenomenon can be a disadvantage in cold regions but in hot areas that are also subjected to constant wind, sulphur phytotoxicity is prevented from occurring.

(*v*) *Sulphur type*. Amorphous sulphur having a higher vapour pressure than crystalline sulphur produces fumes more quickly and in greater abundance than the latter.

(*vi*) *Particle size*. The activity of powdered sulphur increases with the fineness of the particles. This is due to the better covering power and the increased surface of contact with the air, heat and light. These advantages are limited however below 5 μm as the particles tend to stick together. The adjuvants included in the formulation prevent this from occurring in wettable sulphur although there is still a limit to the desired particle size. Below 1 μm (ultra-micronized sulphur) activity no longer increases, the persistance diminishes and the risks of phytotoxicity are augmented.

4. Method and time of application

A method, similar to that employed for downy mildew, to determine the application timing, does not exist for powdery mildew. The vine-grower must make the decision himself according to the following general principles:

1. The action exerted by sulphur on the fungus is both preventive and curative. Preventive action is dependent on long persistence, sulphur residues being present on the vines should a risk of infection occur. Curative action is improved if treatment is made to the young mycelial filaments at the beginning of their development. At a later stage powdery mildew forms a very tight mycelial mat which impedes contact of the sulphur fumes with all the hyphae.

2. Early treatments are the most effective, the easiest to apply and the least expensive. The fungal mycelium is younger, penetration of the

sulphur to the interior of the vine is facilitated by the low foliage density and the quantity of sulphur used is consequently less.

3. Powdery mildew forms long-lasting foci of infection which tend to become more and more dangerous the longer they are left without being destroyed. Effective control threfore requires such foci to be destroyed by repeated sulphur treatment. It is estimated that six annual sulphur treatments (in powder form) will get rid of the most important foci within two years.

4. Dusting is carried out at the critical periods in the development of powdery mildew as the activity is very powerful.

5. Spraying with wettable sulphur should be used as a preventive measure in the general upkeep of the vineyard.

The following treatments are recommended in France (Gaudineau, 1946; Lafon et al., 1966):

(a) Dusting

(i) *Treatment of young shoots:* 15–20 kg sulphur ha^{-1}. Advantage is taken of a warm, calm, sunny day when the first three leaves are well differentiated. The application should be repeated if rain occurs immediately after teatment. This application serves to check the initial attack resulting from the buds infected during the previous year.

(ii) *Treatment at the flowering stage:* 20–30 kg sulphur ha^{-1}. The treatment is applied in the middle of flowering, partly because the racemes are highly susceptible at this time and partly because sulphur treatment has a beneficial effect on fertilization by dispersing the pollen. This particular sulphur treatment is applied more or less systematically in all vineyards.

(iii) *Treatment after fruit-set.* 30–50 kg sulphur ha^{-1}. If powdery mildew fails to appear, treatment can be delayed until a few days before ripening but in the event of attack, an application should be made immediately.

These three treatments are essential in all situations favourable to powdery mildew development. Where necessary, (that is to say if attacks occur despite these precautions) the treatments should be repeated.

(b) Spraying with wettable sulphur

These treatments serve either to enhance the effect of sulphur dusting in the cases cited above, or to ensure the prevention of powdery mildew attack in less susceptible vineyards.

The net cost may be decreased if wettable sulphur is used in association with anti-downy mildew mixtures in vineyards where these latter treatments are routine.

The amount of wettable sulphur employed varies from 3–10 kg ha^{-1} depending on the temperature. The highest dose is applied initially with 7 kg ha^{-1} around flowering time, followed by 5 and 3 kg after fruit set during the summer.

Either mechanical spraying with a pressurized jet or low volume pneumatic spraying may be employed.

(c) Secondary effects of sulphur treatment

Apart from their fungicidal action on powdery mildew, sulphur treatments may also have other effects which must be taken into consideration.

Amongst the favourable effects, the following should be mentioned:

1. The activity of early dusting against mites (*Eriophyes vitis* Pgst., *Phyllocoptes vitis* Nal.) and tetranychids (*Panonychus ulmi* Koch and *Eotetranychus carpini* Ond.).
2. The indirect action on grey mould (*Botrytis cinerea* Pers.) through the prevention of lesion formation by powdery mildew on the berries and peduncles of the inflorescences. Grey mould frequently colonizes such lesions.
3. The beneficial effect on the vine vegetation, probably a trophic effect but due especially to improved fertilization of the grapes by the dust applied at flowering, This is possibly due to a specific action on the pollen but more probably to its better dispersal.

Unfortunately sulphur also has negative effects on the vine:

Burning (similar to solar damage) may occur at high temperatures. This phenomenon may be observed from 35°C whereas it occurs only above 42°C in the absence of sulphur. The phytotoxicity of sulphur is due to production of sulphurous anhydride, sulphureted hydrogen and various sulphates. The affected plant cells become acidified, their protoplasm coagulates and they dry up and die. Wettable sulphur has a cumulative phytotoxic action due to its long persistence. Powdered sulphur is less dangerous if it is applied dry and in a thin uniform layer on the plant. Dusting of wet plants is not to be recommended as the moistened sulphur sticks together forming persistent crusts which are phytotoxic in sunlight.

Sulphur may transfer a bad taste to the alcohol if it is used too near to harvest time. In those regions producing refined spirits (Cognac, Armagnac) stopping of sulphur treatment is recommended 70–80 days before harvest time or even sooner where the more persistent wettable sulphur is employed (J. Lafon *et al.*, 1966).

During this period sulphur may be replaced where necessary with dinocap or another synthetic fungicide.

C. Other fungicides

A number of effective synthetic fungicides exist for the control of grape powdery mildew but very few have been commercialized up until now because of their cost and their phytotoxicity.

1. Potassium permanganate

This was used previously for its curative effect at a rate of 125 g ha^{-1}. A special wetting agent is required and the fungus must be treated by contact for the compound to be active. Unfortunately the suckers are not affected and mycelial activity can start up again immediately after treatment. It is recommended to prolong the action of permanganate by following it up with a sulphur treatment.

2. Dinocap

This is marketed in France for use at 30 g. a. i. hl^{-1}. Like wettable sulphur, this compound is employed preferably with a wetting agent which improves its curative action. It is active at a lower temperature than sulphur but presents similar risks of phytotoxicity above 34°–35°C.

Hardly any of the other products authorized in France for the control of powdery mildew are actually used directly against this disease. These include the vine fungicides used to control grey mould (benomyl 25 g. a. i. hl^{-1}, methylthiophanate 140 g. a. i. hl^{-1}) or simultaneously against grey mould and downy mildew (dichlofluanid 125 g. a. i. hl^{-1}) and which can exert a considerable secondary effect against powdery mildew.

References

Arnaud, G. et Arnaud, M. (1931). "Traité de Pathologie Végétale". P. Lechevalier et Fils, Paris.
Berkeley, M. C. (1847). *Gdnrs' Chron.* 779.
Boubals, D. (1961). *Ann. Amélior. Plant.* **11**, 401–500.
Couderc, G. (1893). *C. r. hebd. Séanc. Acad. Sci., Paris* **116**, 210–211.
Delp, C. J. (1954). *Phytopathology* **44**, 615–626.
Duchartre, P. (1851). *Ann. agron.* **1**, 1–173.
Enjalbert, H., Marquette, J. B., Cavignac, J., Huetz De Lemps, Chr., Butel, P., Pijassou, R. et Guillaume, P. (1974). "La Seigneurie et le vignoble du Château Latour. Histoire d'un grand cru du Médoc (XIVe–XXe siècle)". Fédération historique du Sud-Ouest, Bordeaux.
Galet, P. (1973). *Fr. Vitic.* **8**, 217–229; **9**, 235–241; **10**, 265–267.
Galet, P. (1974). *Fr. Vitic.* **4**, 107–112.
Galet, P. (1975). *Fr. Vitic.* **4**, 113–115; **5**, 141–153; **6**, 169–186; **7**, 214–219; **8**, 232–233.

Galloway, G. T. (1895). *Bot. Gaz.* **20**, 486–491.

Gaudineau, M. (1946). *1er Congrès International de Phytopharmacie*, Heverle, Belgique.

Geoffrion, R. (1976). *Phytoma* **276**, 15–25.

Lafon, J., Couillaud, P. et Hude, R. (1966). "Maladies et Parasites de la Vigne". J. B. Baillière, Paris.

Lhoste, J. et Lambert, J. (1970). "Les Fongicides". Rullière Libeccio, 84–Morières, France.

Oku, H., Hatamoto, M., Ouchi, S. and Fujii, S. (1975). *Sci. Rep. Fac. Agric.*, Okayama Univ. **45**, 16–20.

Schweinitz, L. D. (1834). *Trans. Am. phil. Soc.*, *N.S.* **IV**, 141–316.

Toma, N. (1974). Thèse (résumé), Fac. Agron., Inst. Agron. N. Balcescu, Bucuresti.

Torgeson, D. C. (1969). "Fungicides". Academic Press, New York and London.

Viala, P. (1893). "Les maladies de la Vigne". Masson, Paris.

Viennot-Bourgin, G. (1949). "Les champignons parasites des plantes cultivées". Masson, Paris.

Yossifovitch, M. (1923). *Thèse Fac. Sciences*. Bonnet, Toulouse.

Subject Index

Dithiocarbamate fungicides, 407
Diurnal cycle, 417, 434
Diurnal periodicity, 292
  in sporulation, 451–452
Diuron, 197
Dodemorph, 272–273, 429
Dormant mycelium, 294
Drazoxolon, 310
  as surface fungicide and seed treatment, 264
Drought, 460
Dusting
  with sulphur to control *Uncinula necator*, 545
Dynamics of pathogen response, 130–136

E

Egg plant, 514
Elm *see Ulmus*, 491
Endive, 518
Environment
  effect on vine powdery mildew, 537–538
Environmental conditions, 108, 288
  effects on disease, 293–294
Environmental factors,
  affecting rose-mildew development, 426–429
Epidemics, 336
  temporal and spatial aspects, 70–72
  in vine powdery mildew, 535
Epidemic spread, 43
Epidemiology, 51–82, 398–400
  of cereal mildews, 285–296
  definition, 52
  effect of temperature on, 54
  of *Erysiphe betae*, 338–340
Eradication of inoculum, 352, 357
Erysiphaceae, 8, 16, 19
Erysiphe, 8, 12, 15, 17, 18, 25, 26, 27, 29, 67, 474, 496
Erysiphe acaciae, 475
Erysiphe aggregata, 476
Erysiphe aquilegiae, 433
Erysiphe bertolini, 432
Erysiphe betae, 46–48, 61, 63, 65, 323–346
  biology, 326–335
  course of disease, 332–333
  distribution, 323–324

economic significance and control, 340–343
epidemiology, 335–340
first records, 335
high temperature requirements of, 332
morphology and taxonomy, 324–326
range of host plants, 333
varietal susceptibility, 333
xerophytic properties of, 329
Erysiphe cichoracearum, 7, 8, 12, 15, 27, 43, 54–70, 202, 361, 363, 430–438, 447, 451, 462, 464, 480, 481, 499, 515–516
  comparison with *Sphaerotheca fuliginea*, 362
  occurrence on Compositae, 516–519
  sporulation and dispersal on tobacco, 451–453
Erysiphe communis, 361, 497, 499
Erysiphe cruciferarum, 497–499
Erysiphe graminis, 7, 8, 12, 15, 16, 25, 27, 30, 48, 54–59, 61–68, 73, 83–98, 101, 108, 124, 126, 174, 185, 200, 215–221, 225, 237, 244, 255, 284–312
  distribution and importance, 83–84
  haustorium structure and development, 202–203
  host range of cultures, 93–96.
  isolates from small grain crops, 96
  specialization on wild grasses, 93
Erysiphe graminis f. sp. *agropyri*, 103, 104, 116, 126, 290–291
Erysiphe graminis f. sp. *avenae*, 83, 92, 95–96, 106, 247, 250
Erysiphe graminis f. sp. *hordei*, 83, 85–91, 102, 103, 104, 177, 180, 238, 239, 245, 250, 251, 290–291
  development of hyphae and conidia, 199
  inheritance studies with resistant cultivars, 163
Erysiphe graminis f. sp. *tritici*, 83, 92, 95–96, 101–127, 173, 175, 177, 179, 242, 246, 251, 253, 290–291, 451
Erysiphe heraclei, 483, 512
Erysiphe martii, 438, 439
Erysiphe pisi, 212, 214, 225, 438, 502–506
  conidia tolerant to pisatin, 505
  crops other than pea, attacked by, 505
  formae speciales in, 502